Nuclear Medicine Physics

Series in Medical Physics and Biomedical Engineering

Series Editors: John G Webster, Slavik Tabakov, Kwan-Hoong Ng

Series in Medical Physics and Biomedical Engineering

Nuclear Medicine Physics

Edited by

J. J. Pedroso de Lima

IBILI Biophysics Institute
Coimbra, Portugal

IMPRENSA DA UNIVERSIDADE DE COIMBRA
COIMBRA UNIVERSITY PRESS

CRC Press
Taylor & Francis Group
Boca Raton London New York

CRC Press is an imprint of the
Taylor & Francis Group, an **informa** business
A TAYLOR & FRANCIS BOOK

CRC Press
Taylor & Francis Group
6000 Broken Sound Parkway NW, Suite 300
Boca Raton, FL 33487-2742

First issued in paperback 2019

ISBN-13: 978-1-58488-795-9 (hbk)
ISBN-13: 978-1-138-37496-6 (pbk)

Visit the Taylor & Francis Web site at
http://www.taylorandfrancis.com

and the CRC Press Web site at
http://www.crcpress.com

Contents

About the Series

The *Series in Medical Physics and Biomedical Engineering* describes the applications of physical sciences, engineering, and mathematics in medicine and clinical research.

The series seeks (but is not restricted to) publications in the following topics:

- Artificial organs
- Assistive technology
- Bioinformatics
- Bioinstrumentation
- Biomaterials
- Biomechanics
- Biomedical engineering
- Clinical engineering
- Imaging
- Implants
- Medical computing and mathematics
- Medical/surgical devices

- Patient monitoring
- Physiological measurement
- Prosthetics
- Radiation protection, health physics, and dosimetry
- Regulatory issues
- Rehabilitation engineering
- Sports medicine
- Systems physiology
- Telemedicine
- Tissue engineering
- Treatment

The *Series in Medical Physics and Biomedical Engineering* is an international series that meets the need for up-to-date texts in this rapidly developing field. Books in the series range in level from introductory graduate textbooks and practical handbooks to more advanced expositions of current research.

The *Series in Medical Physics and Biomedical Engineering* is the official book series of the International Organization for Medical Physics.

The International Organization for Medical Physics

The International Organization for Medical Physics (IOMP), founded in 1963, is a scientific, educational, and professional organization of 76 national adhering organizations, more than 16,500 individual members, several corporate members, and four international regional organizations.

IOMP is administered by a council, which includes delegates from each of the adhering national organizations. Regular meetings of the council are held electronically, as well as every three years at the World Congress on Medical Physics and Biomedical Engineering. The president and other officers form the executive committee, and there are also committees covering the main areas of activity, including education and training, scientific, professional relations, and publications.

Objectives

- To contribute to the advancement of medical physics in all its aspects
- To organize international cooperation in medical physics, especially in developing countries
- To encourage and advise on the formation of national organizations of medical physics in those countries that lack such organizations

Activities

Official journals of the IOMP are *Physics in Medicine and Biology* and *Medical Physics and Physiological Measurement*. The IOMP publishes a bulletin, *Medical Physics World*, twice a year, which is distributed to all members.

A World Congress on Medical Physics and Biomedical Engineering is held every three years in cooperation with the International Federation for Medical and Biological Engineering (IFMBE) through the International Union for Physics and Engineering Sciences in Medicine (IUPESM). A regionally based international conference on medical physics is held between world congresses. IOMP also sponsors international conferences, workshops, and courses. IOMP representatives contribute to various international committees and working groups.

The IOMP has several programs to assist medical physicists in developing countries. The joint IOMP Library Programme supports 69 active libraries in 42 developing countries, and the Used Equipment Programme coordinates equipment donations. The Travel Assistance Programme provides a limited number of grants to enable physicists to attend the world congresses. The IOMP Web site is being developed to include a scientific database of international standards in medical physics and a virtual education and resource center.

Information on the activities of the IOMP can be found on its Web site at www.iomp.org.

Editor

J. J. Pedroso de Lima was awarded a degree in physics and chemistry in 1957 from the University of Coimbra, Portugal; a diploma for specialization in methodology of radioisotopes in 1958 from the University of Brazil, Rio de Janeiro; a diploma for advanced studies in research in science in physics in 1968 from the University of Manchester, UK; and PhD in physics in 1970 from the University of Manchester, UK. He was nominated associate professor in 1980 at the Faculty of Sciences and Technology, University of Coimbra; aggregate professor in 1982 at the Faculty of Sciences, University of Coimbra; full professor in 1982 at the Faculty of Sciences, University of Coimbra; full professor in 1986 at the Institute of Biophysics/Biomathematics, Faculty of Medicine, University of Coimbra; "Doctor Honoris causa" in 2004, at the University of Aveiro. He is a professor since September 2004. He was head of the radioisotopes laboratory of the University of Lourenço Marques, Mozambique, 1965–1967 and 1970–1973; head of the Department of Physics, University of Coimbra, 1974–1977; head of the Department of Biophysics/Biomathematics of the Faculty of Medicine, University of Coimbra, 1986–2004; president of the Executive Board of IBILI (Institute of Biomedical Research on Light and Image, University of Coimbra) 1993–2004; and president of the installation committee of ICNAS (Institute for Nuclear Sciences Applied to Health) 2000–2004.

Dr. De Lima was president of Portuguese Society of Nuclear Medicine, president of Portuguese Society of Physiology, vice president of Portuguese Society of Protection against Radiations, member of the executive board of European Association of Nuclear Medicine (EANM) and of the Portuguese Society of Physics. He was the president of the European Association of Nuclear Medicine Congress in 1992 at Lisbon. He was awarded 22 scientific prizes (special reference to Marie Curie Prize of EANM in 1992). He is a member of the Portuguese Medical Sciences Academy (1997), honorary member of ATARP (Portuguese Technologists Association) since 2004 and honorary member of SPPCR (Portuguese Society of Protection against Radiations) since 2008. He has published 5 books as a single author, 8 books in collaboration, 83 papers in international journals/chapters of books, and 127 papers in national journals. He has presented 120 communications at scientific meetings or congresses and was an invited lecturer at more than 50 scientific events.

Contributors

Antero J. Abrunhosa
Institute for Nuclear
    Sciences Applied to Health
University of Coimbra
Coimbra, Portugal

Francisco J. C. Alves
College of Health Technology
Polytechnic Institute of
    Coimbra

and

Institute for Nuclear
    Sciences Applied to
    Health
University of Coimbra
Coimbra, Portugal

Helder Araújo
Institute for Systems and
    Robotics
University of Coimbra
Coimbra, Portugal

M. Filomena Botelho
Institute of Biophysics/
    Biomathematics
University of Coimbra
Coimbra, Portugal

Francisco J. Caramelo
Faculty of Medicine
Institute of Biomedical Research
    on Light and Image
University of Coimbra
Coimbra, Portugal

Durval C. Costa
Faculty of Medicine
Institute of Biomedical Research
    on Light and Image
University of Coimbra
Coimbra, Portugal

Paulo Crespo
Department of Physics
University of Coimbra
Coimbra, Portugal

António Dourado
Department of Informatics
    Engineering
University of Coimbra
Coimbra, Portugal

Nuno C. Ferreira
Faculty of Medicine
Institute of Biomedical Research
    on Light and Image
University of Coimbra
Coimbra, Portugal

Paulo M. Gordo
Department of Physics
University of Coimbra
Coimbra, Portugal

Carina Guerreiro
Faculty of Medicine
Institute of Biomedical Research
    on Light and Image
University of Coimbra
Coimbra, Portugal

Augusto D. Oliveira
Nuclear and Technological
    Institute
Sacavem, Portugal

Adriano Pedroso de Lima
Department of Physics
University of Coimbra
Coimbra, Portugal

J. J. Pedroso de Lima
Faculty of Medicine
Institute of Biomedical Research
    on Light and Image
University of Coimbra
Coimbra, Portugal

Maria Isabel Prata
Faculty of Medicine
Institute of Biomedical Research
    on Light and Image
University of Coimbra
Coimbra, Portugal

1

Introduction

J. J. Pedroso de Lima

In 1973, the WHO defined nuclear medicine (NM) as a discipline

> "...embracing all applications of radioactive materials in diagnosis and treatment or in medical research, with the exception of the use of sealed radiation sources in radiotherapy."

Apart from medical knowledge, NM includes physics, pharmacy, biology, chemistry, mathematics, computer science, and all branches of engineering concerned with the development of devices used in NM.

NM embraces a vast set of applications such as imaging methodologies using radionuclides (ranging from autoradiography to positron emission tomography [PET]), *in vitro* analysis, metabolic radiotherapy, and medical research studies and techniques involving radioactive tracers.

The relative contribution of these applications has varied considerably over the years, although at present imaging methodologies play a leading role. Despite this, it is still inaccurate to reduce the definition of NM to nuclear imaging applications, as is frequently claimed, even when this includes all the functional information resulting from imaging studies. NM is not simply a medical imaging technique, although in this area it does have very special characteristics.

It is perhaps important, at this stage, to raise the following question: Why NM?

There are several important reasons:

1. Of all the available methodologies, NM is best at supplying functional information.

2. It has the most sensitive detection methods (detecting masses below picomole level), enabling studies to be performed in physiological conditions without any interference from the processes being studied.

3. Most biological molecules can be labeled and used as radiotracers, making the information supplied by this technique multiparametric and unique in comparison with other techniques that provide information about just one or a small number of properties.*

---

* Fluorescence techniques rival NM in this area but are generally only applicable to surface organs or tissues.

4. Direct image analysis can provide functional results in relative terms.
5. With some additional steps, it can provide quantitative results.
6. Studies can be repeated within short periods of time.

Certain negative aspects of NM techniques may also be cited:

1. They require a nonconventional environment and certain precautions (radioactivity cannot be switched off as an x-ray machine can).
2. In NM imaging, it is always necessary to administer radiopharmaceuticals, which emit ionizing radiation.
3. The spatial resolution of NM images is almost always worse than that of morphological images.

Focusing our attention on the second point, the effective doses received by patients in the vast majority of NM exams fall within the medium dose range for radiological studies, that is, 0.5–6 mSv, corresponding to a few months to about 2 years of average background irradiation.

The various types of NM images can be classified as

- Static planar scintigraphy
- Dynamic planar scintigraphy
- Whole-body scintigraphy
- Single photon emission computerized tomography (SPECT)
- Gamma camera coincidence detection
- Computerized PET
- Hybrid systems

Generally speaking, the current NM imaging techniques using external detection provide local functional information that is specific, easily presented in relative terms, and very useful in a large number of situations. Quantitative analysis is also possible if the appropriate corrections are applied and, eventually, additional data are obtained.

In some situations, the limitations associated with the unavailability of appropriate tracers (if existing ones fail to enhance the structures being studied) or the inadequacy of the spatial or temporal resolution of the available NM imaging systems can make nuclear images an unattractive option in comparison with direct sampling of biological material and measurement of radioactivity, or the possible application of other techniques.

Important developments are currently taking place in most areas of NM. New radio-labeled biological drugs, new cell labeling techniques, new technical concepts in radiation detection, improvements in instrumentation, access to multimodal techniques, the rapid increase in computer power, and a much better understood partnership with the clinical disciplines are all combining to create a new profile for NM.

In addition, the sequencing of the human genome and the ever-increasing knowledge of proteomics, systems biology, and the pathogenesis of human disease have created unprecedented opportunities for molecular imaging in NM and have also provided opportunities for understanding the molecular basis of normal and diseased cellular functions.

These developments have had a very positive effect in terms of improved diagnostic quality and, it is hoped, in reducing patient exposure to radiation.

This book, *Nuclear Medicine Imaging Physics*, aims at providing a series of contributions, mainly in areas of physics that are related to the theoretical bases of NM and its applications, some of which cannot be currently found in text books or, at least, in the perspective and detail that new or advanced approaches may require.

The subjects have been selected according to their direct potential interest to NM today or to disciplines particularly concerned with the applications of this science.

A reasonable background in mathematics and radiation physics; the basics of biology and human physiology; and an understanding of linear and non-linear system theory, model building, and simulation (linear and nonlinear) are recommended.

The book covers the following themes: the production of radioisotopes and radiopharmaceuticals, positron physics and positron applications in medicine and biology, modern radiation detectors and measuring methods, systems in NM, imaging methodologies, dosimetry, and the biological effects of radiation.

It is divided into the following chapters:

1. Introduction
2. Cyclotron and Radionuclide Production
   a. The Quantitative Aspects of Radionuclide Production
   b. The Cyclotron: Physics and Acceleration Principles
3. Positron Physics and Positron Applications in Medicine and Biology
   a. The Physical and Chemical Aspects of the Positron and Positronium in Matter
   b. Perspectives on the Medical Applications of Positrons
4. Radiopharmaceuticals: Development and Main Applications
   a. Radiochemistry and Radiopharmacy in Conventional NM and PET
   b. Advances in Radiopharmacy and New Trends in Radiochemistry
5. Radiation Detectors and Measuring Methods
   a. The Physics of Detection Methods

Chapter 1 serves as an introduction and aims at explaining the idea and philosophy of the book and the reasons for the choice of material. It also provides a short description of the subjects contained in each chapter.

Chapter 2, on radioisotope production, deals with the physics of the production of radioisotopes suitable for use in conventional NM and PET as well as the machines used for this purpose. At the beginning of the chapter, the quantitative theoretical aspects of radionuclide production, nuclear reactions, cross sections, and excitation functions are considered.

The physical principles of the cyclotron and matters relating to its performance, such as resonance equations, defocusing effects, and the maximum energy problem, are studied.

The two types of focusing (magnetic and electric) and their properties are then analyzed, taking into account points such as the conjunction of the two focusing effects, phase relationships and maximum energy, trajectory, phase at first revolution, and relativistic limit.

The different processes leading to the transmutation of target material by cyclotron beam interaction, the cross section of nuclear reactions, the corresponding excitation functions, the energy or mass balance, and so on are considered next.

Positron physics and the applications of positrons in medicine and biology are the subjects of Chapter 3. Symmetry properties, the processes of positron interaction with matter, loss of energy and thermalization, ranging and track pattern, diffusion and mobility, positron lifetime, annihilation, and finally, momentum conservation are initially considered.

Positrons have been used as a probe in structural studies of different materials. The experimental techniques used are based on the detection of annihilation radiation, as the processes occurring in the medium involving the positron and the positronium (Ps) can provide significant information. In liquids and also in tissues, the probability of Ps formation and the physical and chemical processes that occur depend considerably on the environment, for example, chemical composition and the presence of free volumes.

The most well-known positron application in biomedical science is PET, in which the distribution of the two gamma annihilation sites is obtained in patient sections or volumes. However, other possibilities for positron medical applications may open up when detailed knowledge of positron and Ps behavior in biological tissues becomes available.

The necessary requirements for establishing the interest of a particular radiopharmaceutical for medical use, such as specific activity, yield, purity, and so on, are considered in Chapter 4. Within an analysis of the advances in radiopharmacy and modern trends in radiochemistry, important issues concerning the development of new tracers and ligands for both conventional NM and PET are also discussed.

First, the selection of molecular targets to be used in oncology, cardiology, and neuroscience is analyzed. Radionuclide alternatives obviously have to be considered (Fluorine 18, Carbon 11, etc.), in addition to the position of the tracer in the labeled molecule and the possibilities for radiosynthesis, given the available reactions. The deoxyglucose model is specifically studied. Some aspects and properties of the most commonly used radionuclides in conventional NM, namely the coordination chemistry of Tc, Ga, In, and Cu and the production and radiopharmaceuticals that can be usefully produced with them, are discussed. The use of antibodies and peptides in molecular imaging is then considered. Radiopharmaceutical quality control, including chemical purity, specific activity, the effects of radiolysis, etc., is also discussed in this chapter. Particular attention is paid to the paths opened up by new approaches such as molecular modeling, the use of knowledge-based models, and advanced compartmental analysis.

Finally, cases of radiopharmaceuticals recently developed or under development for clinical or research applications are discussed.

Modern radiation detectors and measuring methods are covered in Chapter 5, which includes a physical review of the modern detectors used in nuclear imaging. The properties of semiconductor, solid state, and gas detectors are analyzed in the context of the specific needs of the different NM applications. The reasons that particular detector types are preferred in specific fields are analyzed, and the characteristics and methods used in modern imaging devices are discussed. Acquisition and processing electronics and trends in the evolution of detection technologies for these applications are also briefly considered.

Chapter 6, "Imaging Methodologies," begins with a review of the important physics concepts in NM functional imaging, their generation, and applications.

The effects of image degradation (loss of contrast, loss of quantification, loss of counts, and partial volume effect) owing to physical phenomena (e.g., scattering, attenuation, and dead time) or to other sources such as patient movement are also considered and some solutions are indicated.

The techniques most commonly used to process data in NM are presented, including the preprocessing of raw data acquired before image reconstruction (e.g., corrections for intrinsic efficiency variations, geometry, dead time, decay, arc, scatter, attenuation, and various other corrections).

Both two- and three-dimensional analytic and iterative image reconstruction algorithms are described, and a short description of rebinning methods is included. The chain of different data processing sequences, from raw data to reconstructed images, is explained in relation to NM techniques (PET and SPECT). The real possibilities of extracting quantitative information in NM functional imaging studies and the necessary correction techniques, particularly in PET, are analyzed.

The new challenge of PET–CT is considered, particularly within the context of new nuclear oncology methodologies. The evolution of NM methods has led to an increasing number of studies in nuclear oncology, related to an improved understanding of the molecular mechanisms that are the basis of malignant transformation, from the alterations associated with innumerable metabolic steps to angiogenesis, hypoxia, and various genetic alterations and modifications in the cells. This chapter demonstrates how every step in this process can be evaluated with the aid of NM, obtaining information not only from the metabolic situation of the tumor tissue being studied but also from the therapeutic response.

The application of physiological models to the central nervous system (CNS) is the focus of the final part of this chapter, which also considers the anatomy, physiology, and pharmacology of the CNS, particularly the brain, in terms of understanding self-regulating mechanisms (including the individual–environment relationship) and the progression of pathologies that afflict brain tissues in all phases of human life. The anatomy of the CNS is briefly described, and the crossing of the blood and neuronal tissue barriers is explained, with important historical references that reflect the evolution of the concept of the hematoencephalic barrier. The neuronal intercommunication systems are introduced in their common form among neurotransmitters, with reference to the synthesis, liberation, and neuronal recapture of the neurotransmitter as well as its action at postsynaptic receptor level. As an interesting and practical supplement, references are made to the most important clinical applications of methods using radionuclides for the purpose of collecting information on neurological disease mechanisms and evaluating the response to existing and new medication. Finally, physiological models are considered in relation to clinical applications, in general NM and CNS diseases.

Chapter 7 presents an introduction to biological systems theory and NM methods as systems theory procedures as well as aspects of kinetic modeling.

This chapter first considers the general properties of biological systems, their elements and relationships with the external medium, feedback and self-regulation, open and closed systems, homeostasis, and entropy and negentropy. Processes such as autopoiesis, adaptation, feedback, and hierarchy are introduced for analysis. After this, the mathematical modeling of biological systems is considered as well as the integral–differential paradigm specified by transfer functions or state equations and, in particular, compartmental analysis.

The representation and analysis of population models, the growth of biological species, and the Lotka–Volterra model for the predator–prey pair are also discussed.

Aspects of the resolution of the linear state equation in temporal domain, nonlinear systems, singular points, and linearization are then considered; and finally, bifurcations and chaos are reviewed.

Next, the main general properties of some of the more representative techniques for the mathematical modeling of systems in NM are described. In NM, an input function is used to study the properties of objects through an induced response. Relevant aspects of the kinetic modeling of radiotracers and their transport and localization mechanisms are described.

The most frequently used compartmental models are reviewed: the single-compartment model, two-compartment model, series two-compartment model, closed two-compartment system in equilibrium, three-compartment model, three-compartment systems with no equilibria, open mamillary systems, and mamillary systems with multiple equilibria.

Concepts associated with dilution theory using radionuclides are emphasized, and volume and flow evaluation in dilution systems are considered.

The Stewart–Hamilton principle and the behavior of tracers in serial dilution systems are reviewed; and, in addition, convolution integral is introduced and some applications are analyzed.

The different approaches in data analysis and the application of models in studies of tracers in examinations using PET images are reviewed, including those that are data driven, that is, assume no models or compartments, and those which are model driven, that is, assume a compartmental system. In the case of the former, multiple time graphic analysis (MTGA), spectral analysis (SA), standardized uptake value (SUV), and compartmental models are generally used to quantify the dynamics of the processes being studied. The most common MTGA methods (Patlak and Logan plots) that enable exchange constant values to be obtained are also introduced. Finally, parametric analysis and Monte Carlo simulation are briefly considered.

The main subjects of Chapter 8 are dosimetry and the biological effects of ionizing radiation. The chapter begins with an introduction to the interaction of photons with matter within the context of their application to radiological protection problems. Five types of interaction are considered: the Compton effect, photoelectric effect, pair production effect, coherent (Rayleigh) dispersion, and photonuclear interaction. Quantities such as the cross section for the interaction of photons with electrons, loss of energy per collision, linear energy transfer (LET), stopping power and range, and mean energy dispensed per ion pair formed are defined.

Fundamental dosimetric quantities are defined: particle and energy fluence, kerma, and absorbed dose. Due to the biological risks, more specific quantities, known as radiological quantities, are required, namely the radiation weighted dose (previously the equivalent dose) and the effective dose. Other radiological quantities with biophysical relevance are presented, such as LET, relative biological effectiveness (RBE), and the quality factor as a function of LET; and reference is made to radiation and tissue weighting factors.

From a theoretical point of view, radioactive decay, radiation attenuation, and radionuclide biokinetics have a formal and mathematical identity resulting from the presence of first-order equations. Radionuclide dosimetry is basically developed by considering topics such as activity and strength of source and the linear differential equation model of attenuation and scattering. The concept of point source is generalized to the point-kernel concept with application to a linear source as a superposition of point sources. Surface and volume sources are also considered. For internal dose assessment, basic concepts such as the air–kerma rate constant, source and target organs, and the committed effective dose are presented; and one-compartment radionuclide biokinetic models are used. Procedures for internal dose assessment, using Medical Internal Radiation Dose (MIRD) and International Commission on Radiological Protection (ICRP) models, are presented. The overall dosimetry for carbon-11-labeled tracers is estimated, which can be generalized to cover individual probes. Mathematical methods of cellular dosimetry, which are mainly concerned with electron dosimetry and the application of electron point kernels, are presented. In addition, the concepts of cross-dose and self-dose at cellular level will be introduced, together with other relevant mathematical aspects such as the geometric factor. As a modern approach to radiation dosimetry, the Monte Carlo approach is briefly reviewed.

The mechanisms of the biological action of ionizing radiation, specifically on DNA actions that will induce double-strand breaks and on the consequences of these events on the development of chromosome alterations, are then analyzed.

The following target-hypotheses (target-theory models) are considered:

1. One sensitive region, n hits
2. Multiple sensitive sublethal regions/one hit
3. Mixed model
4. Linear–quadratic model (L–Q)

Target theory and L–Q models are compared.

Finally, low-dose nontargeted effects complementary to the direct action of ionizing radiation are introduced.

2

Cyclotron and Radionuclide Production

Francisco J. C. Alves

CONTENTS

2.1   The Quantitative Aspects of Radionuclide Production

The radionuclei that lead to different nuclear medicine techniques are not
readily available in nature and need to be produced through nuclear reactions.
The main fundamental physics mechanisms used in radionuclide production
are fission, neutron activation, and particle irradiation. Fission and neutron
activation are performed in nuclear reactors. Exposing Uranium-235 to thermal neutrons produced in a nuclear reactor can induce the fission of the
uranium isotope, resulting in low (usually between 30 and 65) atomic number
nuclei. Some of the nuclei produced in this way can be chemically separated
from other fission fragments and are widely used in biomedical sciences, particularly in nuclear medicine. Fission is used, for instance, in Molybdenum-99
production, through the nuclear reaction

$$^{235}U + n \rightarrow {}^{236}U \rightarrow {}^{99}Mo + {}^{132}Sn + 4n.$$

Neutron capture activation is another production process that can be performed in a nuclear reactor. The following nuclear reactions are important
examples of this mechanism, as used in the production of Molybdenum-99
and Phosphor-34, respectively:

$$n + {}^{98}Mo \rightarrow {}^{99}Mo + \gamma,$$

$$n + {}^{32}S \rightarrow {}^{32}P + p.$$

Particle irradiation, leading to nuclide transmutation, is performed in a
cyclotron. A target material is irradiated with accelerated particles (protons,
deuterons, and in some cases, α particles), leading to a nuclear reaction. This
type of nuclear reaction, induced by cyclotron beam particle irradiation, can
be generically described using Bothe's notation:

$$A(a, b)B$$

The target nucleus appears before the first bracket and the resulting nucleus
after the final bracket. Inside the brackets, the irradiating and emitted particles
are separated by commas.

This chapter focuses mainly on the physical mechanisms involved in cyclotron radionuclide production, given the widespread use of this accelerator in nuclear medicine centers—particularly those using PET—as opposed to nuclear reactors, which continue to be located in specialist central laboratories outside the clinical or hospital environment. Nevertheless, the main concepts and principles, for example, cross section or excitation function, apply broadly to nuclear reactions, regardless of the type of reactor or accelerator used.

2.1.1 Nuclear Reactions

A process involving interaction with a particular nucleus leading to the alteration of its original state is known, in general terms, as a nuclear reaction.

A set consisting of a projectile and a target nucleus and their energy characteristics is called the *entrance channel* of a given nuclear reaction. The *exit channel* is the name given to the set of products of a nuclear reaction that are characterized by their respective energy or internal excitation status. If a certain channel is not physically possible (e.g., if not enough energy is available), it is said to be *closed*. Otherwise, it is *open*.

The total relativistic energy, momentum, angular momentum, total charge, and number of nucleons are conserved in a nuclear reaction. Parity is also conserved, as interactions in a nuclear reaction are defined by the strong nuclear interaction, which conserves this nuclear state wave-function property.

One important characteristic of a nuclear reaction is the difference between the kinetic energy of the initial participants and final products of the reaction in the laboratory referential system (in which the target nucleus is considered at rest). This difference is the Q *value* of the nuclear reaction. Representing the kinetic energy in the laboratory reference system as K^{lab} and the rest mass as m, in a generic nuclear reaction it can be stated that

$$Q = K_B^{lab} + K_b^{lab} - K_a^{lab} = [(m_a + m_A) - (m_B + m_b)]c^2. \qquad (2.1)$$

Second equality results from relativist energy conservation and demonstrates that Q is a characteristic of the reaction, regardless of the coordinate system used.

A nuclear reaction can be *exothermic* or *endothermic* depending on the positive or negative value of Q. In an endothermic reaction, $|Q|$ represents the minimum energy given to the initial reagents in the center-of-mass system to enable the nuclear reaction to take place.

In a real experiment setup, direct measurement of K^{lab} is not always simple, and momentum conservation is often used in a nonrelativistic approach to deduce an expression that enables Q to be calculated from K_a^{lab} and K_b^{lab} only:

$$Q = K_b^{lab}\left(1 + \frac{m_b}{m_B}\right) - K_a^{lab}\left(1 - \frac{m_a}{m_B}\right) - \frac{2}{m_B}\sqrt{K_a^{lab}K_b^{lab}m_am_b}\cos\theta_{lab} \qquad (2.2)$$

From this expression, an important intrinsic characteristic of endothermic nuclear reactions can be inferred: For each b product emission angle θ_{lab}, measured in the laboratory reference system in relation to the projectile incidence direction, a minimum projectile kinetic energy K_a^{lab} is required to allow the nuclear reaction to take place. This kinetic energy reaches its minimum value, called *threshold energy* (E_T), when $\theta_{lab} = 0°$.

In the laboratory reference system with A at rest, a projectile speed (v_a^{lab}) corresponding to a kinetic energy of $|Q|$ is not enough to induce a nuclear reaction. In fact, not all the projectile's kinetic energy will be available for the nuclear reaction, due to conservation of the center-of-mass momentum:

$$E_r = |Q|\frac{m_A + m_a}{m_A} \Rightarrow (v_a^{lab})_{min} = \sqrt{\frac{2|Q|(m_A + m_a)}{m_A m_a}} \tag{2.3}$$

Two main mechanisms can be observed in typical nuclear reactions induced by protons or deuterons in a cyclotron: *direct reactions* and *compound nucleus* mechanisms.

A nuclear reaction is called *direct* (or sometimes, peripheral) when it involves interaction leading to energy and/or particle transfer between the projectile and the outer (peripheral) nucleons of the target nuclei, without interference to other nucleons. Figures 2.1 and 2.2 illustrate some reaction types that can be considered direct.

A direct reaction involves only a small number of system degrees of freedom and is characterized by a significant overlap of the initial and final wave functions. Therefore, the transition from the initial to the final state takes

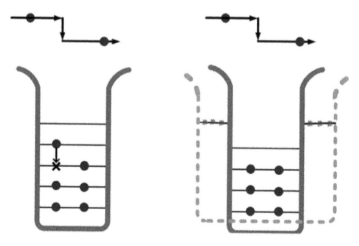

FIGURE 2.1
Diagram of examples of direct reactions, involving only energy transfer (elastic dispersion): excitation of a single nucleon (left) and collective excitation of a rotational or vibrational state (right).

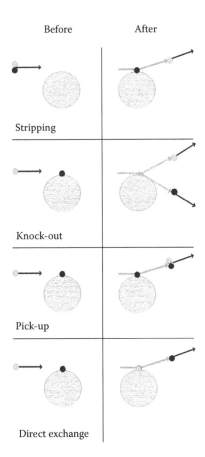

Before   After

Stripping

Knock-out

Pick-up

Direct exchange

FIGURE 2.2
Diagram of some types of direct reactions involving the exchange of nucleons.

place in a very short time (about 10^{-22} s, the order of magnitude of the transit time of a nucleon through a nucleus) and with minimum rearrange processes. Consequently, a strong interdependence between the pre- and postreaction energy states of the projectile, emitted particles and target nucleus can be observed. This interdependence causes (and is seen in) the anisotropic angular distribution of the emitted particles, resulting from a strong correlation with the projectile incident direction.

In a compound nucleus reaction mechanism, the projectile is captured in the target-nucleus potential; and a highly excited system, the compound nucleus, is formed. Projectile energy is distributed throughout all the compound nucleus nucleons, reaching a state comparable to thermal equilibrium. Even though the average energy per nucleon is insufficient to overcome the binding potential, since the system particle number is relatively small, important fluctuations in energy distribution will occur, until one or more nucleons gathers enough energy to exit the nucleus. The typical time interval between projectile target penetration and particle emission is in the order of 10^{-16} s.

It may be possible that, in a compound nucleus reaction, no particle is emitted and all the excess energy is released through γ radiation emission. This is the case in *capture reactions*, a mechanism predominant in thermal neutron irradiation (and used in nuclear reactor radionuclide production), but it is relatively infrequent in charged particle irradiation.

The process of forming the compound nucleus and the energy states that are assumed characterize and define the specific properties of this type of nuclear reaction.

Since the compound nucleus is essentially a nuclear excited state in which energy is distributed through many particles, it can assume many different states, called *many-particle states*. These excited states result from possible energy distributions among the different nucleons (which should not be confused with individual nucleon energy states) and are quantified states defined by quantic numbers and properties, such as spin or parity. However, their energy state value is associated with an intrinsic uncertainty resulting from Heisenberg's principle, corresponding to approximately an electron-volt (calculated on the basis of an energy state lifetime in the order of 10^{-16} s).

If the energy available from the nuclear reaction equals one of the many particles' energy states in the compound nucleus to be formed, a resonance phenomenon will occur and there will be maximum probability of compound nucleus formation. *Compound nucleus resonances* are characteristic of this type of nuclear reaction. However, these resonances are easy to observe only when the projectile is a relatively low-energy nucleon, because in many particles, the separation of energy states rapidly decreases as energy increases, and energy width increases at the same rate.

An important specific property of compound nucleus reactions is the verification of the *independence hypothesis*, according to which the formation and decay of the compound nucleus are independent processes [1]. As a consequence, the relative probabilities of possible decay mechanisms will be independent of each other and independent of the process leading to compound nucleus formation. Therefore, the emission angular distribution is expected to be isotropic.

Although the clear and total separation of direct and compound nucleus reactions may be pedagogically convenient, these mechanisms are not mutually exclusive over the whole projectile energy range for a given reaction.

Usually, even for the same projectile energy value, mechanisms of both types and intermediate processes can be observed, even when, as in most cases, one of the mechanisms is predominant. The relative importance of the different mechanisms to a given nuclear reaction depends on projectile energy and on the Q value, as can be observed in Figure 2.3. In this example, the upward trend (corresponding to a compound nucleus mechanism) is inverted when the (p,n) channel is opened, as this channel statistically competes with the (p,p$'$) channel for compound nucleus decay. The corresponding effect

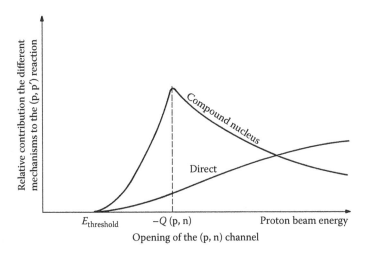

FIGURE 2.3
Relative contribution of direct and compund nucleus mechanisms to a (p, p′) reaction, as a function of projectile energy.

is particularly important, as, for emitted neutrons, the Coulomb barrier is transparent.

Experimental evidence of combined mechanisms can be seen in the degree of spatial anisotropy (even if sometimes very slight) that can always be found in the nuclear reaction emission. This anisotropy is present even in reactions characterized by compound nucleus formation, as a result of the interdependence between the entrance and exit channels due to the laws of conservation, as the initial constants are determined by the entrance channel.

2.1.2 Cross Section

The cross section quantifies the probability, per projectile current-density unit, of the occurrence of a given nuclear reaction. It is defined from the number N_{NR} of nuclear reactions induced when a beam of N_p particles of a given energy hits a surface with an N_{target} target nucleus per unit area. The incident beam is assumed to be parallel, monoenergetic, and composed of small-sized particles (compared with the target nucleus). It is also assumed that the nuclear reaction is induced only in a small proportion of the total number of targets, which can, therefore, be considered constant. In these conditions, the cross section σ for the production of a nuclear reaction is given by

$$N_{NR} = N_p N_{target} \sigma. \tag{2.4}$$

Cross-section dimensions are area dimensions and measure occurrence probability. This apparent ambiguity is explained by its physical significance, which can be understood by considering a simple classic collision problem between a point particle and a target particle with a given radius. Classically, every time the impact parameter is such that the incident particle hits the target particle, a collision will occur. However, the collision of a certain projectile, with a given energy against a specific target, does not always lead to the (expected) nuclear reaction. First, this is because it is a nondeterministic quantic phenomenon, for which a strict definition of radius is not logical and in which the nuclear force can be felt at a finite distance beyond it. Moreover, even if the collision is 100% probable for a certain impact parameter, as in classical mechanics, it should be noted that a nuclear reaction entrance channel can produce different exit channels. The physical phenomenon inherent to these concepts can be interpreted as a proportional diminution of the target particle area in a classic collision. The cross section then represents the effective area for the nuclear reaction to which it refers. However, it is not a property of the target alone; but it is a compound property of both target and projectile, reflected in the interaction between them.

In nuclear physics, it is convenient to use barn, b, as the cross-section unit. It is related to the standard area unit (square meter) as follows:

$$1 \, \text{barn} = 10^{-28} \, \text{m}^2. \tag{2.5}$$

2.1.3   Excitation Function

In any given nuclear reaction, each energy value for a projectile particle beam hitting a target has a corresponding cross-section value. The set of cross-section values in relation to particle beam energy is called the *excitation function* of the given nuclear reaction.

As an example, Figure 2.4 shows the excitation function of the ^{10}B(p,n)^{10}C reaction, obtained from cross-section values experimentally measured for several energies [2].

2.1.4   Excitation Function and Radionuclide Production

In a study aimed at establishing a production process for a given radionuclide, two key steps can be identified in which knowledge of excitation functions is vital: the choice of nuclear reaction to be used and the target incoming and outgoing projectile beam energy values for a given reaction.

Since several different nuclear reactions can lead to production of the same nuclide, knowledge of their excitation functions will determine which reaction has the better yield. In practice, experimental conditions limit the choice to the reactions induced by the projectiles available and the acceleration energy range. The respective excitation function is, therefore, an important selection criterion for the available reactions, initially, to prevent the choice of any reaction with a zero excitation function over the range of energies for which the

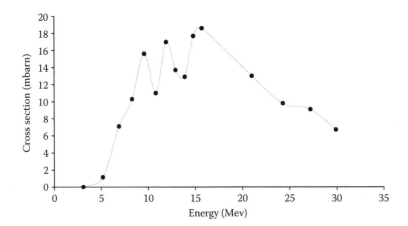

FIGURE 2.4

Excitation function for the ^{10}B(p, n)^{10}C reaction. The line shown is an interpolation from the experiment values, shown as black dots.

projectile beam can be accelerated and, second, to rank reactions in terms of their efficiency in producing the desired nuclide.

However, the process of establishing a protocol for a given radionuclide production is not confined to the choice of the nuclear reaction that leads to a greater physical production yield. A key aspect to consider is the physical and chemical characteristics of the constituent material of the target, which determine the physical feasibility of implementation and its physical and chemical stability during and between irradiation. These include mechanical properties; heat capacity and thermal conductivity; the melting, evaporation, and sublimation points; and chemical reactivity under the conditions contemplated. These features also influence the produced isotope extraction or separation process, in which the practicality, efficiency, and timing of the process are decisive factors. Moreover, in terms of the raw material of the target, it is inevitable that the relationship between commercial availability and price must be considered, which are, to a large extent, dependent on natural occurrence.

The process of choosing the nuclear reaction for the production of a given radionuclide involves, in most cases, a compromise between the excitation functions of the possible reactions and their implementability. In practical terms, the physical and chemical characteristics and the price or commercial availability of the target material often become the predominant factors. One example of this [3] is the preference for the proton irradiation of a pure water (H$_2$O) target, using the ^{16}O(p,α)^{13}N reaction, in the production of nitrogen-13 (^{13}N) in PET-dedicated cyclotrons. The cross-section values of this reaction for the typical range of energies (up to a maximum value in the order of two dozen MeV for protons) in this type of cyclotron are about half of those for the $^{13}C(p,n)^{13}N$ reaction. However, in practice, the higher excitation function of this possible alternative reaction is not sufficient, in itself, to make it the main factor when compared with the price difference between H_2O and highly enriched carbon-13 (^{13}C, only 1.10% in natural carbon) targets. Ease of implementation and cooling during the irradiation of liquid targets are added advantages in the use of a H_2O target.

Once a nuclear reaction has been selected for production, the next step is to choose the incident energy of projectiles on the target and the target thickness (which determines the value to which the incoming energy will be degraded). The resulting range of projectile energies across the target is of extreme importance in achieving the best possible compromise in terms of maximizing production efficiency in the desired radioisotope and minimizing the percentage of radioisotopic impurities in the final product.

The fact that, in any given nuclear reaction, the same entrance channel (characterized by a given projectile and target) can lead to different exit channels, depending on the available energy, is an important physical justification for the need for an appropriate choice of energy range. The number of open channels is generally low for low projectile energy values, but the number of nuclear reactions increases as the projectile energy increases, to compete with the reaction that generates the radioisotope and create impurities in the final product. Nonisotopic impurities can be removed chemically, whereas the percentage of isotopic impurities can only be maintained at acceptable levels by studying the range of energies that should be used. Any failure to maintain these levels makes a production protocol based on a particular nuclear reaction unviable, regardless of its physical yield. In this type of study, the knowledge of the excitation functions of the nuclear reactions involved is fundamental.

The process of choosing the parameters for the production of selenium-73 (^{73}Se), a sulphur analog with potential use in PET [4,5], is illustrative of the need to choose the incident energy and target thickness with a view to mitigating isotopic impurities and of the role played by the excitation functions in this choice. ^{73}Se, whose half-life is 7.15 h, may be feasibly produced by the reaction $^{75}As(p, 3n)^{73}Se$. However, the irradiation of Arsenic-75 (^{75}As) with protons leads to the formation of other selenium isotopes, such as Selenium-75 (^{75}Se; $T_{1/2} = 119.8$ days) and Selenium-72 (^{72}Se; $T_{1/2} = 8.4$ days), which can, due to their half-life, become significant isotopic impurities. The best compromise between maximization of ^{73}Se production and reduction of these impurities can be inferred from a comparative analysis of the excitation function for the $^{75}As(p,3n)^{73}Se$ reaction and the excitation functions for the reactions leading to the formation of isotopic impurities, namely $^{75}As(p, n)^{75}Se$ and $^{75}As(p,4n)^{72}Se$. Since these two reactions show higher cross-section values for energies below 20 MeV and above 40 MeV and the reaction that leads to ^{73}Se shows an excitation function with a maximum level of 30–50 MeV, it can be concluded [5] that a proton beam of 40 MeV should be used and an appropriate ^{75}As target thickness as well to degrade the beam to 30 MeV.

Apart from impurities due to a diversity of exit channels resulting from a single entrance channel in a nuclear reaction, other impurities are common in the final product: the products of nuclear reactions induced in impurities in the target material. Again, special importance should be given to isotopic impurities (in both senses, since in most cases, isotopic impurities in the target will lead to isotopic impurities in the final product), which cannot be chemically separated. To avoid this type of impurity, the logical solution is the use of isotopically enriched target materials. However, it is not always possible (or affordable) to obtain 100% enrichment. In cases where the target material is not a pure isotope, the choice of incoming and outgoing target beam energy, based on knowledge of the excitation functions, can once again prove a major contribution toward minimizing levels of isotopic impurities in the final product. A good example is the optimization of the production parameters of carbon-10 (^{10}C), an isotope used (as ^{10}CO$_2$) as a tracer for regional cerebral blood flow in PET, through the irradiation of boron targets [2,6]. Natural boron consists of two stable isotopes, ^{10}B and ^{11}B, with 19.9% and 80.1% natural abundance, respectively. The irradiation of ^{10}B with a proton beam of (at least) 30 MeV leads to the production of ^{10}C through the ^{10}B(p, n)^{10}C reaction whose excitation function is illustrated in Figure 2.4. The irradiation of ^{11}B with the same beam would lead to the formation of ^{11}C (unstable with β^+ decay and a 20.39 min half-life) through the ^{11}B(p,n)^{11}C reaction [7–9]. In the production of ^{10}C for the above mentioned PET application, the simultaneous production of ^{11}C is highly undesirable, given the comparatively longer half-life of this isotope (with inevitable consequences for the dose administered) and the impossibility of its chemical separation from ^{10}C. The use of 100% ^{10}B targets would be the solution to the problem, thus eliminating ^{11}C production. However, such targets are impossible to obtain, at least commercially. In a real-life situation involving the use of highly enriched targets of ^{10}B, but with some percentage of ^{11}B, the production of ^{11}C can be mitigated by careful choice of the irradiation parameters, including the proton beam energy and target thickness. Indeed, with an increase in beam energy (at least in the range where both the excitation functions are known) followed by proper (but not complete) target beam energy degradation, the relative percentage of ^{11}C can be reduced by more than 50%, while still maintaining a reasonable yield in the absolute production of ^{10}C [2].

2.1.5 Ion Kinetic Energy Degradation in Matter Interaction

In the previous sections, the possibility of ions accelerated in a cyclotron interacting with a target nucleus to produce nuclear reactions (which, in a broad sense, includes the phenomena of elastic and inelastic scattering) was demonstrated. However, these projectiles may also interact with the electrons in the electronic cloud of the target material or, in general terms, the matter. Coulomb interaction then plays such a key role that the degradation of the beam of particle energy from interaction with the nucleus becomes negligible in comparison with the result of interaction with the electronic cloud. Thus, given the use of ion beams, especially hydrogen ion beams, in the production of radionuclides in PET-dedicated cyclotrons, a knowledge of how beam energy degrades in crossing the target becomes very relevant.

The degradation of the energy of a beam of ions is almost entirely due to inelastic collisions with the bound electrons of the atoms forming the material crossed by the beam. The atoms of the target material will be ionized or excited with the energy that comes from the kinetic energy of the beam, which consequently, decreases. The decrease in beam energy during this process is almost continuous as it crosses the target material. The energy reduction rate decreases when it becomes small enough for the beam particles to capture electrons, but it continues until the energy of the beam is equivalent to the thermal energy of the atoms in the medium.

2.1.6 Stopping Power: Bethe's Formula

For ion velocities that are much greater than the typical velocities of electrons in the electronic cloud of atoms in the medium traversed by a projectile beam, it is possible to calculate the average energy degradation of the beam along its path in this medium. This degradation reflects the *stopping power* of this matter, for this beam. Under these conditions (up to a tenth of MeV for protons and one MeV for α particles), in which it is unlikely that the particle beam will capture electrons, it has been demonstrated that for nonrelativistic velocities [10,11]:

$$-\frac{dE}{dx} = \frac{4\pi z^2 q_e^4}{m_e v^2} NZ \log_e \left(\frac{2m_e v^2}{I} \right). \tag{2.6}$$

The left side of the equation is called the linear stopping power of the material, corresponding to the degradation of the average beam-energy per unit length of its path through the medium. In accordance with the international system, it should be expressed in joules per meter.

In nuclear physics, it is common to use MeV as the energy unit and CGS units, multiplied by the density of the medium material, for distance. In this context, the stopping power is expressed in MeV g^{-1} cm^2. It, therefore, corresponds to the linear stopping power divided by the density of the material (and thus only makes sense when the material is specified) and should be termed *massic stopping power*.

On the right-hand side of the equation, m_e and q_e are the mass and charge of the electron, z and v are the atomic number and velocity of the particles constituting the beam, and Z and N are the atomic number and number of atoms per volume of the material crossed by the beam, respectively. I is the average excitation potential of an atom of the material crossed by the beam. It represents the average energy, extended to all connected electrons, that can be transferred to an electron during the process of excitation, including ionization. It is substantially larger than the ionization potential of the atom and should not be confused with it. The value for most elements can be found in the literature that has emerged from experimental measurements. The precise theoretical calculation is complex, but some semiempirical expressions produce a good approximation, for example, [11]:

$$\frac{I}{Z} = 9.1(1 + 1.9Z^{-2/3}) \text{ [eV]}. \tag{2.7}$$

The expression for the linear stopping power presented above can be demonstrated and is, therefore, valid [12–14] in quantum mechanics or classical mechanics. For relativistic velocities of the beam, it should be corrected [15,16]:

$$-\frac{dE}{dx} = \frac{4\pi z^2 q_e^4}{m_e v^2} NZ \left(\log_e \left(\frac{2m_e v^2}{I} \right) - \log_e (1 - \beta^2) - \beta^2 \right). \tag{2.8}$$

This expression, in which β is the ratio between the speed of the beam projectiles and the speed of light in the vacuum, is known as the *Bethe formula*.

The relativistic correction terms express the effect of the increased rate of beam energy degradation with increasing beam particle speed for relativistic beam velocities. This effect [17] comes from the increase in the maximum energy that can be transferred to an electron (1st parcel) and from the Lorentz contraction of the particle beam Coulombian field, which allows energy to be transferred to electrons located further away (2nd parcel).

In the upper limit of relativistic velocities, when the beam particle energy, E_{beam}, verifies the condition

$$E_{beam} \geq \frac{m_{proj}}{m_e} m_{proj} c^2 \tag{2.9}$$

the spin of beam particles with mass m_{proj} becomes important [18,19], and even the relativistic expression for the linear stopping power loses validity. The energy degradation rate becomes increasingly dependent on the spin of the particle, and considerably higher energy degradation can be observed for spin-1 particle beams than for spin-$\frac{1}{2}$ (of protons) or spin-0 particle beams. In the same energy range, the density and dielectric properties of the medium should be considered. The electric field of the projectile felt in an atom that is off course decreases due to the polarization of the atoms in the medium, causing a decrease in the transfer of energy [20–24], an effect that becomes progressively more important as the effect of the interaction distance increases, due to the Lorentz contraction.

In the range of energies used in the production of radionuclides for biomedical use, if all these ultra-relativistic effects are disregarded, the relativistic expression for the linear stopping power or Bethe formula will remain valid.

In fact, for a proton beam, for example, the ultra-relativistic effects only become significant for energies in the order of 10^6 MeV.

In the lower limit, as previously stated, the expressions for the stopping power lose their validity for projectile velocities low enough for significant electron capture and the consequent neutralization of the beam. The probability of electron capture by completely ionized beam nuclei depends strongly on the ratio between their speeds. The physical principles involved were discussed in relation to a proton beam by Niels Bohr [25], who showed that the process can be completely different depending on the medium crossed by the beam and also calculated the probability of its occurrence in media with a high atomic number. He found, in this case, a proportionality to the sixth power of the ratio between the speed of the electron in the medium (considered the first Bohr orbit) and the speed of the proton beam, as opposed to the probability corresponding to media with low atomic numbers, which is proportional to the 12th power of the ratio of the velocities [26]. As a consequence of the neutralization process, the effective charge of the beam is reduced, leading to a reduction in the energy degradation rate over distance. Therefore, both the Bethe formula and the classic expression for linear stopping power overestimate the rate of spatial energy degradation in the low-energy range of the incident beam.

Given the assumptions contained in the definition of linear stopping power, it seems logical that the linear stopping power of a compound is given by the sum of the linear stopping power of each of the individual atoms that constitute it. This principle, which reflects Bragg's rule, simplifies to a large extent the calculation of the linear stopping power of compounds. However, the results are only an approximation (although usually very good). When a compound is formed, there is a change in the electron wave functions of the more peripheral layers. Therefore, the excitation potential will change, generally corresponding to a greater binding of the electrons involved. An increase in the value of the electron excitation potential (which is expected to be in the same order of magnitude as the energies corresponding to chemical bonds, i.e., a few units of electron volts) will, of course, result in a slight change in the value of the mean excitation potential for all electrons, I. This can be up to 2%, corresponding to a linear decrease in stopping power in the order of tenths of a percentage. The effect is less evident in a compound made of heavy elements, where the percentage of electrons in the element involved in chemical connections is smaller.

2.1.7 Range

The distance that a beam of a given energy has to travel through a given material, if this energy is to degrade completely, is called *range*. The range R of a beam of given energy in this material is, therefore, the integral of the inverse of the linear stopping power for the beam in this material, where the integration limits are the value of the energy of the beam when entering the material, E_i, and zero:

$$R = \int\limits_0^{E_i} -\frac{\mathrm{d}x}{\mathrm{d}E}\,\mathrm{d}E. \qquad (2.10)$$

2.1.8 Straggling: Statistical Fluctuation in Energy Degradation

Bethe's formula calculates only an average rate of linear energy degradation. In fact, the energy of a charged projectile does not continuously decrease as the formula may suggest; instead, there are a large number of small but finite decreases, corresponding to interactions with the electrons in the medium. There are two implications of the consequent fluctuation around the average value that occurs for each projectile-medium pair. First, for a given trajectory of the projectile in the medium, the degradation of energy fluctuates around a mean value. In addition, the path of the beam that corresponds to the energy degradation also oscillates around a statistical average. The phenomenon that is manifested by these statistical fluctuations is called *straggling*.

It is possible to theoretically study the variance in the amount of energy degradation in projectiles with the same initial energy crossing a thickness x of a given medium,

$$\left(\sigma_E^2\right)_x = \left(\langle\varepsilon^2\rangle - \varepsilon_0^2\right)_x, \qquad (2.11)$$

where ε_0 denotes the average energy degradation and $\langle\varepsilon^2\rangle$ denotes the expected value for the squared value of energy degradation in a single event (note that the expected energy value for a single event $\langle\varepsilon\rangle$ will correspond to the average energy ε_0). Livingston and Bethe [27], using quantum assumptions, arrived at the expression:

$$\frac{\mathrm{d}}{\mathrm{d}x}\left(\sigma_E^2\right)_x = 4\pi q_e^4 z^2 N\left(Z' + \sum_n k_n \frac{I_n Z_n}{m_e v^2}\log_e \frac{2m_e v^2}{I_n}\right). \qquad (2.12)$$

The notation is consistent with the one used in Bethe's formula. Z' denotes the number of electrons in atoms that are not present in the more internal layers for which the average energy of excitation is greater than $2m_e v^2$. The sum is performed for all layers not excluded by this criterion, weighted by the constant k_n, with Z_n being the number of electrons in the nth layer, whose average excitation energy is I_n.

For high energies (compared with $2m_e v^2$), the sum can be ignored and Z' can be replaced by the total number of electrons in the atom, to arrive at the expression deduced by Niels Bohr [17] using only classical assumptions [28]:

$$\frac{\mathrm{d}}{\mathrm{d}x}\left(\sigma_E^2\right)_x = 4\pi q_e^4 z^2 N Z. \qquad (2.13)$$

Consider, moreover, that a beam of projectiles from the same initial energy experiences the energy degradation when crossing a given medium. The variance in the value of the corresponding distance, using a notation similar to the one above, is,

$$\left(\sigma_x^2\right)_E = \left(\langle x^2\rangle - x_0^2\right)_E. \tag{2.14}$$

Once again, the expected value of the distance travelled in a particular instance $\langle x \rangle$ will correspond to the average distance x_0. Thus, if dx and dE are an infinitesimal distance and energy degradation, respectively, and if dE is the mean energy degradation corresponding to the distance dx:

$$\left(\sigma_E^2\right)_{dx} = \left(\frac{dE}{dx}\right)^2 \left(\sigma_x^2\right)_{dE}. \tag{2.15}$$

If this expression is integrated, the value of σ_x^2 for the degradation of the finite energy E can be obtained:

$$\left(\sigma_x^2\right)_E = \int_0^E \frac{d}{dx}\left(\sigma_E^2\right)\left(\frac{dE}{dx}\right)^{-3} dE. \tag{2.16}$$

Using the Livingston and Bethe expression and integrating from the initial beam energy E_i to total energy degradation, the variance in range value, σ_R^2, is obtained:

$$\sigma_R^2 = \int_0^{E_i} 4\pi q_e^4 z^2 NZ' \left(\frac{dE}{dx}\right)^{-3} \left(1 + \sum_n k_n \frac{I_n Z_n}{m_e v^2 Z'} \log_e \frac{2 m_e v^2}{I_n}\right) dE. \tag{2.17}$$

This expression can also be simplified for high energies in the incident beam leading to

$$\sigma_R^2 = \int_0^{E_i} 4\pi q_e^4 z^2 NZ \left(\frac{dE}{dx}\right)^{-3} dE. \tag{2.18}$$

The study of linear stopping power based on Bethe's formula only considers interactions with the electrons in the medium, resulting in an energy transfer that is small in comparison with the total energy. The energy degradation experienced by each particle of a monoenergetic beam crossing the same distance in the medium can be seen as an independent statistical phenomenon. Considering only interactions with the electronic cloud, it is expected that the distribution of the energy degradation value in each of the large number of beam projectiles shows a Gaussian distribution around the mean value.

However, the energy degradation in each nuclear interaction, although so comparatively rare that its contribution to the stopping power of the medium can be ignored, is significantly higher than that which is found in every electronic interaction, which has a considerable influence on the statistical fluctuations under discussion. A perhaps more important consequence is that the distribution of the energy degradation value is not Gaussian around the average, but it shows a "tail" toward the higher degradation values. This is particularly true for heavy projectiles (e.g., fission products) at distances close to the range value [25].

2.1.9 Energy Degradation versus Ionization

An ion passing through a given medium may interact with an electron, transferring enough energy to overcome the ionization potential of the atom to which it is connected. In this case, it causes an ionization known as *primary* ionization, as it is the direct result of projectile–electron interaction. In most cases, the electron is ejected from the atom with relatively low kinetic energy in comparison with the ionization potential. However, some primary ionizations result in electrons ejected with energies greater than or equal to the ionization potential of the medium. These electrons, called *delta rays*, can lead to the *secondary* ionization of most atoms in the medium. Total ionization is the sum of primary ionization, in which the number of ions produced equals the number of ejected electrons, and secondary ionization, in which the electron resulting from primary ionization ionizes one or more atoms.

One remarkable experimentally demonstrated property of the ionization caused by the energy deposition of charged projectiles is that the energy degradation per ion formed, w, is practically independent of the energy, charge, and mass of the particle that gives rise to ionization, whether primary or secondary. Considering the total energy available for ionization, regardless of whether the carrier particle is a beam particle or a secondary electron, this property can be qualitatively explained. On average, each excitation leads to a W_{exc} reduction of available energy. Similarly, each ionization resulting in a low kinetic energy electron (unable to produce secondary ionizations) reduces, on average, the energy available by a W_{ion1} amount. If an electron with enough kinetic energy to induce secondary ionizations results from this ionization, the accountable reduction in available energy will be equal to the ionization potential of the medium, V_I, as the remaining energy will be available for secondary ionizations (in which the respective degradation will be accounted for). With σ_{exc}, σ_{ion1}, and σ_I, the cross-section values for each of these processes, the average amount of energy required to form an ion will be given by

$$w = \frac{\sigma_{exc} W_{exc} + \sigma_{ion1} W_{ion1} + \sigma_I V_I}{\sigma_{ion1} + \sigma_I}. \qquad (2.19)$$

The approximate constant value of w is then explained by the low (compared to the other cross sections) value of σ_I and the fact that the energies W_{exc} and W_{ion1} and the ratio between the σ_{exc} and σ_{ion1} cross-section values are virtually independent of the ionizing particle energy [29].

The almost constant ion beam energy degradation value per ion formed in the medium crossed and the consequent analogy between the distribution of total ionization along the path of the projectile and the energy degradation rate have been widely used [30,31] to determine the energy of charged particles, using the number of ions created in a known medium as a first approximation.

2.1.10 Thick Target Yield

The amount of radionuclide that is expected to be produced by a nuclear reaction in a target of a given thickness can be calculated by integrating the reaction excitation function over the range of beam energies on target. The limited practical interest in the absolute quantity of radionuclides produced without taking into account the decay during irradiation paved the way for a generalized measure of the efficiency of a nuclear reaction using the so-called *thick target yield*, the radioisotope activity expected to be induced by the irradiation of a thick target, per unit of beam current, in saturation.

Saturation condition is reached when there is a balance between the number of nuclides being produced and decaying in constant rate production. Given the large number of target nuclei and the relatively low cross section of nuclear reactions involved, the reduction in the number of target nuclei during a production process (such as the one carried out in a PET-dedicated cyclotron) can be considered negligible. The rate of production will be constant if the projectile beam intensity is constant, which happens in most protocols for routine production. If P represents the constant rate of production of a radionuclide with a decay constant λ, the temporal variation of the number N of nuclides in the target is given by

$$\frac{dN(t)}{dt} = -\lambda N(t) + P. \tag{2.20}$$

Considering the activity A is nil at the moment production begins, the solution for this differential equation is

$$A(t) = \lambda N(t) = P(1 - e^{-\lambda t}). \tag{2.21}$$

As time t goes on, the activity and the rate of production tend, asymptotically, to equalize, as shown in Figure 2.5, where time is counted in units of the half-life of the radionuclide in question. It should be observed that although, strictly speaking, saturation is only achieved for an infinite time, after a time

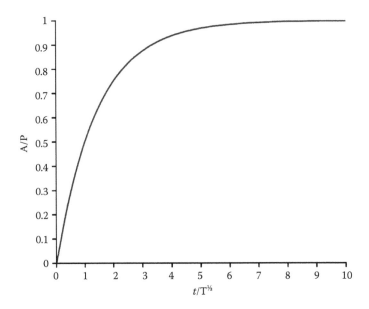

FIGURE 2.5
Evolution of the relationship between the activity and the rate of production, assumed constant, of a radionuclide, in units of its half-life.

equal to about a dozen half-lives it does not make sense to extend irradiation any longer (at this rate of production), because the number of nuclides produced will have already practically compensated for those which have decayed. Thus, in this condition, the activity of the radionuclides produced at the end of irradiation does not depend on irradiation time.

The thick target yield Y can be calculated from

$$Y = \frac{N_A H}{M z q_e} \int_{E_{inc}}^{E_{em}} \left(\frac{dE}{dx}\right)^{-1} \sigma(E)\, dE, \qquad (2.22)$$

N_A is the Avogadro number, z the projectile atomic number, and q_e is the electron charge. M and H are the atomic mass and the target material isotopic enrichment, respectively. The inverse of the linear stopping power multiplied by the excitation function $\sigma(E)$ is integrated along the beam energy values, from entering (E_{inc}) to emerging from the target (E_{em}).

The convenient practical units of thick target yield used in daily production of radionuclides in PET-dedicated cyclotrons are MBq/μA sat, where *sat* is an explicit indication that the measurement refers to saturation condition.

It is also common for the thick target yield to refer not to a range of energies but only to an energy value. This value indicates the energy from the incident beam, which, it is assumed, will be completely stopped in the target. The corresponding thick target yield corresponds to the highest activity that can be produced using the nuclear reaction in question per beam current unit acting on a target with this energy. This explains the use of diagrams of the thick target yield as a function of this energy, which are very common in literature on the subject.

The measurement of the experimental thick target yield for different energies should lead to values consistent with those obtained in theoretical calculations based on the excitation function. It should be noted, however, that an experimental measure refers to a target under certain experimental conditions. The approach, common in literature on the subject, that characterizes a reaction based on thick target yields experimentally measured for various energies of the incident beam, though pragmatic, is still empirical, as the basic parameter in the production of a radionuclide using a nuclear reaction is the cross section and not the yield of a particular target, which reflects the particular conditions during the process.

Further, the thick target yield calculated from the excitation function represents the maximum yield that can be obtained for a target. Consequently, accurate knowledge of the excitation function and, therefore, the thick target yield of nuclear reactions that lead either to the desired radioisotope or to impurities in the final product is crucial in the planning, development, and evaluation of the performance of high-yield targets.

2.1.11 Experimental Measurement of the Excitation Function: The Stacked Foils Methodology

The experimental measurement of excitation functions in nuclear reactions typically produced in a low- or medium-energy cyclotron is traditionally performed using a method commonly termed *stacked foils* [32–38]. This experimental approach can be used independently of the particles of the cyclotron emerging beam [39–43], and its applications also extend to monitoring beam energy and/or intensity [34–46].

The technique involves the irradiation of a number of layers aligned perpendicular to the beam, as shown in Figure 2.6, followed by measurement of the activity induced in each layer, usually by counting the gamma radiation emitted in the decay product of the nuclear reaction induced.

The incident beam is degraded as it crosses the different layers of the stack and, therefore, assumes a different value for each foil. By ensuring the geometry leads to a perpendicular incidence of the beam in the stack of layers, it can be assumed, with good approximation, that the size of the trajectory of any beam particle in each layer—and consequently the energy degradation—simply corresponds to the thickness of the foil, avoiding additional calculations for the angle of incidence.

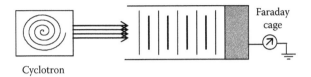

Cyclotron

FIGURE 2.6
Diagram showing the traditional configuration for stacked foil irradiation. The beam of particles from the cyclotron perpendicularly enters the stack consisting of several target sheets (tall and thin in the diagram) interspersed with degrading layers (short and thick). After the last foil, a thick metal block ensures that the beam is completely stopped. The layers form a Faraday cage, which enables the intensity of the beam current to be measured.

Thus, for a given incident beam, it is possible to establish a direct relationship between the beam energy in each foil and the number of nuclear reactions induced (proportional to the activity) in the same foil. If the incident beam intensity is known, the nuclear reaction cross section can be calculated from the activity induced in a given foil for the value of beam energy at that foil. Combining different pairs of energy values—the cross section for the different layers of the stack—allows the excitation function of the nuclear reaction being studied to be obtained.

Conversely, if the cross section of the nuclear reaction is known, the induced activity measurement in each of the stack foils can be used to calculate the energy and/or intensity values of the incident beam.

The effective beam energy in each foil can be calculated from the energy value of the incident beam (corresponding to the energy of the beam leaving the cyclotron toward the first layer), using experimental tables such as Williamson's [47] or calculations from the theoretical expressions of the linear stopping power.

The cross-section values are determined, according to their definition, from the number of nuclear reactions produced and the number of particles in the beam passing through the surface of each foil. It is assumed that only a negligible fraction of beam particles are absorbed (to induce the nuclear reaction) on target. A beam transverse section entirely projected within the target area (and, therefore, smaller than this area) is assumed—an important issue to be assured when performing the experiment.

The number of projectiles, N_P, charge, q and atomic number, z on the surface formed by the target particles is related to the intensity of the beam that passes through this area by

$$N_P = \frac{1}{qz} \int_{\Delta t_i} i(t)\, dt, \qquad (2.23)$$

where $i(t)$ is the beam current value at time t and the integral is extended over the irradiation time, Δt_i. For a given beam current value I, constant over the irradiation time, this expression is simplified to

$$N_P = \frac{I \Delta t_i}{qz}.$$ (2.24)

Usually, foils are electrically connected to each other. Behind the last one, a block driver thick enough to completely degrade the energy of the emerging beam is placed. This set is a Faraday cage (represented in Figure 2.6), where the total collected charge is recorded. When the intensity of the beam current is constant, it enables its value to be measured directly.

The measurement of the beam current obtained by the Faraday cage is often confirmed or completed with the one obtained by a monitored reaction: between the blades of the primary battery, other blades are included (represented as shorter and thicker lines in Figure 2.6), for which the cross section for irradiation with the incident beam particles is known. Given the energy of the incident beam in the first foil (and hence calculating the effective energy in each of the following primary and monitor foils), the activity induced in the monitor foils enables the value of the integrated intensity of the beam current to be calculated.

The intensity of the beam current, measured by the Faraday cage or monitor foil reaction, is assumed to be constant throughout the stack. This assumption loses validity when the cross section of the reaction is relatively high and the stack consists of a large number and/or thick foils. In this case, since a large number of particles in the incident beam is lost in the stack to produce the nuclear reaction under study, the degradation of current intensity along the stack must be known or, at least, estimated [48,33].

The stacked foil technique can also be used to calculate the beam energy if the excitation function of the reaction involved is known. Since the shape of the curve of the excitation function is independent of the beam intensity, the curve obtained from measuring the foils can be used as a reference. In this procedure, the beam current can be assumed constant along the stack, and measurement is not necessary. Moreover, an even simpler reference can be used: the energy threshold, the minimum energy of the incident beam that induces the nuclear reaction, which, in practice, leads to a good approximation, due to the typical exponential growth of the cross section with the increasing energy of the incident beam above the threshold.

The energy resolution of such measurements is mainly determined by the thickness of the foils and is typically in the order of one tenth of an MeV.

To obtain the highest possible number of points in the excitation function in a single irradiation while ensuring a constant beam current along the stack as a valid approximation and avoiding excessive degradation of energy in each foil (with the consequent loss of resolution in energy and the risk of masking a possible fine structure), it is fundamental that the thickness of the layers is kept as low as possible. This premise is even more important for low energies, as confirmed by the Bethe formula, and low energy threshold reactions, as it typically corresponds to a region of sharp variation in the excitation function (with a sharply sloping increase). However, the foils must be thick enough to ensure statistical relevance in the measurement of the activity produced.

Further, an excessive number of layers increases the size of the stack, which can lead to excessive dispersal of the beam (deflecting it beyond the limits of the target area) and progressive deviation from the perpendicularity of the beam-foil geometry. In each reaction studied, it will therefore be necessary to find a good compromise between the number and the thickness of the foils.

In the case of ductile metals (aluminum, titanium, copper, silver, gold, etc.), foils of different thicknesses are commercially available. The same is not true in the case of nonmetallic or amorphous materials, which require the use of special preparation techniques to obtain a solution similar to a thin foil. These techniques [49–55] include sublimation and vacuum evaporation (with which it is possible to obtain thin films in a suitable medium, similar to a metal blade), uniform suspension of the target material in self-sustained polystyrene film, and electrolytic deposition. The most common practical method for substances in the form of powder or small particles is compression against a strip of metal at a pressure of several metric tons per cm^2 to obtain a thin layer.

In the evaluation of thin foil (or foil analog) production techniques, particular attention must be paid to certain characteristics of the final product. First of all, with regard to the thickness, it should be of an appropriate size and uniform throughout the target area. A nonuniform thickness would result in a variation in energy degradation along the transverse section of the incident beam, with the consequent loss of simplicity (and probably the inviability) of the method. Second, but equally important, the chemical purity and degree of isotopic enrichment of the material should also be taken into consideration: immeasurable and uncontrollable changes in these characteristics during the manufacturing process may also make the technique unviable.

There are three major sources of uncertainty when applying the stack foil technique: determination of the intensity of the beam current, absolute determination of the activity induced in each of the foils, and beam energy calculation in each of the foils. Usually, measuring the beam current through a Faraday cage would be more precise than using the monitor reactions, although the secondary electrons produced are a source of inaccuracy. Measurement using monitor reactions is subject to accurate knowledge of the excitation function and, since the monitor foils are themselves a stack of foils, to the same sources of uncertainty as the primary reaction studied. Moreover, despite several proposals for monitor reactions [44–46] and the fact that they are widely used, there is no standard [56], making it difficult to compare and critically evaluate the excitation functions measured by different research groups.

In assessing the results of measurements of beam current intensity, errors inherent in the constant current approximation or in the current decrease along the stack estimate or calculation must be also considered. Even with a suitable choice of foil thickness, there will always be uncertainty in determining the absolute activity induced in the foils. This uncertainty originates in the decay data in question (e.g., uncertainty regarding the amount of gamma rays produced in decay) and, especially, uncertainty in the efficiency of counting statistics. Calibrated sources are commonly used, allowing knowledge of the counting efficiency energy curve of the gamma rays to be detected. The most common examples are Europium-153 (^{153}Eu), Sodium-22 (^{22}Na), Cobalt-57 (^{57}Co), Manganese-54 (^{54}Mn), Barium-133 (^{133}Ba), Cesium-137 (^{137}Cs), Yttrium-88 (^{88}Y), and Mercury-203 (^{203}Hg). The contribution made by statistical counting error to the overall uncertainty can be diminished by a suitable choice of time length for stack foil irradiation and measurement of activity.

Calculation of the effective energy in each of the stack foils is based on knowledge of the decay energy of the beam in the material. Thus, the precision of the table values for energy degradation or errors associated with their theoretical calculation, together with the precision of foil thickness measurement are directly reflected in the precision of the excitation function energy scale when using the stacked foil technique.

Uncertainty regarding foil thickness depends heavily on the technique used to manufacture the foil. In commercially available ductile metal foils, uniform thickness is guaranteed by the supplier with a very good level of accuracy and reduced uncertainty. Otherwise, the accuracy and uncertainty regarding the measurement and uniformity of thickness depend on the technique used in their production. These sources of error and uncertainty may become more significant than beam energy fluctuations and/or variations along its transverse area.

The total uncertainty in cross-section measurements using the stacked foil technique is typically in the order of 10–15%, rising to 25% in low-yield reactions [32–38,57].

2.2 The Cyclotron: Physics and Acceleration Principles

The cyclotron is perhaps the most convenient accelerator within the energy range that is of interest to the biomedical sciences, with the added benefits of its compact and relatively small size. It has been regularly used in the production of radionuclides for biomedical purposes since the 1950s and became unquestionably important in the 1980s, with the increased use of PET and the growth of respective centers, where it is the accelerator of choice for the dedicated production of positron emitters.

The most common variant is the isochronous cyclotron, which enables high-intensity beams with uniform energy to be produced, simultaneously overcoming both the energy limitations of conventional cyclotrons and the technological complexity of frequency modulation, the significant lower time rate of accelerated particles, and the higher costs of synchrocyclotrons. Further, the theoretical and technical difficulties associated with the complexity of the magnetic field of the isochronous cyclotron, often leading to fine adjustments and empirical models, is becoming less of a disadvantage, due to mass production as a result of increasing demand (to a large extent due to the expansion of PET). The fact that nowadays the serial production and installation of isochronous cyclotrons designed for proton acceleration to energies below the relativistic limit of a conventional cyclotron is becoming more and more common is proof of this.

It is important to mention the two main types of cyclotron typically installed in PET centers that are almost exclusively dedicated to the production of positron emitters used in this technique, generally known as PET-dedicated cyclotrons: the cyclotrons designed for the acceleration of protons and deuterons up to 18 and 9 MeV, respectively (depending on the model), and the models usually known as *baby-cyclotrons*, which are able to accelerate (only) protons up to an energy value of approximately 11 MeV, depending on the model.

In this subchapter, the physical working principles of the cyclotron are described, ranging from the conventional models to the most common variants.

2.2.1 Introduction and Historical Background

A nucleus can be passively studied by observation of the radiation emitted from nuclear decay or its interaction with the electronic surroundings. It was studied in this way until the pioneering work of Lord Rutherford, who used natural emitted α particles to irradiate the nuclei of different elements, demonstrating the importance of studying the nucleus through its interaction with accelerated projectiles. After his work, many efforts were made to develop artificial acceleration methods for nuclei and subatomic particles at similar or higher energies than those obtainable from natural radioactive sources.

Cockroft and Walton accelerated protons by applying an electrostatic potential of around 500 kV, aiming the beam directly at Lithium nuclides and observing the emission of two α particles, the first nuclear transmutation induced by artificially accelerated particles [58]. Several methods using the same acceleration principle and applying an electrostatic potential to obtain kinetic energy from charged particles in a vacuum tube were then attempted. The most successful was Van de Graaff's model: using a moving belt as physical charge carrier, it was capable of creating electrostatic potentials corresponding to energies higher than MeV in an insulated electrode. However, this design created technical problems in terms of implementation related to control and insulation, which increased when higher energies and potentials were needed.

Aiming at avoiding the use of extremely high potentials, Wideröe proposed and demonstrated an acceleration method based on the use of small, repeated acceleration forces [59]. The method, which is the principle of linear accelerators, involves the use of several hollow cylindrical electrodes along a glass pipe axis in a vacuum (Figure 2.7). The electrodes are alternately connected to the poles of an alternating constant frequency current generator, with the first, third, fifth, and so on set at positive potential when the second, fourth, sixth, and so on are at negative potential.

Assuming a source emits a positive ion at the end of the vacuum tube when the first electrode is at its highest negative potential, the electrical field between the ion source and the electrode will accelerate the ion, which will move into the inside of the cylinder, provided that the source is placed at the assemblage symmetry axis. Inside the cylinder, a region without the interference of an electrical field, the ion will maintain a constant speed. With the appropriate relationship between the cylinder length and the ion speed established inside, the ion will reach the electrode end at exactly the same moment the applied potentials (for all the electrodes) are changed. Therefore, the ion will be subjected to a new acceleration along the gap between this electrode and the next, where it will assume a new speed value. This process is repeated along the accelerator, and the resulting final energy is the sum of all the small increases acquired between electrodes. Moreover, if the frequency of the generator alternating current is constant, the time the ion takes to go from one electrode to the other is always the same. Since the ion speed is always increasing, the pipe length must increase proportionally.

The biggest technical difficulty in the practical implementation of high energy accelerators based on Wideröe's method is their length, as the electrical field frequency applied cannot be made infinitely high.*

A skilful solution to the length problem was proposed by Ernest O. Lawrence using only two large electrodes facing each other and a magnetic field to deflect the path of the ions into circular orbits, forcing them to return again and again to the acceleration gap between the electrodes in opposite directions in consecutive passages, in such a way that the energy increase builds up. As a principle, it is similar to a Wideröe accelerator rolled up in a spiral. At each accelerating zone passage, the magnetic field creates a bigger circular ion path radius, due to the energy increase.

Lawrence found that the motion equations predicted a period of constant revolution, enabling the ions to be accelerated in resonance with the oscillating electric field between the electrodes. He published his idea [60] before it was verified in experiments carried out by Livingston, his PhD student [61].

---

* Historically, by the time the first linear accelerators had been developed, the generation of high radio frequency values was a major limitation. By then, the upper limit was close to tens of MHz; and this was only overcome by enormous technological advances during World War II, essentially related to the development of radar, finally giving rise to the production of GHz radio frequency generators.

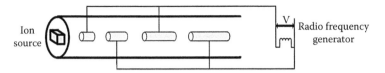

FIGURE 2.7
Simplified diagram of a Wideröe linear accelerator.

After that, they installed [62,63], the first *magnetic resonance accelerator*, which would become known as the *cyclotron*, the jargon term used by the authors for their accelerator. Inside a vacuum chamber, two hollow copper electrodes are installed, called Ds due to their shape,* which resemble a thin cylinder cut in two along the diameter. A radio frequency power supply provides a sinusoidal alternating electric field between both electrodes, although the electric field inside them is null. A vacuum chamber is placed between the poles of a big magnet, fed by a coil that produces an almost uniform magnetic field, perpendicular to the radial plane of the Ds. The ion source is located at the center of the vacuum chamber in such a way that the low-speed ions leaving the source are accelerated in the gap between electrodes and enter the D charged with the opposite potential. Inside the D, the electric field is no longer felt by the ions; and the uniform magnetic field determines their motion, describing a semicircular orbit on a plane perpendicular to the field, until they reach the gap between the Ds again. Since the direction of the electric field in the acceleration region is inverted in the meantime, the ions will be accelerated again, entering the other D and describing a semicircular path with a greater radius. This process continues following to the path outlined in Figure 2.8.

Lawrence was awarded the 1939 Nobel Prize for Physics for the invention and development of the cyclotron.

2.2.2 The Resonance Condition

The cyclotron magnetic field does not assist in increasing the kinetic energy of the accelerating ions.† Its aim is to drive them to the region between the electrodes, where the accelerating electrical field is felt. However, its value must be such that the time required for the ion to complete the semicircle (inside the D) is half the electrical field oscillation period, that is, the alternating electrical field inversion time. If this condition (the so-called *resonance*

---

* The name D is still used, even when a different shaped electrode is used.
† In principle, the cyclotron could be used to accelerate any charged particle. However, practical limitations (relating to the source or to particle stability) or inherent to the accelerator principle (such as the relativistic limit, which will be discussed later) restrict it to ion acceleration. Therefore, in this text, we use this term to describe the body that is to be accelerated.

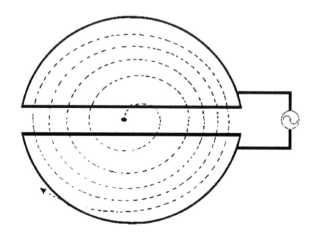

FIGURE 2.8
Simplified diagram of a cyclotron. Outline view of the ion acceleration path.

condition) is met, the ion speed increases every time it crosses the gap between the Ds.

An ion with a given mass m and a given charge q, moving on a plane perpendicular to the magnetic field with a speed modulus v, experiences a force in the direction given by the vector product of the speed and the magnetic field vectors. Its modulus is given by

$$\left| \vec{F} \right| = qvB, \tag{2.25}$$

where B is the magnetic field vector modulus. Under the effect of this force, the ion describes a circular path with radius r. Since the centripetal force equals the magnetic force:

$$\frac{mv^2}{r} = qvB \Leftrightarrow mv = qBr. \tag{2.26}$$

The frequency of the circular path revolution, f_r, is given by

$$f_r = \frac{v}{2\pi r} = \frac{qB}{2\pi m}. \tag{2.27}$$

This expression shows that the revolution period depends only on the magnetic field and on the relation between the accelerating ion charge and mass. The time interval during which it is inside the D is independent of its speed and the radius of its path.

The resonance condition implies that the alternating electrical field frequency f applied to the electrodes is equal to the circular path revolution frequency, f_r:

$$f = \frac{qB}{2\pi m}. \tag{2.28}$$

This expression is the *cyclotron resonance fundamental equation* and shows the relation between the supposedly uniform magnetic field and the alternating electrical field frequency applied to the electrodes (often called *cyclotron frequency*), which enables the acceleration of an ion with a given charge or mass ratio to take place.

An important intrinsic characteristic of cyclotron resonance is that all the ions from a source with a certain charge to mass ratio are accelerated and not only the ones that "find" the maximum value of the alternating electric field. The entire half cycle with the same polarity is used, which makes an important contribution to high performance in terms of the number of accelerated particles per time interval in this accelerator. The ions reaching the gap between the electrodes in the phase corresponding to the highest potential value will go through it in the same phase, reaching maximum energy level within a smaller number of revolutions. The ones that find the accelerating potential in different phases of the same polarity will experience smaller energy growth (depending on the corresponding potential phase) but will remain in resonance, as the angular frequency will be the same. They will continue to cross the accelerating gap in the same phase, reaching the maximum radius (and corresponding energy) after a greater number of revolutions. In this way, a large number of ions are kept in resonance within the spatial region between the Ds, with an azimuthal length corresponding to the resonance condition phase band. In each radio frequency cycle, a group of ions starts its orbital path from the source, corresponding typically to several tens of revolutions and meters covered. These groups are only discrete for the first revolutions and end up overlapping in a radial distribution of all possible energies that is almost continuous. Each ion's path in the cyclotron depends on its initial conditions, especially the position and time relative to the radio frequency cycle. On the other hand, using the cyclotron resonance fundamental equation and the revolution frequency expression, it is possible to relate the accelerating ion kinetic energy, E_c, with the magnetic field and the orbit radius r described in a given time:

$$E_c = \frac{1}{2}mv^2 = \frac{1}{2}\frac{q^2}{m}B^2r^2. \tag{2.29}$$

This relationship helps explain the preference for cyclotron implementation with a high magnetic field (and, therefore, high frequency) and high applied voltage amplitude. The use of magnetic fields with Tesla magnitude and corresponding frequencies with an order of magnitude of tens of MHz allows high kinetic energies to be obtained, minimizing the technical difficulties in large high vacuum chambers under a uniform magnetic field. On the other hand, the increase in voltage amplitude applied to the Ds allows for an average reduction in the accelerating ion revolution numbers, reducing the time they stay inside the cyclotron and, consequently, the probability of loss by collision with any atoms or molecules that remain, due to the impossibility of absolute vacuum.

2.2.3 Magnetic Focusing

2.2.3.1 Axial Focusing

The basic principle of cyclotron resonance applies only to ions in an ideal orbital movement in the cyclotron chamber median plane, crossing the acceleration gap synchronously with the electrical field. In a uniform magnetic field with perfectly parallel field lines, the ions travelling with a small deviation in the axial angle in relation to the *equilibrium orbit* in the cyclotron median plane will follow a helicoidal path, colliding with the inner surface of the *D*. The number of ions axially deviating from the equilibrium orbit can be quite large, given the chance of scattering collisions, which are significant in the first accelerating orbits (due to lower accelerating ion energy, high density of local source residual molecules, and difficulty in achieving perfect vertical alignment of the ion source). Therefore, the probability of an ion reaching the last orbit and leaving the cyclotron with the desired energy would be very low.

The effect that enables ions with small axial deviations in relation to the equilibrium orbit to be accelerated to the cyclotron extracting orbit is termed *axial magnetic focusing* and is achieved when the magnetic field flux lines are concave on the inside. This magnetic field shape exists naturally near the outer surface of the cylindrical poles, and the corresponding focusing effect was identified very early on [63].

In the inner regions of the cyclotron, it can be achieved by a small decrease in the magnetic field amplitude along the cyclotron radius.

A magnetic field that slightly decreases with the radius (Figure 2.9) has an axial component (the only one that would be present in a uniform field) in any place outside the median plane and a radial component with an amplitude proportional to the deviation in the equilibrium orbit lying on the median plane. The axial component corresponds to a horizontal centripetal force that is responsible for the circular motion of the ions. The radial component corresponds to a *vertical restore force* that drives them back to the median plane. In

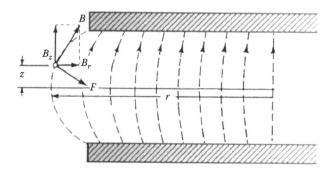

FIGURE 2.9

Representation of the radial and axial components of a magnetic field slightly decreasing along the cyclotron radius, and the corresponding net force.

this way, every ion in resonance away from the median plane experiences not only horizontal but also vertical acceleration. Consequently, those ions will accelerate toward the median plane, cross the plane at a certain speed, and then experience acceleration in the opposite direction, thus oscillating around the plane.

The action of the vertical restore force on an ion of mass m and charge q, moving at an axial distance z from the median plane, can be studied quantitatively using the resulting axial motion equation. Assuming a constant angular frequency ω corresponding to the cyclotron resonance frequency and a fixed radial position, we have

$$\frac{d}{dt}\left(m\frac{dz}{dt}\right) = qvB_r, \tag{2.30}$$

where B_r is the modulus of the magnetic field vector component along the radial coordinate r. This expression can be shown as a power series with coefficients referring to the median plane (where the coefficient values are constant over a certain radius):

$$B_r = z\left(\frac{\delta B_r}{\delta z}\right)_{z=0} + \frac{z^2}{2}\left(\frac{\delta^2 B_r}{\delta z^2}\right)_{z=0} + \cdots. \tag{2.31}$$

For small values of z (a good approach for a typical cyclotron, where the D inner height is small compared with the distance between the magnetic field poles [64]), all the series terms, except the first, can be disregarded.

Considering Maxwell's equation for the magnetic field curl operator, it can be written thus, for the cyclotron inner D field:

$$\left(\frac{\delta B_r}{\delta z}\right)_{z=0} = \left(\frac{\delta B_z}{\delta r}\right)_{z=0}, \tag{2.32}$$

where B_z is the magnetic field axial component modulus, equal to the magnetic field modulus B in the medium plan where the radial component is zero. Thus, in the classical approach where the mass m can be considered constant (which is especially valid for the first revolutions), the motion equation can be written as follows:

$$\frac{d^2z}{dt^2} - \frac{qv}{m}\frac{\delta B}{\delta r}z = 0. \tag{2.33}$$

Considering the relation between the speed v and the cyclotron (resonance) frequency:

$$\frac{d^2z}{dt^2} + \omega^2\left(-\frac{r}{B}\frac{\delta B}{\delta r}\right)z = 0. \tag{2.34}$$

This equation corresponds to a harmonic motion around the medium plan defined by the angular frequency ω_z such that

$$\omega_z = \omega\sqrt{n}. \tag{2.35}$$

This equation is known as the axial oscillation *Kerst–Serber equation*, named after the researchers who first described it [65].

The n index is called the *field index* and stands for the relationship between the decrease rate of the magnetic field and the increase rate of the cyclotron radius:

$$n = -\frac{r}{B}\frac{\delta B}{\delta r}. \tag{2.36}$$

The integration of this expression between two radius values, r and r', corresponding to two magnetic field values, B and B', respectively, is another way of describing the field radial profile:

$$B = B'\left(\frac{r'}{r}\right)^n. \tag{2.37}$$

The oscillations around the equilibrium orbit described by the Kerst–Serber equation are called *free axial oscillations*, because after starting they continue freely, only influenced by the action of the accelerating magnetic field. Since they were previously predicted for another circular cyclic accelerator (the betatron), they are commonly known as *betatron oscillations*, although this is incorrect because it is not a phenomenon exclusive to this accelerator.

2.2.3.2 Radial Focusing

In addition to, and separately from, axial stability, it is also important to analyze the radial stability of the orbits, that is, to determine whether an ion diverting slightly from the equilibrium orbit along a radial direction will tend to return to the orbit or move irreversibly away from it, leaving the resonance condition. As with axial deviations, radial deviations can occur through the scattering of residual gas atoms and molecules in the cyclotron chamber or by the ion source deviating from the center.

Radial stability can also be studied from the motion equation of a particle moving in the cyclotron median plane, using the radial acceleration (a_r) expression in cylindrical coordinates:

$$a_r = \frac{d^2 r}{dt^2} - r\left(\frac{d\theta}{dt}\right)^2. \tag{2.38}$$

Since the time derivative of the azimuthal angle θ is, by definition, the angular frequency for a particle moving with velocity in an axial magnetic field with B modulus, in the nonrelativistic limit:

$$m\frac{d^2r}{dt^2} - \frac{mv^2}{r} = -qvB \tag{2.39}$$

The negative sign on the right-hand side signifies the resulting force reducing the value of r. For a constant radius (a circular path), the simple solution to this is the fundamental cyclotron resonance equation. For a particle with a radial coordinate $(r_e + x)$ (where x is the radial deviation relative to the circular orbit equilibrium of the radius r_e), the motion equation is

$$m\frac{d^2(r_e + x)}{dt^2} - \frac{mv^2}{(r_e + x)} + qvB_x = 0, \tag{2.40}$$

where B_x is the magnetic field at the radius $(r_e + x)$. Since the deviation x is much smaller than the equilibrium radius, it can be written:

$$\frac{1}{r_e + x} \approx \frac{1}{r_e}\left(1 - \frac{x}{r_e}\right). \tag{2.41}$$

Thus, since r_e is constant:

$$m\frac{d^2x}{dt^2} - \frac{mv^2}{r_e}\left(1 - \frac{x}{r_e}\right) + qvB_x = 0 \Leftrightarrow$$

$$m\frac{d^2x}{dt^2} + (B_x - B_e)qv + qvB_e\frac{x}{r_e} = 0, \tag{2.42}$$

where B_e is the magnetic field value for the radius r_e:

$$B_e = \frac{1}{qv}\frac{mv^2}{r_e}. \tag{2.43}$$

For small displacements, the field radial dependence can be expressed using the Taylor series:

$$B_x = B_e + x\frac{dB}{dr} \ldots \Rightarrow B_x - B_e \approx x\frac{dB}{dr}. \tag{2.44}$$

Considering that a good approximation to particle velocity is given by multiplying the angular frequency by the radius r_e and given the magnetic field index in the vicinity of the equilibrium orbit:

$$n = -\frac{r_e}{B_e}\frac{dB}{dr}, \tag{2.45}$$

and, finally:

$$\frac{d^2x}{dt^2} + \omega^2(1 - n)x = 0. \tag{2.46}$$

This equation shows that an ion with little radial deviation from the equilibrium orbit will oscillate harmonically around it with an angular frequency that is related to the angular frequency corresponding to the cyclotron resonance condition through the expression:

$$\omega_x = \omega\sqrt{1 - n}. \tag{2.47}$$

This is the Kerst–Serber equation for free radial oscillations.

2.2.3.3 Stability Criteria

From the Kerst–Serber criteria, we can deduce the possible magnetic field index n value range so that the stability of the orbital path of the accelerated ions in the cyclotron is ensured. With regard to stability after small axial deviations, their relation to the magnetic field radial decrease has been established, implying that n should be positive. Examining only the equation, there is no maximum limit for the value of n, which is an indicator of restore force efficiency, in the sense that the axial oscillation frequency increases (without limit) when n increases.

The upper limit is determined by the equation referring to the radial oscillations, which only gives a real solution for n values of less than 1. If this condition is not verified, in physical terms it would result in the exponential growth of the oscillation amplitude and, therefore, radial deviation instability. Qualitatively, the physical meaning can be understood by examining the balance between the magnetic force and the centrifugal force experienced by particles along the radial coordinate. Stability will only be possible if, in a radius smaller than the equilibrium radius in which both forces are equal, the magnetic force is greater than the centrifugal force, and the reverse is true for a bigger radius. This will only happen if B_z does not decrease faster then the radius reverse (i.e., corresponding to n positive but smaller than one) as illustrated in Figure 2.10, where both forces are expressed as a function of the cyclotron orbital radius. When this condition can be verified (the left-hand side of Figure 2.10) for a radius larger than the equilibrium radius (r_e in the figure), the magnetic force will be greater than the centrifugal force and the path radius will tend to get smaller, whereas for a smaller radius the centrifugal force will tend to get bigger. The resulting force is always a restore force, contrary to the effect when the magnetic field index is greater than one (Figure 2.10 on the right). In the first case, the magnetic force prevails for a radius smaller than the equilibrium radius, whereas the centrifugal force prevails for a bigger radius.

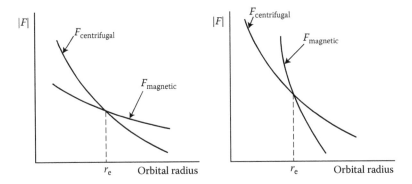

FIGURE 2.10
Relation between the centrifugal force and the magnetic force acting on the accelerating ion in the cyclotron median plane as a function of its orbit radius when the radial deviation related to the equilibrium orbit leads to free oscillations around it (on the left) and instability with ion loss (on the right).

The Kerst–Serber equation for radial oscillations would result in an oscillating motion and, therefore, instability even for negative values of n. The corresponding magnetic field, increasing with the radius, would be desirable from the point of view of radial stability, because it would correspond to a greater radial oscillation frequency. However, it would also correspond to axial deviation instability.

The orbital motion stability of the accelerating ions in the cyclotron, essential to acceleration efficiency (in terms of the number of accelerating particles per time unit) in this accelerator, is, therefore, ensured for the magnetic field index range, which corresponds, at the same time, to the stability of small axial deviations and to the radial orbit equilibrium. In conclusion, it can be said that the magnetic field index must obey the stability criteria:

$$0 < n < 1 \tag{2.48}$$

The need for a magnetic field that decreases over a greater radius is a difficult problem in terms of the practical planning and implementation of cyclotrons, above all due to the limitations imposed by the resonance conditions. It is necessary to make a compromise in such a way that the radial decrease in the magnetic field is kept small enough, allowing resonance until the maximum energy orbit is reached. The usual solution is to use a magnetic field at the cyclotron center, which is slightly bigger than the corresponding fundamental resonance equation and decreases very slightly at the periphery to a lower value than the corresponding fundamental resonance equation. In this way, the ions will receive a phase advance shift at the beginning each time they go through the acceleration gap between the *D*s, while during the final revolutions, after the magnetic field value corresponds exactly to the resonance condition, they experience a small phase delay when going through the acceleration gap. The magnetic field must be such that phase divergence is never enough, either individually or as a whole, to break the acceleration process.

In practice, modulation of the cyclotron magnetic field can be achieved by changing the shape of the pole faces (wearing them down progressively by increasing the radius to slowly increase the distance between them) and/or by reducing the effective distance between them in the central region by inserting concentric cyclotron shim disks [66].

Clearly the latter offer easy reversibility and greater versatility and are, therefore, used very frequently, mainly in the semiempirical fine-fit, using an ion beam current maximizing criterion that reaches the extraction orbit, although the correct magnetic field shape that is obtained is not easily described in theory [67].

High efficiency ion acceleration and high-intensity beam cyclotrons typically have a decreasing radial magnetic field in an almost linear mode, up to the zone where the edge effect is clear, making the decrease more pronounced. Therefore, the n value is not constant and typically has almost zero value in the central region, rising to two or three tenths on the outside after an almost linear variation.

In cyclotrons with a maximum proton kinetic energy of around 20 MeV, the total decrease in the magnetic field from the central region to the outside is around 2%. The radial decrease can be higher (3% or 4%) in small cyclotrons where the voltage applied to the Ds is relatively high and the number of revolutions is low. It should be smaller (around 1%) in higher energy cyclotrons, where the relativistic accelerating ion mass increase makes it difficult to maintain the resonance condition, as we will see later.

2.2.3.4 Free Oscillation Amplitude

Axial and radial magnetic focusing mechanisms are essentially explained by the previous process, which leads to the Kerst–Serber equations and acknowledgment of radial and axial harmonic oscillations. Nevertheless, strictly speaking, this is only a rough description. In fact, a constant value is assumed for the magnetic field index and the classical limit in which the accelerating ion mass does not increase. However, this supposition is justifiable, because in cases where the relativistic effects (and the resulting implications for the resonance condition) may be rejected it provides good solutions for combinations of successive revolutions in which the field index does not vary considerably. If these approximations are not considered, this implies that the motion equations must be written in more generic terms. For the axial and radial motion, respectively:

$$\frac{d}{dt}\left(m\frac{dz}{dt}\right) + m\omega^2 nz = 0 \Leftrightarrow \frac{d^2z}{dt^2} + \frac{1}{m}\frac{dm}{dt}\frac{dz}{dt} + \omega^2 nz = 0, \qquad (2.49)$$

$$\frac{d}{dt}\left(m\frac{dx}{dt}\right) + m\omega^2(n-1)x = 0 \Leftrightarrow \frac{d^2x}{dt^2} + \frac{1}{m}\frac{dm}{dt}\frac{dx}{dt} + \omega^2(n-1)x = 0. \quad (2.50)$$

The respective solutions are of the type [68]:

$$z = K\frac{1}{\sqrt{m\omega\sqrt{n}}}\text{sen}\left(\int\omega\sqrt{n}\,dt\right), \quad (2.51)$$

$$x = K'\frac{1}{\sqrt{m\omega\sqrt{1-n}}}\text{sen}\left(\int\omega\sqrt{1-n}\,dt\right). \quad (2.52)$$

These solutions, in which K and K' are time-independent constants, show that oscillation amplitude is proportional to $(m\omega^2)^{-1/2}$ or (from the resonance condition) to the inverse magnetic field square root. The axial oscillation amplitude is proportional to the inverse of $n^{1/4}$, whereas the radial oscillation amplitude is proportional to the inverse of $(1-n)^{1/4}$.

In a typical cyclotron with a magnetic field that decreases slightly with the radius and a magnetic field index n increasing from almost zero to some tenths, the axial oscillations are dampened whereas the radial oscillation amplitude increases [64,69].

Figure 2.11 depicts a D radial profile showing the axial oscillation decrease known as *adiabatic damping*. The initial amplitude of the oscillations is determined by the inner opening. This is why the biggest amplitude curve begins at the inner edge of the D. This curve corresponds to the hypothetical case of an ion that starts its motion in the cyclotron at a critical distance from the median plane and is pushed in the direction of this plane in its orbit, after which the magnetic focusing effect begins to exert its influence. Any ion beginning its motion beyond this critical distance will be lost, whereas the ions beginning the motion at closer distances will oscillate with proportionally smaller amplitudes. As a consequence, the outer limits, represented by a dotted line, show the axial width of the beam, where the resulting condensation, axial focusing, can be seen. Limiting the amplitude of the axial oscillations, the D interior height is, therefore, very important to the beam intensity that can be extracted

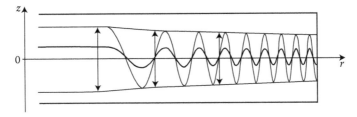

FIGURE 2.11
A D section showing the ion axial motion under the influence of the restoring force resulting from axial magnetic focusing. The dotted lines and vertical arrows represent the evolution of the beam axial dimension.

from the cyclotron. Further, it sets the dimensions (and cost) of almost all the other main components in the accelerator, clearly showing the importance of studying axial oscillations around the equilibrium orbit in the cyclotron construction project. This study is only complete when the acceleration in the gap between the Ds is taken into account, leading to the electrostatic focusing effect, which will be discussed in the next section. As far as radial oscillations are concerned, the cyclotron geometry itself makes the increase in amplitude resulting from the increasing magnetic field radial value less critical. Nevertheless, the interaction between the magnetic field radial profile and the evolution of oscillation along the acceleration process is interesting. At the start of acceleration, when the n value is almost zero, the radial oscillation frequency is almost equal to the ion orbital frequency. One orbit displaced in relation to the cyclotron geometric center (and, therefore, to the magnetic field) will continue to expand in semicircles, increasing the radius around the same deviation point from the center. However, as the n value grows, the radial oscillation frequency decreases, resulting in a precession motion such that the maximum amplitude azimuth moves ahead around the circle. Ions with essentially the same energy will also present azimuthal angular scattering related to the equilibrium orbit, in addition to radial scattering that is directly related to the amplitude of free radial oscillation.

2.2.4 Electric Focusing

2.2.4.1 Introduction

A complete study of accelerating ion motion in a cyclotron, especially the axial components of the respective path, must consider the accelerating gap between the Ds where they are under the influence of an accelerator electrical field. This field exerts a force over the ions that divert from the ideal equilibrium orbit, depending on the shape of the electric field and the radio frequency cycle phase value when they cross the acceleration gap. We can find the explanation for this by considering the gaps between the D faces as an electrical bidimensional lens [69], disregarding the magnetic field action over the accelerating ions. This approach is correct, because the magnetic field can be considered uniform along the short distance between the Ds. The approach remains valid for the first revolutions, when crossing the acceleration gap is still a considerable part of the ion path, as this happens in the central region of the cyclotron where the magnetic field is almost uniform, ensuring only circular motion. This is why, when tracing the accelerating ion path on a plane, we can consider the electrical field lines of force between the Ds at a given moment as a function of the axial coordinate z and a coordinate s, representing the motion direction ordinate when passing through the accelerating region (Figure 2.12).

When moving in the plane in which the axial coordinate is zero, the ion only experiences the action of the electrical field component along the s coordinate, responsible for its energy increase. When its axial coordinate is not zero, it also

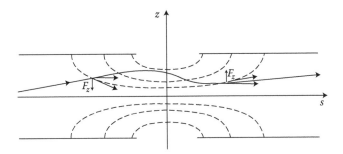

FIGURE 2.12
Diagram representing the electric focusing effects when the ion goes through the acceleration gap between the Ds. The electrical field lines of force are shown as dotted lines.

experiences the electrical field axial component effect. In this case, approaching the lens makes it deflect toward the median plane, experiencing a focusing effect proportional to the initial distance to the plane and to the ratio between the energy acquired in its passage through the acceleration gap and the energy it was carrying. When it moves away from the lens, it is deflected far away from the median plane and becomes out of focus. This focusing and unfocusing would cancel each other if the electric field and the ion energy were constant, regardless of the D geometry. Neither of these two conditions is true for the cyclotron, resulting in two different electrical focusing mechanisms: *static focusing* and *dynamic focusing*.

2.2.4.2 Static Focusing

Since the ions are accelerated as they go through the gap between the Ds, the time they are under the converging field effect at the beginning of the accelerating gap is greater than the time they are under the diverging field effect. The overall result is a focusing action that is static, because it is independent of the electrical field time variation. This effect is greater during the first accelerations, because the ion energy is low and the speed variation is more significant. The corresponding analytical expression for the path deflection suffered by an ion with charge q, axial coordinate z, and radial coordinate r is [69,70]:

$$(\Delta\chi)_{static} \approx -C\left(\frac{qV_0}{E_c}\right)^2 \frac{z}{r}\cos^2\varphi, \qquad (2.53)$$

where V_0 is the electric potential amplitude applied to the Ds, and E_c and φ are the kinetic energy and the ion phase (related to the electric field), respectively, when passing through the gap. C is a constant with an approximate value of 1, dependent on the spatial geometry of the Ds, reflecting the fact that the energy variation effect is the reason that deflection is more significant for larger D apertures and smaller distances between them [64,69].

2.2.4.3 Dynamic Focusing

In addition to energy variations, the particles passing through the gap between the *Ds* experience a variable electrical field during the transit time. The accelerator field time variation is responsible for dynamic focusing. In this portion of the radio frequency cycle, when the field decreases its amplitude the converging force at the beginning of the gap has a greater amplitude than the diverging force at the end, resulting in an overall converging effect. During the other acceleration half-cycle quadrant, the effect is inverted and the final result is diverging.

The amplitude of the dynamic focusing effect decreases as the particle energy (and therefore the orbital radius) increases, as the time the particle needs to go through the gap decreases and, as a consequence, the electrical field variation amplitude decreases. The analytical expression for the resulting deflection [69] using the same notation is

$$(\Delta \chi)_{\text{dynamic}} \approx -\frac{q \, V_0 \, \text{sen} \, \varphi}{E_c r} z. \tag{2.54}$$

It is, therefore desirable for the ions to go through the acceleration gap when the potential difference between electrodes reaches maximum amplitude and starts to decrease. However, this advantage cannot be realized in revolutions corresponding to the inner cyclotron zone, where the need to ensure the resonance condition with the best possible benefit in terms of magnetic focusing leads to a magnetic field value that is slightly larger than the resonance value. In the first stages of acceleration, this dynamic unfocusing is, therefore, responsible for the loss (by collision with the *D* surface) of many ions with considerable axial components.

2.2.4.4 Combining the Focusing Effects

Having considered the two electric focusing aspects separately, it is now important to make a combined assessment considering a number of acceleration gaps. This study can be achieved by analogy with an electrostatic lens sequence combination [64,69,70].

Examining the expressions, it can be seen that the static focusing effect only prevails over the dynamic—resulting in efficient focusing—in the initial passages through the acceleration gap and over a restricted phase range corresponding to very small phases. The static focusing effect resulting from a change in speed over the acceleration gap quickly decreases with energy. Therefore, the electric field variation effect—dynamic focusing—tends to be stronger in most revolutions. For that reason, the equivalent lens focal distance will be relatively short for low-energy ions in the appropriate phase but will become very long for high-energy ions. Taking into account the expression that relates kinetic energy to the cyclotron orbit square of the radius and considering that the ion is experiencing the average potential V_m effect, for each one of the k times it passes through the acceleration gap to gain this energy:

$$E_c = kqV_m = \frac{1}{2}\frac{q^2}{m}B^2 r^2 \Leftrightarrow r = \frac{1}{B}\sqrt{2k\frac{m}{q}V_m}. \tag{2.55}$$

When observing the spiral path transferred to a plane, this informative expression helps us understand that the gap between successive lenses (with an increasing focal distance) is proportional to the square root of an integer series. We can formally express axial motion under the electrical field influence between the electrodes mathematically as a differential equation that is a function of the independent variable k, the number of times the ion goes through the acceleration gap [69]:

$$\frac{d^2 z}{dk^2} + \frac{\pi q V_0 \operatorname{sen} \varphi}{E_c} z = 0. \tag{2.56}$$

Bearing in mind that the axial coordinates and their variation rate (corresponding to the ion deflection) slowly change, a possible solution is [68,69], for positive φ_i:

$$z = A\left(\frac{E_c}{\operatorname{sen}\varphi}\right)^{1/4} \operatorname{sen}\left(\sqrt{\pi}\int_0^k \sqrt{\frac{q V_0 \operatorname{sen}\varphi}{E_c}}\, dk + \delta\right). \tag{2.57}$$

This expression, in which A and δ are integral constants, shows that even if there is electric focusing, the axial oscillation amplitude increases in proportion to the 4th root of the ratio between the kinetic energy and the phase sine (therefore, very slowly). It also grows with $k^{1/4}$ when the phase is constant. In the initial phase of acceleration, it is important to emphasize this electric focusing resulting effect, otherwise, the axial coordinate would change a lot faster, in line with the kinetic energy or with k.

In negative phases, the motion is no longer oscillatory but exponential, either increasing or decreasing. Regardless of the initial conditions, exponential growth will prevail, resulting in an increase in the absolute value of the ion axial coordinate and, therefore, a defocus situation [69].

Figure 2.13 shows the profile of some ion paths along a D taking electric focusing only into account.

As the radius increases, focusing becomes more and more important, due to the nonuniformity of the magnetic field. This is why it cannot be disregarded and, in fact, becomes predominant. The final combined result of magnetic and electric focusing is efficient ion beam axial focusing for the ions that are not lost in the first revolutions where the electric effect prevails. A combined consideration of the various expressions for axial deflections resulting from the aforementioned electric and magnetic focusing effects leads to the prediction (experimentally confirmed [64]) of a beam axial profile (transferred to a

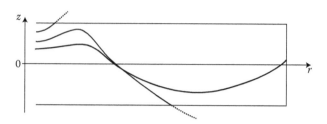

FIGURE 2.13
Bidimensional diagram of a D section of some ion paths under the electric (de)focusing effect.

plane) resembling an envelope, whose cross thickness increases until it is limited by the internal D opening and afterward decreases in amplitude with the increase in ion energy (and increasing orbital radius), due to the decreasing magnetic field (Figure 2.14).

2.2.5 Phase Relations and Maximum Energy

2.2.5.1 *Path and Phase in the First Revolutions*

Even though ions can be emitted from the source during the entire electric field half cycle, corresponding to an accelerating force between the ion source and the D, the resulting wide spatial distribution of ions quickly narrows, making all the ions cross the acceleration gap in approximate phases distributed around the electric field peak.

Exact ion path analysis is complex, because it depends to a great extent on detailed knowledge of the electric and magnetic fields, especially the electric field between the ion source and the Ds. However, a good approximation can be achieved, illustrating the conclusion of the previous paragraph, by studying the initial ion path—the first revolutions—on the basis of some simplifying assumptions.

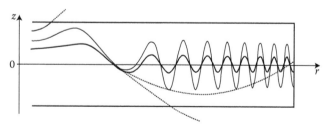

FIGURE 2.14
Bidimensional diagram of a D section of cyclotron ion vertical motion. The lines are the ion paths, considering only the electric field effects. The dotted lines represent the combined effects of the magnetic and electric fields after the magnetic field has become predominant.

First, a point-like ion source and a uniform electric field between the *Ds* can be assumed. A gap between the *Ds* big enough to prevent the ions from entering during the first revolutions can also be considered (which is completely true as the gap is usually several centimeters long to allow for ion source positioning and the insertion of testing probes and beam diagnostics). Finally, it can be assumed that the ions do not suffer any radial or axial deflection in relation to the ideal orbit, thus preventing the radial and axial focusing effects from masking the phase relations to be explained.

Using a coordinate system with its origin in the ion source and with the *x* and *z*-axes lying along the electrical field and magnetic field, respectively (the open *D* surfaces are parallel to the *y*-axes), the motion in the *xy* plane can be described by the equations:

$$m\frac{d^2x}{dt^2} = qE + qB\frac{dy}{dt}, \tag{2.58}$$

$$m\frac{d^2y}{dt^2} = -qB\frac{dx}{dt}, \tag{2.59}$$

in which *q* and *m* are the accelerating ion and mass charge, respectively, and *B* is the magnetic field modulus (assumed to be uniform for the purposes of simplification and because it is the cyclotron central zone, which is not usually modulated to secure axial focusing); whereas the electrical field ε with the amplitude ε_0 and angular frequency ω is given by

$$\varepsilon = \varepsilon_0 \cos(\omega(t + t_0)). \tag{2.60}$$

The possibility of the ions beginning their path in different accelerator cycle phases of the electrical field is considered by the introduction of t_0.

By introducing a (unique) magnetic field axial component value corresponding to the fundamental condition of the cyclotron resonance, the motion equations can be written as

$$\frac{d^2x}{dt^2} = \frac{q}{m}\varepsilon_0 \cos(\omega(t + t_0)) + \omega\frac{dy}{dt}, \tag{2.61}$$

$$\frac{d^2y}{dt^2} = -\omega\frac{dx}{dt}. \tag{2.62}$$

Considering that the ions are resting at source in the initial moment, the motion equation solutions are

$$x = \frac{q\varepsilon_0}{2m\omega^2}[\text{sen}(\omega t_0)\,\text{sen}(\omega t) - t\,\text{sen}(\omega(t + t_0))], \tag{2.63}$$

$$y = \frac{q\varepsilon_0}{2m\omega^2}[\cos(\omega t_0)\,\text{sen}(\omega t) - t\cos(\omega(t + t_0)) - 2\,\text{sen}(\omega t_0)(1 - \cos(\omega t))]. \tag{2.64}$$

From these expressions, the initial ion paths can be illustrated in diagrams that show the different phase values at the moment of emission, as in Figure 2.15. The position of the various ions (emitted in different phases) for the same ωt value is represented by the series of white points in the center connected by dotted lines.

The ion depicted by the corresponding curve $\varphi = 0$ leaves the source when the electric field reaches maximum value. It is in perfect resonance with the electric field, crossing the y-axes to $\omega \cdot t = \pi, 2\pi, 3\pi, 4\pi$, etc. An initial phase equal to $\varphi = -\pi/2$ corresponds to an ion emitted from the source at the exact moment that the field accelerates in the positive x-axis direction, so that it is under maximum acceleration in this direction, describing a larger orbit than the one emitted in phase zero. Figure 2.15 also illustrates several intermediate phases, and it is significant that the increasingly positive phases correspond to ions emitted when the field is progressively lower on emission and nearer to the point of inverting its direction, thus initially describing only a small arc.

The limit phase for acceleration in this x-axis direction is $\varphi = +\pi/2$. For a hypothetic point-like source that allows for emission in all directions

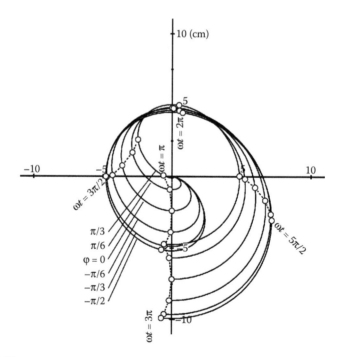

FIGURE 2.15
Initial ion paths in a cyclotron, numerically calculated, for a sinusoidal electrical field that is uniform along the xx axis (maximum amplitude 25 kV/cm, 20 MHz frequency) and a 13.05 kGauss axial magnetic field. Revolutions start from the center, in different phases of the accelerating electrical field.

from this phase value, the ions will be accelerated in the opposite direction and the respective path will be symmetrical to the one corresponding to $\varphi = -\pi/2$.

After three-fourths of the cycle (assuming it began the moment the ions corresponding to $\varphi = -\pi/2$ were emitted), all the ions are near the median line between the Ds (corresponding to $x = 0$), almost independently of the field value at the moment of emission. This "gathering" is repeated at each half cycle with an asymptotic approximation to the median line between the Ds, corresponding to the position of the emitted ions in phase zero. As acceleration increases, the motion of the ions approaches what it would be if they had all been emitted at the moment when the accelerating electric field had reached its maximum value, in what can be termed the *phase focusing* around the zero phase.

Resonant acceleration would take place even if the electric field did not reach zero inside the Ds. However, in practice, the D surfaces are very close to each other, and the electric fields cannot penetrate very deeply inside the Ds. Therefore, the larger orbit shapes are controlled only by the magnetic field (which is closer to the arc of a circle than a spiral) (Figure 2.16).

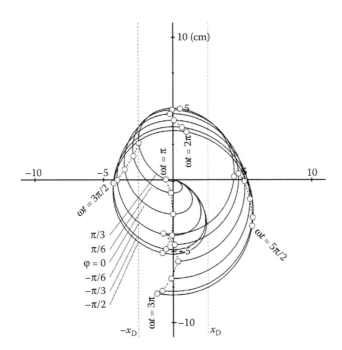

FIGURE 2.16
First revolutions cyclotron ion path evaluated under the same conditions as Figure 2.15 but assuming the electric field is zero in the planes $x = \pm x_D$. The spiral orbits turn into circle segments outside x_D. Phase focusing and spatial concentration are not as complete as in the uniform field situation but are still observable.

2.2.6 Maximum Kinetic Energy: The Relativistic Limit

The fundamental equation for cyclotron resonance, as previously expressed, and the resulting basic working principle are based on equal applied electric field frequency and ion revolution frequency, resulting in a direct proportional relationship between this frequency value and the magnetic field (both considered as constants). In practice, the electrical field frequency is kept constant but this is not the case with ion revolution frequency. When the accelerated ion kinetic energy surpasses its rest mass by some percentage units, the relativistic effects can no longer be disregarded. Among these, the increasing mass directly interferes with the validity of the resonance equation.

A transit time increase inside the Ds corresponding to this increase implies a phase change for the ions entering the acceleration gap, resulting in progressive delay in relation to the existing electrical field, until they are eventually slowed down by this field.

More than loss by residual molecule dispersion inside the cyclotron vacuum chamber, the advent of relativistic effects is an important factor supporting the option of a high accelerating electrical field amplitude. Obviously, if the resulting phase jump from the revolution and resonance frequencies were different, there would be no consequences if the applied potential to the Ds corresponded to the maximum energy reached after one acceleration only. The smaller the maximum accelerating electrical field value, the quicker the effect of the difference will become evident. In a quantitative analysis, considering a uniform and constant magnetic field with modulus B, this becomes clear when we relate the phase change to the kinetic energy increase for each revolution.

Let us consider the electrical field frequency f applied to the Ds (corresponding to the cyclotron resonance condition expressed) and the accelerating (charge q) ion revolution frequency f_r given by

$$f = \frac{qB}{m_0},$$ (2.65)

$$f_r = \frac{qB}{m},$$ (2.66)

where m_0 and m are the ion rest mass and the ion relativistic mass, respectively. The phase variation φ for each revolution N will be given by [70]:

$$\frac{d\varphi}{dN} = 2\pi\frac{f - f_r}{f_r} = 2\pi\left(\frac{m}{m_0} - 1\right).$$ (2.67)

From the relativistic equivalence between mass and energy:

$$\frac{m}{m_0} = \frac{E}{E_0} = \frac{E_0 + E_c}{E_0} = 1 + \frac{E_c}{E_0},$$ (2.68)

where E_c and E_0 are the ion kinetic energy and the ion rest energy, respectively, defined as relativistic. The previous expressions lead to

$$\frac{d\varphi}{dN} = 2\pi \frac{E_c}{E_0}. \tag{2.69}$$

On the other hand, the kinetic energy increase in each revolution can be related to the maximum applied potential to the Ds by

$$\frac{dE_c}{dN} = 2qV_M \, \text{sen} \, \varphi. \tag{2.70}$$

The factor 2 results from the fact that in each revolution the ion goes through the accelerating gap between the Ds twice. From this expression:

$$\frac{d\varphi}{dN} = 2qV_M \, \text{sen} \, \varphi \frac{d\varphi}{dE_c} = -2qV_M \frac{d(\cos \varphi)}{dE_c}. \tag{2.71}$$

Comparison of this expression of the phase variation in each revolution with the previous one leads to

$$\frac{d(\cos \varphi)}{dE_c} = -\frac{\pi}{qV_M} \frac{E_c}{E_0}. \tag{2.72}$$

where φ_0 is the ion initial phase. Taking into account the fact that the kinetic energy at the moment of emission can be considered zero, integrating this expression results in

$$\cos \varphi_0 - \cos \varphi = \frac{\pi}{qV_M E_0} \frac{(E_c)^2}{2}. \tag{2.73}$$

Even though this equation emerges after some simplifying assumptions, it enables us to conclude that for a certain phase change in which the ions find the electrical field when they arrive at the acceleration gap, the greater the potential amplitude applied to the Ds, the greater the kinetic energy obtained will be. As previously noted, in the first revolutions (even before the relativistic effects start to become relevant), independently of the phase at the moment of emission, all the ions go through the acceleration gap with a phase value very similar to the corresponding potential peak ($\varphi_0 = 0$). The greatest acceptable total phase deflection must be one that permits the ions to arrive at the gap between the Ds and still find the accelerating half cycle, that is, the phase limit $\varphi = \pi/2$. Substitution of these values would enable us to discover the maximum kinetic energy in a cyclotron, with the specifications that are the basis of the simplifying assumptions leading to this expression, such as uniform and constant magnetic field considerations.

Technical problems relating to the production of high electric potential values and the related insulation impose limitations on the use of this method of obtaining the desired kinetic energy in a cyclotron, based on an increase in the maximum electrical field applied to the Ds in such a way that the number of revolutions N is small enough to verify the condition:

$$N \frac{d\varphi}{dN} \leq \frac{\pi}{2}.$$ (2.74)

A different (and complementary) solution is to increase the angular range of the phase variation, which can be achieved if the magnetic field value is such that in the first revolutions the ion revolution period is less than that of the accelerating electrical field frequency. The phase will move toward a maximum limit of $-\pi/2$. Meanwhile, the relativistic effect of mass increase will gradually be felt, making the phase jump smaller each time in the negative direction, until finally (when the electrical field and revolution frequency values are the same, in a transient resonance condition only reached when the ions are at a considerable distance from the cyclotron center) it becomes positive, that is, the ions arrive at the acceleration gap with a time delay. It still remains a phase range below the $\pi/2$ limit in which acceleration is possible.

This procedure permits the phase range to be increased by a factor of 3 and is mathematically reflected by the appearance of a negative term in the expression that quantifies the phase variation in terms of number of revolutions.

On the other hand, in discussing magnetic focusing, the decrease in revolution frequency as a consequence of the small magnetic field radial decrease necessary for axial focusing has already been described (and the implications of attempting a practical solution discussed). This (decreasing) variation in revolution frequency in relation to the cyclotron orbit radius, which emphasizes the relativistic effect of increasing mass, should also be considered in a quantitative analysis of phase variation.

The combined consideration of these factors results in the expression [71]:

$$\frac{d\varphi}{dN} = 2\pi \frac{B_c}{B} \left(\frac{B_0}{B_c} - \frac{B}{B_c} + \frac{B_0 E_c}{B_c E_0} \right),$$ (2.75)

where the kinetic energy E_c and the magnetic field B are functions of the cyclotron path radius. The magnetic field value for radiuses equal to zero, that is, the cyclotron center, is represented by B_c. It is a constant value similar to B_0, which is the magnetic field value corresponding to the fundamental cyclotron resonance condition initially defined, that is, for the oscillation frequency of the electrical field applied to the Ds and the ion rest mass (corresponding to energy E_0). Despite the implied complexity of the integration of this expression, it is possible to reach a compromise between a magnetic field radial profile (needed for axial focusing), an electric potential applied to the Ds that is technically feasible, and maximum accelerating ion kinetic energy, with regard to the kinetic energy variation in each revolution and along the radius [70,71]. This limit explains why the cyclotron is not used to accelerating light particles such as electrons. For protons and deuterons it amounts, in practice, to around tens of MeVs. The limitation of the fundamental cyclotron principle to nonrelativistic energies is due to the assumption of a constant magnetic field and a constant applied electrical field oscillating frequency. Technologically, it is possible to implement the variation in both frequency and magnetic field adequately, enabling acceleration to take place with relativistic energies. This implementation is the basis of the particle accelerators that appeared after the first cyclotrons, the *synchrocyclotron* and the *isochronous cyclotron*.

2.2.7 The Synchrocyclotron

2.2.7.1 Working Principle

The relativistic mass increase limit imposed on maximum cyclotron kinetic energy can be overcome when the electrical field oscillating frequency applied to the electrodes decreases during the acceleration process, after a decrease in revolution frequency as a result of the relativistic mass increase. This is the basis of the synchrocyclotron working mode.

In this modulated frequency cyclotron, the yield in terms of the number of accelerated particles per time unit is much lower—by about two orders of magnitude—than the conventional cyclotron, because the particles are not distributed continuously and almost uniformly along the path. Instead, they form groups or clusters, which are synchronously accelerated from the center to the outside with the frequency variation. The process can only restart for a new particle cluster once a cluster has been accelerated to a high energy level.

2.2.7.2 Phase Stability

Phase stability [72,73] is an important feature of the synchrocyclotron and is crucial to the success of the acceleration process. It prevents the loss of ions in a group being accelerated that arrive at the acceleration gap at slightly different times; and it instead enables them to group themselves in the center of the cluster, in a kind of longitudinal focusing with space and phase features.

Figure 2.17 illustrates phase stability in a synchrocyclotron. An ion in the center of the accelerating cluster that reaches the accelerating gap at the moment the accelerator electric field moves to zero value (point *a* in the figure) will always circle in this equilibrium orbit, which is termed *synchronous*. Another ion from the same cluster reaching the accelerating gap slightly before this (illustrated as point *b* in the figure) will find an accelerator field that will increase its orbit radius and energy, consequently decreasing its revolution frequency. Thus, in the next passage through the accelerating gap, it will be closer to the cluster center. In a similar manner, the ions reaching the

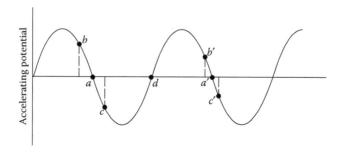

FIGURE 2.17

Phase stability in a synchrocyclotron. The points *a* and *a'* correspond to successive passages through the ion accelerating gap in the same direction in a synchronous orbit. Ions reaching the acceleration gap slightly later (*b*) will be accelerated, bringing their phase in the next equivalent acceleration gap passage (*b'*) closer to the synchronous orbit. A similar effect occurs with the ions arriving at the synchronous orbit in advance (*c* and *c'*). The point *d* corresponds to instability, as the delayed or in advance ions in the vicinity of this point tend to increase.

accelerating gap later than the cluster center (corresponding to point *c* in the example) will be decelerated, and the corresponding mass decrease will lead to an increase in revolution frequency, bringing them closer to the cluster center for the next passage. As a consequence, the ions in the same cluster oscillate around a synchronous orbit. In each passage through the accelerating gap, they are not only accelerated but also grouped together by this phase stability effect.

The accelerating process and phase stability restrict the accelerating ion phase in relation to the accelerating magnetic field in such a way that, unlike the conventional cyclotron, it is not absolutely necessary to use high accelerating electrical field values. Therefore, with the aim of reducing high electric power production and insulation problems, the individual accelerations (i.e., the accelerating electrical field amplitude) are typically lower than in a conventional cyclotron, resulting in a much higher number of revolutions [74].

2.2.8 The Isochronous Cyclotron

2.2.8.1 Thomas Focusing and Working Principle

An alternative method of overcoming the relativistic mass increase limit imposed on maximum kinetic energy, which does not involve modulation of the accelerating electrical field frequency (and the consequent decrease in beam intensity), can be inferred from the cyclotron resonance condition: Increasing the cyclotron magnetic field at the rate determined by the relativistic mass increase enables the revolution period to be maintained (isochronism).

This magnetic field radial increase violates the stability condition resulting from the Kerst–Serber equations or, in other words, results in a loss of the beam magnetic focusing effect, making the acceleration process impossible. The proposed solution by J. H. Thomas [75] is magnetic field azimuthal periodic modulation, which creates an axial focusing mechanism even in the presence of an increasing radial magnetic field. In addition, the magnetic field radial increase corresponds to a negative field index, thus increasing radial focusing, as predicted by the relevant Kerst–Serber equation. The result is the *isochronous cyclotron*, in which the magnetic field is, therefore, highly nonhomogeneous, both radially and azimuthally (which is why it is also known as the azimuthally varying field (AVF) cyclotron).

Thomas focusing can be understood with the aid of Figure 2.18, which illustrates the azimuthal cut (for a constant radius) perpendicular to the median plane between the cyclotron magnetic field poles, transferred to a paper plane. The distance between the magnetic poles has two alternating values, a larger one (corresponding to the *top* region) and a smaller one (the *valley* region) along the azimuth around the cyclotron center. For each top-valley pair, the azimuthal periodicity element is termed the *sector*, thus explaining the alternative name for this cyclotron: *sector focused*.

The bend in the magnetic field lines in the transition zone between a top and a valley implies that there is a magnetic field azimuthal component. Figure 2.18 depicts the magnetic field lines for this geometry, allowing for qualitative variation analysis of this component (B_θ) along the same field line and between different field lines. In both cases, its respective direction is inverted when crossing the medium cyclotron plan and the top and valley axial geometric center, where it is zero.

Even inside a *D*, far away from the electrical field interaction, the equilibrium orbit of an ion in this magnetic field is not a perfect circle arc. The magnetic field axial component and ion velocity vector component along the cyclotron medium plan produce a Lorentz force equivalent to a centripetal force. However, variations in the magnetic field axial component mean that the values are different in tops and valleys, corresponding to variations in the bending radius, which will be bigger when this component is smaller and vice versa, as illustrated in Figure 2.19. The oscillating orbit around a circle

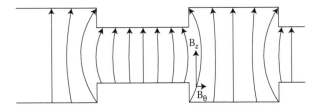

FIGURE 2.18
Isochronous cyclotron azimuthal cut transferred to a paper plane showing the magnetic field lines (and an example of the field axial and azimuthal component in a space point) resulting from the tops and valleys.

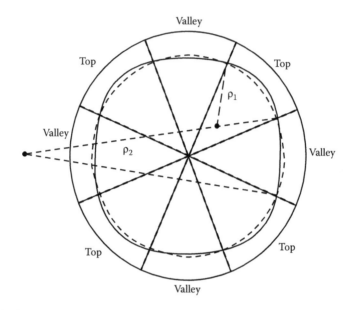

FIGURE 2.19
Diagram showing bending radius, assuming a completely closed hypothetical orbit in an isochronous cyclotron with a four-sector profile, as depicted in Figure 2.18.

and the ion velocity vector present a radial component alternately pointing toward the cyclotron center or the outside.

A vertical Lorentz force results from the magnetic field azimuthal component and the velocity vector radial component, which always point toward the cyclotron median plane, providing axial focusing.

This behavior of the magnetic field azimuthal component, ion velocity vector, and, therefore, the Thomas focusing occurs not only in relation to azimuthal modulation in this field but also for any other geometry with the same periodicity and symmetry type, such as sinusoidal modulation. In any example, it is greater, as the relative field variation between top and valley increases (directly associated with the bending of field lines) and can be configured to ensure an overall focusing effect even in an increasing radial field.

2.2.9 Focusing Reinforcement Using the Alternating Gradient Principle

Thomas axial focusing can be reinforced if the tops and valleys do not occur in the same azimuthal position in different radii but in a spiral-shaped distribution from the center (Figure 2.20). Thus, in each top-valley transition, the magnetic field will have a radial component whose gradient in successive transitions is alternately centripetal or centrifugal. Crossing these spatial regions where the magnetic field radial components are felt, the ion, given

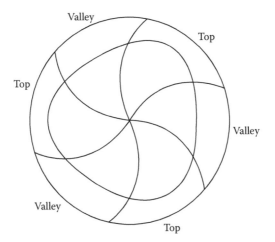

FIGURE 2.20
Three isochronous cyclotron sectors viewed from above, where the borders between tops and valleys are spiral shaped according to the alternating gradient principle. A closed hypothetical path is also represented.

its azimuthal speed component, will be under the axial Lorentz force in a direction that is alternately in and out of focus. The resulting general effect is efficient focusing, as established in the *alternating gradient principle* [76]. This principle is simply the application to the charged particle beam dynamics of the geometrical optics principle, which establishes that the focal distance of a converging and diverging lens group with an equal absolute focal distance value is always positive (i.e., corresponds to a general focusing effect), regardless of the order of the lens pair (Figure 2.21). In addition to being proportional to the difference between the field values in tops and valleys, the amplitude of this focusing effect depends on the limit line profile between them, increasing in proportion to the increase in the angle that the tangent to the limit line makes with the radial direction, as can be seen from the relevant geometry.

From this analysis, it can also be concluded that the ion path in the region where it is under an out-of-focus force is more nearly perpendicular to the transition boundary than when it crosses a focusing region, so that it remains for a comparatively longer time in this region under a focusing force. This fact is reflected in a second overall axial focusing effect associated with the spiral shape of the tops and valleys [77], reinforcing the alternating gradient principle (which would be felt even if the time taken to cross the in-focus and out-of-focus regions were the same).

Due to these effects, an isochronous cyclotron with spiral tops and valleys—also called an FFAG (fixed-field* alternating-gradient) cyclotron—can easily

---

* With regard to magnetic field time invariance.

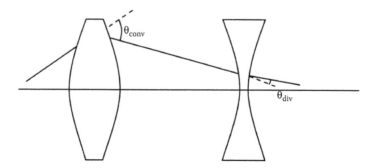

FIGURE 2.21
Illustration of the alternating gradient principle using an optical analogy: the overall effect of the path through a converging lens and a diverging lens of the same distance (in modulus) is always converging, no matter which lens is first. The explanation relies on the proportionality between the deviation angle and the distance from the radius to the optical axis.

generate axial focusing forces that are much greater than those usually obtained through a magnetic field radial decrease in a conventional cyclotron. These forces are strong enough to create a pronounced axial focusing effect, even with a radial increase in the magnetic field in order to maintain the resonance condition for the relativistic energies of the accelerating ion beams. For this reason, the axial focusing characteristic of the FFAG cyclotron is known in jargon terms as *strong focusing*, as opposed to the *weak focusing* obtained in the conventional cyclotron.

2.2.10 Aspects of Quantitative Characterization

An analogous but symmetrical magnitude to the field index is usually used in the quantitative characterization of the magnetic field in an isochronous cyclotron, known as the *average field index*, $\bar{n}$, given by

$$\bar{n} = \frac{r}{\bar{B}_z} \frac{d\bar{B}_z}{dr},\tag{2.76}$$

where r is the radial distance to the isochronous cyclotron center and $\bar{B}_z$ represents the average value of the magnetic field axial coordinate (B_z) for a course along the path in the median plane:

$$\bar{B}_z = \frac{1}{S} \oint B_z \, ds,\tag{2.77}$$

where s is the coordinate corresponding to a course along the path with total length S.

The resonance condition implies that the revolution angular frequency ω_r is equal to the applied electrical field constant angular frequency ω. For an ion with charge q, mass m, and rest mass m_0, the latter is given by

$$\omega = \frac{qB_c}{m_0},$$

(2.78)

where B_c is the axial magnetic field in the cyclotron center in which the radial coordinate is zero. The bend radius variation constant that is present in the accelerating ion path in this kind of cyclotron leads to the definition of ω_r in average terms, with the implicit concept of *equivalent radius* equal to the arc described, divided by the covered azimuthal angle, which can be approached by the radial coordinate relating to the cyclotron center, except in the first revolutions. Thus, if m is the relativistic mass:

$$\omega_r = \frac{q\bar{B}_z}{m}.$$

(2.79)

The resonance condition is maintained for each radial coordinate value by the average axial magnetic field radial modulation in such a way that the revolution frequency is kept equal to the accelerating electrical field frequency, implying that

$$\bar{B}_z = B_c\gamma,$$

(2.80)

where γ characterizes the relativistic mass increase:

$$\gamma = \left(1 - \frac{v^2}{c^2}\right)^{-1/2} = \frac{m}{m_0}.$$

(2.81)

From the differentiation of the expression that relates the magnetic field axial coordinate average value to the value it assumes in the cyclotron center, the average field index needed to maintain the isochronous resonance condition can be written in terms of γ:

$$\bar{n} = -\frac{r}{\gamma}\frac{d\gamma}{dr}.$$

(2.82)

In FFAG cyclotrons, magnetic field characterization is produced with the aid of the *flutter function*, ϕ. This is a function of the radial coordinate (only) that describes the variation in magnetic field intensity between the tops and valleys. It is defined as an average quadratic deviation along a circle (corresponding to a radial coordinate value) of a function Φ, chosen so that its average value along a revolution is the unit and conforms to the equation:

$$B_{z0} = \bar{B}_z \Phi. \tag{2.83}$$

Similar to B_{z0}, Φ is a function of the radial and azimuthal θ coordinates. The flutter function is then given by

$$\varphi = \frac{1}{2\pi} \int_0^{2\pi} (\Phi - \bar{\Phi})^2 d\theta, \tag{2.84}$$

$$\bar{\Phi} = \frac{1}{2\pi} \int_0^{2\pi} \Phi \, d\theta = 1. \tag{2.85}$$

The complete description of betatron oscillations in a generic isochronous cyclotron is a complex problem demanding the use of approximations in its resolution and often presenting the results as series expansion. A large part of this complexity is related to the fact that the average field index, the flutter function, and the angle ζ that the tangent to the limit line between the tops and the valleys makes with the radial direction (crucial, not only to axial focusing but also to the corresponding free oscillations) are functions of the radial coordinate, which does not correspond to the (variable) bending radius of the equilibrium orbit. Moreover, as a function of these two variables, the magnetic field gradients have clearly different values and orientations. As a first approach, the result for free radial oscillations is the same as the one described by the corresponding conventional cyclotron Kerst–Serber equation, that is (using the average field index), the proportionality coefficient between the free radial oscillation frequency and the resonance frequency Ω_r is given by

$$\Omega_r = \sqrt{1 + \bar{n}}. \tag{2.86}$$

As far as axial free oscillations are concerned, as a first approximation the proportionality coefficient between the oscillating frequency and the resonance frequency Ω_z will be given by

$$\Omega_z^2 = -\bar{n} + \varphi(1 + 2 \tan^2 \zeta) + \cdots . \tag{2.87}$$

The first term in this expression is analogous to the one expressed by the Kerst–Serber equation for free axial oscillations in a conventional cyclotron, whereas the term that follows is the diverging effect compensation of a positive average field index, given by the magnetic field azimuthal variation (quantified by the flutter function) and the shape of the transition line between the tops and valleys.

References

1. N. Bohr. Neutron capture and nuclear constitution. *Nature* 137, 344, 1936.
2. F. Alves, M. Jensen, H.J Jensen, R.J. Nickles, S. Holm. Determination of the excitation function for the $^{10}B(p,n)^{10}C$ reaction implications for the production of [^{10}C]carbon dioxide for use as a PET tracer. *Appl. Radiat. Isot.* 52, 899, 2000.
3. S.M. Qaim, J.C. Clark, C. Crouzel, M. Guillaume, H.J. Helmeke, B. Nebeling, V.W. Pike, G. Stöcklin. PET radionuclide production. In *Radiopharmaceuticals for Positron Emission Tomography—Methodological Aspects* (Eds. Stöcklin, G.; Pike, V.W.). Kluwer Academic Publishers, Dordrecht, Netherlands, 1993.
4. T. Nozaki, Y. Itoh, K. Ogawa. Yield of ^{73}Se for various reactions and its chemical processing. *Int. J. Appl. Radiat. Isot.* 30, 595, 1979.
5. A. Mushtaq, S.M. Qaim, G. Stöcklin. Production of ^{73}Se via (p,3n) and (d,4n) reactions on arsenic. *Appl. Radiat. Isot.* 39, 1085, 1988.
6. F. Alves, J.J.P. Lima, R.J. Nickles, M. Jensen. Carbon-10: Example of cyclotron production of positron emitters as an open research field. *Rad. Phys. Chem.* 76, 343, 2007.
7. N.M. Hintz, N.F. Ramsey. Excitation functions to 100 MeV. *Phys Rev.* 88, 19, 1952.
8. M.L. Firouzbakht, D.J. Schyler, A.P. Wolf. Yield measurements for the $^{11}B(p,n)^{11}C$ and the $^{10}B(d,n)^{11}C$ nuclear reactions. *Nucl. Med. Biol.* 25, 161, 1998.
9. B. Anders, P. Herges, W. Scobel. Excitation functions of nuclear reactions producing $^{11}C^*$. *Z. Phys. A* 301, 353, 1981.
10. H.A. Bethe. Theory of the passage of fast corpuscular rays through matter. *Ann. Physik* 5, 325, 1930.
11. E. Segrè. *Nuclei and Particles.* W.A. Benjamin Inc., New York, 1964.
12. N. Bohr. On the theory of the decrease of velocity of moving electrified particles on passing through matter. *Philos. Mag.* 25, 10, 1913.
13. F. Bloch. Stopping power of atoms several electrons. *Z. Physik* 81, 363, 1933.
14. E.J. Williams. Application of ordinary space–time concepts in collision problems and relation of classical theory to Born's approximation. *Rev. Mod. Phys.* 17, 217, 1945.
15. H.A. Bethe. Bremsformel für Elektronen relativistischer Geschwindigkeit. *Z. Physik* 76, 293, 1932.
16. C. Møller. Zur theorie des durchgang schneller elektronen durch materie. *Ann. Phys.* 14, 531, 1932.
17. N. Bohr. On the decrease of velocity of swiftly moving electrified particles in passing through matter. *Philos. Mag.* 30, 581, 1915.
18. J.R. Oppenheimer, H. Snyder, R. Serber. The production of soft secondaries by mesotrons. *Phys. Rev.* 57, 75, 1940.
19. B.B. Rossi, K.I. Greisen. Cosmic ray theory. *Rev. Mod. Phys.* 13, 240, 1941.
20. E. Fermi. The absorption of mesotrons in air and in condensed materials. *Phys. Rev.* 56, 1242, 1939.
21. O. Halpern, H. Hall. Energy losses of fast mesotrons and electrons in condensed materials. *Phys. Rev.* 57, 459, 1940.
22. E. Fermi. The ionization loss of energy in gases and in condensed materials. *Phys. Rev.* 57, 485, 1940.
23. G.C. Wick. Sul frenamento delle particelle veloci. *Nuovo Cimento* 1, 302, 1943.

24. O. Halpern, H. Hall. The ionization loss of energy of fast charged particles in gases and condensed bodies. *Phys. Rev.* 73, 477, 1948.
25. N. Bohr. Scattering and stopping of fission fragments. *Phys. Rev.* 58, 654 & *Kgl. Danske Videnskab. Selskab, Mat.-Fys. Medd.* 18, 8, 1948.
26. H.C. Brinkman, H.A. Kramers. *Proc. Acad. Sci. Amsterdam* 33, 973, 1930.
27. M.S. Livingston, H.A. Bethe. Nuclear physics C. Nuclear dynamics, experimental. *Rev. Mod. Phys.* 9, 245, 1937.
28. N. Boembergen, J. Van Heerden. The range and straggling of protons between 35 and 120 MeV. *Phys. Rev.* 83, 561, 1951.
29. U. Fano. On the theory of ionization yield of radiations in different substances. *Phys. Rev.* 70, 44, 1946.
30. M.G. Holloway, M.S. Livingston. Range and specific ionization of alpha-particles. *Phys. Rev.* 54, 18, 1938.
31. W.P. Jesse, J. Sadauskis. Alpha-particle ionization in argon and in air and the range-energy curves. *Phys. Rev.* 75, 1110, 1949.
32. R. Collé, R. Kishore, J.B. Cumming. Excitation functions for (p,n) reactions to 25 MeV on ^{63}Cu, ^{65}Cu and ^{107}Ag. *Phys. Rev. C* 9, 1819, 1974.
33. R. Weinreich, H.J. Probst, S.M. Qaim. Production of chromium-48 for applications in live sciences. *Int. J. Appl. Radiat. Isot.* 31, 223, 1980.
34. S.M. Qaim. Nuclear data relevant to cyclotron produced short-lived medical radioisotopes. *Radiochim. Acta* 30, 147, 1982.
35. A. Grütter. Excitation functions for radioactive isotopes produced by proton bombardment of Cu and Al in the energy range from 16 to 70 MeV. *Nucl. Phys. A* 383, 98, 1982.
36. B. Scholten, S.M. Qaim, G. Stöcklin. Excitation functions of proton induced nuclear reactions on natural tellurium and enriched ^{123}Te: Production of ^{123}I via the ^{123}Te (p,n) ^{123}I—process at a low-energy cyclotron. *Appl. Radiat. Isot.* 40, 127, 1989.
37. G. Blessing, W. Bräutigam, H.G. Böge, N. Gad, B. Scholten, S.M. Qaim. Internal irradiation system for excitation function measurement via the stacked foil technique. *Appl. Radiat. Isot.* 46, 955, 1995.
38. A. Honh, B. Scholten, H.H. Coenen, S.M. Qaim. Excitation functions of (p,xn) reactions on highly enriched ^{122}Te: Relevance to the production of ^{120g}I. *Appl. Radiat. Isot.* 49, 93, 1998.
39. M. Hille, P. Hille, M. Uhl, W. Weisz. Excitation functions of (p,n) and (α,n) reactions on Ni, Cu and Zn. *Nucl. Phys. A* 198, 625, 1972.
40. P. Jahn, H.J. Probst, A. Djaloeis, W.F. Davidson, C. Mayer-Böricke. Measurement and interpretation of ^{197}Au (d, xnyp) excitation functions in the energy range from 25 to 86 MeV. *Nucl. Phys. A* 209, 333, 1973.
41. J.H. Zaidi, S.M. Qaim, G. Stöcklin. Excitation functions of deuteron induced nuclear reactions on natural tellurium and enriched ^{122}Te: Production of ^{123}I via the ^{122}Te (d,n) ^{123}I—Process at a low energy cyclotron. *Appl. Radiat. Isot.* 34, 1425, 1983.
42. N.G. Zaitseva, C. Deptula, O. Knotek, Kim Sem Khan, S. Mikolaewski, P. Mikec, E. Rurarz, V.A. Khalkin, V.A. Konov, L.M. Popinenkova. Cross sections for the 100 MeV proton-induced nuclear reactions and yields of some radionuclides used in nuclear medicine. *Radiochim. Acta* 54, 57, 1991.

43. M.L. Firouzbakht, D.J. Schyler, R.D. Finn, G. Laguzzi, A.P. Wolf. Iodine-124 production: Excitation function for the ^{124}Te(d,2n)^{124}I and ^{124}Te(d,3n)^{123}I reactions from 7 to 24 MeV. *Nucl. Instr. Method B* 79, 909, 1993.
44. P. Kopecký. Proton beam monitoring via the Cu(p,x)^{58}Co, ^{63}Cu(p,2n)^{62}Zn and ^{65}Cu(p,n)^{65}Zn reactions in copper. *Int. J. Appl. Radiat. Isot.* 36, 657, 1985.
45. F. Tárkányi, F. Szelecsényi, P. Kopecký. Excitation functions of proton induced nuclear reactions on natural nickel for monitoring beam energy and intensity. *Appl. Radiat. Isot.* 42, 513, 1991.
46. P. Kopecký, F. Szelecsényi, T. Molnar, P. Mikecz, F. Tárkányi. Excitation functions of (p,xn) reactions on ^{nat}Ti: Monitoring of bombarding proton beams. *Appl. Radiat. Isot.* 44, 687, 1993.
47. C.F. Williamson, J.P. Boujot, J. Pickard. Tables of range and stopping power of chemical elements for charged particles of energy 0.05 to 500 MeV. *Report CEA-R* 3042, 1966.
48. H.J. Probst, S.M. Qaim, R. Weinreich. Excitation functions of high-energy α-particle induced nuclear reactions on aluminium and magnesium: Production of ^{28}Mg. *Int. J. Appl. Radiat. Isot.* 27, 431, 1976.
49. D. Basile, C. Birattari, M. Bonardi, L. Goetz, E. Sabbioni, A. Salomone. Excitation functions and production of arsenic radioisotopes for environmental toxicology and biomedical purposes. *Int. J. Appl. Radiat. Isot.* 32, 403, 1981.
50. K. Kondo, R.M. Lambrecht, A.P. Wolf. ^{123}I production for radiopharmaceuticals XX. Excitation functions for the ^{124}Te(p,n)^{123}I and ^{124}Te(p,n)^{124}I reactions and the effect of target enrichment on radionuclinic purity. *Int. J. Appl. Radiat. Isot.* 28, 395, 1977.
51. M. Guillaume, R.M. Lambrecht, A.P. Wolf. Cyclotron isotopes and radiopharmaceuticals XXVII. ^{73}Se. *Int. J. Appl. Radiat. Isot.* 29, 411, 1978.
52. He Youfeng, S.M. Qaim, G. Stöcklin. Excitation functions for ^{3}He-particle induced nuclear reactions on ^{76}Se, ^{77}Se and ^{nat}Se: Possibilities of production of ^{77}Kr. *Int. J. Appl. Radiat. Isot.* 33, 13, 1982.
53. R. Weinreich, O. Schultz, G. Stöcklin. Production of ^{123}I via the ^{127}I(d,6n)^{123}Xe (β$^+$,EC)^{123}I process. *Int. J. Appl. Radiat. Isot.* 25, 535, 1974.
54. S.M. Qaim, G. Stöcklin, R. Weinreich. Excitation functions for the formation of neutron deficient isotopes of bromine and krypton via high-energy deuteron induced reactions on bromine: Production of ^{77}Br, ^{76}Br and ^{79}Kr. *Int. J. Appl. Radiat. Isot.* 28, 947, 1977.
55. D. De Jong, G.A. Brinkman, L. Lindner. Excitation functions for the production of ^{76}Kr and ^{77}Kr. *Int. J. Appl. Radiat. Isot.* 30, 188, 1979.
56. K. Okamoto, Ed. *IAEA Consultants' Meeting on Nuclear Data for Medical Radioisotope Production*, Vienna, April 1981, INDC (NDS) – 123/G, 1–22, 1981.
57. B. Scholten, S.M. Qaim, G. Stöcklin. Excitation functions of proton induced nuclear reactions on natural tellurium and enriched ^{123}Te: Production of ^{123}I via the ^{123}Te (p,n) ^{123}I—process at a low-energy cyclotron. *Appl. Radiat. Isot.* 40, 127, 1989.
58. J.D. Cockroft, E.T.S. Walton. Experiments high velocity positive ions II. Disintegration of elements by high velocity protons. *Proc. Roy. Soc.* A 137, 229, 1932.

59. R. Wideröe. Über ein neues prinzip zur herstellung hoher spannungen. *Arch. f. Elektotech.* 21, 387, 1929.

60. E.O. Lawrence, N.E. Edlefsen. On the production of high speed protons. *Science* 72, 376, 1930.

61. M.S. Livingston. The production of high-velocity hydrogen ions without the use of high voltages. Ph.D. thesis, University of California, 1931.

62. E.O. Lawrence, M.S. Livingston. The production of high speed protons without the use of high voltages. *Phys. Rev.* 38, 834, 1931.

63. E.O. Lawrence, M.S. Livingston. The production of high speed light ions without the use of high voltages. *Phys. Rev.* 40, 19, 1932.

64. R.R. Wilson. Magnetic and electrostatic focusing in the cyclotron. *Phys. Rev.* 53, 408, 1938.

65. D.W. Kerst, R. Serber. Electronic orbits in the induction accelerator. *Phys. Rev.* 60, 53, 1941.

66. M.S. Livingston. The cyclotron. *J. Appl. Phys.* 15, 2, 1944.

67. E.O. Lawrence, D. Cooksey. On the apparatus for the multiple acceleration of light ions to high speeds. *Phys. Rev.* 50, 1131, 1936.

68. P. Debye. Nährungsformeln für die zylinderfunktionen für große werte des arguments und unbeschränkt veränderliche werte des index. *Math. Ann.* 67, 535, 1909.

69. M.E. Rose. Focusing and maximum energy of ions in the cyclotron. *Phys. Rev.* 53, 392, 1938.

70. B.L. Cohen. The theory of the fixed frequency cyclotron. *Rev. Sci. Instr.* 24, 589, 1953.

71. J.J. Livingood. *Principles of Cyclic Particle Accelerators.* D. Van Nostrand Company, New York, 1961.

72. E.M. McMillan. The synchrotron—A proposed high energy particle accelerator. *Phys. Rev.* 68, 143, 1945.

73. J.R. Richardson, K.R. MacKenzie, E.J. Lofgren, B.T. Wright. Frequency modulated cyclotron. *Phys. Rev.* 69, 669, 1946.

74. D. Bohm, L. Foldy. Theory of the synchro-cyclotron. *Phys. Rev.* 72, 649, 1947.

75. L.H. Thomas. The paths of ions in the cyclotron. *Phys. Rev.* 54, 580, 1938.

76. E.D. Courant, M.S. Livingston, H.S. Snyder. The strong focusing synchrotron—a new high energy accelerator. *Phys. Rev.* 88, 1190, 1952.

77. L.J. Laslett. Fixed-field alternating-gradient accelerators. *Science* 124, 781, 1956.

78. M.S. Livingston, J.P. Blewett. *Particle Accelerators.* McGraw-Hill, New York, 1962.

3

Positron Physics

Adriano Pedroso de Lima and Paulo M. Gordo

CONTENTS

3.1 Positron and Positronium: General Remarks

In this chapter, the different processes involving the interaction of the positron with matter and the way that information about the annihilation site can be transported by the emitted radiation are briefly described. The experimental setups used to extract this information will also be discussed.

At present, the medical applications using positrons are mainly concentrated in positron emission tomography (PET), where the annihilation and subsequent detection of emitted radiation are used to define tissue or organ images, as described in detail in Chapter 6. However, we will show that the possibility of combining information on the chemical environment at the position where the annihilation takes place may offer new and important opportunities for applications in clinical diagnostics.

3.1.1 History and General Concepts

The history of the positron starts in 1928 with Dirac's prediction of the existence of the antiparticle to the electron [1]. The first experimental evidence of the existence of the positron was reported by Anderson in 1932 through cosmic ray pictures taken in a cloud chamber [2]. Studies based on the annihilation of the positrons with the electrons in matter were initiated in 1949 after the discovery by DeBenedetti et al. [3] that, when the positron annihilates in solid matter, the two annihilation gamma rays are not exactly collinear. This was interpreted as resulting from the effect of the momentum of the electron taking part in the annihilation. It was then suggested that, for the annihilation process, both the momentum and energy conservation laws of the electron–positron pair could be used to provide information on the properties of the matter.

The first experimental studies using positrons on the electronic structure of matter were devoted to the identification of Fermi surfaces in metals. The dramatic developments in nuclear spectroscopy equipment that took place in the two decades starting 1945 allowed the unequivocal establishment, by the end of the 1960s, that

- Positron annihilation parameters were sensitive to crystal lattice imperfections.
- Positrons could annihilate after being captured in a defect, which means that their wave functions were confined in the defect location (with the positron described by a localized state).

This positron behavior was clearly demonstrated in various papers, for example, thermal holes studies in metals [4] and ionic crystals [5], and elastic deformations in semiconductors [6].

Until the mid-1980s, positron studies of defects in solids were performed, preferentially, in metals and metal alloys. The experience obtained in this field was then applied to semiconductor studies, and soon the majority of published papers using positrons were directed toward studies on simple and compound semiconductors.

In parallel with the progress on the understanding of the physics involved in the positron annihilation in matter, important developments were also made with the experimental techniques. Around 1980, the first variable energy positron beams specially developed for material studies were constructed. These systems have opened new fields on depth profile defect studies (from the surface to the bulk material), interfaces, and multilayer systems, which are of crucial importance on metallic and semiconductor films used in modern devices. More recently, a new generation of pulsed positron beams, being able to focus the beam in a micrometer scale spot on the target, have become the most advanced positron beams. These systems can also scan the material surface, allowing studies of well-defined regions of the material.

Very recently, already in the new millennium, research groups have started using positrons in matter studies in biological systems. The first results are very promising, as variations in the annihilation radiation characteristics in biological systems have been confirmed.

3.1.2 Properties of Symmetry

The annihilation of the electron–positron pair is the process through which a particle–antiparticle pair is transformed into energy (photons). Since the total energy of the system involved in this annihilation (the pair formed by the free positron and one electron of the solid) has a value $2 \times m_0 c^2 - E_b \cong 1022 \, \text{keV}$, where m_0 represents the rest mass of the electron and the positron and E_b represents the binding energy of the bulk electron, the energy of the emitted photons are in the same energy range as the γ photons emitted by atomic nuclei. Since this process is invariant under charge conjugation, a general selection rule can be applied to the annihilation process. In particular, depending on the relative orientations of the electron and positron spins, the annihilation is only allowed with formation of an even number of photons when spins have antiparallel orientations ($S = 0$) or with an odd number of photons when spins are parallel ($S = 1$). Following this selection rule, the number of photons emitted during the annihilation of a randomly oriented electron–positron pair is not completely defined. However, in practice, only the annihilations involving the emission of 2γ or 3γ photons have an important role, as the cross sections for emission of higher number of photons are several orders of magnitude smaller than the annihilations with emission of 2γ or 3γ. The single photon annihilation of the electron–positron pair is forbidden by the momentum conservation law and can only occur in the presence of a third body that absorbs the recoil momentum, as was experimentally observed in 1991 by Palathingal et al. [7]. The probability for occurrence of this annihilation process is, in general, five orders of magnitude lower than the 2γ emission probability. The ratio between the probabilities for 3γ and 2γ annihilation for collisions between electrons and positrons with random orientations is given by the cross-section ratio $\sigma_{3\gamma}/\sigma_{2\gamma} = 1/372$, showing that the 2γ annihilation is, indeed, the dominant process.

The probability rate, per unit time, for annihilation of the positron with an electron from the medium Γ, is given by the product of the collision cross section σ, with the electron speed v, and the probability of finding one electron at the positron position, which is expressed by the overlap of the electron and positron wave functions. For 2γ annihilations in the nonrelativistic limit (since thermalized positrons have very low kinetic energy), the collision cross section, as calculated by Dirac [1], is given by

$$\sigma_{2\gamma} = \frac{\pi r_0^2 c}{v}, \tag{3.1}$$

and the probability rate is given by

$$\Gamma \equiv \Gamma_{2\gamma} = \sigma_{2\gamma} \, v n_e = \pi r_0^2 \, c n_e = \frac{12}{r_s^3} \, (\text{ns}^{-1}), \tag{3.2}$$

where r_0 $(r_0 = e^2/m_0c^2 = 2.8 \times 10^{-15} \text{ m})$ is the classical radius of the electron or positron, c is the speed of light in vacuum, n_e is the electronic density at the annihilation site, m_0 is the rest mass of the electron or positron, and r_s is the electronic density parameter given by

$$r_s = \frac{1}{r_0} \sqrt[3]{\frac{3}{4\pi n_e}}. \tag{3.3}$$

The expected values for the average positron lifetime are naturally dependent on the electronic density of the medium. Lifetimes of the order of 100 ps are expected for positrons in metals, but the average lifetime will increase with the decrease of the r_s parameter (semiconductors, polymers, and liquids).

3.1.3 Positronium

Positronium (Ps) is a quasistable bound system of a positron and an electron, similar to the hydrogen-like atom (and can be described in the same way), with a reduced mass $m_0/2$ (where m_0 is the electron mass). The Schrödinger equation for Ps is then identical to that for the hydrogen atom, when we replace the reduced mass of the hydrogen by one half of the electron mass. The Ps binding energy in vacuum is 6.8 eV and its diameter is approximately 1.6 Å.

The angular momentum sum for the electron and the positron spins can have two possible results leading to two fundamental Ps states: the singlet state, $^1S_0$, named as *para-positronium*, p-Ps (configuration with antiparallel spins); or the triplet state, $^3S_1$, referred to as *ortho-positronium*, o-Ps (configuration with parallel spins). In the absence of external magnetic fields, the formation probabilities for these two states are 25% and 75%, respectively. The difference in energy between the different spin states, the so-called hyperfine structure, is only 8.4×10^{-4} eV, which leads to all the possible states having approximately the same formation probability. As a consequence, there will be, on average, three o-Ps for each p-Ps formed.

Following the parity conservation rules, the Ps annihilation in vacuum from the p-Ps state can only be through the emission of an even number of photons, whereas the o-Ps state annihilation occurs with the emission of an odd number of photons.

As was previously referred to, if we consider the collisions between positrons and electrons with randomly oriented spins, the number of annihilations with emission of three photons is considerably less than the number of annihilations with emission of two photons. However, the electron–positron bound states, p-Ps and o-Ps, where the spins have definite relative orientations before annihilation, present quite different annihilation rates. With no external perturbation, the p-Ps and the o-Ps cannot be modified from their formation states before their annihilations and, consequently, the ratio of 2γ

to 3γ Ps annihilation should be 1/3. The lower annihilation probability, per unit time, for the 3γ emission is also indicated by the longer lifetime of the o-Ps state. The experimental values of p-Ps and o-Ps lifetimes, 125 ps and 142 ns [8], respectively, are in excellent agreement with the values previously calculated by West [9].

Nevertheless, within a material where Ps formation is possible, the frequent collision of o-Ps with atoms and molecules of the medium results in a fast exchange of its spin state. This spin exchange can occur either through the change of the electron in the Ps by an electron from a neighboring molecule, followed by annihilation (pick-off process); or, in special cases such as in the presence of paramagnetic molecules, through the change of spin of the Ps electron (spin conversion). In this way, the initial 1/3 ratio of the formation of the two possible positron states is not preserved in time; and the ratio of the 2γ to 3γ emission is, in this case, larger. As a consequence of this, a substantial reduction in the o-Ps lifetime in materials is observed (typically, the reduction is two orders of magnitude, for example, reaching values of few nanoseconds). The o-Ps lifetime then reflects the ortho- to para-transition probability and is one of the basic properties used to study the Ps interactions with neighboring atoms or molecules inside materials.

Four models are presently used to describe Ps formation in matter:

- The Ore model [10, 11]: The Ps may be formed from the ionization produced by the energetic positron with the subsequent binding to the free electron produced. This process can occur if the positron kinetic energy satisfies the condition: $(E_i-6.8 \text{ eV}) \leq E \leq E^*$, where E_i represents the ionization energy of the molecule and E^* represents the energy needed for electronic excitation of the molecule.
- Resonant mechanism [12]: The Ps formation process involving the last inelastic collision of the positron with bound electrons resulting, through energy resonant absorption, in the formation of an intermediary excited state of the positron–molecule system.
- The Spur mechanism [13]: through the capture of one of the many electrons, created by ionization, during the last steps of the positron energy loss process.
- Blob model [14] is an extension of the Spur mechanism, taking into account the distribution of free electrons formed by ionization during all the paths of the positron inside the medium, in order to explain the change of Ps formation probability under the influence of an applied external electric field [15].

The first two models are applied to the gaseous state whereas the last two are valid for condensed matter (liquids and solids).

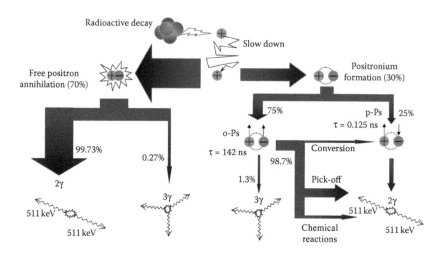

FIGURE 3.1
History of a positron emitted by a radioactive source since its creation until annihilation in water.

Positronium formation frequently occurs in gases, liquids, solid polymers, and living tissues, that is, in materials with low electronic density and with enough free space to contain the Ps. In ionic crystals, its formation is associated with the existence of defects. Positronium formation in metals and semiconductors is inhibited by their compact structure, by the high electronic density in the interstitial space between the atoms, and by the Ps atom dimensions. In metals and semiconductors, the Ps formation can only occur on the surface, either the external surface of the material or the surface of large dimension cavities inside the material.

In Figure 3.1, a schematic representation is shown of the physical processes as seen by the positron implanted in water, from which it can be observed that annihilation occurs preferentially through the emission of two gamma rays; nevertheless, the 3γ annihilation is also present, although with a small percentage (0.5%) and is mainly due to the formation of Ps.

3.2   Interaction of Positrons with Matter

3.2.1   Energy Loss and Thermalization

The different interaction processes that occur when energetic positrons enter a solid are schematically represented in Figure 3.2. Some positrons are backscattered in elastic collisions at the surface, and those entering into the solid rapidly lose their kinetic energy in a set of ionizations and excitations with

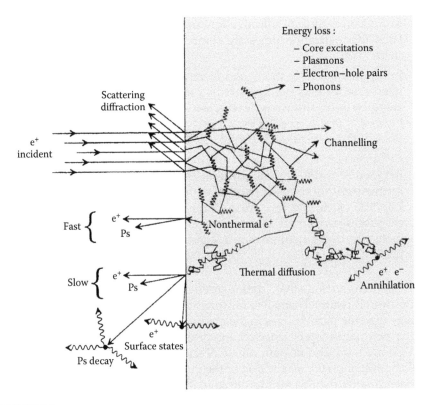

FIGURE 3.2
Positron interactions with a solid and a surface.

the electrons of the solid. In the case where the radioactive source is placed inside the material, the backscattering process cannot take place.

The energy loss rates are typically 1 MeV/ps when the positron energy is in the range from 100 keV to 1 MeV and 100 keV/ps in the energy range from 100 eV to 100 keV, independent of the material where the positron is moving [16]. For kinetic energies less than 100 eV, the positron energy loss rate strongly depends on the environmental characteristics:

- In metals, the excitation of free electrons, with the creation of plasmons and electron–hole pairs, is the dominant mechanism for positron energy loss. In the energy range below 1 eV, the dominant process is phonon dispersion. The total positron thermalization time, as calculated at room temperature, is of the order 1–3 ps, much smaller than the typical lifetime of a positron in a metal, 100–200 ps [16,17].

- In semiconductors, when the positron energy is lower than the semiconductor energy gap, electron–hole excitations are not possible, and the positron energy loss is due to phonon dispersion. This

dispersion is already effective for energies around 1 eV. The calculated thermalization times are comparable with those obtained in
metals [18].

- In liquids, the ionizations and electronic excitations represent the
main role for energies greater than the ionization limit. For lower
energies, the positron reaches thermal equilibrium primarily through
intra- and intermolecular excitations and through rotational excitations of the molecules. The total duration of the process is less than
1 ps [19, 20].

- In gaseous media, particularly in noble gases, the time scale needed
for the positron to reach thermal equilibrium is much longer, from
1 to 100 ns. In the low energy range (<100 eV), through chemical processes, various interactions of the positron with atoms and
molecules of the medium may be considered, including Ps formation
[21, 22].

We can then conclude that, in condensed matter, the thermalization time of
the positron is relatively small (a few picoseconds) when compared with the
average time the particle lives in the medium and so can be neglected.

After thermalization is achieved, the positron attains thermal equilibrium
with the medium, and its path can be described through a diffusion process
until annihilation with an electron of the medium.

Depending on the distance to the solid surface when it gets thermalized,
the positron can reach the surface during the diffusion and escape as a free
particle or as Ps, or be captured as a positron or Ps surface state.

3.2.2   Range and Implantation Profile

The implantation profile (or stopping profile) of the positron in matter
depends on various aspects, in particular the material in which it is implanted
and, especially, on its energy. However, two distinct conditions can be
considered:

- Positrons emitted by radioactive sources having a continuous energy
spectrum, characteristic of the β^+ decay (maximal energy of the order
of the MeV for most radioactive isotopes commonly used).

- Positrons with well-defined energy produced by monoenergetic
beam systems.

The stopping profile of energetic positrons emitted by a radioactive source
can be described by an exponential function:

$$P(z) = \alpha e^{-\alpha z}, \tag{3.4}$$

where $P(z)$ represents the probability of positron penetration to the depth z (measured from the surface in the case where the positron is implanted in matter from outside or the average distance from the radioactive source when the source is located inside matter), with the absorption coefficient α being given by

$$\alpha = 16 \frac{\rho}{E_{max}^{1,4}} \text{ cm}^{-1}, \tag{3.5}$$

where ρ (g/cm^3) is the material density and E_{max} (MeV) is the maximal energy of the emitted positrons. For positrons emitted by a ^{22}Na radioactive source (which is the isotope most used on material studies), the characteristic penetration length, $1/\alpha$, is 0.11 mm in silicon, 0.050 mm in germanium, and 0.014 mm in tungsten. In the case of positrons emitted by ^{11}C, ^{13}N, ^{15}O, and ^{17}F nuclei (with nuclear medicine applications), the characteristic penetration length in water varies between 0.5 and 1.4 mm.

For monoenergetic positrons obtained from variable energy positron systems with energies up to some tens of keV, the positron implantation profile of a positron with energy E, as simulated by Valkealahti and Nieminen [23], to the form originally suggested for electrons by Makhov [24], obeys the relationship:

$$I(z, E) = -\frac{d}{dz} [P(z, E)] = -\frac{d}{dz} \left[e^{-(z/z_0)^m} \right] = \frac{m z^{m-1}}{z_0^m} e^{-(z/z_0)^m}, \tag{3.6}$$

where m is a dimensionless parameter and z_0 is related to the average stopping depth, $\bar{z}$, by

$$z_0 = \frac{\bar{z}}{\Gamma(1 + (1/m))} \quad \text{and} \quad \bar{z} = \frac{AE^n}{\rho}, \tag{3.7}$$

where n and A are empirical parameters, ρ is the material density, Γ is the gamma function, and $P(z, E)$ represents the probability for the incident positron with energy E to stop (thermalize) inside the region between z and $z + dz$ (Makhov distribution). The more commonly used values for these parameters, which are considered to give a better description of the material behavior, are [25]

$$A = 4.0 \,\mu\text{g cm}^{-2} \text{ keV}^{-n}, \quad m = 2 \text{ and } n = 1.6.$$

The average positron stopping depth varies from the nanometer scale up to a few micrometers when its energy is up to $\cong 30$ keV.

This position selectivity along the average stopping depth allows the monoenergetic positrons the possibility of depth investigation from the surface until the bulk material. Using a variable energy positron system, it is then possible to obtain the depth profile of defects close to the surface region and to perform the analysis of interfaces of thin film and multilayer systems.

3.2.3 Diffusion and Mobility

After reaching thermal equilibrium inside the medium, the positron behaves as a quantum-free particle and its movement can be described through the diffusion equation Boltzmann equation. For thermal energies, the phonon dispersion is the dominant process in solids and the diffusion coefficient, as calculated using the relaxation time approximation; and the Nernst–Einstein relation is given by

$$D_+ = \frac{\mu_+}{e} k_B T = \frac{k_B T}{m^*} \tau_{rel}, \tag{3.8}$$

where m^* is the effective positron mass, k_B the Boltzmann constant, T the temperature, e the positron charge, and τ_{rel} the relaxation time, which is the inverse of the dispersion rate.

The mobility and the average free path of the positron are, respectively, given by

$$\mu_+ = \frac{e\tau_{rel}}{m^*} = \frac{D_+ e}{k_B T}, \tag{3.9}$$

and,

$$l_+ = \left(\frac{3k_B T}{m^*}\right)^{1/2} \tau_{rel}. \tag{3.10}$$

It has been experimentally confirmed through the observation of a $T^{-1/2}$ dependence on the diffusion coefficient [26] that, in metals and semiconductors, the dominant dispersion process is due to acoustical phonons, leading to

$$D_+ = D_0 \left(\frac{300\,K}{T}\right)^{1/2}. \tag{3.11}$$

The values of the positron diffusion coefficient in metals and semiconductors at 300 K is within the interval $D_0 = 1\text{–}3$ cm^2 s^{-1}. Combining these values of the diffusion coefficient with the typical positron lifetime values ($\tau_b \approx 100 - 200$ ps), the calculated average diffusion length of the positron is

$$L_+ = \sqrt{D_+ \tau_b} \approx 100 - 200 \text{ nm}. \tag{3.12}$$

These values of the positron diffusion length are, usually, much larger than the average free path (Equation 3.10), $l_+ \approx 5 - 10$ nm, and the Broglie wavelength, $\lambda_{term} = 6.2(300\,K/T)^{1/2}$ nm, of the thermal positrons, which are the main reasons that justify the application of the diffusion theory to describe the motion, after thermalization, of the positron inside metals and semiconductors.

In liquids, the positron diffusion length is also of the same order of magnitude and is still much greater than interatomic distances, so that diffusion theory also remains valid for these materials.

In summary, the thermalized positrons in condensed matter can be considered as waves, and the typical values of the characteristic lengths are within the following scale:

$$1/\alpha \geq 50\,\mu m > L_+ \approx 100\,nm > \lambda_{term} \approx 10\,nm > l_+ \approx 5\,nm. \qquad (3.13)$$

From the analysis of these values, the following considerations can be made:

- In the so-called conventional positron systems, when positrons emitted from radioactive sources are used, the implications of positron diffusion are not so important, as the implanted positrons are distributed through a large region ($\gg L_+$).
- In the variable energy positron systems, and particularly in the low energy region (≤ 10 keV), where the average implantation depth is small and the distribution is sharp, the motion of the positron after thermalization has to be considered and, in this case, is well described by the diffusion theory.

3.2.4 Annihilation

The positron annihilates preferentially after thermal equilibrium with the medium has been reached. Under this condition, the positron wave function in a solid can be calculated through the Schrödinger equation:

$$-\frac{\hbar^2}{2m}\nabla^2\Psi_+(\vec{r}) + V(\vec{r})\Psi_+(\vec{r}) = E\Psi_+(\vec{r}). \qquad (3.14)$$

In a perfect periodic lattice, the wave functions of thermalized positrons are described by delocalized states, that is, Bloch wave functions. The potential has two terms:

$$V(\vec{r}) = V_{Coul}(\vec{r}) + V_{corr}(\vec{r}), \qquad (3.15)$$

where the first term represents the Coulomb electrostatic potential and the second term takes into consideration the electron–positron correlation effects within the local density approximation.

Due to the strong Coulomb repulsion of positive ions inside the solid, the positron wave function is concentrated in the interstitial space between atoms. The positron wave function changes drastically in the neighborhood of a lattice site where the crystal periodicity has a gap, for example, due to the presence of one vacancy in the lattice (one atom is missing): this can be easily generalized to more complex defects. If an atom is removed from the periodic crystal lattice, the local electrostatic potential changes and the local electronic density decreases, as the core electrons are missing. The delocalized electrons of the solid that are present in the lattice vacancy create a negative charge in this region, which means an attractive potential for the positron. This potential can be strong enough to create a positron bound state (localized state) in the lattice vacancy site. The positron transition to localized states is referred to as *trapping*, and the defects are called *trapping centers*. In principle, positrons are trapped by any type of lattice defect that has an attractive electronic potential. In metals and semiconductors, the more general cases are open-volume-type defects: vacancies, vacancy aggregates, impurity–vacancy complexes, dislocations, grain boundaries, cavities, interfaces, surfaces, etc. Impurities and precipitates can also give rise to attractive potentials, in this way forming trapping centers to the positron. The trapping rate and the degree of localization of the positron are strongly dependent on the defect type and on its physical properties.

The high values of trapping rates and the dependence of the annihilation gamma ray characteristics on the defect type where the positron is trapped are the main reasons for the high sensitivity and selectivity of positron annihilation techniques in the characterization of defects in condensed matter. As examples, single vacancy concentrations in the range 0.1–200 at. ppm can be detected by this technique, whereas cavities with radius of some micrometers can also be observed at concentrations as low as 0.02 at. ppm. The dependence of both the positron trapping and the annihilation characteristics on the size of the defects, the charge states, and electronic configurations of defects makes possible, in most cases, the identification of the type and concentration of the defects present in the matter. These sensitivity and selectivity properties have already awarded positron annihilation techniques an important role in defect detection in condensed matter studies. The technique is nowadays established as positron annihilation spectroscopy.

In the majority of cases, as was previously indicated, the electron–positron pair annihilates mostly through the emission of 2γ photons. The study of the positron annihilation characteristics starts with the calculation of the probability per unit time, $\Gamma(\vec{p})$, that an electron–positron pair annihilates through the emission of a photon pair with total angular momentum $\vec{p} = \hbar\vec{k}$.

Equation 3.2 provides a first step to calculate this probability. However, in this equation, neither the interaction of the positron with the electrons from the medium nor the momentum of the pair involved in the annihilation is taken into consideration. One frequently used model is the independent particle model where an n particle system is described by n systems of individual particles, with each of them seeing the average field created by all the other particles. The Coulomb attraction between the electron and positron (as well as the correlations between all the electrons of the system), which implies an increase of the annihilation rate due to the higher electronic density in the neighborhood of the positron, is taken into account separately by the inclusion of an additional term. After these approximations, the probability density per unit time for the annihilation of the electron–positron pair by two photons of momentum $\vec{p} = \hbar\vec{k}$, usually referred to as momentum density, is given by

$$\rho_{2\gamma}(\vec{p}) = \frac{\pi r_0^2 c}{(2\pi)^3} \sum_{i,j} n_i^+ n_j^- \left| \int d\vec{r} \cdot e^{-i\vec{p}\cdot\vec{r}} \Psi_i^+(\vec{r}) \Psi_j^-(\vec{r}) \sqrt{\gamma[n(\vec{r})]} \right|^2 , \qquad (3.16)$$

where r_0 is the Bohr electron radius, c is the speed of light in vacuum, $\Psi_i^+(\vec{r})$ and $\Psi_j^-(\vec{r})$ are the wave functions of the positron and the electron, respectively; n_i^+ and n_j^- are the distribution functions describing the occupation of the positron and electron states, respectively; and $\gamma[n(\vec{r})]$ represents the electronic density amplification factor at the annihilation site. The summation includes all the states occupied by the positron and the electron. For valence electrons in a crystalline solid, n_j^- is the Fermi–Dirac distribution function. In the positron system generally used, n_i^+ is, to a good approximation, described by a delta function, as only one positron with negligible kinetic energy (thermalized positron) is present in the sample at a particular time. The amplification factor, $\gamma[n(\vec{r})]$, describes the effect of the positron–electron attractive force and of the resulting electron cloud accumulated around the positron. This has the effect of changing the potential seen by the positron and, consequently, the wave functions of both positron and electron. The effect of the higher electron density around the positron, due to the electron cloud, is to produce an increase of the annihilation probability and, obviously, a decrease in the positron lifetime. Different expressions for $\gamma[n(\vec{r})]$ based on many body calculations can be found in literature [27,28]. Naturally, electron–positron correlations are different for metals and semiconductors [29].

Equation 3.16 shows that positron annihilation characteristics in matter are strongly related to the electronic state at the site where the annihilation occurs. Since, in the majority of cases, the positron annihilates from thermalized states, the total momentum of the process contains information of the momentum of the electron at the annihilation site. The experimental measurement of the electronic density and of the electronic momentum distribution obtained from the information transported by the annihilation gamma photons is then possible.

Using the expression (Equation 3.16) for the momentum density, one can calculate the measurable physical quantities of positron annihilation experiments:

- The integration of $\rho_{2\gamma}(\vec{p})$ over a single momentum component gives the two-dimensional angular distribution of the annihilation gamma rays emitted, the so-called 2D-ACAR (two-dimensional angular correlation of the annihilation radiation):

$$N(\theta, \varphi) \propto N(p_y, p_z) \propto \int \rho_{2\gamma}(\vec{p}) \, dp_x; \qquad (3.17)$$

- The integration of $\rho(\vec{p})$ over two momentum components gives the one-dimensional angular distribution, the so-called 1D-ACAR, and the broadening of the annihilation radiation energy distribution due to the Doppler effect:

$$N(\theta) \propto N(E) \propto \int\int \rho_{2\gamma}(\vec{p}) \, \mathrm{d}p_x \, \mathrm{d}p_y; \qquad (3.18)$$

- The integration of $\rho(\vec{p})$ over all three momentum components gives the annihilation probability, per unit time, of a photon pair of arbitrary total momentum. This value corresponds to the inverse of the positron lifetime, τ:

$$\tau^{-1} = \lambda \equiv \Gamma = \int\int\int \rho_{2\gamma}(\vec{p}) \, \mathrm{d}p_x \, \mathrm{d}p_y \, \mathrm{d}p_z. \qquad (3.19)$$

The positron lifetime is, therefore, inversely proportional to the electronic density sampled by the positron at the annihilation site. For a delocalized positron, its lifetime reflects the average electron density in the interstitial space between atoms. For a localized positron, the annihilation can occur from different localized states, for each state i, the positron experiences a particular value of the electronic density corresponding to a particular value of the lifetime, $\tau_i = 1/\Gamma_i \equiv 1/\lambda_i$.

These three measurable observables constitute the basic quantities for the positron spectroscopy in material studies, using the techniques

- 2D-ACAR
- 1D-ACAR and Doppler broadening measurements
- Lifetime measurements

In the media where Ps formation can take place (e.g., in liquids, polymers, porous materials, living tissues, and organs), it is mainly the information related to this particle (lifetime and intensity) that can lead to useful information concerning the chemical surrounding at the Ps trapping sites. In particular, in polymers and porous materials, the Ps behavior constitutes a powerful way of obtaining information, for example, on porosimetry (measurements of dimensions and size distribution of porous and testing connectivity between porous) and on surface properties of porous materials including thin films. The Ps formed inside the bulk of polymeric or porous materials may reach, by diffusion, the surface of an open volume or porous and get trapped there. Inside the porous material, the Ps loses its initial kinetic energy, of some eV, through collisions with the cavity wall. Since it has a negative work function, the Ps, after losing its energy excess, cannot go back to the bulk material and stays localized, that is, trapped inside the porous material. In this state, the Ps can have different interactions:

1. In addition to self annihilation, it can annihilate with one porous surface electron by a pick-off process. The pick-off annihilation rate is proportional to the probability of Ps surface overlap. In this way, the Ps lifetime decreases when the porous dimension is reduced. Due to this lifetime dependence on the dimensions of the porous, the Ps lifetime measurements are, indeed, a very important technique for the measurements of average porous dimensions and size distributions of porous materials with dimensions from some tenths until a few tens of nm (mainly in the range 0.2–50 nm).

2. If the porous material is partially or completely occupied with liquid or gas molecules, the Ps annihilation rate by the pick-off process with electrons of the molecules filling the porous, a significant reduction of the Ps lifetime is then observed, and the value obtained may be used to identify chemical adsorption or molecular occupation of the pores.

3. If there are negative ions or free electrons inside the porous material, Ps may be involved in chemical reactions or spin conversion processes, respectively.

These situations can quite often be differentiated, which brings to Ps spectroscopy unique characteristics in surface chemistry, especially in catalysis investigations.

3.3 Fundamentals of the Experimental Techniques

The different positron techniques mentioned above are based on the analysis of the annihilation radiation due to the mass energy transformation of the electron–positron pair, which, in general, produces 2γ photons, each with energy approximately 511 keV. In special cases, a bound state electron–positron is formed, the so-called Ps atom and, in this case, the transformation can occur with the emission of either two or three photons, as previously discussed in Section 3.1.3.

The basic principles of each technique are shown, schematically, in Figure 3.3. These can be classified in two main groups that can be distinguished either by the positron sensitivity to the electronic density (positron lifetime measurements) or to the distribution of the electron momentum (Doppler broadening and ACAR measurements) in the sample to be studied.

Positrons from a radioactive source (e.g., ^{22}Na) enter the sample, thermalize in a few picoseconds, and diffuse through the medium before final annihilation. The maximum implantation depth is of the order 100 μm, which is much larger than the typical value of the positron diffusion length (of the order 100 nm). The positron lifetime can be measured as the temporal difference, Δt, between the emission of the 1.27 MeV photon, which corresponds

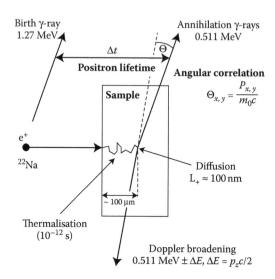

FIGURE 3.3
Schematic representation of the different experimental techniques in positron studies.

to the "birth" of the positron and the 511 keV annihilation photon, which corresponds to its "death." The Doppler broadening technique records the broadening ΔE of the 511 keV line. ΔE can be calculated through the electron momentum component along the propagation direction, p_z, $\Delta E = p_z\,c/2$. The angular deviation Θ is measured by the angular correlation of the annihilation radiation method. The angular deviation in the x–y plane, $\Theta_{x,y}$ is related to the components $p_{x,y}$ of the electron momentum, with $\Theta_{x,y} = p_{x,y}/(m_0c)$, where m_0 is the electron rest mass and c is the light velocity in vacuum.

3.3.1 Positron Lifetime

The probability that a positron annihilates in the time interval between t and $t + \mathrm{d}t$ is given by

$$p(t)\,\mathrm{d}t = \frac{\mathrm{d}}{\mathrm{d}t}[1 - e^{-t/\tau}]\mathrm{d}t = \frac{1}{\tau}e^{-t/\tau}\,\mathrm{d}t, \qquad (3.20)$$

where $e^{-t/\tau}$ represents the probability that in the instant t, after its production, the positron exists in a state with average lifetime τ.

In some cases, the positron can exist in different possible states, each with a characteristic lifetime, τ_i, and a probability, I_i, to be in that state. The lifetime spectrum is then a weighted average (with the respective probabilities) of the different exponential terms:

$$P(t) = \sum_i I_i \frac{1}{\tau_i} e^{-t/\tau_i}, \tag{3.21}$$

and, this way, is equivalent to the absolute value of the time derivative of the probability that the positron exists at the instant t, $D(t)$:

$$D(t) = \sum_i I_i e^{-t/\tau_i}. \tag{3.22}$$

In an experimental lifetime system, the measured spectrum, $N(t)$, is a convolution of $P(t)$ with the time resolution of the system, $R(t)$:

$$N(t) = \int_{-\infty}^{+\infty} P(t - t') R(t') \, dt'. \tag{3.23}$$

From this expression, and knowing the resolution function of the spectrometer, $R(t)$, it is possible to obtain, experimentally, the τ_i and I_i values by performing a numerical fit of Equation 3.23 to the measured time spectrum. However, it should be noted that, due to the correlations between the parameters to be adjusted, decomposition up to two components ($i = 2$) is only possible when the lifetimes are relatively short (100–200 ps), as in the case of metals and semiconductors; and up to four components when the lifetimes present in the material being studied are longer (>1 ns), as in the case of materials where Ps formation can occur (as in polymers). Even so, in metals and semiconductors, the unambiguous identification of two lifetimes in experimental time spectra is only possible if $\tau_2/\tau_1 > 1.5$.

Since the average lifetime is calculated through the weighted average of the observed lifetimes,

$$\bar{\tau} = \sum_{i=1}^{N} I_i \tau_i, \tag{3.24}$$

is then a parameter that can be obtained with high statistical precision (variations of 1–2 ps may be observed) and related to the time spectrum centrum; variations in this parameter of the order of a few picoseconds can be considered as a reasonably strong indication that some change has occurred in the region seen by the positron (e.g., a new type of defect has appeared in the material under study).

The experimental measurement of the average positron lifetime can be made if we obtain good precision the instants the positron enters the material and it annihilates and if we measure the time gap between these two events. The signal corresponding to the annihilation can easily be obtained through

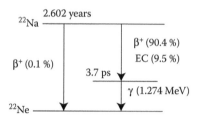

FIGURE 3.4
Schematic diagram of the ^{22}Na radioactive decay.

the detection of one of the two 511 keV emitted in the positron annihilation. The signal corresponding to the instant the positron enters the material can be obtained in different ways depending on the type of experimental system in use.

In the so-called "conventional systems," that is, those using the positrons directly from a radioactive source, such as ^{22}Na, the signal corresponding to the positron emission is obtained by the detection of the 1.27 MeV gamma photon due to the decay of the excited state in ^{22}Ne, which was populated by the ^{22}Na positron emission (see Figure 3.4 for the schematic diagram of the ^{22}Na radioactive decay).

In the pulsed beam positron systems, the signal indicating the positron arrival to the sample can be electronically generated by the beam control system.

In Figure 3.5a, a system for positron lifetime measurements is schematically represented; and in Figure 3.5b, typical positron lifetime spectra are compared.

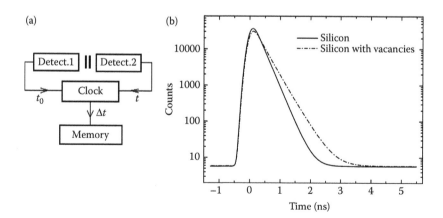

FIGURE 3.5
(a) Block diagram of a system for positron lifetime measurements. (b) Simulated lifetime spectra of positrons in crystalline Si and in Si with vacancy-type defects.

3.3.2 Momentum Distribution

In the annihilation process of the positron–electron pair, the energy and momentum conservation laws imply that, in the mass center reference frame, the 2γ photons are emitted in exactly opposite directions with energy 511 keV per gamma. However, in the laboratory frame, this pair has a nonvanishing momentum that has to be transported by the annihilation photons. As a consequence, a deviation of colinearity is observed, with a value θ, and also a Doppler shift of the 2γ photons in relation to the energy m_0c^2.

A geometric interpretation of the momentum conservation law in the positron–electron annihilation as viewed in the laboratory frame is represented in Figure 3.6, where

$\vec{p}$ represents the momentum of the positron–electron pair at the instant of annihilation.
$\vec{p}_1$ and $\vec{p}_2$ represent the momentum of the annihilation photons.
$\vec{p}_t$ and $\vec{p}_l$ represent the transversal and longitudinal components of $\vec{p}$ in relation to the direction taken as reference.

The energy and linear momentum conservation imply that

$$p_1c + p_2c = 2m_0c^2 \tag{3.25}$$

and

$$\vec{p}_1 + \vec{p}_2 = \vec{p}, \tag{3.26}$$

where m_0 is the rest mass of the electron and of the positron.

The angular shift from the colinearity θ depends on the magnitude of $\vec{p}_t$, whereas the magnitude of $\vec{p}_l$ determines the energy shift (Doppler broadening) of the photons.

Since the kinetic energy of the annihilating pair is a few eV, a simple calculation shows that, in the laboratory frame, the colinearity shift is

$$\theta \cong \frac{p_t}{m_0c} \tag{3.27}$$

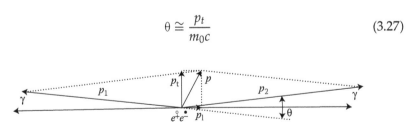

FIGURE 3.6
Vector diagram of the momentum conservation law in the annihilation of the positron–electron pair as viewed in the laboratory frame.

and the Doppler broadening, relative to the energy m_0c^2, is given by

$$\Delta E \cong \frac{cp_l}{2}. \tag{3.28}$$

The momentum of the thermalized positron is very small (almost zero) when compared with the momentum of the electrons of the material; and so, the momentum of the pair and also $\vec{p}_t$ and $\vec{p}_l$ are, in reality, very nearly the transversal and longitudinal components of the electron linear momentum relative to the emission direction of the annihilation radiation. In this way, the experimental systems based on the measurement of the one-dimensional angular shift (as referred before by 1D-ACAR) or of Doppler broadening of the energy, ΔE, give information on momentum distribution of the electrons involved in the annihilations.

A typical value for the momentum of the electrons in metals based on the free electron model is 1.5×10^{-24} kg m s^{-1}, which corresponds to an energy spread of 1.4 keV. This value is, in general, larger than the resolution that can be achieved with high-purity germanium detectors (HPGe), and so it is possible to obtain information on the electron momentum distribution for one material through Doppler broadening measurement of the annihilation radiation. The corresponding maximum angular shift θ is 5.6 mrad, which is naturally much larger than the angular resolution of the systems where angular correlation is currently observed.

In Figure 3.7, a classic angular correlation system (Figure 3.7a) and a typical Doppler spectrometer (Figure 3.7b) are schematically represented. The angular correlation system records the count rate $N(\theta)$, as expressed by Equation 3.18, which contains the linear momentum density of the 511 keV photons as a function of the angle θ. The need to obtain good resolutions (0.1–0.5 mrad) with these experimental systems implies the use of thin collimators and large sample-to-detector distances. Consequently, the coincidence count rates are very low and the acquisition time becomes too much long. More recent angular correlation systems (2D-ACAR) [30] use position-sensitive gamma radiation detectors with good angular resolution. In this case, a two-dimensional projection of the electron momentum distribution, as expressed in Equation 3.17, is obtained. The very low count rate of these detection systems is compensated by multiple coincidence techniques using multiple detector systems or large dimension (area) position-sensitive detectors (Anger chambers and multiwire proportional counters).

As described before, the information extracted from the energy spectrum of the annihilation radiation obtained in Doppler broadening measurements is equivalent to that obtained by the 1D-ACAR system. Effectively, using Equations 3.27 and 3.28 and considering that the expected values of the transversal and longitudinal components of the momentum must be the same, we can write:

$$\theta \cong \frac{2}{m_0c^2}\Delta E. \tag{3.29}$$

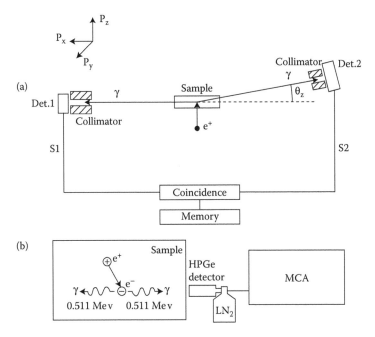

FIGURE 3.7
Experimental configuration for (a) 1D-ACAR. (b) Doppler broadening measurements.

The Doppler broadening systems have an equivalent angular resolution that is poorer than the angular correlation systems, but they are quite often an interesting alternative solution due to their considerable simplicity and high efficiency.

Experimentally, the Doppler broadening system uses a high-purity germanium gamma ray detector (HPGe) to obtain the energy spectrum of the annihilation gamma radiation.

The Doppler broadening spectra contain, as previously indicated, information on the contributions of each individual electron moment involved in annihilations. This information can, in principle, be obtained from the spectrum as so on as the resolution function of the detection system for the measured energy is well known. Although there are different deconvolution techniques specially adapted to the analysis of Doppler spectra, for example in references [31–34], there exist some reluctance on their applications, which are fundamentally due to two reasons: (1) the system resolution is intrinsically low and the identification of the different components is only possible in very simple cases (when higher precision on the determination of the electron momentum distribution is needed, another technique, ACAR, with one order of magnitude better resolution is normally used); and (2) in the majority of the experiments in materials, the research is only concerned with the presence and evolution of defects and these can be achieved using simpler techniques and methods of analysis.

As an alternative to the deconvolution methods, parameters related to the shape of the 511 keV peak of the annihilation radiation can be defined. These form parameters that present good sensitivity to shape variations of the Doppler spectra, are integrals of counts obtained in defined regions of the spectrum, are normalized to the total area of the 511 keV peak, and can be calculated after background subtraction.

The results of Doppler broadening measurements are, in general, discussed in terms of these parameters. The form parameters more usually accepted are the parameter S (abbreviation of "sharp"), introduced by MacKenzie [35], and the parameter W (abbreviation of "wing"). The parameter S is defined by the ratio between the area of the peak central part, A_s, and the total area of the 511 keV peak, A_o, after background subtraction,

$$S = \frac{A_s}{A_0}, \quad A_s = \int_{E_0 - E_s}^{E_0 + E_s} N(E)\, dE. \tag{3.30}$$

The parameter W corresponds to the sum of the count integrals in two regions away from the central part but symmetrically placed in relation to the peak centrum, which are also normalized to the total area of the 511 keV peak:

$$S = \frac{A_w}{A_0}, \quad A_w = \int_{E_1}^{E'_1} N(E)\, dE + \int_{E_2}^{E'_2} N(E)\, dE \tag{3.31}$$

In Figure 3.8, the integration regions for definition of each parameter are schematically represented. The energy limits for each region are symmetrically chosen around the central energy $E_0 = 511$ keV, taking $E_0 \pm E_s$ for the parameter S calculation. The energy limits E_1, E'_1, E_2, and E'_2 for the calculation of W should be defined to avoid effects related to the S parameter, considering that $E_0 = (E'_1 + E_2)/2 = (E_1 + E'_2)/2$. The choice of these limits should be made to obtain greater sensitivity of the parameter to variations of the moment distribution of the electrons involved in the annihilations associated with the defects. However, it should be remembered here that positron annihilations with low moment electrons, that is, the valence electrons, contribute most strongly to the value of the parameter S. In contrast, positron annihilations with high moment electrons, that is, the core electrons, contribute predominantly to the value of the parameter W. Due to this, these parameters are also called *valence parameter*, I_v, and *core parameter*, I_c, respectively. Considering the main type of annihilations contributing to the values of the form parameters, it is possible to conclude that both parameters are sensitive to concentration and type of defects in the material, but the W parameter is more sensitive to the chemical ambient around the annihilation site. This sensitivity to the chemical environment at the annihilation site is, indeed, taken as an advantage

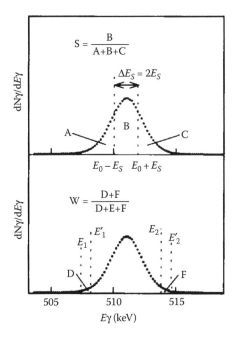

FIGURE 3.8
Definitions of the integration regions in the form parameters S and W. The energy windows are placed symmetrically in relation to the value $E_0 = 511$ keV.

of the techniques measuring the electron momentum (Doppler and ACAR) when the chemical environment of defects is under study. It should be mentioned here that the positron lifetime technique is a powerful technique that is very sensitive to the electronic density on the defect environment and is able to identify univocally the defects; however, it is not sensitive to electron momentum densities.

The absolute values of the form parameters do not have, *per se*, a clear physical meaning, as they depend on the position of the integration limits ($E_0 \pm E_s$, E_1, E_1', E_2, and E_2', which are not normalized) and are also strongly dependent on the resolution function of the experimental system. In material studies, only the changes in the parameter values (e.g., during a sample treatment process) are relevant and considered as indicators of modifications of the electron linear momentum distributions at the annihilation sites.

The positron annihilation techniques have been used with great success in the identification and characterization of defects in metals and semiconductors as well as in chemistry, in particular in the characterization of new materials with industrial applications, for example, polymers. This fact is naturally associated with the high sensitivity and selectivity of the positron to open-volume-type defects from the atomic scale (Angstrom) up to microcavities (some nanometers), as previously discussed in this chapter. An important aspect of the selectivity of the technique is the possibility of distinguishing different chemical environments at the positron annihilation sites. At this point, it becomes pertinent to raise the question as to how these remarkable possibilities can be used in clinical diagnosis.

It was only very recently that the positron annihilation techniques had started being applied to biological systems. Although the information is still limited, there are clear indications of the applicability of positron techniques on these systems. Shape parameters, S and W, of the annihilation radiation observed in tissue samples (rat skin in this example) that have been exposed to successive ultraviolet light irradiations [36] show a clear dependence of the irradiation dose. In addition, it is well known that certain types of cancerous tumors present a significant deficit in the amount of oxygen. Such chemical changes in the tissues lead to variations in the annihilation radiation characteristics as well as in the Ps pick-off rate, which imply a change in the observed $3\gamma/2\gamma$ fraction. Factors such as these are likely, in future, to stimulate the use of positron annihilation gamma radiation in medical diagnostics. However, quite a few changes and optimizations have to be made to the current positron systems to achieve the prompt response needed for medicine.

Let us now discuss the annihilation techniques that could be applied:

- Positron lifetime measurements: This technique will need a relatively short-lived positron emitter radioactive isotope, which, at the same time, would also emit a γ photon to indicate the birth of the positron. In addition, fast-response detectors would be needed (typical time resolution 250 ps) and acquisition times would be very long. With the inclusion of such features, the PET system would become quite exotic but, in practice, impossible to apply to living bodies.

- Annihilation radiation Doppler broadening measurements: This technique needs a gamma detector with high energy resolution, typically with full width at half maximum (FWHM) lower than 2 keV for the 511 keV peak. Actually, such energy resolutions are achieved with HPGe semiconductor detectors (which need to be cooled to liquid nitrogen temperature) and with CdTeZn (which operates at room temperature). With this technique, simultaneous information on the chemical environment at the annihilation site and on the $3\gamma/2\gamma$ fraction can be achieved.

As discussed in Chapter 6 on the image reconstruction of tissues and organs in PET, only the time coincidence of the two annihilation gammas with approximate energy of 511 keV is required; and it is not necessary to record any other information produced by the radiation. In general, in the conventional PET systems, the gamma ray detectors are based on the coupling of BGO scintillators to photomultipliers. With this configuration, excellent radiation detection efficiency and good spatial resolution are achieved. Unfortunately, these detectors have poor energy resolution; typically, the

511 keV peak has 12% FWHM, which is not enough to provide the detailed information that can be achieved through the annihilation radiation Doppler broadening technique.

With the development of PET systems, and particularly with the inclusion of better energy resolution detectors, it will be possible to achieve a more complete characterization of the chemical environment at the annihilation site and also of the $3\gamma/2\gamma$ fraction, with a significant improvement in the clinical diagnostics given by PET.

References

1. P.A.M. Dirac, On the annihilation of electrons and protons. *Proc. Cambridge Philos. Soc.* 26:361, 1930.
2. C.D. Anderson, The apparent existence of easily deflectable positives. *Science* 76:238, 1932.
3. S. DeBenedetti, C.E. Cowan, W.R. Konneker, Angular distribution of annihilation radiation. *Phys. Rev.* 76:440, 1949.
4. I.K. MacKenzie, T.L. Khoo, A.D. MacDonald, B.T.A. McKee, Temperature dependence positron mean lives in metals. *Phys. Rev. Lett.* 19:946, 1967.
5. W. Brandt, H.F. Waung, P.W. Levy, *Proc. Int. Symp. Color Centers in Alkali Halides* Rome, 48, 1968.
6. I.Y. Dekhtyar, V.S. Mikhalenkov, S.G. Sakharova, *Fiz. Tverd. Tela* 11:3322, 1969.
7. J.C. Palathingal, P. Asoka-Kumar, K.G. Lynn, Y. Posada, X.Y. Wu, Single-quantum annihilation of positrons with shell-bound atomic electrons. *Phys. Rev. Lett.* 67:3491, 1991.
8. R.N. West, Positron studies of condensed matter. *Adv. Phys.*, 22, 3:263, 1973.
9. C.I. Westbrook, D.W. Gidley, R.S. Conti, A. Rich, Precision measurement of the orthopositronium vacuum decay rate using the gas technique. *Phys. Rev. A*, 40 5498, 1989.
10. J. Green, J. Lee, (Eds.). *Positronium Chemistry*, Academic Press, NY, 1964.
11. V.I. Goldanskii, Physical chemistry of positron and positronium. *At. Energy Rev.*, 6:3, 1968.
12. Z. Zhang, Y. Ito, A new model of positronium formation: Resonant positronium formation. *J. Chem. Phys.*, 93:1021, 1990.
13. O.E. Mogensen, Spur reaction model of positronium formation. *J. Chem. Phys.*, 60:998, 1974.
14. S.V. Stepabov and V.M. Byakov, Physical and radiation chemistry of the positron and positronium. In: Jean, Y.C., Mallon, P.E., Schrader, D.M. (Eds.), *Principles and Applications of Positron and Positronium Chemistry*. World Scientific Publishers, Singapore, pp. 117–149, 2003.
15. S.V. Stepabov, V.M. Byakov, Y. Kobayashi, Positronium formation in molecular media: The effect of the external electric field. *Phys. Rev. B* 72 054205, 2005.
16. R.M. Nieminen, J. Oliva, Theory of positronium formation and positron emission at metal surfaces. *Phys. Rev. B* 22:2226, 1980.
17. K.O. Jensen, A.B. Walker, Positron thermalization and non-thermal trapping in metals. *J. Phys. Condens. Matter* 2:9757, 1990.

18. H.H. Jorch, K.G. Lynn, T. McMullen, Positron diffusion in germanium. *Phys. Rev. B* 33:93, 1984.
19. H.G. Paretzke, Radiation track structure theory. In: Freeman, G.R. (Ed.), *Kinetics of Nonhomogeneous Processes.* John Wiley & Sons, New York, pp. 89–170, 1987.
20. T. Tabata, Y. Ito, S. Tagawa, (Eds.). *Handbook of Radiation Chemistry,* CRC Press, Boca Raton, FL, 1991.
21. M. Charlton and J.W. Humberston, *Positron Physics,* Cambridge University Press, Cambridge, 2001.
22. M. Kimura, O. Sueoka, A. Hamada, Y. Itikawa, A comparative study of electron- and positron-polyatomic molecule scattering. *Adv. Chem. Phys.* 111:537, 2000.
23. S. Valkealahti, R.M. Nieminen, Monte Carlo calculations of keV electron and positron slowing down in solids. II. *Appl. Phys. A* 35:51, 1984.
24. A.F. Mahkov, *Sov. Phys. Sol. State* 2:1934, 1961.
25. A. Vehanen, K. Saarinen, P. Hautojärvi, H. Huomo, Positron diffusion in Mo: The role of epithermal positrons. *Phys. Rev. B*35:163, 1987.
26. B. Bergensen, E. Pajanne, P. Kubica, M.J. Stott, C.H. Hodges, Positron diffusion in metals. *Solid State Commun.* 15:1377, 1974.
27. M.J. Puska, R.M. Nieminen, Theory of positrons in solids and on solid surfaces. *Rev. Mod. Phys.* 66:841, 1994.
28. E. Boronski, R.M. Nieminen, Electron-positron density-functional theory. *Phys. Rev. B* 34:3820, 1986.
29. M.J. Puska, S. Mäkinen, M. Manninen, R.M Nieminen, Screening of positrons in semiconductors and insulators. *Phys. Rev. B*39:7666, 1989.
30. S.Tanigawa, A. Uedono, L.Wei, R. Suzuki, Defects in semiconductors observed by 2D-ACAR and by slow positron-beam. In: Dupasquier, A., Mills Jr., A.P. (Eds.), *Positron Spectroscopy of Solids.* IOS Press, Cambridge, 1986.
31. V.I. Goldanskii, K. Petersen, V.P. Sharantarovich, A.V. Shishkin, Another method of deconvoluting positron annihilation spectra obtained by the solid-state detector. *Appl. Phys.* 16:413, 1978.
32. K. Shizuma, Iterative unfolding method for positron annihilation radiation spectra measured with Ge(Li) detectors. *Nucl. Inst. Methods* 173:395, 1980.
33. P. Verkerk, Resolution correction: A simple and efficient algorithm with error analysis. *Comp. Phys. Commun.* 25:325, 1982.
34. W.H. Press, B.P. Flannery, S.A. Teukolsky, W.T. Vetterling, In: *Numerical Recipes, the Art of Scientific Computing,* Cambridge University Press, Cambridge, 1986.
35. I.K. MacKenzie, J.A. Eady, R.R. Gingerich, The interaction between positrons and dislocations in copper and in an aluminum alloy. *Phys. Lett. A* 33:279, 1970.
36. Y.C. Jean, H. Chen, G. Liu, J.E. Garcia, Life science research using positron annihilation spectroscopy: UV-irradiated mouse skin. *Rad. Phys. Chem.* 76:70, 2007.

4

Radiopharmaceuticals: Development and Main Applications

Antero Abrunhosa and M. Isabel Prata

CONTENTS

4.1 Overview

One of the bases of NM is the ability to produce molecules that are capable of acting as selective markers for the biophysical, biochemical, or pharmacological processes that are of interest. Over recent decades, thousands of molecules have been labeled with emitting nuclides for studying more than a hundred different processes in the human body. The unparalleled ability of these molecules to depict *in vivo* metabolic pathways and interactions has led to the development of a new scientific area: *molecular imaging*. In this chapter, we will describe the main questions related to the use of radiopharmaceuticals in molecular imaging studies and indicate some of their principal applications in clinical and research studies.

A molecular imaging study with radiopharmaceuticals usually begins with a clinical or research question that needs to be answered (Figure 4.1), frequently placed in the form of a model (paradigm) that is then confronted with the experimental results ("proof of action" approach). To answer this question, we have to select a molecular target, that is, a physiological process

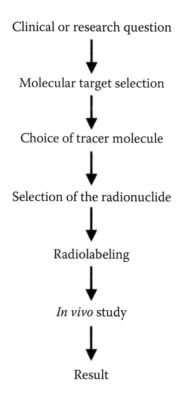

FIGURE 4.1
Main steps of a molecular imaging study with radiopharmaceuticals.

that is altered by the pathology that we want to study. For this molecular target, we select a tracer molecule that has the right physicochemical characteristics to map that process. The molecule is then labeled with an appropriate emitting nuclide which will enable us to perform the *in vivo* study that will allow us to obtain, if completely successful, a final image, which is frequently quantitative, and that will enable us to answer the initial question.

4.2 Molecular Target Selection

The quantification of a parameter of interest related to a certain physiological process imposes certain conditions on the chosen molecular target. First, we need a sufficiently high concentration of target molecules (e.g., receptors, enzymes, and transporters) in the region of interest, with minimal target occupation to avoid interfering with the process, to allow a sufficiently potent *in vivo* signal for the study to be made.

This potency of the signal is normally represented by the kinetic parameter binding potential (BP), defined as

$$BP = \frac{\beta_{max}}{K_d} \tag{4.1}$$

in which β_{max} is the density of molecular targets in the region of interest, and K_d is the kinetic dissociation constant between the chosen tracer molecule and the target [1].

Second, we need the structure we intend to visualize to be anatomically separable, bearing in mind the spatial resolution of the scanner used. This can be a problem for certain applications, for example, in neurology or with small metastases in oncology. Nevertheless, it is important to note that sometimes, if the signal-to-noise ratio is high, it is possible to observe structures with dimensions below the spatial resolution of the camera used, although, in this case, quantification would be impossible due to the so-called partial volume effect (see Chapter 6).

In the majority of studies, steady-state conditions would also be required with regard to the occupation of the active sites by endogenous molecules, in order to avoid interference concerning the affinity of the radiopharmaceutical for its active site. However, in some cases, this capacity of an endogenous molecule to displace the radiopharmaceutical from its active site can be exploited in order to measure, indirectly, a physiological effect that promotes its variation after an internal or external stimulus [2].

Table 4.1 lists some of the radiopharmaceuticals developed for the three main clinical areas of application: oncology, cardiology, and neuroscience, with the indication of the physiological process or molecular target they use.

TABLE 4.1

Examples of Some Radiopharmaceuticals Used in Molecular Imaging Studies and Their Application

Radiopharmaceutical	Physiological Process/Molecular Target	Reference
$H_2{}^{15}O$	Blood flow (neurology)	[3]
$C^{15}O$	Blood volume	[4]
[^{18}F]fluoromisonidazole (FMISO)	Cellular hypoxia	[5]
[^{18}F]annexin V	Apoptosis	[6]
2-[^{18}F]fluoro-2-deoxy-D-glucose (FDG)	Oxidative metabolism	[7]
3'-deoxy-3'-[^{18}F]fluorothymidine (FLT)	Proliferation (cell replication)	[8]
[^{11}C]choline	Proliferation (membrane synthesis)	[9]
2'-deoxy-2'-[^{18}F]fluoro-5'-iodo-1β-D-arabinofuranosyluracil (FIAU)	Gene expression	[10]
[^{13}N]NH$_3$	Blood flow (cardiology)	[11]
[^{11}C]acetate	Oxidative metabolism (cardiology)	[12]
[^{11}C]palmitate	Fatty acid metabolism (cardiology)	[13]
14-[^{18}F]fluoro-6-thia-heptadecanoic acid (FTHA)	Fatty acid metabolism (cardiology)	[14]
6-[^{18}F]fluoro-L-DOPA (FDOPA)	Dopamine biosynthesis	[15]
[^{11}C]methyl-methionine (MET)	Proliferation (protein synthesis)	[16]
O-(2-[^{18}F]fluoroethyl)tyrosine (FET)	Proliferation (protein synthesis)	[17]
L-[3-^{18}F]-α-methyl-tyrosine (FMT)	Proliferation (protein synthesis)	[18]
[^{11}C]WAY1000635	Serotonin 5-HT$_{1A}$ receptor	[19]
[^{11}C-methyl]SCH 23390	Dopamine D$_1$ receptor	[20]
[^{11}C]raclopride (RAC)	Dopamine D$_2$ receptor	[21]

4.3 Tracer Molecule

A fundamental characteristic of a radiopharmaceutical is that it can access the physiological compartment where the target is located. To achieve this, two fundamental characteristics are needed: the availability of the tracer in blood plasma and its distribution across the relevant biological barriers. The majority of drugs are transported in blood bound to plasma fractions, either proteic or lipidic. The extent of binding as well as the kinetics of the binding process (fast or slow) determine the availability of the radiopharmaceutical to enter the tissue. In addition, the passage of the radiopharmaceuticals into the tissue depends on the existence of a transport mechanism or the presence of favorable physicochemical characteristics of the molecules for their passive diffusion across the barriers. For example, in the central nervous system, the main physical barrier involves the tight junctions of the capillary endothelium (blood–brain barrier). The extent by which a certain compound can cross this barrier can be estimated by multiplying the permeability of the compound by the capillary surface (the so-called PS product). In this context, the permeability of the barrier to a certain compound is a characteristic value easily related by structure–function studies with physicochemical properties such as lipophilicity, charge, and molecular size [22].

The maximization of the signal measured *in vivo* and, consequently, the viability of a radiopharmaceutical for molecular imaging (Figure 4.2), ultimately depend on the balance between the molecular interactions established with the active site (specific signal) and all other interactions (nonspecific signal).

We can therefore write, assuming first-order kinetics that the tissue concentration of a radiopharmaceutical C_t is a function of its plasma concentration C_p over time:

$$C_t(t) = \sum_{i=1}^{n} \alpha_i \int_0^t C_p(\tau) e^{-\beta_i(t-\tau)} \, d\tau. \tag{4.2}$$

In this expression, components with a high value of α_i correspond to significant contributions to the observed signal, whereas β_i values for the specific signal that are significantly different from the β_i values of the other components allow us to separate the specific signal. A common situation is when the β value of the specific process is much lower than those of all the others ($\beta \rightarrow 0$). In this case, the protocol of the exam can allow sufficient time for the activity due to nonspecific processes to decay before image acquisition takes place. This is, for example, the case of the studies with 2-fluoro-deoxy-D-glucose (see Section 4.7.2.1), in which an essentially irreversible specific

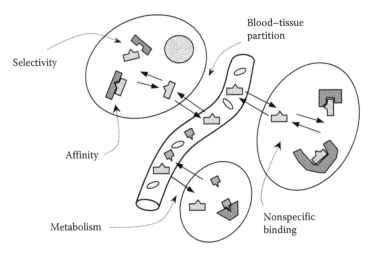

FIGURE 4.2
Overview of the main physiological processes involved in a molecular imaging study with radio-pharmaceuticals. (Adapted from *Physiological Imaging of the Brain with PET*, Academic Press, San Diego, pp. 51–56, 2001.)

process (phosphorylation by hexokinase) and the definition of an appropriate waiting time after the injection allow the quantification of the specific process (oxidative metabolism).

Metabolism of the radiopharmaceuticals can be another obstacle for the quantification of a specific component of the signal. In fact, the formation, during the study, of metabolites that maintain the labeling are able to cross relevant biological barriers and present affinity for molecular targets in the region of interest can preclude the quantification of the specific binding of the radiopharmaceutical to its specific site. The development of metabolism-resistant radiopharmaceuticals has been one of the main concerns, especially in the case of molecules that are present in several metabolic pathways (e.g., carbohydrates or amino acids).

4.4 Radionuclide Selection

The success of a radiopharmaceutical for *in vivo* imaging requires that the labeling process has a minimal interference in the physicochemical characteristics of the molecule. This requires, ideally, an isotopic substitution or, in case this is not possible, the substitution by an atom or chemical group that does not significantly change the *in vivo* behavior of the labeled molecule.

In the case of PET, the existence of radioactive isotopes of carbon (^{11}C, $t_{1/2} = 20.4$ min), nitrogen (^{13}N, $t_{1/2} = 9.97$ min), and oxygen (^{15}O, $t_{1/2} = 122$ s) that are positron emitters allows the labeling of virtually any organic molecule. However, the relatively short half-lives of these nuclides significantly limit the complexity of the molecules to be labeled as well as the range of processes that can be studied *in vivo*. In contrast, halogen positron emitters present half-lives that are more appropriate for complex labeling procedures and more prolonged studies ($^{18}F, t_{1/2} = 109.8$ min, $^{76}Br, t_{1/2} = 16$ h, or ^{124}I, $t_{1/2} = 4.2$ days); but their presence is not very common in organic molecules, and the introduction of a halogen atom frequently causes significant changes in the physicochemical properties of the compounds. A notable exception is fluorine-18, which has been used with great success as a substitute for the hydrogen atom that has a close Van der Waals radius and, sometimes, even for hydroxyl groups with which it shares a similar electronic configuration.

The substitution of hydrogen by Fluorine-18 is, as a consequence, the most common form of nonisotopic labeling in PET, as the changes introduced in a molecule by this substitution are minimal. The physicochemical changes introduced in a molecule by a change of a certain chemical group are usually evaluated in a diagram similar to that presented in Figure 4.3 (Craig plot) in which a series of chemical groups are plotted in terms of two main physicochemical characteristics: lipophilicity (σ) and polarity (π). This dia-

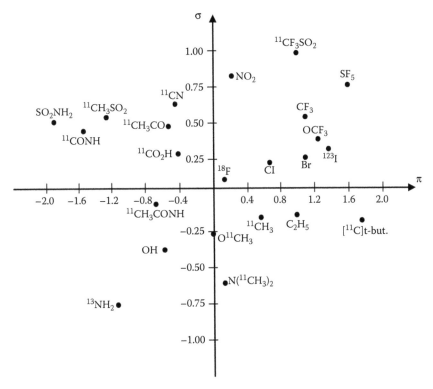

FIGURE 4.3
Craig plot for the two main physicochemical properties, charge (σ) and lipophilicity (π). Points that are close in the diagram represent chemical groups with similar physicochemical profile. H, by definition, is the origin. (Adapted from P. N. Craig. *J. Med. Chem.*, 14(8): 682, 1971.)

gram allows us to identify which chemical group is the best to substitute at a certain position of a molecule to minimize the effect of the substitution over the biological activity of the compound.

As can easily be seen in Figure 4.3, fluorine is the element closest to the origin, which makes it an excellent substitute for hydrogen [by definition, $\sigma(H) = \pi(H) = 0$], a fact that is the basis for the development of many PET radiopharmaceuticals such as [^{18}F]fluoro-2-deoxy-D-glucose ([^{18}F]FDG). This is also another considerable advantage of molecular imaging with PET compared with that by single photon emission tomography (SPECT) as the most appropriate γ emitter, ^{123}I, for use in this would undoubtedly lead to considerable modifications in the physicochemical characteristics of a molecule.

Similarly, we can conclude that the possible substitution of the OH group by fluorine-18 would not be very favorable. As we can see, these two groups are present in opposite quadrants of the diagram in Figure 4.3 ($+\sigma/+\pi$ for F—$\sigma/-\pi$ for OH). The hydroxyl group is, therefore, much better substituted by the O-methyl group and in fact, the labeling by the O-[^{11}C]methyl group is one

TABLE 4.2

Positron Emitting Nuclides with Indication of Half-lives and an Example of a Production Reaction

Nuclide	$t_{1/2}$	%β^+	Reaction	Nuclide	$t_{1/2}$	%β^+	Reaction
^{11}C	20.4 m	99.8	^{14}N(p,α)^{11}C	^{68}Ga	1.13 h	90	(Generator)
^{13}N	9.97 m	100	^{12}C(d,n)^{13}N	^{68}Ge	270.8 d		^{69}Ga(p,2n)^{68}Ge
^{15}O	122.24 s	99.9	^{14}N(d,n)^{15}O	^{73}Se	7.1 h	65	^{75}As(p,3n)^{73}Se
^{18}F	109.8 m	97	^{20}Ne(d,α)^{18}F	^{75}Br	1.62 h	75.5	^{76}Se(p,2n)^{75}Br
^{38}K	7.6 m	100	^{35}Cl(α,n)^{38}K	^{76}Br	16 h	57	^{75}As(^{3}He,2n)^{76}Br
^{52}Fe	8.28 h	57	^{52}Kr(^{3}He,3n)^{52}Fe	^{82}Rb	1.26 m	96	(Generator)
^{55}Co	17.5 h	77	^{56}Fe(p,2n)^{55}Co	^{82}Sr	25.6 d	100	Mo(p,spall)^{82}Sr
^{62}Cu	9.74 m	98	(Generator)	^{86}Y	14.74 m	34	^{88}Sr(p,3n)^{86}Y
^{62}Zn	9.22 h	93	^{63}Cu(p,2n)^{62}Zn	^{89}Zr	3.27 d	25	^{89}Y(p,n)^{89}Zr
^{64}Cu	12.7 h	18	^{64}Ni(p,n)^{64}Cu	^{94m}Tc	53 m	72	^{94}Mo(p,n)^{94m}Tc
^{66}Ga	9.5 h	57	^{67}Zn(p,2n)^{66}Ga	^{124}I	4.18 d	25	^{124}Te(p,n)^{124}I

Note: d: days; h: hours; m: minutes; s: seconds.

of the most widely used systems for the development of radiopharmaceuticals labeled with carbon-11.

Table 4.2 presents a list of some of the main positron emitting nuclides used in PET. Although the majority of the PET tracers have been developed around the first four of these, there are positron emitters from elements across the entire periodic table including, for example, isotopes of technetium and iodine, elements that are usually associated with SPECT studies, as well as nuclides available from reactors such as gallium-68 and rubidium-82.

Other important considerations involved in the selection of a positron emitting nuclide include the percentage of β emission, the energy of the positron (that influences its range in tissue), and its half-life, which should be long enough to permit chemical synthesis and the exam to take place but not too long to avoid exposing the patient to unnecessary doses of radiation after the exam is finished.

4.5 Labeling Position

The selection of the labeling position in a radiopharmaceutical is critical, as it can determine the fate of the emitting nuclide as the molecule is metabolized in the body. As an example, Figure 4.4 indicates two possible labeling positions for the molecule dihydroxyphenylalanine (DOPA), a marker of dopaminergic function in Parkinson's disease: [1-^{11}C]DOPA [23] and [2-^{11}C]DOPA [24]. In the first case (labeling on carbon 1), the action of the enzyme L-amino acid decarboxylase present in the target tissue causes the loss of the labeling

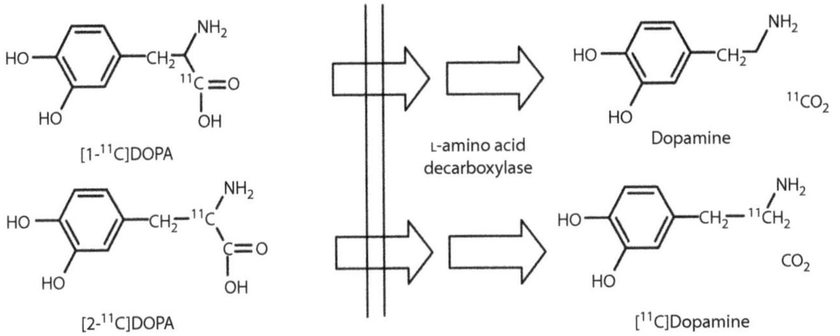

FIGURE 4.4
Fate of the emitting nuclide for two carbon-11 labeling positions of DOPA. Labeling in carbon 1 causes the loss of the labeling through $^{11}CO_2$ whereas labeling in carbon 2 leads to the production of [^{11}C]dopamine.

with the production of $^{11}CO_2$. In the second case (labeling on carbon 2), the same metabolic mechanism leads to the production of the dopaminergic neurotransmitter [^{11}C]dopamine.

4.6 Radiosynthesis

The specific nature of the radiopharmaceuticals imposes some conditions on the synthesis process as well as on the necessary quality control procedures. Additionally, considering that it will be applied in clinical studies, the final product should also be sterile, apyrogenic, and suitably formulated for human use. Also for clinical reasons, the compounds have to be produced with very high specific activity, which means that the entire synthesis process has to be made within a very short time. In the case of nuclides with very short half-lives, such as oxygen-15 and nitrogen-13 ($t_{1/2} < 10$ min), their chemistry is limited to very simple molecules (CO, H_2O, and NH_3) that do not have significant requirements in terms of quality control. In the case of nuclides with longer half-lives, such as carbon-11 ($t_{1/2} = 20.4$ min), it is possible to synthesize small organic molecules such as amino acids, carbohydrates, or fatty acids and perform typical quality control procedures. For radionuclides with longer half-lives, such as fluorine-18 ($t_{1/2} = 109.8$ min), it is possible to have a more complex chemistry and to include detailed quality control procedures. In this latter case, it is also possible to distribute the radiopharmaceuticals to centers that do not have a cyclotron available, a feature that facilitates the clinical use of these compounds.

In order to minimize synthesis time, the process is optimized to include the radionuclide in the last possible step. The process relies, therefore, on the use of (cold) prevalidated precursor molecules that are added to the labeling agent (radionuclide). It is also common to use a large stoichiometric excess of the cold reagents to compensate for the relative lack of the labeling agent and to improve the general yield of the reaction.

In compounds labeled with positron emitters, the short half-lives also require working with very high initial activities. The principles of radio-protection, particularly considering the high-energy γ rays that result from annihilation of the positrons (511 keV), require the use of automated synthesis and dispensing modules housed in heavily shielded hot cells with a minimum thickness of 75 mm of lead in their walls. This increasing automation of the processes also helps in establishing reproducibility, allowing the optimization of the synthesis conditions and facilitating production according to good manufacturing practice (GMP) regulations.

4.7 PET Nuclides: Reactions and Radiopharmaceuticals

4.7.1 Carbon-11

Carbon-11 is usually produced by proton irradiation of a nitrogen target through the reaction $^{14}N(p,\alpha)^{11}C$. The presence of small quantities of oxygen leads to the formation of $^{11}CO_2$, whereas the presence of hydrogen leads to the formation of $^{11}CH_4$. After a purification stage, each of these gases is propelled by a helium overpressure and is cryogenically trapped for the radiolabeling of a variety of important chemical precursors (Table 4.3). Of these, the most important is methyl iodide ($^{11}CH_3I$), which is used for the radiolabeling of numerous radiopharmaceuticals that are based on this nuclide.

TABLE 4.3

Examples of Chemical Species Labeled with Carbon-11 as Radiopharmaceuticals *Per Se* or as Precursors for Other More Complex Molecules

Reaction	Via		Labeled Species	
$^{14}N(p,\alpha)^{11}C$	$\rightarrow ^{11}CO_2$	$\rightarrow ^{11}CH_3OH$	$\rightarrow ^{11}CH_3I$	$\rightarrow ^{11}CH_3OSO_2CF_3$
			$\rightarrow H^{11}CHO$	
		$\rightarrow R^{11}COOMgX$	$\rightarrow R^{11}COOH$	
			$\rightarrow R^{11}COCl$	
		$\rightarrow CH_3{}^{11}COOLi$	$\rightarrow CH_3{}^{11}COOH$	
		$\rightarrow ^{11}CO$		
	$\rightarrow ^{11}CH_4$	$\rightarrow ^{11}CHCl_3$	$\rightarrow ^{11}CH_2N_2$	
		$\rightarrow H^{11}CN$	$\rightarrow ^{11}CH_3NH_2$	
			$\rightarrow ^{11}CO(NH_2)_2$	
		$\rightarrow ^{11}CCl_4$	$\rightarrow ^{11}COCl_2$	

4.7.2 Fluorine-18

Two distinct nuclear reactions are used to produce the two chemical main forms of fluorine-18 used in PET (Table 4.4): $^{18}F^-$ (nucleophilic fluorine) and $^{18}F_2$ (electrophilic fluorine).

Fluoride ion is normally produced by the reaction $^{18}O(p,n)^{18}F$ through the cyclotron irradiation of ^{18}O-enriched water. The resulting $^{18}F^-$ is then trapped in an anionic exchange resin and extracted from the aqueous medium. Since the fluoride ion is not very reactive in aqueous solution, it is taken to dryness and resuspended in an aprotic solvent such as DMSO or acetonitrile.

A key factor for the production of reactive nucleophilic fluoride is the choice of the counter ion. Potassium is normally used, although it is usually chelated in a poli-ether macrocyclic complex (Cryptand 222), which increases the distance between both ions and therefore, helps the release of the fluoride for nucleophilic substitution reactions. This is, in fact, the most common radiolabeling mechanism with ^{18}F used, for example, for the synthesis of [^{18}F]FDG through the nucleophilic attack of F$^-$ on the precursor tetra-O-acetyl-2-O-trifluoro-methanesulfonyl-β-D-mannopyranose [25].

In contrast, the production of molecular fluorine is traditionally done by the reaction $^{20}Ne(d,n)^{18}F$ by deuteron bombardment of neon in a nickel target. Molecular fluorine is a very reactive species, and the activity produced is trapped by adsorption onto the target walls. To remove this, we need to flush the target with cold fluorine, which, by competition, will remove the $^{18}F_2$ adsorbed onto the nickel walls of the target. The fluorine-18 obtained is consequently heavily diluted with nonradiolabeled fluorine, which leads to a very low specific activity of the radiopharmaceuticals produced [26]. This constraint limits the use of this synthesis pathway for the labeling of most radiopharmaceuticals, although it is still used as the major path for the

TABLE 4.4

Examples of Chemical Species Labeled with [^{18}F]fluoride and [^{18}F]fluorine, in Most Cases with the Objective of being Used as Hot Precursors for the Labeling of More Complex Molecules

Reaction	Via	Labeled species
$^{20}Ne(d,n)^{18}F$	$\rightarrow{}^{18}F_2$	$\rightarrow CH_3COO^{18}F$
		$\rightarrow RSO_2N^{18}FR$
$^{18}O(p,n)^{18}F$	$\rightarrow{}^{18}F^-$	$\rightarrow Br^{18}F$
		$\rightarrow I(CH_2)_n{}^{18}F$
		$\rightarrow CH_3CH^{18}FCOOCH_3$
		$\rightarrow{}^{18}FArCHO$

production of a few radiotracers (e.g., [^{18}F]DOPA) for which the specific activity factor is not so critical.

4.7.2.1 Deoxyglucose Method

The method is based on the properties of 2-deoxy-D-glucose (deoxyglucose, DG), and it was originally developed to determine local glucose utilization *ex vivo* in animals by autoradiography with carbon-14 (2-deoxy-D-[1-^{14}C]glucose) [27]. This method was later adapted for *in vivo* imaging by labeling DG with positron emitters. Carbon-11 was tested (2-deoxy-D-[1-^{11}C]glucose) [28], but it was the nonisotopic substitution of one of the two hydrogens in position 2 by fluorine-18 (2-[^{18}F]fluoro-2-deoxy-D-glucose [^{18}F]FDG) [6] that showed ideal characteristics to enable the determination *in vivo*, in humans, the local energy metabolism (Figure 4.5).

Deoxyglucose shares with glucose the same passive transport mechanism for entering cells (GLUT transporters). In the cells, deoxyglucose is also metabolized by hexokinase through phosphorylation to produce deoxyglucose-6-phosphate. This compound is effectively trapped within the cells due to its high polarity and is considered, within a compartmental model, to be in a different compartment with no direct access back to the blood pool (Figure 4.6).

In normal glucose, phosphorylation by hexokinase is only the first step of glycolysis, the initial step of the process of oxidative metabolism. The glycolytic pathway then proceeds with the action of an enzyme called phosphoglucose isomerase, which catalyzes the reaction that transforms glucose-6-phosphate into fructose-6-phosphate. This reaction, which can be easily understood using Fischer projections (Figure 4.7), relies on the formation of an enediol intermediate, which is essential for the conversion of the aldol group of glucose into the keto group of fructose.

In the case of deoxyglucose, the absence of the hydroxyl group in the β position makes this reaction impossible; and as a consequence, there is accumulation of the metabolic intermediate deoxyglucose-6-phosphate. It

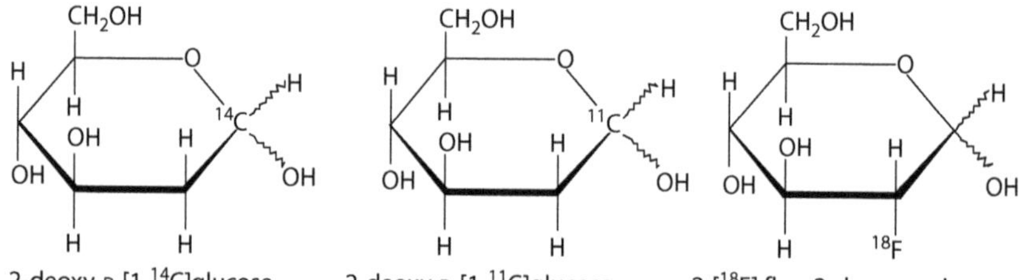

2-deoxy-D-[1-^{14}C]glucose   2-deoxy-D-[1-^{11}C]glucose   2-[^{18}F] fluro2-deoxy-D-glucose

FIGURE 4.5

Structural formulas of the different forms of deoxyglucose labeled with β-emitters: 2-deoxy-D-[1-^{14}C]glucose ([^{14}C]DG) and β+: 2-deoxy-D-[1-^{11}C]glucose ([^{11}C]DG) and 2-[^{18}F]fluoro-2-deoxy-D-glucose ([^{18}F]FDG).

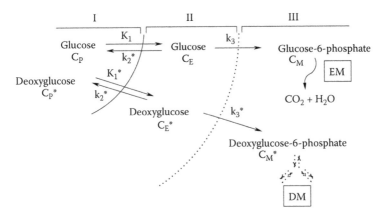

FIGURE 4.6
Three-compartment model for deoxyglucose labeled as a tracer for the determination of oxidative metabolism. Compartment I: capillary lumen; compartment II: intact compounds in tissue; compartment III: metabolic products. Cp, CE, and CM glucose concentrations in compartments I, II, and III respectively. K1-k3, kinetic constants of passage between compartments. Variables labeled with '*' are the same for deoxyglucose. EM, energetic metabolism; DM, deoxyglucose metabolites.

should be noted that, although this chemical reaction is not possible, there is evidence that deoxyglucose-6-phosphate binds effectively to phosphoglucose isomerase [29], leading to a competitive inhibition of the binding of glucose. This is, in fact, one of the mechanisms proposed for the toxic effect of deoxyglucose when administered in higher doses, which is suggested to cause an inhibition of glycolysis and the consequential development of symptoms similar to those of hypoglycemia. Although the simple reversion of the phosphorylation reaction of deoxyglucose by hexoquinase is, as in the case of glucose, highly unfavorable in terms of the energy balance, several

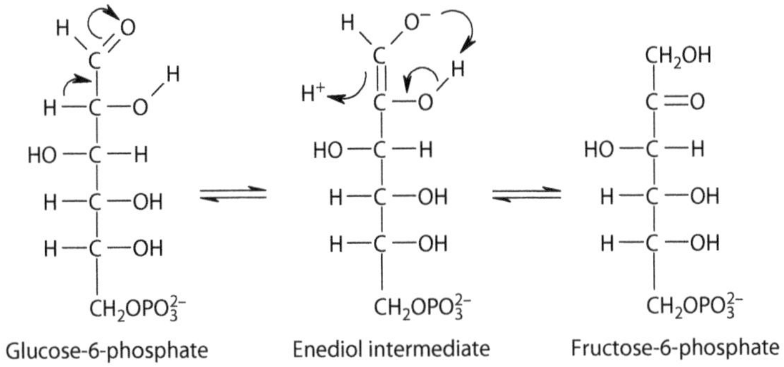

FIGURE 4.7
Schematic representation of the reaction catalyzed by the enzyme phosphoglucose isomerase that converts glucose-6-phosphate in fructose-6-phosphate evidencing the enediol intermediate fundamental to the formation of the final product.

mechanisms may occur to diminish the accumulation of deoxyglucose-6-phosphate in tissue.

In cancer cells, there is usually an increase in the GLUT expression that favors an increased uptake of glucose into the tissue. There are also several key enzymes in glycolysis that have their activity enhanced in tumor processes, including hexokinase, the first glycolytic enzyme. These two processes lead to an increase in deoxyglucose uptake and consequent entrapment, which provide the physiological basis for the widely successful use of [^{18}F]FDG in oncology.

4.8 SPECT Radionuclides

Currently, ^{99m}Tc is the most widely used diagnostic radionuclide and is responsible for more than 80% of diagnostic scans done each year in hospitals and clinics. However, a large number of other radioisotopes have found application, both in diagnostics and in therapy [30–34]. Table 4.5 summarizes some clinically used radionuclides, their physical characteristics, and their production methods. The most important properties of a radionuclide to be used in clinical applications are its half-life and decay mode, the energy of the emitted radiation, and, at least as important, the cost and ease of its production. Radionuclide half-lives should be long enough to allow production of the corresponding radiopharmaceutical and to perform the diagnostic scan but not so long as to increase the patient dosimetry without diagnostic benefits. Its energy should be appropriate for the detection system; for example, for SPECT cameras the energy of the gamma rays should be between 100 and 250 keV. Outside this range, the images will have poor quality. For lower energies, most of the photons will be stopped inside the patient's body, resulting in poor statistics, while at the same time increasing the patient's radiation dose. At high energies, as a consequence of the low selectivity of the photons

TABLE 4.5

Physical Data for the Most Widely Used Nuclear Medicine Radionuclides and Their Production Processes

Nuclide	$t_{1/2}$	Main Emissions (keV)	Production
Techenetium-99m	6.01 h	γ 141	Generator (^{99}Mo/^{99m}Tc)
Iodine-123	13.2 h	γ 159	Cyclotron (^{124}Te(p, 2n)^{123}I)
Iodine-131	8.04 d	γ 364, β 606	Nuclear fission
Thalium-201	73.1 h	γ 167, 135; X68 − 82	^{203}Tl(p,3n)^{201}Pb → ^{201}Tl
Gallium-67	78.3 h	γ 300, 181, 93	Cyclotron (^{68}Zn(p, 2n)^{67}Ga)
Indium-111	2.81 d	γ 245, 171	Cyclotron ^{111}Cd(p,n)^{111}In
Xenon-133	5.25 d	γ 81; β 46; X 30–36	Nuclear fission
Krypton-81m	13.3 s	γ 190	Generator (^{81}Rb/^{81}Kr)

by the collimator, there will be an increase in the scattering and, thus, a loss of image quality.

The decay scheme is also a very important criterion to choose an appropriate radionuclide: For therapeutic purposes, β^- or α emission is desirable; whereas for imaging diagnosis, particle emission is a big disadvantage, as it only contributes to the absorbed radiation dose without providing any benefit for the imaging. A good example is ^{131}I, which decays by β^- emission to an excited nuclear state of ^{131}Xe, which, in turn, decays by emitting gamma radiation of various energies, including 384, 637, 284, and 80 keV.

4.8.1 Technetium-99m: Production, Coordination Chemistry, and Radiopharmaceuticals

In the Periodic Table of elements, technetium (Tc) is the element 43, a transition metal. Its name comes from the Greek *technetos*, meaning artificial.

As indicated, technetium in the form of one of its isotopes (Tc-99m) is, by far, the most widely used radioisotope in NM. This is due both to its excellent nuclear properties and also to the fact that it can be produced *in situ* and daily by a generator (see below).

Radionuclide generators are based on a relatively long-lived parent that decays to a daughter nuclide, which is itself radioactive but with a short half-life [72]. The system will be useful for routine applications only if there is a straightforward method to separate the daughter radionuclide from its parent. The most common generator system is the Mo-Tc-99m [73], which is used daily in every NM Service around the world. However, there are other generator systems, such as Rb-81/Kr-81m, Sr-82/Rb-82, and of growing importance due to the increasing use of Ga-68 in PET imaging, the generator system Ge-68/Ga-68 [75–77].

The commercial generator Mo-99/Tc-99m (Figure 4.8) contains radioactive molybdate (sodium molybdate), adsorbed in an aluminum oxide column. The ^{99m}Tc, which is continuously formed inside the column, and the molybdate have different chemical affinities for the alumina, which allows their separation by column elution with saline solution. The reaction inside the column follows the scheme:

$$^{99}\text{MoO}_4^- \text{ (adsorbed in the column)} \rightarrow {}^{99m}\text{TcO}_4^- \text{ (free in solution)} \quad (4.3)$$

From consideration of the half-life ratio between ^{99}Mo ($t_{1/2} = 66$ h) and ^{99m}Tc ($t_{1/2} = 6$ h), we can easily conclude that this leads to a transient equilibrium, with the time or activity curves for the parent or daughter radionuclides given by

$$A_{[\text{Mo}]} = A_{[\text{Mo}]}^0 e^{-\lambda_{[\text{Mo}]}t} \quad (4.4)$$

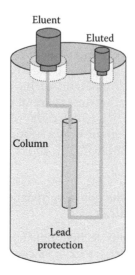

FIGURE 4.8
Schematic representation of the ^{99}Mo-^{99m}Tc generator.

$$A_{[Tc]} = A^0_{[Mo]} \frac{\lambda_{[Mo]}}{\lambda_{[Tc]} - \lambda_{[Mo]}} (e^{-\lambda_{[Mo]}t} - e^{-\lambda_{[Tc]}t}) + A^0_{[Tc]} e^{-\lambda_{[Tc]}t} \qquad (4.5)$$

For the case where the initial technetium-99m activity equals zero (after one elution), the above equations take the form:

$$A_{[Tc]} = A^0_{[Mo]} \frac{\lambda_{[Mo]}}{\lambda_{[Tc]} - \lambda_{[Mo]}} (e^{-\lambda_{[Mo]}t} - e^{-\lambda_{[Tc]}t}) \qquad (4.6)$$

The two curves involved can be represented by the diagram given in Figure 4.9.

The upper curve for ^{99m}Tc corresponds to the direct application of these equations; whereas the lower curve takes into account the fact that only 87% of the decays produce 140 keV γ rays.

It is straightforward to demonstrate that the time at which the maximum technetium activity is obtained corresponds to

$$t_{[Tc]max} = \left[1.44 \frac{t_{1/2[Mo]} t_{1/2[Tc]}}{t_{1/2[Mo]} - t_{1/2[Tc]}} \right] \ln \left(\frac{t_{1/2[Mo]}}{t_{1/2[Tc]}} \right). \qquad (4.7)$$

This corresponds to ca. 22.8 h, which is the main reason that the ^{99}Mo–^{99m}Tc generators are normally eluted the same time each day to allow an almost ideal accumulation of pertechnetate. Combining this process with the decay scheme of the ^{99m}Tc permits determination of the conditions of maximum activity in the presence of ^{99}Tc (Figure 4.10).

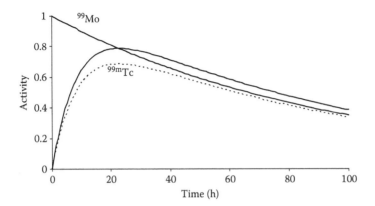

FIGURE 4.9
Activity/time diagrams for ^{99}Mo and ^{99m}Tc in the ^{99}Mo–^{99m}Tc generator.

Tc-99m emits 140 keV γ rays with 89% abundance, which makes it almost ideally suited for imaging with gamma cameras while being safe from the point of view of the radiation dose given to the patient. It has a very convenient half-life of 6 h, which allows the preparation and quality control of the ^{99m}Tc-radiopharmaceuticals and the possibility of performing even a complicated imaging protocol.

The existence of lyophilized cold kits containing appropriate formulations for ^{99m}Tc complexation makes it easy to obtain ^{99m}Tc radiopharmaceuticals just by addition of ^{99m}Tc in the pertechnetate form (^{99m}TcO$_4^-$) to the kit.

The ^{99m}Tc is eluted from the generator as an aqueous solution of sodium pertechnetate (Na ^{99m}TcO$_4$) and has an oxidation number of +7. The ^{99m}Tc^{7+} is unreactive with most of the ligands and needs to be reduced. Tin chloride and HCl are among the most commonly used agents to reduce the pertechnetate ion. After being reduced in the presence of suitable ligands, the technetium bears an oxidation number ranging from +1 to +6, depending on the reductant and ligand characteristics and the coordination reaction conditions.

Although the possibility that technetium can exist in all these oxidation states may make it difficult to control the reactions and also increases the lability of the complexes formed, on the other hand, it does offer more possibilities for the modification of the structure and properties of complexes

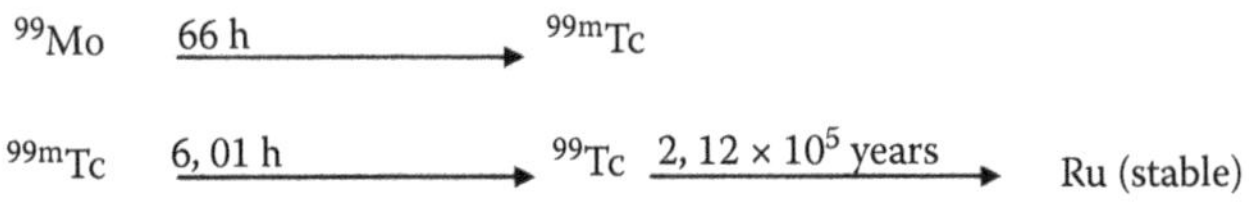

^{99}Mo $\xrightarrow{\text{66 h}}$ ^{99m}Tc

^{99m}Tc $\xrightarrow{\text{6, 01 h}}$ ^{99}Tc $\xrightarrow{2,12 \times 10^5 \text{ years}}$ Ru (stable)

FIGURE 4.10
Production and decay scheme of ^{99m}Tc.

through the choice of the appropriate ligands [35]. A further important characteristic of these technetium compounds is the possibility of the presence of isomeric structures: geometric isomers, epimers, enantiomers, and diastereoisomers. The presence of isomers, which is more frequent with technetium oxo-complexes, may have an important impact on the biological properties of the radiopharmaceuticals, as these isomers often have differences in their lipophilicity and, as a consequence, in their biodistribution profile [35].

To sum up, technetium can produce a large number of complexes displaying different structures, oxidation states, with coordination numbers ranging from 4 to 7, which introduces a very rich chemistry and a huge number of synthetic pathways.

A large variety of ^{99m}Tc-radiopharmaceuticals have been developed and approved by FDA for diagnosis, whereas a myriad of new ones are at the preclinical phase. There is virtually no organ function or disease process, from neuroreceptor imaging to oncology that cannot be visualized in NM using a ^{99m}Tc-based tracer [57]. Table 4.6 summarizes some of the new developments in this area. For more information about ^{99m}Tc-based radiopharmaceuticals and ^{99m}Tc chemistry, the reader is referred to excellent comprehensive reviews and textbooks [31,35,36].

As with other metal-based radiopharmaceuticals, ^{99m}Tc radiopharmaceuticals can be classified in two categories: "technetium essential" and "technetium tagged." Technetium essential radiopharmaceuticals are those in which the Tc is an integral part of the radiopharmaceutical and for which the molecule would not be delivered to its target in the absence of the Tc. Their *in vivo* behavior depends only on their chemical–physical properties, such as size, charge, and lipophilic or hydrophilic character. The latter class, technetium tagged radiopharmaceuticals, includes those in which the targeting moiety (e.g., antibody, peptide, and hormone) has been labeled with ^{99m}Tc, either directly or by using a bifunctional chelate; and their *in vivo* localization is mediated by biological interactions with, for example, receptors or proteins. A bifunctional chelator bears chelator groups that are able to coordinate the metal, while, at the same time, having functional groups which allow the radiopharmaceutical to be recognized by a target biological molecule. The introduction of a chelator in a bioconjugated molecule can have a profound effect on its biodistribution, and this will increase with a decrease of the size of the biomolecule; this will be more pronounced for small peptides than for larger antibodies. To minimize this effect, it is usual to introduce a "spacer" to separate the chelator from the bioactive part of the molecule [45].

Considering ^{99m}Tc compounds, we can include the following radiopharmaceuticals in the first group (where their *in vivo* behavior depends only on the complex structure): ^{99m}Tc-D,L-HM-PAO (Ceretec) and ^{99m}Tc-LL-ECD (^{99m}Tc coordinated to N,N'-1,2-ethylenediylbis-L-cysteine diethyl ester—Neurolite), both of which are used to determine brain–blood flow; the complexes ^{99m}Tc-DTPA and ^{99m}Tc-MAG3 (Technescan) used to evaluate kidney function; and

TABLE 4.6
Examples of ^{99m}Tc-Labeled Ligands as Diagnostic Radiopharmaceuticals

Radiopharmaceutical	Chemical and Biological characteristics	Application	Reference
^{99m}Tc-TRODAT-1	^{99m}Tc complex in which a diaminodithiol ligand is complexed with the metal and a tropane analog is derivatized from one nitrogen. It presents two diastereoisomers syn on the coordination to TcO^{3+}. The human images of ^{99m}Tc-TRODAT-1 showed localization at the basal ganglia, which correlates with the binding to the dopamine transporters (DAT).	Imaging of dopamine transporters	[37]
^{99m}Tc BMS 181321	Nitroimidazole-derivatized TcO^{3+} amine oxime complexes. Hypoxic tissues can be differentiated from normal tissues on the basis of their redox profile inside the cell. The radiopharmaceutical should enter the cell and be reduced in a hypoxic environment.	Hypoxia targeting	[38]
^{99m}Tc-sestamibi	A monocationic hexakis-isonitrile (2-methoxy-2-methyl-1-propyl) isonitrile, where technetium is present with oxidation number+1. This is a very rare Tc oxidation state, but the complex is very kinetically stable. The ^{99m}Tc-sestamibi is a myocardium perfusion agent, but it was also demonstrated that the ^{99m}Tc-sestamibi is transported out of tumor cells expressing MDR by the Pgp glycoprotein, allowing its use for MDR evaluation.	Evaluation of myocardium perfusion; MDR detection	[39,40]
^{99m}Tc-annexin V	Annexin V labeled with ^{99m}Tc(CO)_3(OH_2)_3$. The annexin V is a protein that binds to the phosphatidylserine residues present at the membrane of apoptotic cells. It was demonstrated that the ^{99m}Tc labeling using the tricarbonile does not modify the cell affinity of annexin V.	Apoptosis tracer	[41]
^{99m}Tc-P829	Radiopharmaceutical that binds to the somatostatin receptors overexpressed in a large variety of tumors. The labeling using the precursor ^{99m}Tc(CO)_3(OH_2)_3$ does not need the presence of a reductor species, an advantage as it protects the disulfide peptide bond that is essential to its receptor recognition.	Neuroendocrin tumor detection	[42]
^{99m}Tc- RP419/DMP444	Peptide bearing the RGD (-Arg-Gly-Asp) sequence; it binds to the l GPIIb/IIIa receptors expressed in the activated platelets, the first step in the thrombus formation.	Thrombosis detection	[43]
^{99m}Tc-ciprofloxacin	The ciprofloxacin is an antibiotic. It is not known the coordination mechanism of the technetium to this ligand and the radiolabeling yield is only ca. 40%.	Bacterial infection diagnosis	[44]

Note: MDR, multidrug resistance.

^{99m}Tc-sestamibi (^{99m}Tc-hexakis-isonitrile) (Cardiolite) and ^{99m}Tc-tetrafosmin ([^{99m}Tc-(1,2-bis((bis-ethoxyethyl)phosphino)ethane)]$^+$) (Myoview) as myocardium perfusion agents. Pertechnetate itself (as eluted from the generator) has diverse uses in NM, such as thyroid and parathyroid scintigraphy and imaging of the salivary glands. Figure 4.11 shows the structures of some of these technetium-99m complexes. As examples of ^{99m}Tc-bioconjugates (the second group mentioned), we can include ^{99m}Tc-annexin V, for apoptosis evaluation [38], the ^{99m}Tc-monoclonal antibody used in CEA-Scan [46], or some peptide somatostatin analogs labeled with ^{99m}Tc [42].

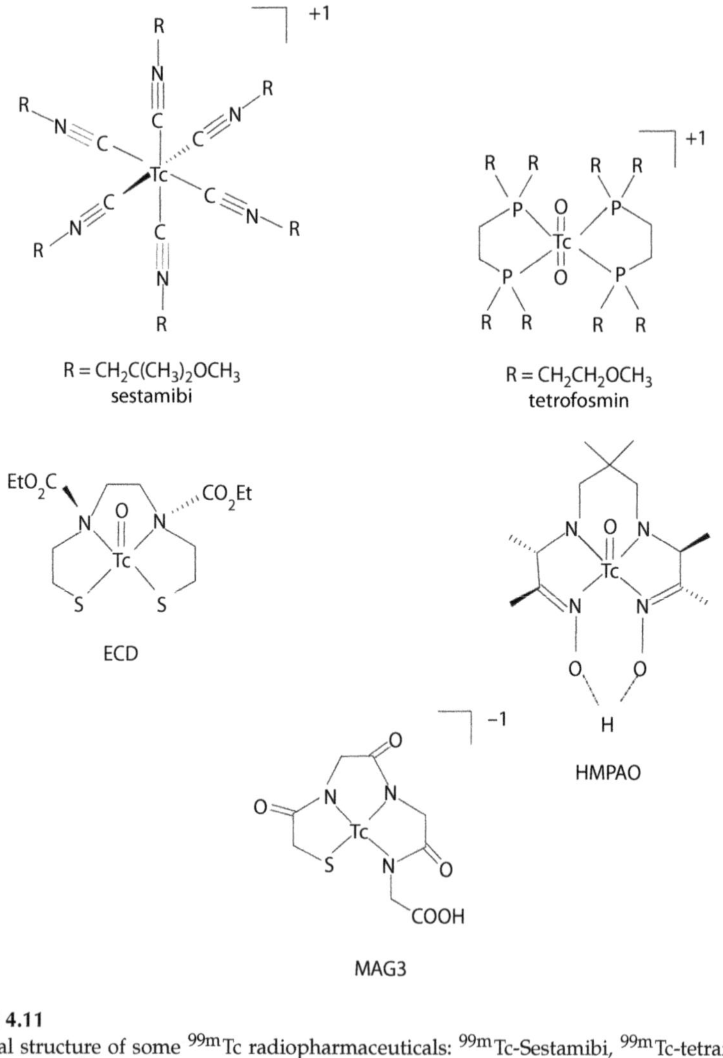

FIGURE 4.11
Chemical structure of some ^{99m}Tc radiopharmaceuticals: ^{99m}Tc-Sestamibi, ^{99m}Tc-tetrafosmin, ^{99m}Tc-L,L-ECD, ^{99m}Tc-D,L-HM-PAO, and ^{99m}Tc-MAG3.

4.8.2   Gallium and Indium: Radioisotopes and Their Coordination Chemistry

There are three gallium radionuclides (^{66}Ga, ^{67}Ga, and ^{68}Ga) and two indium radionuclides (^{111}In, ^{113}In) with suitable physical characteristics for use in gamma scintigraphy or PET [30].

The isotope ^{67}Ga ($t_{1/2} = 78.1$ h) is produced in a cyclotron by the nuclear reaction ^{68}Zn(p,2n)^{67}Ga starting with a ^{68}Zn-enriched sample. The ^{67}Ga obtained is then separated either by solvent extraction or by an ionic exchange process.

^{68}Ga, a positron emitter, is produced by a ^{68}Ge or ^{68}Ga generator and decays by positron emission in 89% yield. The maximum energy of this positron is 1.899 keV, whereas the average energy per disintegration is 740 keV. The long half-life of ^{68}Ge ($t_{1/2} = 280$ days) makes it possible to obtain ^{68}Ga *in situ*, without a cyclotron, although it is necessary to have an efficient method of separation of ^{68}Ga from ^{68}Ge. ^{68}Ga has a very convenient half-life of 68 min compatible with the biokinetics and biological half-lives of a lot of low or average molecular weight radiopharmaceuticals, such as peptides or oligonucleotides [47], thus opening the possibility of developing lyophilized cold kit formulations. The chemical properties of ^{68}Ge and ^{68}Ga are different enough to permit an efficient separation of the two radionuclides. In the literature, two main strategies are found to isolate ^{68}Ga from ^{68}Ge:

1. Using organic matrices bearing phenolic groups that form very stable complexes with Ge(IV) and which allow the elution of ^{68}Ga^{3+} as ^{68}GaCl$_4^-$ by using HCl as eluent. An alternative methodology uses an *N*-methylglucamine-based polymer and a 0.1 M trisodium citrate solution as eluent [48].

2. With inorganic oxides matrices, such as Al$_2$O$_3$, SnO$_2$, Sb$_2$O$_5$, ZrO$_2$, and TiO$_2$, eluted with HCl or EDTA [49]. Recently, a generator has been introduced in the market where the ^{68}Ge or ^{68}Ga separation is done on a TiO$_2$ column using a 0.1 M HCl solution as eluent (Cyclotron Co, Obninsk, Russia). The reader is directed to excellent reviews concerning this subject [50,51].

^{66}Ga is also a positron emitter ($t_{1/2} = 9.45$ h) and is produced in a cyclotron by the ^{66}Zn(p,2n)^{66}Ga reaction. However, there are very few examples of application of this radionuclide.

The radioisotope ^{111}In ($t_{1/2} = 67.2$ h) is produced in a cyclotron, starting from ^{111}Cd, ^{111}Cd(p,n)^{111}In and decays by electron capture producing 173 keV (89%) and 247 keV (95%) photons. The separation of the ^{111}In from the starting material, ^{111}Cd, follows processes similar to those used to isolate ^{67}Ga.

^{113m}In ($t_{1/2} = 1.7$ h) obtained with the generator ^{113}Sn/^{113m}In emits γ radiation with energy 393 keV. However, nowadays this radionuclide has been replaced for most of its applications by ^{99m}Tc.

The coordination chemistry of gallium and indium is of considerable great interest, mostly due to the potential use of their radioisotopes in radio-pharmacy [31,45,52]. Gallium and indium are metals belonging to the III B group of the periodic table and under physiological conditions they only exist in the +3 oxidation state. This fact is determinant for their radiochem-istry. These two metals, together with boron and aluminum, are classified as "hard acids" [53], and they prefer to form chemical bonds with ionic and nonpolarizable Lewis bases such as nitrogen and oxygen atoms (carboxylate, phosphonate, and amino groups). The coordination chemistries of gallium (III) and iron (III) are very similar; and, as we will see, this fact is crucial in the use and preparation of gallium radiopharmaceuticals. The coordination chemistry of In (III), having a larger ionic radius, is comparable with that of Y (III) and lanthanides.

In solution, the hydrated cations Ga (III) and In (III) are only enough stable under acidic conditions, hydrolyzing at higher pH values and leading to the insoluble hydroxides Ga(OH)$_3$ and In(OH)$_3$:

$$M(H_2O)_6^{3+} + H_2O \rightarrow M(OH)(H_2O)_5^{2+} + H_3O^+; \tag{4.8}$$

$$M(OH)(H_2O)_5^{2+} + H_2O \rightarrow M(OH)_2 + (H_2O)_4^+ + H_3O^+; \tag{4.9}$$

$$M(OH)_2 + (H_2O)_4^+ \rightarrow M(OH)_3(s) + H_3O^+ + 3H_2O. \tag{4.10}$$

As an example, in a gallium(III) solution containing 1 mCi/mL (which corresponds to a ^{68}Ga concentration of 4×10^{-10} M and to a ^{67}Ga concen-tration of 2.5×10^{-8} M), Ga(OH)$_3$ precipitation occurs for pH ≥ 3 if there are no stabilizing ligands present in solution [54]. Even though it is possible to use gallium-68 hydroxide-based colloids for liver–spleen PET imaging [55], normally these hydrolysis reactions prevent the preparation of gallium radio-pharmaceuticals. In contrast with what happens with In(OH)$_3$, the hydroxide Ga(OH)$_3$ is amphoteric; and in addition to being soluble in acids, it is soluble and at basic pH through the reaction,

$$Ga(OH)_3(s) + OH^- \rightarrow Ga(OH)_4^-, \tag{4.11}$$

which allows the redissolution of Ga(OH)$_3$.

For their use as radiopharmaceuticals, the compounds of gallium and indium must be thermodynamically stable at physiological pH values and/or kinetically inert on the timescale of the procedures used in NM.

4.8.2.1 Gallium and Indium Ligands

On the basis of their structural properties, one can establish two main classes of ligands suitable for coordinating In^{3+} or Ga^{3+}: linear chain ligands and macrocyclic ligands. In addition to the metal coordinating groups, many of the ligands belonging to the two classes possess other functional groups

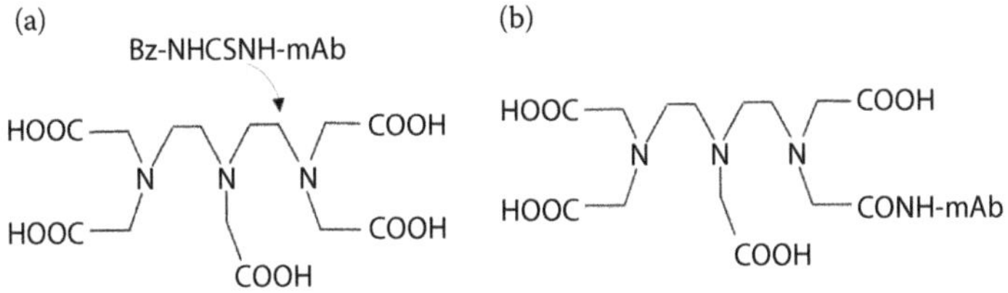

FIGURE 4.12
Structures of the bifunctional chelators (a) SCN-BzDTPA (octadentate) and (b) cDTPA (hepta-dentate) (Adapted from M. W. et al., *Inorg. Chem.*, 25:2772–2781, 1986.). The arrow shows the coupling site.

(e.g., –NH$_2$ or –COOH), which allow the functionalization with a macro-molecule. DTPA (diethylenetriaminepentaacetate) is an example of a widely used linear chain chelator in radiopharmacy, such as for metal radiolabeling of monoclonal antibodies [56] and peptides [57]. The amino acid conjugation is normally achieved through the cyclical bis-anhydride derivative of the DTPA (cDTPA) [58]. This method has the inconvenience of sacrificing one of the car-boxylate groups of the DTPA to form an amide bond with the lysine residue of the lateral chain of the antibody such that only seven coordinate sites of the DTPA stay available for coordination to the metal, thus lowering the ther-modynamic stability of the chelate. For this reason, the use of another deriva-tive of DTPA, 1-(p-isothiocyanatobenzyl)-diethylenetriaminepentaacetate (p-SCN-Bz-DTPA), has been proposed, where the linking group to the antibody uses one carbon of the skeleton of the DTPA (Figure 4.12) [59]. However, DTPA only forms a complex that is sufficiently stable to permit its use *in vivo* with the In^{3+}. Another linear chelator has, therefore, been proposed for studies with gallium: the desferrioxamine-B (DFO) (Figure 4.13), which is used as a bifunc-tional chelator with high yield for $^{67/68}$Ga^{3+} labeling [60]. In addition, DFO presents a –NH$_2$ group, which is available for conjugation with biomolecules, such as peptides or antibodies.

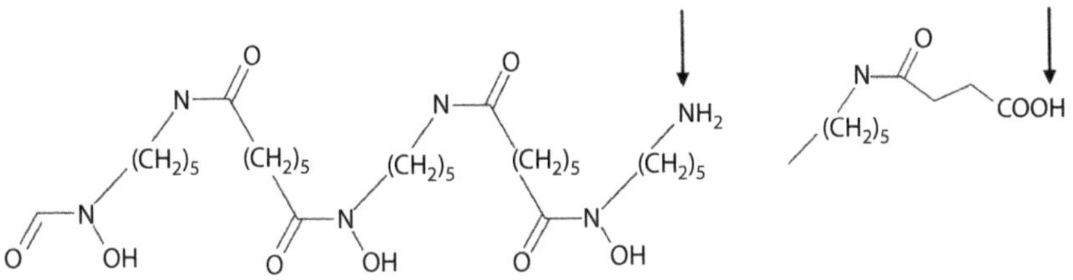

FIGURE 4.13
Structure of the desferrioxamine-B (DFO), a bifunctional chelator coupled to biomolecules via the –NH$_2$ group or a succinyl spacer (Adapted from P.M. Smith–Jones, et al., *J. Nucl. Med.*, 35: 317–325, 1994.).

Macrocyclic ligands are, by definition, cyclic polydentate ligands where the donor atoms belong to the macrocyclic ring and/or to their pendant arms. The main characteristics that differentiate macrocyclics from their linear analogs are the high stability of the complexes formed, their specificity and selectivity for the cations that they coordinate, and the slower formation and dissociation kinetics of their complexes. In fact, polyfunctional macrocyclic ligands show a very selective coordination profile, which is strongly pH dependent, and form very stable complexes with several metal ions. The chelate stability depends on the relationship between the cavity size, the metal ionic radius, the rigidity of the ligands, and the nature of the coordinating groups. In particular, the ligand 1,4,7-triazacyclononane-N,N',N''-triacetate (NOTA) forms chelates of high stability with Ga^{3+} [61]. In fact, the Ga(NOTA) complex has proved to stay intact in a 1 M nitric acid solution for a period of at least 6 months [62]. The high stability of the complex of Ga^{3+} is a consequence of the excellent fitting of the small metal ion (ionic radius 0.76 Å) into the cavity of the NOTA ligand. This class of ligands binds the metal ion efficiently [63], protecting it from competitor ligands, in particular transferrin (the blood serum protein that transports iron), which is the main competitor for Ga^{3+} in the blood stream.

As has already been stated, gallium and iron share a very similar chemistry: charge, ionic radius (62 pm for Ga^{3+} and 65 pm for Fe^{3+}), and the preference for the coordination number 6. Transferrin, a plasma protein, has two sites of binding to iron that also have high affinities for Ga^{3+}, and this protein is present in high concentrations in plasma, 2.5×10^{-3} M. The binding constants for gallium and indium to transferrin are log $K(Ga\text{-}tf) = 20.3$ and log $K(In\text{-}tf) = 18.74$ [63], respectively. Thus, when injecting $^{67}Ga^{3+}$ in the form of gallium citrate (or as another low stability complex), more than 90% of this metal is complexed by the transferrin.

In ^{111}In radiopharmaceuticals, due to its larger ionic radius, 81 pm, and also its higher coordination number (normally it forms complexes with coordination number 7), it is preferable to use the octadentate macrocyclic DOTA (1,4,7,10-tetraaza-cyclododecane, N,N',N'',N'''-tetraacetate), which forms more stable complexes with this metal [45] (Figure 4.14).

4.8.2.2 Gallium and Indium Radiopharmaceuticals

^{67}Ga, in the chemical form of gallium citrate, was initially used in the detection of inflammation, but this technique became less important when methods of radiolabeling leucocytes (with ^{99m}Tc-HMPAO or with ^{111}In-oxine) were developed. However, [^{67}Ga]Ga-citrate is still in current use in the diagnosis of some chronic inflammations, in the diagnosis of some inflammatory processes such as sarcoidosis, and in patients with low level of leucocytes, such as AIDS patients.

In oncology, the citrate of gallium-67 still has a role in the diagnosis of some tumors of soft tissues, especially in some types of lymphomas. Although it

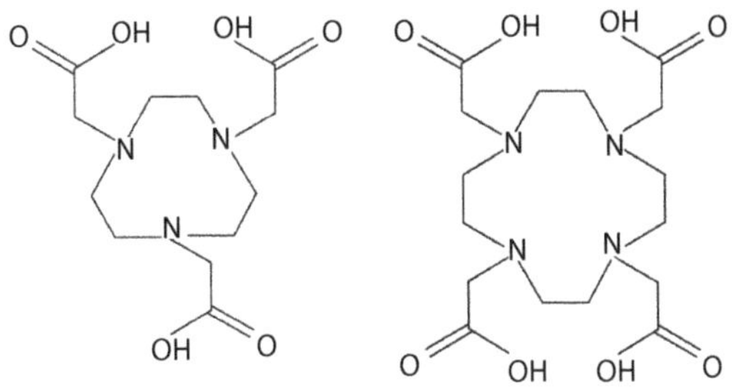

FIGURE 4.14
Structure of the macrocyclics NOTA (1,4,7-triazacyclononane-*N*,*N'*,*N''*-triacetate) and DOTA (1,4,7,10-tetraazacyclododecane-*N*,*N'*,*N''*,*N'''*-tetracetate).

is known that 90% of gallium injected in the blood stream in the form of gallium citrate is complexed by transferrin, the mechanism of incorporation of gallium into tumor cells is not yet known. One hypothesis states that this can be due to the uptake of the complex ^{67}Ga-transferrin by the transferrin receptors of tumor cells, but various other mechanisms can be present or may coexist with that one.

With the increased number of PET centers and with the establishment of the importance that ^{68}Ga could assume in molecular imaging, the interest in new ligands for gallium has reemerged and even increased, particularly those involving bifunctional chelators [64,65].

Until recently, ^{111}In was almost exclusively used for labeling of leukocytes and platelets through the complex between ^{111}In^{3+} and 8-hydroxyquinoline (oxine) [66]. One of the few exceptions listed in the literature was the use of ^{111}In-DTPA to evaluate alterations of the blood–brain barrier [67]. Today, this radioisotope has acquired additional importance in the labeling of antibodies and peptides (see Section 4.9) [30,45].

^{111}In^{3+} forms a liposoluble complex with oxine (8-hydroxyquinoline)$_3$ and penetrates in this form into the cell membrane. After diffusion, the complex ^{111}In-oxine (Figure 4.15) dissociates and the ^{111}In^{3+} binds to cytoplasmatic and nuclear proteins [66]. Although the complex ^{111}In-leukocyte suffers liver and spleen uptake, the absence of uptake in the kidneys, bladder, and gallbladder makes this method of great applicability in middle-low abdomen and chest scintigraphy.

Reference has also been made, although more rarely, to the use of immunoglobulin G (IgG) labeled with ^{111}In in the diagnosis of certain particular types of inflammation (e.g., inflammation of the joints) [68]. The complex ^{111}In-HIG presents the great advantage that it is not taken up in the bone marrow. The complex ^{111}In-oxine is still used to label platelets with a view of the detection of thrombi [69]. The method developed to label platelets with ^{111}In may also be used in their labeling with ^{68}Ga [70].

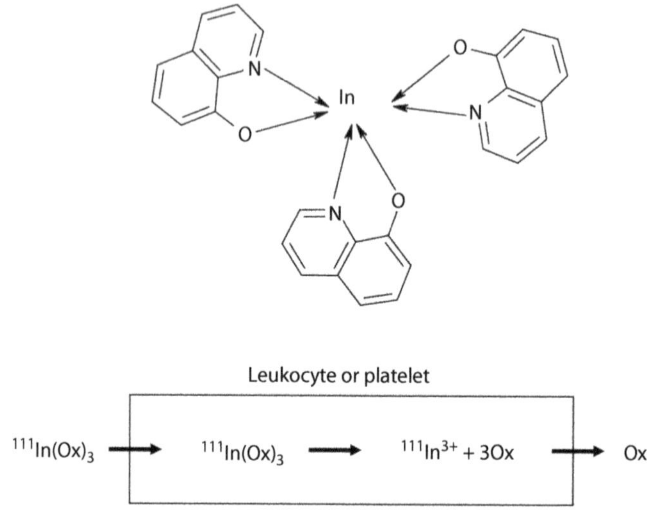

FIGURE 4.15
Structure of the In-oxine complex and the cell labeling mechanism.

4.8.3 Copper Radioisotopes and Copper Coordination Chemistry

There are several copper radioisotopes with suitable characteristics both for diagnostics and therapy: ^{67}Cu, ^{64}Cu, ^{62}Cu, ^{61}Cu, and ^{60}Cu. These radio-isotopes have very different nuclear properties, with half-lives ranging from 9.6 min (^{62}Cu) to 62 h (^{67}Cu), β^+ and/or β^t decay. ^{67}Cu is a β^- emitter with 40% γ emission and hence a suitable therapeutic radionuclide. However, for diagnostic purposes the most interesting copper radioisotope is ^{64}Cu. Due to its relatively long half-life (12.8 h), ^{64}Cu is mainly used in PET imaging of biomolecules with very long biological half-lives, in particular monoclonal antibodies and peptides [71]. Its decay scheme combines β^+(19%), β^-(40%), and electron capture (41%), allowing pretherapeutic dosimetry estimation. The ^{64}Cu radionuclide can be produced in a nuclear reactor by a ^{64}Zn(n,p)^{64}Cu reaction or by the ^{64}Ni(p,n)^{64}Cu mechanism in a cyclotron.

The copper ion exists in only two relevant oxidation states, Cu(I) and (II). The Cu(II), the most used in the production of radiopharmaceuticals, prefers to coordinate to compounds containing atoms of nitrogen (amines, Schiff bases, and pyridines) or sulfur. The chelating agents more commonly used to complex copper, particularly in the form of bifunctional complexes, include the 4-[(1,4,8,11-tetraazacyclotetradec-1-yl)-methyl]benzoic acid (CPTA), the 1,4,8,11-tetraazacyclotetradecane-1,4,8,11-tetraacetic acid (TETA), and the 1,4,7,10-tetraazacyclododecane-1,4,7,10-tetraacetic acid (DOTA) (Figure 4.16) [30].

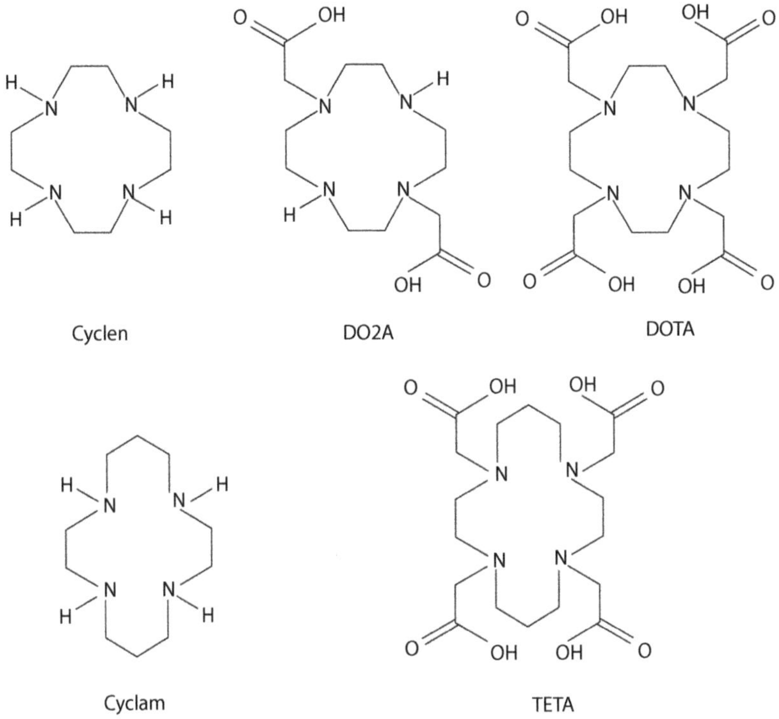

FIGURE 4.16
Macrocyclics used to coordinate copper radioisotopes in radiopharmaceuticals.

4.9 Monoclonal Antibodies and Peptides in Molecular Imaging

4.9.1 Monoclonal Antibodies

Monoclonal antibodies (mAbs) are proteins with molecular weights averaging 160 kDa. In the early 1970s, these antibodies were initially labeled with iodine radioisotopes [72]; but today it is more common to use a bifunctional chelator conjugated to the antibody to complex the radioisotope of choice. The majority of the mAbs used for tumor targeted belong to the IgG (immunoglobulin G) family and are formed by two large heavy chains and two small light chains bounded by disulfide bonds.

The usefulness of mAbs in tumor targeting depends on several factors: (1) the antibodies cannot induce cross reactivity with the normal tissues; (2) their antigens should be present in a good distribution at the cell membrane; (3) the antibodies should present a high affinity for the antigen; (4) no free antigens should be present in the plasma; (5) the tumor should be well vascularized; (6) the antibody should have rapid clearance from the nonspecific locals factors;

and (7) the chelate and the conjugate chelate-mAb should present high *in vivo* stability.

The high molecular weight of the antibodies implies a slow blood hepato-biliary clearance. To reduce these disadvantages, mAbs fragments have been produced with molecular weights 10–100 kDa, which maintain, at the same time, the biological properties of their intact precursors; these fragments received the designation of Fab [73]. However, although Fabs reduce the background due to a more rapid blood clearance and excretion, at the same time, their tumor uptake is also reduced when compared with the intact mAbs.

For use in radiolabeled mAb, a ligand should fulfill the following criteria: (1) the ligand, covalently coupled to the protein, should rapidly complex the radioisotope at physiological conditions and (2) the complex formed should be kinetically and thermodynamically stable under *in vivo* conditions, in particular relation to demetalation promoted by other cations that are present in the blood (e.g., Ca^{2+}, Zn^{2+}, and Mg^{2+}).

Macrocyclic ligands accomplish these requirements with regard to the radioisotopes mainly used for mAbs radiolabeling, ^{111}In, $^{67/68}Ga$, and ^{64}Cu [59,73]. There are several examples in the literature of bifunctional derivatives of NOTA and DOTA coupled to antibodies and radiolabeled with ^{111}In, ^{67}Ga, ^{68}Ga, ^{64}Cu, and ^{67}Cu [62,74].

Although much enthusiasm and developments have been devoted to this area, until now the FDA has only approved three monoclonal antibodies for human use: ^{111}In-DTPA-B72.3 (ONCOScint), ^{111}In-DTPA-7E11.C53 (Proctascint), and IMMU-4 labeled with ^{99m}Tc (CEA-Scan) [75].

4.9.2   Peptides

Although mAbs show high tissue affinity and specificity, they have several disadvantages: slow blood clearance, slow delivery to the target, and a low target to nontarget ratio. Their relatively high molecular weight limits their localization and diffusion, increasing the length of time between their administration and imaging acquisition. For these reasons, the introduction of radiolabeled bioactive peptides in molecular imaging sounds promising, particularly for oncology [76,77]. Peptides, due to their small size, allow rapid blood clearance and a high tumor-to-background ratio.

Many peptides are known to have a regulatory, inhibitory, or stimulatory role in a variety of cellular functions, in both normal and tumor tissues. Peptides act as hormones, neurotransmitters, neuromodulators, and growth factors, in addition to performing other functions. They act at very low concentrations (from nanomolar to femtomolar), and their action is mediated by binding to cellular membrane receptors. Most of these receptors belong to the super-family of G protein coupled receptors, membrane single chain proteins with seven transmembrane domains, The extracellular parts of the receptor contain two highly conserved cysteine residues, which form disulfide bonds to stabilize the receptor structure. The extracellular parts are responsible for the ligand binding, whereas the intracellular domain couples to the G protein.

Many tumors express receptors for different peptides, often a large number, and many of these receptors are known to mediate regulatory growth effects *in vitro*. Some types of tumors also respond to an inhibitory effect or to a growth-promoting effect induced *in vivo* by peptides [45]. This effect constitutes an important approach to the clinical treatment of tumors in humans. An important example is the use of analogs of somatostatin, whose receptors are overexpressed in neoplastic tissues. Since 1979, radiolabeled somatostatin analogs have been successfully used both in diagnosis and in internal radiotherapy [78].

The radiolabeling techniques of small peptides are very similar to those used for MAb labeling. A natural peptide or an analog is covalently bound through a spacer to a chelator, which allows the radiometal complexation. In an alternative approach, the peptide itself contains a prosthetic group able to be labeled with an iodine radioisotope or with ^{18}F for PET imaging [79]. In most of the natural peptides, the molecular recognition site is limited to very specific locations, which provides a clear advantage by allowing modification of other peptide regions to make the radiolabeling possible. However, due to the small size of most of the peptides, the introduction of a prosthetic group or a chelator may cause relevant changes in the peptide biodistribution, and this can also dramatically decrease its receptor affinity. Strategies to diminish the interaction between the chelate and the biological binding site include the introduction of a spacer between the chelate and the biological active portion of the peptide to prevent steric or electronic hindrance in the specific binding to the receptor. Considering the possible pharmacological effects of a peptide, even if injected in very small quantities, it is crucial that the radiopeptide is produced with very high specific activity to guarantee that the radiopharmaceutical will not have any biological activity. This is the reason that it is of fundamental importance to improve the labeling efficiency, which depends on factors such as ligand concentration, chelation kinetics, and the presence of trace concentrations of other metals inside the labeling solution. Another drawback of the *in vivo* utilization is the rapid proteolysis of peptides in plasma by endogenous peptidases and proteases. To circumvent this problem, peptides must be modified in order to decrease their enzymatic recognition, in particular by introducing D-aminoacids or nonusual aminoacids [80].

The majority of peptides developed for molecular imaging are analogs of somatostatin. The naturally occurring somatostatin consists of two peptides, one with 14 amino acids, SS14, and another with 28 amino acids, SS28. There are cellular receptors of somatostatin in several organs and tissues, as this neuropeptide presents multiple physiological functions, which are mainly inhibitory: secretion inhibition of the growth hormone, glucagon, insulin, gastrin, and other peptides. It also has an inhibitory function of the immune system together with a neuromodulator role in the central nervous system [81].

Since the endogenous somatostatins cannot be used in imaging due to their low plasma and tissue stability, there has been a need to synthesize somatostatin analogs that are more resistant to biological degradation. One such

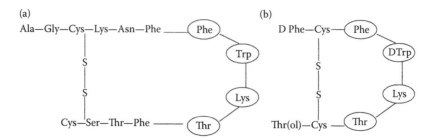

FIGURE 4.17
Structures of (a) somatostatin-14 and (b) the somatostatin analog octreotide. The amino acids within the oval constitute the recognition site of the receptor.

analog is octreotide (^{111}In-DTPA-octreotide, Octreo-Scan) [82]. Figure 4.17 represents the amino acid sequence of somatostatin and its analog octreotide. Other somatostatin analogs, such as DOTATOC [45] (Figure 4.18), have been successfully used in the visualization (labeled with ^{111}In, $^{67/68}$Ga, ^{99m}Tc, and ^{64}Cu) [45,83] and therapy (labeled with β^- emitters such as ^{90}Y or ^{174}Lu) of neuroendrocrine tumors and their metastasis [84].

Other regulatory peptides for which it is known that there are receptors in several types of tumors have also been the subject of considerable investigation with the objective of obtaining new radiopeptides. Among these peptides, we should include the α melanocyte-stimulating hormone (αtMSH), the vasoactive intestinal peptide (VIP), the substance P (SP), the cholecystokinin B (CCK-B), gastrin, neurotensin (NT), and bombesin (BN) [85]. All these peptides mentioned have multiple regulatory functions, either at the central nervous system or at the peripheral nervous system. They also have in

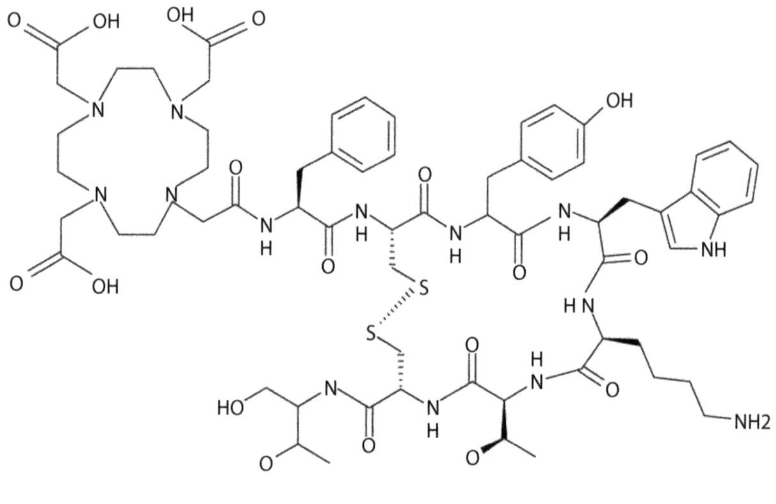

FIGURE 4.18
Chemical structure of DOTATOC.

common the fact that their action is mediated by specific receptors belonging to the super-family of receptors associated with the G protein.

4.10 Radiopharmaceutical Quality Control: Radiochemical Purity, Specific Activity, and Radiolysis

The chemical purity of a radiopharmaceutical can be defined as the percentage of the total sample radioactivity that is present in the desired chemical form of the radiopharmaceutical. For instance, if the radiopharmaceutical is the ^{99m}Tc-HMPAO, the chemical purity measures the quantity of ^{99m}Tc coordinated to HMPAO and compares this value with other forms of ^{99m}Tc present in the sample (e.g., pertechnetate). The radiochemical purity of a radiopharmaceutical should be close to 95%, as the presence of impurities with a different biodistribution invalidates a diagnostic test based on scintigraphic images. Radiochemical purity in the majority of cases is determined by separation methods, such as chromatographic techniques, radio-HPLC and TLC (thin-layer chromatography), or electrophoresis.

The chemical purity of a radiopharmaceutical can also be compromised by a radiolytic breakdown process releasing unbound radionuclides. For this reason, it is necessary to establish a compromise between the maximum activity desirable for biologically active radiopharmaceuticals and the possibility that this amount of radiation promotes the radiopharmaceutical dissociation by radiolysis.

Further, the ionizing radiation can break the water molecules down to produce hydrogen peroxide and free radicals. The presence of these oxidizing species can compromise the radiolabeling with ^{99m}Tc, because, as described in Section 4.8.1, the ^{99m}Tc obtained from the generator needs to be reduced before radiolabeling. This is one of the reasons that the reductor (usually tin chloride) is present in large excess in the labeling kits. It is necessary to point out within this context that specific activity is defined by the ratio between the radionuclide activity and the quantity of ligand present in the sample (MBq/μmol).

Since most of the radiopharmaceuticals are administrated by injection (IV), the product should be prepared under sterile and apyrogenic conditions and regularly tested for these two parameters.

4.11 Advances in Radiopharmaceutical Development and New Trends in Radiochemistry

Molecular imaging (using both PET and SPET radiotracers) will encompass in the future imaging of gene expression, protein expression, protein function, and physiological function [86]. There is increasing enthusiasm and intense research interest in the determination of patterns of gene expression that encode normal processes and in understanding how these patterns change with disease. In addition to spatiotemporal information, a radiolabeled PET tracer can provide information as to whether a particular gene product, such as a receptor or enzyme, is active. The latter information, combined with the capability for the concurrent analysis of normal and disease tissues and the assessment of regional variations, makes PET and NM techniques exciting imaging tools for diagnostics and evaluation of patient response to therapy. In oncology, the noninvasive monitoring of the dynamics of cell proliferation by nucleoside tracers will someday become a prerequisite for the development of molecular-targeted drugs and their postmarketing evaluation [87]. In conclusion, in the near future we will use more radiopharmaceuticals that will allow us to image the basic processes underlying diseases rather than detecting the effects of an installed pathology.

The PET and other molecular imaging methods are being increasingly used in the pharmaceutical industry. The appropriate use of molecular imaging in drug discovery and development could significantly speed up the development process, thus saving money for patients and health care systems by helping the industry in the extrapolation of *in vitro* data to *in vivo* application [88]. In addition, radiopharmaceutical development will benefit from methods already established in the pharmaceutical industry, such as combinatorial techniques, rational design, and high-throughput testing. These approaches provide a modern alternative to the rational conception of ligands, and very large libraries of candidate molecules can be randomly synthesized and simultaneously screened to identify those with the desired property.

References

1. V.W. Pike. Positron emitting radioligands for studies *in vivo*—probes for human psychopharmacology. *J. Psychopharmacol.*, 7:139–158, 1993.
2. M.J. Koepp, R.N. Gunn, A.D. Lawrence, V.J. Cunningham, A. Dagher, T. Jones, D.J. Brooks, C.J. Bench, and P.M. Grasby. Evidence for striatal dopamine release during a video game. *Nature*, 393:266–268, 1998.
3. M.E. Raichle, W.R. Martin, P. Herscovitch, M.A. Mintun, and J. Markham. Brain blood flow measured with intravenous $H_2^{15}O$. II. Implementation and validation. *J. Nucl. Med.*, 24:790–798, 1983.
4. R. Subramanyam, N.M. Alpert, B. Hoop Jr, G.L. Brownell, and J.M. Taveras. A model for regional cerebral oxygen distribution during continuous inhalation of $^{15}O_2$, $C^{15}O$ and $C^{15}O_2$. *J. Nucl. Med.*, 19:48–53, 1978.
5. S.J. Read, T. Hirano, D.F. Abbott, J.I. Sachinidis, H.J. Tochon-Danguy, J.G. Chan, G.F. Egan et al. Identifying hypoxic tissue after acute ischemic stroke using PET and 18F-fluoromisonidazole. *Neurology*, 51:1617–1621, 1998.

6. K.J. Yagle, J.F. Eary, J.F. Tait, J.R. Grierson, J.M. Link, B. Lewellen, D.F. Gibson and K.A. Krohn. Evaluation of 18F-Annexin V as a PET imaging agent in an animal model of apoptosis. *J. Nucl. Med.*, 46(4):658–666, 2005.
7. M.E. Phelps, S.C. Huang, E.J. Hoffman, C. Selin, L. Sokoloff and D.E. Kuhl. Tomographic measurement of local cerebral metabolic rate in humans with (F18) 2-fluoro-2-deoxy-D-glucose: Validation of the method. *Ann. Neurol.*, 6:371–388, 1979.
8. A.K. Buck, G. Halter, H. Schirrmeister, J. Kotzerke, I. Wurziger, G. Glatting, T. Mattfeldt, B. Neumaier, S.N. Reske, and M. Hetzel. Imaging proliferation in lung tumors with PET: [18F]FLT versus [18F]FDG. *J. Nucl. Med.*, 44:1426–1431, 2003.
9. T. Hara, N. Kosaka, and H. Kishi. PET imaging of prostate cancer using carbon-11-choline. *J. Nucl. Med.*, 39:990–995, 1998.
10. T.J. Mangner, R.W. Klecker, L. Anderson, and A.F. Shields. Synthesis of 2'-deoxy-2'-[(18)F]fluoro-beta-D-arabinofuranosyl nucleosides, [(18)F]FAU, [(18)F]FMAU, [(18)F]FBAU and [(18)F]FIAU, as potential PET agents for imaging cellular proliferation. *Nucl. Med. Biol.*, 30:215–224, 2003.
11. H.R. Schelbert, M.E. Phelps, S.C. Huang, N.S. MacDonald, H. Hansen, C. Selin, and D.E. Kuhl. N-13 ammonia as an indicator of myocardial blood flow. *Circulation*, 63:1259–1272, 1981.
12. J.J. Armbrecht, D.B. Buxton, R.C. Brunken, M.E. Phelps, and H.R. Schelbert. Regional myocardial oxygen consumption determined noninvasively in humans with [1–11C]acetate and dynamic positron tomography. *Circulation*, 80:863–872, 1998.
13. E.M. Geltman. Assessment of myocardial fatty acid metabolism with 1–11C-palmitate. *J. Nucl. Cardiol.*, 1:S15–S22, 1994.
14. L. Guiducci, T. Grönroos, M.J. Järvisalo, J. Kiss, A. Viljanen, A.G. Naum, T. Viljanen et al. Biodistribution of the fatty acid analogue 18F-FTHA: Plasma and tissue partitioning between lipid pools during fasting and hyperinsulinemia. *J. Nucl. Med.*, 48:455–462, 2007.
15. J.C. Wu, K. Bell, A. Najafi, C. Widmark, D. Keator, C. Tang, E. Klein, B.G. Bunney, J. Fallon, and W.E. Bunney. Decreasing striatal 6-FDOPA uptake with increasing duration of cocaine withdrawal. *Neuropsychopharmacology*, 17:402–409, 1997.
16. J. Hatazawa, K. Ishiwata, M. Itoh, M. Kameyama, K. Kubota, T. Ido, T. Matsuzawa, T. Yoshimoto, S. Watanuki, and S. Seo. Quantitative evaluation of L-[methyl-C-11] methionine uptake in tumor using positron emission tomography. *J. Nucl. Med.*, 30:1809–1813, 1989.
17. D. Salber, G. Stoffels, D. Pauleit, G. Reifenberger, M. Sabel, N. Shah, K. Hamacher, H. Coenen, and K. Langen. Differential uptake of [18F]FET and [3H]l-methionine in focal cortical ischemia. *Nucl. Med. Biol.*, 33:1029–1035, 2006.
18. T. Yoshioka, T. Fukuda, K Yamaguchi, M. Suzuki, S. Furumoto, R. Iwata, and C. Ishioka. O-[18F]fluoromethyl-l-tyrosine is a potential tracer for monitoring tumour response to chemotherapy using PET: An initial comparative *in vivo* study with deoxyglucose and thymidine. *Eur. J. Nucl. Med. Mol. Imag.*, 33: 1134–1139, 2006.
19. R.N. Gunn, A.A. Lammertsma, and P.M. Grasby. [carbonyl-(11)C]- WAY1000635 PET studies. *Nucl. Med. Biol.*, 27:477–482, 2000.
20. L. Farde, C. Halldin, S. Stone-Elander, and G. Sedvall. PET analysis of human dopamine receptor subtypes using 11C-SCH 23390 and 11C-raclopride. *Psychopharmacology*, 92:278–284, 1987.

21. P.M. Grasby, M.J. Koepp, R.N. Gunn, V.J. Cunningham, A. Lawrence, and C.J. Bench. [11C]raclopride pet detects dopamine release induced by behavioural manipulation. *Schizophrenia Res.*, 29:94–94(1), 1991.
22. W.M. Pardridge. CNS drug design based on principles of blood–brain barrier transport. *J. Neurochem.*, 70: 1781–1792, 1998.
23. J. Korf, S. Reiffers, H.D. Beerling-van der Molen, J.P. Lakke, A.M. Paans, W. Vaalburg, and M.G. Woldring. Rapid decarboxylation of carbon-11 labelled DL–DOPA in the brain: A potential approach for external detection of nervous structures. *Brain Res.*, 145:59–67, 1978.
24. P. Hartvig, H. Agren, L. Reibring, J. Tedroff, P. Bjurling, T. Kihlberg, and B. Langstrom. Brain kinetics of L-[beta-11C]DOPA in humans studied by positron emission tomography. *J. Neural Transm. Gen. Sect.*, 86:25–41, 1996.
25. K. Hamacher, H.H. Coenhen, and G. Stocklin. Efficient stereospecific synthesis of no-carrier-added 2-[18F]-fluoro-2-deoxy-D-glucose using aminopolyether supported nucleophilic substitution. *J. Nucl. Med.*, 27:235–238, 1986.
26. M. Guillaume, A. Luxen, B. Nebeling, M. Argentini, J.C. Clark, and V.W. Pike. Recommendations for fluorine-18 production. *Appl. Radiat. Isot.*, 42:749–762, 1991.
27. L. Sokoloff, M. Reivich, C. Kennedy, M.H. Des Rosiers, C.S. Pratlak, K.D. Pettigrew, O. Sakurada and M. Shinohara. The [14C]deoxyglucose method for the measurement of local cerebral glucose utilization: Theory, procedure, and normal values in the conscious and anesthesized albino rat. *J. Neurochem.*, 28:897–916, 1977.
28. M. Reivich, A. Alavi, A. Wolf, J.H. Greenberg, J. Fowler, D. Christman, R. MacGregor, S.C. Jones, J. London, C. Shiue, and Y. Yonekura. Use of 2-deoxy-D[1–11C]glucose for the determination of local cerebral glucose metabolism in humans: Variation within and between subjects. *J. Cereb. Blood Flow Metab.*, 2:307–319, 1979.
29. R.W. Horton, B.S. Meldrum, and H.S. Bachelard. Enzymic and cerebral metabolic effects of 2-deoxy-D-glucose. *J. Neurochem.*, 21:507–520, 1973.
30. C.J. Anderson and M.J. Welch. Radiometal-labeled agents (non-technetium) for diagnostic imaging. *Chem. Rev.*, 99:2219–2234, 1999.
31. S. Jurisson and J.D. Lydon. Potential technetium small molecule radiopharmaceuticals. *Chem. Rev.*, 99:2206–2218, 1999.
32. W.A. Volkert and T.J. Hoffman. Therapeutic radiopharma-ceuticals. *Chem. Rev.*, 99:2269–2292, 1999.
33. W.A.P. Breeman, M. de Jong, E. de Blois, B.F. Bernard, M. Konijnenberg, and E.P. Krenning. Radiolabelling DOTA-peptides with [68]Ga. *Eur. J. Nucl. Med. Mol. Imag.*, 32:478–485, 2005.
34. G.J. Ehrhardt and M.J. Welch. Gallium 66 and Gallium 68, a new germanium-68/gallium-68 generator. *J. Nucl. Med.*, 19:925–929, 1978.
35. S. Liu and D.E. Scott. ^{99m}Tc-Labeled small peptides as diagnostic radiopharmaceuticals. *Chem. Rev.*, 99:2235–2268, 1999.
36. M.A. Mendez-Rojas, B.I. Kharisov, and A.Y. Tsivades. Recent advances on technetium complexes: Coordination chemistry and medical applications. *J. Coord. Chem.*, 59:1–63, 2006.
37. H.F. Kung, H.-J. Kim, M.-P. Kung, S.K. Meegalla, and H.K. Lee. Imaging of dopamine transporters in humans with technetium-99m TRODAT-1. *Eur. J. Nucl. Med.*, 23:1527–1530, 1996.

38. R.J. Di Rocco, A.A. Bauer, J.P. Pirro, B.L. Kuczynski, L. Belnavis, Y.-W. Chan, K.E. Linder, R.K. Narra, D.P. Nowotnik, and A.D. Nunn. Delineation of the border zone of ischemic rabbit myocardium by a technetium-labelled nitroimidazole, *Nucl. Med. Biol.*, 24:201–207, 1997.

39. R. Taillefer, L. Laflamme, G. Dupras, M. Picard, D.-C. Phaneuf, and J. Léveillé. Myocardial perfusion imaging with ^{99m}Tc-methoxy-isobutyl-isonitrile (MIBI) comparison of short and long intervals between rest and stress injections. Preliminary results. *Eur. J. Nucl. Med.*, 13:515–522, 1998.

40. C.M.F. Gomes, M. Welling, I. Que, N.V. Henriquez, G. Pluijm, S. Romeo, A.J. Abrunhosa et al. Functional imaging of multidrug resistance in an orthotopic model of osteosarcoma using ^{99m}Tc-sestamibi. *Eur. J. Nucl. Med.*, 34:1793–1803, 2007.

41. G.G. Blankerberg, P.D. Katsikis, J.F. Tait, E.R. Davis, I. Naumovski, K. Ohtsuki, S. Kopiwoda et al. *In vivo* detection and imaging of phosphatidylserine expression during programmed cell death. *Proc. Natl. Acad. Sci. USA*, 95:6349–6354, 1998.

42. M. Leimer, A. Kurtaran, M. Raderer, P. Smith-Jones, C. Bischof, W. Petcov et al. Somatostatin receptor *in vitro* and *in vivo* binding of Tc-99m-P829, *J. Nucl Med.*, 39(5, suppl):39–39P, 1998.

43. J.A. Barrett, A.C. Crocker, D.J. Damphousse, S.J. Heminway, S. Liu, D.S. Edwards, A.R. Harris et al. Biological evaluation of thrombus imaging agents utilizing water soluble phosphines and tricine as coligands to label a hydrazinonicotinamide-modified cyclic glycoprotein IIb/IIIa receptor antagonist with [99m]Tc. *Bioconj. Chem.*, 8:155–160, 1997.

44. K.E. Britton, S. Vinjamuri, A.V. Hall, K. Solanki, Q.H. Siraj, J. Bomanji, and S. Das. Clinical evaluation of techenetium-99m infection for the localization of bacterial infection. *Eur. J. Nucl. Med.*, 24:553–556, 1997.

45. A. Hepeller, S. Froidevaux, A.N. Eberle, and H.R. Maecke. Receptor targeting for tumor localisation and therapy with radiopeptides. *Curr. Med. Chem.*, 7:971–994, 2000.

46. B.J. Krause, R.P. Baum, E. Staib-Sebler, M. Lorenz, A. Niesen, and G. Hör. Human monoclonal antibody Tc-99m.88 BV59: Detection of colorectal cancer, recurrent or metastatic disease and immunoscintigraphy assessment. *Eur. J. Nucl. Med.*, 24:72–75, 1997.

47. H.R. Maecke and J.P. André. ^{68}Ga-PET radiopharmacy: A generator-based alternative to ^{18}F-radiopharmacy. *PET Chemistry, the Driving Force in Molecular Imaging*, P.A. Schubiger, L. Lehmann, M. Friebe, Eds., Springer-Verlag, Berlin, 2007.

48. M. Nakayama, M. Haratake, T. Koiso, O. Ishibashi, K. Harada, H. Nakayama, A. Suggii, S. Yahara, and Y. Arano. Separation of ^{68}Ga from [68]Ge using a macroporous organic polymer containing N-methylglucamine groups. *Anal. Chim. Acta*, 453:135–141, 2002.

49. C. Loc'h, B. Maziere, and D. Comar. A new generator for ionic gallium-68. *J. Nucl. Med.*, 21:171–173, 1980.

50. R. Lambrecht and M. Sajjad. Accelerator derived radionuclide generators. *Radiochim. Acta*, 43:171–179, 1978.

51. S. Mirzadeh and R. Lambrecht. Radiochemistry of germanium. *J. Radioanal. Nucl. Chem.*, 202:7–102, 1996.

52. M.A. Green and M.J. Welch. Gallium radiopharmaceutical chemistry. *Nucl. Med. Biol.*, 16:435–448, 1989.
53. R.G. Pearson. Hard and soft acids and bases. *J. Am. Chem. Soc.*, 85: 3533–3539, 1963.
54. A.M. Dymov and A.P. Seventin. *Analytical Chemistry of Gallium*, Ann Arbor Science Publishers, Michigan, 1970.
55. B. Kumar, T.R. Miller, B.A. Siegel, C.J. Mathias, J. Markhan, G.J. Ehrhardt and M.J. Welch. Positron tomographic imaging of the liver Ga-68 iron hydroxide colloid. *Am. J. Roent.*, 136:685–690, 1981.
56. B.A. Khan, J.T. Fallon, H.W. Strauss, and E. Haber. Myocardial infarct imaging of antibodies to canine cardiac myosin with indium-111-diethylenetriamine pentaacetic acid. *Science* 209:295–297, 1980.
57. Y. Arano, T. Uezono, H. Akizawa, M. Ono, K. Wakisava, M. Nakayama, H. Sakahara, J. Konishi, and A. Yokoiama. Reassessment of diethylenetriaminepentaacetic acid (DTPA) as a chelating agent for indium-111 labeling of polypeptides using a newly synthesized monoreactive DTPA derivative. *J. Med. Chem.*, 39: 3451–3460, 1996.
58. D.J. Hnatowich, K.L. Childs, D. Lanteigne, and A. Najafi. The preparation of DTPA-coupled antibodies radiolabeled with metallic radionuclides: An improved method. *J. Immun. Meth.*, 65:147–157, 1983.
59. M.W. Brechbiel, O.A. Gansow, R.W. Atcher, J. Schlom, J. Esteban, D.E. Simpson, and D. Colcher. Synthesis of 1-(p-1sothiocyanatobenzyI) derivatives of DTPA and EDTA. Antibody-labeling and tumor-imaging studies. *Inorg. Chem.*, 25:2772–2781, 1986.
60. P.M. Smith-Jones, B. Stolz, C. Bruns, R. Albert, H.W. Reist, R. Fridrich, and H.R. Maecke. Gallium-67/gallium-68-[DFO]-octreotide—a potential radiopharmaceutical for PET imaging of somatostatin receptor-positive tumors: Synthesis and radiolabeling *in vitro* and preliminary *in vivo* studies. *J. Nucl. Med.*, 35:317–325, 1994.
61. E.T. Clarke and A.E. Martell. Stabilities of the Fe(III), Ga(III) and In(III) chelates of N,N',N''-triazacyclononanetriacetic acid. *Inorg. Chim. Acta*, 181:273–280, 1991.
62. J.P. Broan, L. Cox, A.S. Craig, R. Kataky, D. Parker, A. Harrison, M. Randall, and G. Ferguson. Structure and solution stability of indium and gallium complexes of 1,4,7-triazacyclononanetriacetate and of yttrium complexes of 1,4,7,10-tetraazacyclododecanetetraacetate and related ligands—kinetically stable complexes for use in imaging and radioimmunotherapy X-ray molecular structure of the indium and gallium complexes of 1,4,7-triazacyclononane-1,4,7-triacetic acid. *J. Chem. Soc. Perkin Trans.*, 2:87–99, 1991.
63. W.R. Harris and V.L. Pecoraro. Thermodynamic binding constants for gallium transferrin. *Biochemistry*, 22:292–299, 1983.
64. D. Wild, H.R. Maecke, B. Waser, J.C. Reubi, M. Ginj, H. Rasch, J. Mueller-Brand, and M. Hofmann. ^{68}Ga-DOTANOC: A first compound for PET imaging with high affinity for somatostatin receptor subtypes 2 and 5. *Eur. J. Nucl. Med. Mol. Imag.*, 32:724–724, 2005.
65. W.A.P. Breeman and A.M. Verbruggen. The [68]Ge/^{68}Ga generator has high potential, but when can we use ^{68}Ga-labelled tracers in clinical routine? *Eur. J. Nucl. Med. Mol. Imag.*, 34:978–981, 2007.

66. F.H.M. Corstens and J.W.M. van der Meer. Nuclear medicine's role in infection and inflammation. *The Lancet*, 354:765–770, 1999.
67. F. Servadei, G. Moscatelli, G. Giuliani, A.M. Cremonini, G. Piazza, M. Agostini, and P. Riva, Cisternography in combination with single photon emission tomography for the detection of the leakage site in patients with cerebrospinal fluid Rhinorrhea: Preliminary report. *Acta Neurochir. (Wien)*, 140:1183–1189, 1998.
68. W.J.G. Oyen, R.A.M.J. Claessens, J.W.M. van der Meer, and F.H.M. Coerstens. Detection of subacute infectious foci with indium-111-labeled autologous leukocytes and indium-111-labeled human nonspecific immunoglobulin G: A prospective comparative study. *J. Nucl. Med.*, 32:1854–1860, 1991.
69. M.L. Thakur, L. Walsh, H.L. Malech, and A. Gottschalk. Indium-lll-labeled human platelets: Improved method, efficacy and. evaluation. *J. Nucl. Med.*, 22:381–385, 1981.
70. M.J. Welch, M.L. Thakur, R.E. Coleman, M. Patel, B.A. Siegel, and M. Ter-Pogosian. Gallium-68 labeled red cells and platelets new agents for positron tomography. *J. Nucl. Med.*, 18:558–562, 1977.
71. C.J. Anderson, T.S. Pajeau, W.B. Edwards, E.L.C. Sherman, B.E. Rogers, and M. Welch. *In vitro* and *in vivo* evaluation of copper-64-octreotide conjugates. *J. Nucl. Med.*, 36:2315–2325, 1995.
72. D. Pressman. The development and use of radiolabeled antitumor antibodies. *Cancer Res.*, 40:2960–2964, 1980.
73. K.O. Webber, R.J. Kreitman, and I. Pastan. Rapid and specific uptake of anti-Tac disulfide-stabilized Fv by interleukin-2 receptor-bearing tumors. *Cancer Res.*, 55:318–323, 1995.
74. C. Meares. Chelating agents for the binding of metal ions to proteins. *Nucl. Med. Biol.*, 13:311–318, 1986.
75. A.J. Fishman, J.W. Babich, and H.W. Strauss. A ticket to ride: Peptide radiopharmaceuticals. *J. Nucl. Med.*, 34:2253–2263, 1993.
76. S.M. Okarvi. Recent progress in fluorine-18 labelled peptide radiopharmaceuticals. *Eur. J. Nucl. Med.*, 28:929–938, 2001.
77. A. Signore, A. Annovazzi, M. Chianelli, F. Corsetti, C. van de Wiele, R.N. Watherhouse, and F. Scopinaro. Peptide radiopharmaceuticals for diagnosis and therapy. *Eur. J. Nucl. Med.*, 28:1555–1565, 2001.
78. T. Maack, V. Johnson, T.K. Sen, J. Figueiredo, and D. Sigulem. Renal filtration, transport, and metabolism of low-molecular-weight proteins: A review. *Kidney Int.*, 16:251–270, 1979.
79. H. Lundqvist and V. Tolmachev. Targeting and positron emission tomography. *Biopolymers (Peptide Science)*, 66:381–392, 2002.
80. J.C. Reubi. *In vitro* identification of vasoactive intestinal peptide receptors in human tumors: Implications for tumor imaging. *J. Nucl. Med.*, 36:1846–1853, 1995.
81. L.K. Kvols, C.G. Moertel, M.J. O'Connel, A.J. Schut, J. Rubin, and R.G. Hahn. Treatment of the malignant carcinoid syndrome: Evaluation of a long-acting somatostatin analogue. *New Engl. J. Med.*, 315:663–666, 1986.
82. K.N. Hofer, H.P. Amstutz, H.R. Maecke, R. Scharzbach, K. Zimmermann, J.J. Morgebthaler, and P.A. Schubiger-Hofer. Cellular processing of copper-67-labeled monoclonal antibody chCE7 by human neuroblastoma cells. *Cancer Res.*, 55:46–50, 1995.

83. C.J. Anderson, F. Dehdashti, P.D. Cutler, S.W. Schwarz, R. Laforest, L.A. Bass, J.S. Lewis, and D.W. McCarthy. ^{64}Cu-TETA-octreotide as a PET imaging agent for patients with neuroendocrine tumors. *J. Nucl. Med.*, 42:213–221, 2001.
84. M. de Jong, W.A.P. Breeman, R. Valkema, B.F. Bernard, and E.P. Krenning. Combination radionuclide therapy using ^{177}Lu and ^{90}Y-labeled somatostatin analogs. *J. Nucl. Med.*, 46:13S–17S, 2005.
85. T.M. Behr, N. Jenner, M. Béhé, C. Angersten, S. Gratz, F. Raue, and W.J. Becker. Radiolabeled peptides for targeting cholecystokinin–B/gastrin receptor-expressing tumors. *J. Nucl. Med.*, 40:1029–1044, 1999.
86. R. Weissleder and U. Mahmood. Molecular imaging. *Radiology*, 219:316–333, 2001.
87. J. Toyohara and F. Yasuhisa. Trends in nucleoside tracers for PET imaging of cell proliferation. *Nucl. Med. Biol.*, 30:681–685, 2003.
88. R.M. Weussleder. Molecular imaging in drug discovery and development. *Nat. Rev. Drug Discov.*, 2:123–131, 2003.

Further Reading

1. Sampson, C.B., Ed., *Text Book of Radiopharmacy*, 3rd Edn., Gordon and Breach Science Publishers, Amsterdam, The Netherlands, 1999.
2. Steigman, J., and Eckelman, W.C. (1992) *The Chemistry of Tecnetium in Medicine*, National Academy Press, Washington, DC.

5

Radiation Detectors and Image Formation

Francisco J. Caramelo, Carina Guerreiro, Nuno C. Ferreira,
and Paulo Crespo

CONTENTS

5.1   Methods and Measurement in Nuclear Medicine

Imaging is becoming increasingly important in modern medicine, either as a means of diagnosis or as additional support for planning and controlling clinical procedures. One of the factors that has contributed to this has been the improved quality of medical images. Enhanced contrast and resolution, reduced noise, distortion and artifacts, as well as the possibility of quantification have become key objectives in several areas such as the physics of detectors or image reconstruction mathematics.

Nuclear imaging techniques with radioisotopes include *ex vivo*, *in vitro*, and *in vivo* methods. High-resolution autoradiography with spatial resolutions greater than 100 μm is generally used in *in vitro* studies and is also possible in *ex vivo* tests. Images are obtained on radiological film, nuclear emulsion, or in autoradiography systems in real time [1]. The nuclear medicine (NM) *in vivo* methods can be divided into two groups: single photon techniques and techniques based on positron emission, known as positron emission tomography (PET) (Figure 5.1).

The first group uses radionuclides emitting gamma radiation and includes conventional (planar and whole-body) scintigraphy and single photon emission computed tomography (SPECT).

In general, the detectors used for SPECT imaging are the same as those used in scintigraphy, preferably with two heads, with rotation around the object, allowing for the use of reconstruction algorithms for tomographic images and the possibility of three-dimensional (3D) visualization.

The PET imaging uses radionuclides emitting positrons (β^+ decay). Positrons are emitted with a kinetic energy that has a continuous spectrum, covering unpredictable distances until they thermalize and are annihilated with the electrons of the medium.

When annihilation occurs, the kinetic energy of the positron may not be exactly zero; thus, conservation of the kinetic moment requires two 511 keV gamma photons to be emitted in opposite directions but not along the same line [2]. This near colinearity is exploited in PET to generate a line of response (LOR) corresponding to each coincidence detected, hence obtaining an electronic collimation. Thus, physical collimators, well-known for causing a deterioration in the performance of SPECT systems, can be avoided. However, a number of other favorable factors, including a higher half-life for the radionuclides used in SPECT, contribute to its greater use in NM [3].

Photon energy in PET is 511 keV, whereas for single photon techniques the energy may be less than 400 keV (the photons emitted by ^{99m}Tc, the most commonly used nuclide, have 140 keV of energy). Therefore, suitable, highly efficient systems are required to detect these energies, either directly by collecting the charge produced in a suitable medium or through systems that use electromagnetic converters that produce lower energy radiation, which is then detected by appropriate sensors.

The basic ideas of the technology, regardless of the nature of the detected photons (SPECT or PET), can be summarized as

1. Administration to the patient under study of a radiolabeled molecule with biological interest.
2. Emission of gamma photons, either by radioactive decay (a single photon in planar scintigraphy and SPECT) or positron annihilation, which is emitted by radioactive decay (resulting in two 511 keV photons emitted simultaneously in opposite directions—PET).
3. Interaction of photons with a gamma detector, typically a crystal scintillator emitting secondary photons in the visible or near ultraviolet region.
4. Transduction of the light to electrical signals that are used to determine the position and concentration of the tracer.
5. Image reconstruction from the detected events.

The advantage of combining information of different modalities to assist in clinical diagnosis has led to the development of multimodal PET or CT and SPECT or CT systems, which result in the almost simultaneous acquisition of morphological information from CT and functional information from PET or SPECT. The possibility of using these modalities together is especially helpful in identifying and locating lesions, including tumors. In combination with CT, it brings a second benefit, by allowing attenuation of radiation in the patient to be corrected, which is essential for producing images that are closer to the true distribution of the radiotracer in the patient while avoiding the use of external radioactive sources. This correction is performed in PET by using the transmission data and a simple calculation, in contrast to SPECT, where the correction is more complex and less accurate. It is much easier in PET due to the fact that two photons are detected in coincidence instead of the single photon detected by the SPECT camera. It is unusual to apply the attenuation correction in planar scintigraphy, hence planar images provide essentially qualitative information.

The principal agents used in the detection and measurement of ionizing radiation in NM resulting from the interaction within the medium of the detector are the electric charge (ionization chambers, proportional counters, Geiger tubes, and photoconduction), light from luminescence processes [fluorescence in scintillation counters and thermoluminescence in Thermoluminescent Dosimeter (TLD)], and chemical energy (film autoradiography) [3].

In NM, the measurement of radiation is required for different situations that depend on the source and type of radiation to be measured and need appropriate detectors and instrumentation. The type of radiation involved is mainly gamma, although some situations require the detection of beta particles.

The *in vitro* measurement of the activity of samples of biological fluids is a situation that, at present, requires well counters for gamma emitters and, in rare cases, liquid scintillation detectors for beta emitters.

The *in vivo* detection of activity in specific areas in patients, in order to obtain quantitative or semiquantitative results, can be achieved by using collimated gamma radiation detectors or by defining regions of interest in area detectors.

In most cases, radiation detection aimed at obtaining NM images makes use of scintillation detectors that are embedded in gamma cameras. However, forthcoming major changes are predicted involving the use of semiconductor and photoconductor detectors.

The use and measurement of radiation emitted by γ tracers selectively retained within the body is limited by a set of constraints:

1. The solid angle for the emission radiation of a point source is 4π therefore, with any imaging detector there will always be loss of rays due to the restricted area of detection.

2. A significant proportion of radiation produced by decay is unused, as the energy is outside the acquisition interval, which must be kept as narrow as possible.

3. The counts are affected by statistical variations inherent to the process.

4. The laws of optics cannot be used in γ rays, as they cannot be refracted by lenses and therefore, γ rays that include position information are selected by selective absorption using collimators. Consequently, the origin of emission includes uncertainty that leads to poor spatial resolution and partial volume effect. On the other hand, collimators absorb a significant proportion of useful radiation.

5. There is attenuation of radiation in tissues between the source and detector. Scattered radiation in the surrounding tissues adds background noise, leading to deterioration of the spatial resolution.

6. The intrinsic efficiencies of radiation detectors are, in general, smaller than the unit.

7. Patient movement can lead to deterioration of the quality of the images.

Although the thickness of the absorbent material between a point source and the detector, the attenuation characteristics, and their local variations are, in principle, unknown, it is possible to effectively correct the attenuation effect (referred to in Point 4) for PET imaging and with good approximation for SPECT imaging.

All of the above mentioned points are associated with patient irradiation and have little effect on image formation, with an overall efficiency in the order of 0.001% when a physical collimator is used. Points 3 and 4 may also cause severe deterioration in image information.

Nevertheless, continuous efforts have been made to ensure that NM imaging is, as far as possible, an ideal tool for medical diagnosis: noninvasive and providing functional as well as 3D and quantitative information. Among the two tomographic techniques available in NM, PET is probably closer to this ideal goal than SPECT.

The techniques currently used are capable of creating images with good contrast of the dynamics of labeled molecules (native or functionally similar) that are present in the metabolic processes of the organs under examination.

Isotopes produced in the cyclotron (11C, 18F, etc.) are used in the synthesis of organic molecules (peptides, carbohydrates, steroids, vitamins, etc.) for PET studies.

Soluble or insoluble metal salts are used in the synthesis or labeling of biological molecules used as tracers in SPECT (^{99m}Tc, ^{113}In, etc.) and PET (^{82}Rb, etc.).

Computers are vital in modern NM and are used in data acquisition, data correction in real time, viewing and processing images, mathematical manipulation of images, analysis and reconstruction of images, data storage, networked systems, and multimodality implementation.

5.1.1 The Physics of Detection: Basic Concepts

In a broad sense, the detection of radiation involves changing the nature of the signal to another that is more easily measurable. Regarding the detection of electromagnetic radiation, it is essential that the radiation interacts with the detector through one of the known mechanisms, and it is also crucial that the energy associated with particle radiation is deposited within the volume of the detector. Generally, the interaction of radiation with the detector creates a certain amount of charge that is collected with the help of an electric field. The electric current generated is different from zero for a period equal to the time associated with the process of charge collection. Generally, the current generated is transformed into a voltage signal through an RC equivalent circuit that is outlined in Figure 5.2.

The most common situation is when the time constant of the circuit, $\tau = RC$, is greater than the charge deposition time, which means that the signal presents a typical form due to the charge and discharge of the capacitor through the load resistor (Figure 5.2). The time required for the signal to reach its maximum is mainly determined by the charge collection time in the detector, whereas the decay time is determined by the time constant of the

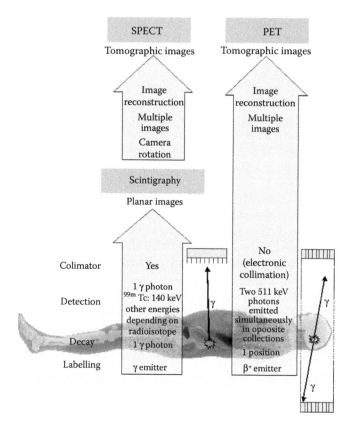

FIGURE 5.1
Comparison between *in vivo* imaging methods in nuclear medicine, with regard to the detection and production of images. The presence of the collimator for planar scintigraphy and SPECT imaging, which is required in order to locate the radioactive source in relation to the detector, causes a very significant loss of sensitivity in comparison with PET and degrades spatial resolution, which deteriorates in proportion to the distance from the radioactive source to the detector.

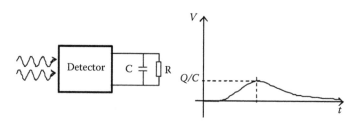

FIGURE 5.2
Diagram of the equivalent circuitry connected to the detector and the typical voltage (V) response of the circuitry.

load circuit. The amplitude of the signal is proportional to the deposit charge in the detector ($V_{max} = Q/C$). For each efficient interaction taking place in the detector, there is a corresponding electrical pulse in the measuring circuit and the pulse rate is, therefore, equivalent to the interaction rate in the detector.

Efficiency, energy resolution, and dead time are important factors in radiation detectors. The efficiency of a detector links the number of photons entering the active volume with the number of electrical impulses generated. For convenience, efficiency is studied in two separate ways: as absolute efficiency and as intrinsic efficiency. Absolute efficiency is defined as the ratio between the number of pulses and the measured number of photons emitted by the source [4]:

$$\varepsilon_{abs} = \frac{\text{number of measured pulses}}{\text{number of emitted photons}}. \tag{5.1}$$

Conversely, intrinsic efficiency is defined by taking into account only the characteristics of the detector and, hence, is independent of the geometrical characteristics of the measuring arrangement. Thus, intrinsic efficiency is defined by

$$\varepsilon_{int} = \frac{\text{number of measured pulses}}{\text{number of incident photons in detector}}. \tag{5.2}$$

Considering a point source emitting isotropically, the relationship between the two efficiencies is based only on geometrical arguments and is given by

$$\varepsilon_{int} = \frac{4\pi}{\Omega}\varepsilon_{abs}, \tag{5.3}$$

where Ω is the solid angle of the detector seen from the position of the point source. Intrinsic efficiency depends on the material of the detector, the thickness (measured in terms of the direction of the incident radiation), and the energy of the incident radiation.

The energy resolution of a detector is defined as the ability to discriminate between two close radiation energies. The response of a detector to a real monoenergetic source is not a narrow peak (mathematically, a delta function) but a Gaussian distribution whose amplitude corresponds to the radiation energy.

Formally, the energy resolution is given by the ratio between the full width at half maximum (FWHM = ΔE) of the energy response of the detector to monoenergetic radiation and the peak energy (E_p) [4]. Two monoenergetic sources are distinguished by a detector if the difference between their energies is greater than the FWHM of the response curve (Figure 5.3).

Several factors contribute toward the deterioration of energy resolution in a detector, such as statistical noise due to the discrete nature of the signal, the noise caused by the detector and by the measuring circuit, and any devi-

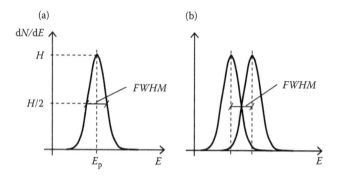

FIGURE 5.3
(a) Response curve for a source point obtained by a real radiation detector. The xx axis represents the energy of the signal, whereas the yy axis is the number of counts per interval of energy. (b) Diagram of the minimum difference in energy that must exist in order to discriminate between two monoenergetic sources.

ations from operating conditions during measurement. Although the latter two sources of error are likely to be minimized by optimizing the detector, the primary cause of noise is always present regardless of the detector. Statistical noise is related to the fact that the charge deposited in the detector is discrete, as it depends on the number of charge carriers. The number of charge carriers is random and follows a certain distribution of probabilities. Assuming that the process follows a Poisson distribution, for a given number N, of created charges, the expected standard deviation is $\sqrt{N}$. Assuming further that the response of the detector is linear to the number of charges generated, the expected deposited energy (E_0) should be equal to KN, where K is a constant of proportionality. For a high number N of charges, the response of the detector follows a Gaussian distribution. For this distribution, the FWHM is equal to 2.35σ, hence;

$$R = \frac{\text{FWHM}}{E_0} = \frac{2.35K\sqrt{N}}{KN} = 2.35\frac{1}{\sqrt{N}}. \tag{5.4}$$

This relationship implies that statistical noise limits the energy resolution, which can be improved if the number of charges per event increases.

The dead time of a detector is usually defined as the time that should exist between two separate events so that they can be measured as two different electrical signals. Depending on the type of detector, the limiting factor is either the detection process or the associated electronic circuit.

5.1.1.1 Interaction of Radiation with Matter

When gamma (or x) rays focus directly on an object and cross it, some photons interact with the particles of the object and their energy can be absorbed or scattered. The process of absorption and dispersion of radiation is generally known as *attenuation* (Figure 5.4).

The intensity (energy per unit area and per unit time) of the radiation emerging from the object is less than the incident intensity, and the fraction of transmission depends on the density, atomic number, and thickness of the material and the radiation energy.

The attenuation resulting from the interaction of gamma radiation with the particles of the medium is not a unique process—a single interaction rarely results in the incident photon changing to another form of energy. Usually, several interactions are needed until all the photon energy is transformed. Therefore, attenuation results from all processes involved, that is, the photoelectric effect, the Compton effect, the Rayleigh–Thomson effect and, for energies above 1.022 MeV, the production of pairs.

Figure 5.5 shows the total attenuation coefficient for water and for lead in terms of incident gamma radiation energy. The usual range of energies for gamma radiation in NM is 80–511 keV, corresponding approximately to the shaded area in the chart. For water (the main constituent of biological tissues), the predominant effect in the energy range in question is the Compton effect, and the occurrence of the photoelectric effect is much smaller. On the other hand, for lead the photoelectric effect is predominant in the same energy range. The total attenuation of water is low, which is fundamental to the success of PET and SPECT imaging, whereas the attenuation characteristics of lead are essential for collimators and the shielding devices used in radioprotection.

The photoelectric effect occurs when the incident photon transfers its total energy to an electron of an atom, which is ejected with a kinetic energy equal to the difference between the energy of the incident photon and the binding energy ($E_c = hv - E_\ell$). The ejection of the electron causes the atom to become an ion for a short period of time. This excited state remains until a more energetic electron occupies the hole created, emitting a photon with energy that is equal to the difference between the energies of the initial and final

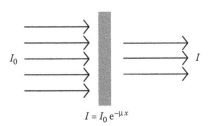

$$I = I_0\, e^{-\mu x}$$

FIGURE 5.4

Illustration of how the attenuation experienced by a monoenergetic beam when crossed by an object of thickness x. I_0 is the intensity of the incident radiation beam, reduced to I after crossing the object. The reduction fraction is given by $e^{-\mu x}$, where μ is the linear attenuation coefficient of the medium crossed by the radiation.

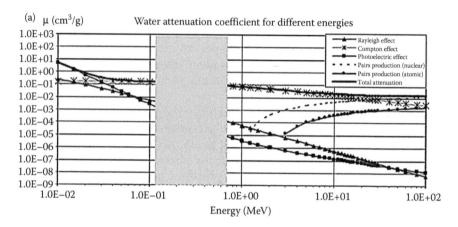

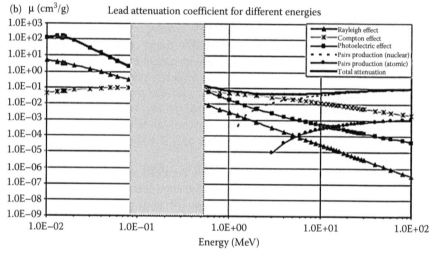

FIGURE 5.5
(a) Variation of the water attenuation coefficient with energy. The total attenuation is the combined effect of the four mechanisms. (Data from: http://physics.nist.gov/PhysRefData/Xcom/html/xcom1.html.) (b) Variation of the lead attenuation coefficient with energy. The total attenuation is the combined effect of the four mechanisms. (Data from: http://physics.nist.gov/PhysRefData/Xcom/html/xcom1.html.)

states of the electron. The radiation emitted, which is generally termed *characteristic x-ray*, is usually absorbed later by the detector. The probability of the occurrence of the photoelectric effect depends on the fourth power of the atomic number (Z), which is easily observable in Figure 5.5.

The Compton effect or Compton scattering, occurs when a photon interacts with an electron, causing deflection of the photon, which sends part of the energy to the electron that recoils (Figure 5.6).

The relationship between the energy of the photon, E_0, before interacting with the electron and the energy of the deflected photon, E', for a given angle

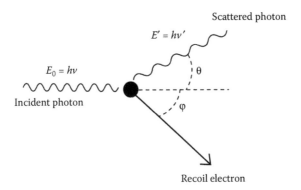

FIGURE 5.6
Diagram showing the Compton effect—the interaction of a photon with an electron results in the deflection of the photon, which transfers energy to the electron. This interaction results in a scattered photon with lower energy (longer wavelength) and the recoil of the electron. (Data from http://physics.nist.gov/PhysRefData/Xcom/html/xcom1.html.)

θ can be obtained from the law of conservation of linear momentum and the law of conservation of energy. The application of these laws results in the expression:

$$E' = \frac{E_0}{1 + (E_0/m_ec^2)(1 - \cos \theta)},$$ (5.5)

where m_ec^2 is the energy of the electron at rest.

Compton scattering is anisotropic, that is, the probability of a photon being deflective is not equal in all directions. The Klein–Nishina formula describes scatter probability in terms of direction:

$$\frac{d\sigma}{d\Omega} = \frac{r_e^2}{2}\left(P_{E_0,\theta} - P_{E_0,\theta}^2 \operatorname{sen}(\theta) + P_{E_\gamma,\theta}^3\right),$$ (5.6)

where r_e is the classical electron radius and $P_{E_0,\theta}$ is the ratio of the energies before and after Compton deflection:

$$P_{E_0,\theta} = \frac{E'}{E_0} = \frac{1}{1 + (E_0/m_ec^2)(1 - \cos \theta)}.$$ (5.7)

The value $d\sigma/d\Omega$ is usually known as the differential cross section and represents the probability of a photon being scattered in the solid angle $d\Omega$.

Rayleigh–Thomson scattering is also known as *elastic scattering*, as the kinetic energy is conserved. The probability of this effect occurring is greater for low-energy photons (<10 keV) that are deflected with no loss of energy when interacting with an atom.

Since pair production is an interaction mechanism that occurs when the energy is greater than 1.022 MeV, it is therefore, not observed in the energy ranges associated with NM. The pair production mechanism can be summarized as the interaction of a high-energy photon with the nucleus field or the electron field in which the photon is annihilated, producing an electron–positron pair.

Photoabsorption is a mechanism that occurs only with high-energy photons, in which a photon is captured by the nucleus, which then de-excites emitting one or more particles.

5.1.2 Gamma Radiation Detectors

The success of detection depends on the efficacy of the radiation interaction with the detector, because what is actually measured are the products of interaction. Measurement is mainly based on the fast electrons that result from the interaction of radiation with matter. The maximum energy of these electrons is equal to the energy of the incident photon; the electrons gradually lose their energy through interaction with the atoms of the medium that are ionized or excited. The charge produced is collected either directly via proportional counters or semiconductor detectors or indirectly via crystal scintillators.

5.1.2.1 *Gas-Filled Detectors*

Gas-filled detectors are composed of two electrodes separated by a gaseous atmosphere. Generally, the container that encloses the gas is one of the electrodes (cathode), and the other electrode (anode) is a thin wire running through the center of the system [4] (Figure 5.7).

Gamma radiation interacts with the gas contained in the detector causing, in certain circumstances, ionization and the consequent ejection of electrons. These are accelerated toward the anode and, if fully collected by the anode, produce a measurable electric current. Hence, the detector operates as an ionization chamber.

When the external voltage transfers enough kinetic energy to electrons, they produce new ionized atoms emitting more electrons—the avalanche effect. The negative charge created during this process is then collected by the anode, causing an electrical impulse which is the signal that is actually measured (see Section 5.1.1). For a certain range of the electric field, the signal obtained is proportional to the number of electron–ion pairs generated by incident radiation. In this case, the detector is designated as a proportional counter and operates in the proportionality region, which represents the conventional mode of operation for these detectors.

In sufficiently high electric fields, the avalanche effect extends to the entire anode, which leads to a loss of proportionality in the relationship between the charge produced and the energy released by the incident radiation. The signal from the detector always has the same amplitude, therefore meaning that the detector operates in the Geiger–Müller region.

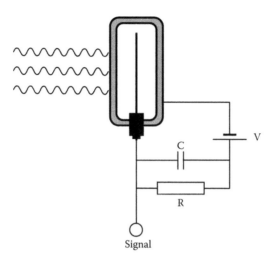

FIGURE 5.7
Equivalent circuit diagram of a gas detector. The electric field created within the detector accelerates the electrons produced by the interaction of radiation with the gas, which are then collected
at the anode (internal wire).

5.1.2.2 Semiconductor Detectors

The semiconductor detector most commonly used in the detection of gamma
radiation is the germanium–lithium detector. This type of detector has been
used in gamma radiation spectroscopy since the 1960s. In comparison with
gas detectors, these detectors offer the advantage of a smaller size and larger
density. An additional interesting feature is the excellent energy resolution.
However, until recently, these detectors required a cooling device in order
to operate efficiently, which limited their use in medical imaging. With the
emergence of detectors such as the CdZnTe (CZT), which can operate at room
temperature, this limitation has disappeared and gamma cameras using these
semiconductor detectors are now available. Nevertheless, the energy resolution of CZT detectors is lower than that of germanium detectors, although
still higher than scintillator detectors. There are also solid-state detectors (silicon) that require no cooling devices but can only be used with low-energy
radiation (tens of keV).

The working principle of semiconductor detectors is based on the formation
of electron–hole pairs whose displacement leads to a measurable electrical
signal.

A semiconductor can be defined as a solid whose valence band is complete
when $T = 0 \, \mathrm{K}$ but whose forbidden band is so small (few eV) that electrons can
be easily thermally excited (at room temperature) and pass to the conduction
band [5]. If the forbidden band is larger, the number of electrons capable
of moving to the conduction band is much smaller and the material is then
classified as an *insulator*.

For each electron excited to the conduction band in a semiconductor, a hole is created in the valence band whose mobility is only a fraction of the mobility of the electrons. In this situation, conduction of an electric current may exist if an electric field is applied. However, the electric conductivity is lower than that of metal materials and depends on temperature.

A semiconductor that is free of impurities has an equal number of electrons in the conduction band and holes in the valence band, as the electrons are thermally excited; and there is a one-to-one relationship between electrons and holes. A semiconductor with these characteristics is known as *intrinsic* or *undoped*. Considering the concentration of electrons (n) in the conduction band and the concentration of holes (p) in the valence band, it can be easily inferred that, for an intrinsic semiconductor:

$$n = p. \tag{5.8}$$

The doping of pure semiconductors with appropriate impurities leads to situations in which the previous relationship is not maintained and, consequently, there may be an excess of electrons or an excess of holes. The materials obtained by this process are known as n-type semiconductors or p-type semiconductors, respectively.

A semiconductor with two adjacent regions, a p-region and a n-region, demonstrates properties suitable for use as radiation detector. In the pn junction, there is diffusion of electrons from the n to the p region and a diffusion of holes in the opposite direction. This process creates a voltage difference in the pn junction. Near the pn junction, a region is created—the depletion region—in which there is balance of charge carriers. The depletion region has favorable features as a means of detecting radiation, as any electron that is created in this region is accelerated into the n side and any created hole is accelerated into the p side. However, in a nonpolarized pn junction, the depletion region is small and the junction capacitance is high; hence, the spontaneous electric field generated is of low intensity and does not allow the charge carriers to move quickly. For these reasons, the pn junction is generally polarized when used as a radiation detector in real applications.

Figure 5.8 shows a diagram of a typical semiconductor detector. The radiation interacts with the detector, perhaps causing ionization and, therefore, free charges, which are accelerated by the imposed electric field. The charge created is then collected by the polarization electrodes, generating an electrical signal.

The incident photons are absorbed at the pn junction of the semiconductor crystal, which is affected by a high-voltage difference, creating a large number of electron–hole pairs. For each electron–hole pair produced, approximately 3–5 eV are expended on average, and the total number of pairs is proportional to the gamma photon energy. With NaI (Tl) crystal, about 30 eV are needed to create one ionization and to produce one scintillation within the crystal. The number of electron–hole pairs produced is about 6–10 times greater than the amount of scintillation.

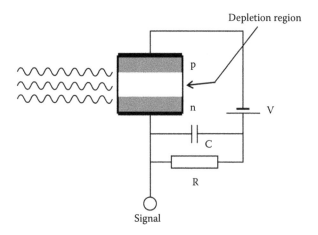

FIGURE 5.8
Equivalent circuit diagram of a semiconductor detector. The detector is an inversely polarized pn junction. When the gamma radiation produces an electron–hole, the charge is conducted through the sensitive region (depletion region) generating a measurable signal.

The energy resolution of the CZT gamma camera is approximately 5%, compared with about 14% for the NaI (Tl) gamma camera. The fact that the CZT gamma camera does not use photomultiplier tubes (PMT) is also an advantage.

The most recent CZT crystals are slightly more efficient than NaI (Tl). The intrinsic spatial resolution of CZT gamma cameras is also greater, being approximately 2–3 mm compared with 3–4 mm for the NaI (Tl) gamma cameras, at 140 keV.

In addition, the collection time for the electron–hole pairs is at least 100 times shorter than the decay time of the scintillation crystal NaI (Tl), which allows for an increase in the counting speed (250,000 count/s).

The CZT camera weighs about 100 times less, is considerably smaller than the scintillation camera, and can, therefore, be easily moved between different hospital departments.

These differences result in several advantages for image quality, such as improved contrast due to greater efficiency, higher energy resolution, and accuracy of position determination (intrinsic resolution), which is also enhanced as an effect of the modular structure [6].

5.1.2.3 Scintillation Detectors

Scintillators convert high-energy gamma photons into photons with wavelength in the visible range, which, in turn, are detected using PMT. The PMT collect the light created in the crystal by the γ rays, generating electrical signals that contain information about the energy of the interactions taking place in the crystal.

Scintillators are the most commonly used detectors in NM. They are generally inorganic substances in the form of solid crystals or organic substances dissolved either in a liquid solution or in a solvent and later polymerized, that is, formed into a solid solution. However, inorganic crystals are more widely used in modern NM equipment for the detection of both x-rays and γ rays.

When γ rays with energy levels of up to 511 keV hit the crystal NaI (Tl), several situations can occur:

1. No interaction
2. Interaction via the Compton effect
3. Interaction via the photoelectric effect

After the interaction of radiation through any of these mechanisms, new events caused by the release of secondary electrons may occur in the crystal, leading to generation of scintillation and the subsequent conversion into electrical impulses in the PMT.

The many possible combinations that could lead to production of scintillation are outlined in Figure 5.9a.

The requirements of an ideal scintillator are highly efficient detection, good energy resolution, good intrinsic spatial resolution, low dead time, and limited cost. These parameters depend on attenuation length, luminosity, the photoelectric fraction, and the decay constant of the detector [7].

Detection efficiency is critical in NM, as the dose administered to patients must be as small as possible, which naturally entails low count rates. This means that the scintillator thickness (measured in the axial direction) should be at least equal to two attenuation lengths and, in fact, is usually equal to three. However, intrinsic spatial resolution deteriorates with increased thickness, because the probability of multiple interactions occurring in the

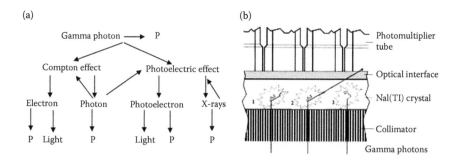

FIGURE 5.9
(a) Possible combinations of processes that may lead to the generation of scintillation after the interaction of the incident photon. (b) Portion of a slice showing a scintillation gamma camera with total absorption of energy (1 and 3), or partial absorption of energy (2) by loss of the Compton photon.

crystal increases, involving the total absorption of photon energy. Thus, a compromise must be reached between efficiency and spatial resolution.

The spatial resolution associated with the Anger logic is proportional to the PMT diameter and inversely proportional to the square root of the number of photoelectrons [7]. Therefore, the greater the yield of the scintillator (see Equation 5.10), the better the intrinsic resolution. In the case of the gamma camera, spatial resolution is mainly affected by the collimator, and intrinsic resolution has a marginal impact. Thus, by using a scintillator with higher luminosity the PMT diameter can be increased, which, in turn, can reduce the cost of the detectors.

As we saw earlier (see Section 5.1.1), increasing the yield has a direct consequence in terms of energy resolution, which can be used to discard the photons subjected to Compton interaction.

When there is total absorption of photon energy, the accuracy of the interaction position within the detector benefits more if this is achieved by photoelectric effect rather than by the Compton effect (see Section 5.1.1.1). Until it is absorbed, a photon subjected to Compton interaction deposits its energy in at least two separate locations, which degrade spatial resolution. Hence, a high photoelectric fraction is desirable, and the atomic number of the scintillator material is a major factor in this context.

Table 5.1 shows the principal properties of the main scintillators used either in NM or in research and development. The table was compiled on the basis of content [4,8–10], and from the Crystal Clear Web site [11] and the references it provides. The density of the scintillator ρ is multiplied by the fourth power of its effective atomic number Z_{eff}, as the absorption by photoelectric effect per unit of length is proportional to ρZ_{eff}^{3-4} [8]. For 511 keV photons, the probability of interaction by photoelectric effect at $Z_{eff} = 80$ is less than 50%, whereas the remaining 50% is due to the Compton effect. Depending on the type and size of the scintillator crystal, the detection of the Compton electron in addition to the scattered photon may significantly contribute to the intensity of the full-energy peak, as demonstrated, for example, in positron emission mammography (PEM) dedicated systems using LuAP scintillators coupled to avalanche photodiode arrays [12]. In PET applications, a scintillator with a small attenuation length is necessary in order to minimize the crystal radial length, typically in the order of thrice the attenuation length. This, in turn, minimizes the parallax effect responsible for radial degradation of the images at voxels further away from the isocenter of the tomograph [13]. The photoeffect fraction together with the photopeak energy resolution are indeed important for high full-energy peak intensity and, consequently, for high-efficiency detection of events valid for reconstruction; a low level of Compton-scattered photons in the patient, which would otherwise show up as a continuous blur in the images; and a high true-to-random coincidence ratio, because random events are more affected by Compton photons than by true events [14]. The photon yield and the decay constant of the scintillator play their role in minimizing the coincidence–time resolution τ achievable by

TABLE 5.1

Properties of Scintillators with Application in Nuclear Medicine

Scintillator	ρ (g/cm³)	Atten. Length (mm)	Photoelectric Fraction (%)	Hygroscopicity	Luminosity (photons/keV)	Decay Const. (ns)	Emission Peak (nm)	$\Delta E/E$ (% FWHM)	Fluorescence (%/ms)	Refractive Index	Clinical Application
NaI: Tl	3.67	29.1	17	Yes	41	230	410	5.6	0.3–5/6	1.85	SPECT
CSI: Na	4.51	22.9	21	Yes	40	630	420	7.4	0.5–5/6		XII
CSI: Tl	4.51	22.9	21	Little	66	>800	420	6.6	0.5–5/6		PET, SPECT, CT
CSF	4.64	20	23	High	2	3	390			1.48	TOF-PET
BaF$_2$	4.89	20.5	17	Little	2	0.7	220	10		1.54	TOF-PET
BGO (Bi$_4$Ge$_3$O$_{12}$)	7.13	10.1	40	No	9	300	480	9	0.005–3	2.15	PET
LSO (Lu$_2$SiO$_5$:Ce)	7.4	11.4	32	No	26	40	420	7.9	<0.1/6	1.82	TOF-PET
Lu$_{1.8}$Y$_{0.2}$SiO$_5$:Ce	7.1	11.5		No	26	41	420	7–9	<0.1/6	1.81	TOF-PET
LuYSiO$_5$: Ce	6	16.7	21	No	26		420	7–9	<0.1/6		TOF-PET
LuAP (LuAlO$_3$:Ce)	8.3	10.5	30	No	12	18	365	~15		1.94	TOF-PET
LPS (Lu$_2$Si$_2$O$_7$:Ce)	6.2	14.1	29	No	30	30	380	~10		1.74	TOF-PET
GSO (Gd$_2$SiO$_5$:Ce)	6.7	14.1	25	No	8	60	440	7.8		1.85	PET
YAP (YalO$_3$)	5.5	21.3	4.2	No	21	30	350	4.3		1.95	PET
LaCl$_3$:Ce	3.86	27.8	14	Yes	46	25 (65%)	353	3.3		1.9	SPECT
LaBr$_3$:Ce	5.3	21.3	13	Yes	61	35 (90%)	358	2.9		1.9	SPECT
CeBr$_3$	5.2	21.5	14	Yes	68	17	370	3.4			TOF-PET
LXe (liquid xenon)	3.06	30.4	21	—	11	27 (30%)	165	22/16			DOI-PET
Ideal (PET)	>6	<12	>30	No	>8	<300	300–500	<10			

the system. A small time resolution is essential to achieve the lowest random coincidence rate C_r possible, as $C_r = 2\tau C_i C_j$, where C_i and C_j are the singles count rates in detectors and i and j form the LOR ij. Further, a small scintillation decay time is also necessary for minimizing dead time, that is, the time in which a coincidence cannot be registered, because the PET system is busy handling a previous coincident event. Several parts of the system contribute to the dead time, and the detector is one of them. In addition to its influence on energy resolution, a high light yield is also important for achieving optimum spatial resolution in the detector in systems using scintillation-light-sharing methods to reduce the number of readout electronic channels [15–17].

By far, the most commonly used scintillator for NM applications is the thallium-activated sodium iodide crystal NaI(Tl), used in combination with a PMT readout mostly not only in single-photon emission systems but also in high-resolution PET scanners [18]. Other scintillators that are particularly suitable for PET at present are BGO, LSO, LuAP, LPS, GSO, and YAP (Table 5.1). The most widely used in commercial PET systems is BGO, which has a high detection efficiency for 511 keV photons (a crystal with 3 cm depth has nearly three attenuation lengths), a high photoeffect fraction, and a relatively low production cost. The main disadvantages of BGO are its rather long decay time of 300 ns and low light yield in comparison with other scintillators such as LSO, leading to inferior time and energy resolutions, respectively. One of the most suitable scintillator materials for PET is LSO. It has a high number ρZ_{eff}^4; and high light yield, a short decay time, and relatively good mechanical properties necessary for high-throughput manufacturing. Due to its short decay constant, a time resolution of 1.2 ns has been already achieved in commercial PET scanners [19], whereas BGO-based tomographs typically present 12 ns. Therefore, renewed interest in profiting from the time-of-flight (TOF) information in LSO-based tomographs has emerged [20]. For completeness, it must be stated that several groups have been actively developing new PET methods and techniques using noble elements, such as liquid xenon (LXe) in scintillation and ionization correlated modes in order to solve the problem of parallax error in PET [21,22].

Current commercial PET scanners generally use detector module readouts in Anger logic, that is, one scintillator detector block is implemented with saw cuts defining *individual* pixels and the scintillation light is read by several PMTs [20]. The analog ratio among the PMT signals yields the coordinate of the hit crystal [23]. This reduces the number of electronic readout channels by a factor proportional to the ratio between the number of individual crystals and the PMT but, on the other hand, increases the detector dead time per front area unit* by the same amount and introduces a spatial resolution degradation typically of 2 mm added in quadrature to the other factors influencing the spatial resolution of the tomography [24].

---

* The detector dead time per unit area is a figure of merit commonly used to describe the effective detector dead time due to the scintillator decay time together with the readout scheme implemented.

5.1.2.4 *Photomultipliers*

The reasons for using PMTs for scintillation light readouts are their much higher gain and better noise characteristics in comparison with other more compact light detectors such as silicon-based avalanche photodiode arrays.

A PMT consists of a photocathode and several dynodes. The photocathode is located near the window of the PMT and is composed of a photosensitive material. The photons coming from the scintillator crystal interact with the photocathode, transferring its energy to the electrons of the photocathode material. The electrons ejected by the photocathode are then accelerated toward the first dynode. The electrons interact with the first dynode causing the ejection of more electrons that are accelerated, by another voltage difference, toward the second dynode. This process is repeated with the charge multiplying at each stage between the dynodes. The charge is finally collected at the anode generating a signal, which, after it is amplified and shaped, is the input of a pulse height analyzer circuit and position circuit.

The final amplitude of the signal is proportional to the light energy of the scintillation within the crystal, which, in turn, is proportional to the energy lost by the incident radiation that produced the scintillation.

Figure 5.10 shows a diagram of a scintillation detector and the assembly that is generally used to obtain the voltage between dynodes.

The steps described above can be analyzed and quantified independently to deduce an equation that reflects the performance of a PMT.

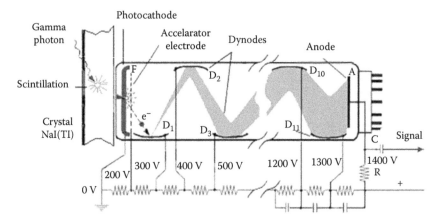

FIGURE 5.10

Diagram of a photomultiplier tube. The photons incident on the window of the PMT cause electrons to be emitted in the photocathode that are accelerated along a chain of electrodes under high tension, leading to multiplication of the charge. The voltage of the dynodes is obtained by dividing the output of the voltage source by a series of high-value resistors (MΩ).

When a photon with energy $E\gamma$ interacts with the crystal and produces N_0 fluorescence photons of individual energy E, the total light energy released is

$$E_\tau = N_0 E. \tag{5.9}$$

The light yield of a scintillator crystal is defined by the relationship

$$\lambda = \frac{E_\tau}{E_\gamma} \tag{5.10}$$

For the NaI (Tl), the light yield can be up to 0.1.

From the N_0 photons of light emitted within the crystal in all directions, only a number N_1 reaches the photocathode. This is called the *optical yield*, for the optical quotient

$$\omega = \frac{N_1}{N_0}, \tag{5.11}$$

which, under optimum conditions, may equal 0.5.

Part of these N_1 photons will produce a photoelectric effect in the photo-cathode.

Let N_2 be the number of ejected photoelectrons. The ratio

$$\xi = \frac{N_2}{N_1}, \tag{5.12}$$

is the quantum yield of the photocathode. For the antimony-cesium this yield is around 0.1.

Only a portion, N_3 of the N_2 photoelectrons, reaches the first dynode. The term *collection yield* is usually employed to describe the relationship

$$\chi = \frac{N_3}{N_2}, \tag{5.13}$$

which may be close to a unit.

Each dynode will multiply the number of electrons up to a certain, very high, factor that depends on the voltage difference between the dynodes and the coating substance. The gain (A) of a dynode is the ratio between the number of electrons emitted and the number of incident electrons.

If n is the number of dynodes, the total gain of the PMT is

$$G = A^n. \tag{5.14}$$

The total charge received by the anode after a photon of energy $E\gamma$ has released N_0 visible photons in the crystal is

$$Q = N_0 \omega \xi \chi G e, \tag{5.15}$$

where e represents the electron charge.

Equation 5.15 can be rewritten as

$$Q = \frac{E_\tau}{E}\omega\xi\chi Ge. \tag{5.16}$$

or

$$Q = \lambda\frac{E_\gamma}{E}\omega\xi\chi Ge. \tag{5.17}$$

In this expression, the values E, λ, ω, ξ, χ, and e are constant and specific for a certain crystal-photomultiplier set.

For a given voltage difference between dynodes, the gain G is constant. Thus, we can write

$$Q = E_\gamma T, \tag{5.18}$$

where

$$T = \lambda\frac{1}{E}\omega\xi\chi Ge. \tag{5.19}$$

The light yield is defined by assuming that all the energy $E\gamma$ of the photon is absorbed, creating N_0 visible photons. Therefore, we can conclude that the charge collected by the anode of a PMT coupled to the crystal is proportional to the energy of the incident photon when it is fully absorbed by photoelectric effect or when multiple interactions occur in the crystal, leading to the loss of the total energy of the incident γ photon.

When the energy of a photon is not totally absorbed, the charge collected at the anode is only proportional to the energy absorbed.

This is the case, for example, when one or more Compton interactions occur and the photon is scattered on leaving the crystal.

5.1.2.5 Variants and Alternatives

The photomultiplier is the photodetector used in the overwhelming majority of current systems used in conventional NM, namely PET and SPECT. Continuous improvements have been made to its technology, but there are some intrinsic limitations that preclude its use in new applications. They are, for example, inappropriateness for use in the new multimodal PET or MRI (magnetic resonance imaging) systems, as their sensitivity to magnetic fields and their relatively large dimensions prohibit use in systems with very high resolution, such as those dedicated to small animals. Fortunately, variants of the traditional PMTs and other alternative devices exist with a performance suitable for the requirements predicted for the near future, which are expected to include systems with the ability to measure TOF (requiring a short response time) and depth of interaction (for which small size and high efficiency are advantageous), possibly with better energy and time resolution. Other photodetectors providing a sufficiently high-gain bandwidth for use in PET applications are multianode PMT [25], metal channel dynode PMT [26], hybrid PMT [27], micro-channel plate PMT (MCP-PMT) [28], visible light photon counters (VLPC) [29], and Geiger-mode avalanche photodiodes [30]. The latter has been gaining in importance due to its low-cost, compactness, ability to operate at room temperature, insensitivity to magnetic field and hadrons, and good timing characteristics in low-light environments, with pixel recovery times of about 20 ns. The avalanche photodiodes (APDs) and variants (HAPDs—Hybrid Avalanche Photodiodes and PSAPDs—Position Sensitive APDs, among others) are compact, have high quantum efficiency and high gain, and are insensitive to magnetic fields. Although presenting high noise levels, the APDs have a performance suited to the development of high-resolution PET systems that are compatible with MRI. The silicon PIN diodes are small, with high quantum efficiency and insensitivity to magnetic fields, but they have a low gain and need preamplifiers with low noise. Other technologies involving silicon-based PMTs are also available and are being rapidly developed, in addition to devices based on CCDs (charge-coupled devices).

5.1.3 Signal Acquisition Electronics

The basic structure of the signal acquisition module is represented in Figure 5.11.

Detectors generate a signal that is a pulse of current lasting a few microseconds. The electric charge contained in these pulses is related to the energy deposited within the detector volume by the ionizing radiation. It is necessary to transform the current pulse to a voltage pulse in such a way that the amplitude is proportional to the deposited energy. The preamplifier, with huge input impedance and low output impedance, is the device responsible for this transformation, while also allowing for the adjustment of impedance. The preamplifier gain is usually small (≈ 1) (Figure 5.12).

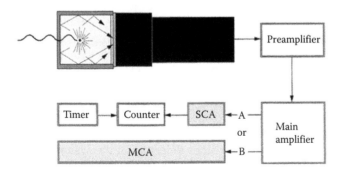

FIGURE 5.11
Basic diagram of the electronic module used in detector signal acquisition. (Adapted from N. T. Ranger, *Radiographics*, 19(2):481–502, 1999.)

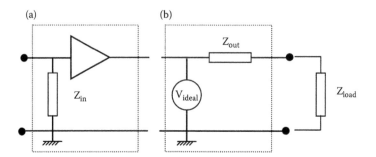

FIGURE 5.12
(a) Signal input stage; the triangle represents an operational amplifier and the input impedance (ideally infinite) is represented by Z_{in}. (b) output stage that can be seen as a voltage generator in series with a resistor; Z_{out} represents the output impedance, which is as small as possible.

The output signal of the preamplifier enters the amplifier where it is amplified and shaped. The typical gain for this stage is 1000, and the appropriate shaping is required for the height pulse analysis. Impulse shaping depends on the electronic circuitry for peak analysis, which is performed by a single or a multichannel analyzer. Since both single and multichannel analyzers measure the pulse amplitude in relation to a voltage reference, it is essential that, between pulses, the output from the amplifier quickly returns to the stable voltage reference [32]. It is, therefore, also essential to implement a restoration baseline circuit.

The output of the amplifier is a symmetrical voltage pulse (usually unipolar) whose amplitude is proportional to the energy deposited in the PMT. Thus, analysis of the signal amplitude provides information on the energy of the radiation detected, which can then be used to accept or reject events. For this purpose, as already noted, single or multichannel analyzers can be used. In the case of the former, only pulses that present amplitudes within a certain range are accepted for counting. In this case, only radiation with a particular energy is measured and the channel presents a certain width that determines energy resolution. With the multichannel analyzer, a histogram of amplitudes is created, corresponding to an energy spectrum. This histogram can be obtained through the use of an analog-to-digital converter (ADC).

The PET and SPECT cameras that perform coincidence require the detectors to operate in coincidence mode. This mode requires a number of detectors operating simultaneously. When a photon is detected, the corresponding output signal of the PMTs is used to generate a time signal that accurately indicates the instant t of photon detection. The time signal is then sent to an electronic unit that processes the measurement of coincidences within a time window Δt with all the other detectors in the system. The unit detects a coincidence if there is a time signal from another crystal in the time range $(t, t + 2\Delta t)$. The time window has to be carefully chosen and should take into account the temporal resolution of the system, which measures the existing uncertainty in determining the instant of the photon detection. If Δt is too small in comparison with the temporal resolution of the system, fewer true coincidences than those that actually exist may be identified; and if, on the other hand, Δt is too large, it increases the probability of detecting random coincidences. When Δt is very small, that is, less than 3 ns, the difference in the time intervals required by the two photons resulting from annihilation to reach the detectors becomes important. There are systems based on scintillators that are very fast and can use TOF information to improve spatial resolution and reduce the noise in the image.

Another important aspect of signal processing that must be taken into account is how to determine the position of radiation interaction inside the crystal. The interaction of a gamma photon with the crystal produces numerous scintillation photons that are detected by several PMTs. The sum of the charge measured by all PMTs is proportional to the energy of the gamma photon energy, and it is reasonable to assume the charge distribution is such that it is more intense near the interaction position and less intense further away from it. Anger logic is based on this assumption and enables the position of the gamma photon in the scintillator to be determined by a weighted average.

Let X and Y be the axes of a coordinate system whose origin is at the centre of each crystal. A position (x_f, y_f) is associated with each PMT, f, which is the position at the center of the PMT window in the coordinate system in question (Figure 5.13). In a given event, each PMT gives a measurement a_f.

The position of a particular event will be given as if it were a center of mass calculation [33], that is,

$$X = \frac{\sum_{f=1}^{N} a_f x_f}{\sum_{f=1}^{N} a_f} \quad \text{and} \quad Y = \frac{\sum_{f=1}^{N} a_f y_f}{\sum_{f=1}^{N} a_f}. \tag{5.20}$$

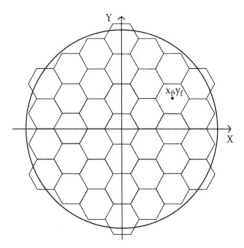

FIGURE 5.13
Diagram showing the arrangement of PMTs in a gamma camera and their relationship in determining position using Anger logic.

5.2 Properties of Image Detection Systems

The quality of medical images is mainly related to the capability of detecting relevant information. The image quality depends on its information content and how accessible the information is.

The properties of imaging systems are usually expressed in physical terms. Several concepts are used to describe imaging systems in terms of quality of the response: spatial resolution, sensitivity, contrast, and noise.

There are complementary physical quantities, which are used together with these basic concepts to enhance their objectivity. The most important complementary quantities are the point spread function (PSF), the modulation transfer function (MTF), the Wiener spectrum, and signal-to-noise ratio (SNR).

In addition, we have the Rose model with the contrast–detail curves and ROC analysis (receiver–operator characteristics) as intermediaries, which also unify concepts that include basic properties.

The detector is obviously one of the most important parts of imaging systems, which, in the case of NM, are gamma radiation detectors.

Some desirable properties of γ-ray detectors are the high effective atomic number, high density, high photopeak fraction, and short decay time.* The first two properties are related to the sensitivity of the detector, the third to selectivity in photon detection, and the fourth to the ability to process high activities [34].

Other properties, such as high luminosity efficiency, large detector area, and the low refractive index, may be important in specific contexts.

Some of the generally accepted parameters for measuring the properties of the image detection systems are briefly considered in the next section.

5.2.1 Distance of Resolution: PSF

The spatial resolution is related to the sharpness or detail of the images.

The performance of the devices for medical imaging, in terms of sensitivity and spatial resolution, can be entirely defined by the PSF, which is the function that describes the image when the object is a point.

If $B(x, y)$ is the PSF for a 2D imaging system and is symmetric about a central axis, then the response of the system can be fully described by the curve $B(x)$ in an axial plane that intersects $B(x, y)$.

The volume under the surface $B(x, y)$ is proportional to the system sensitivity.

For a system that gives an isotropic response, the width at half maximum of the curve $B(x)$ is the distance of resolution, d (Figure 5.14a).

---

* In the case of scintillation detectors, the time decay is the time required for the scintillation emitted to fall by a factor equal to e^{-1}.

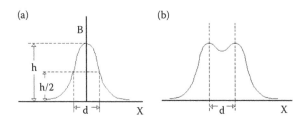

FIGURE 5.14
(a) The distance of resolution, d, is the width at half maximum of the curve $B(x)$, that is, the PSF.
(b) Two points to the distance d are separated with difficulty in the image.

Two points located at distance d are separated with difficulty in the image (Figure 5.14b).

For digital systems, the resolution can be associated with the number of pixels representing the image (typically 128×128, 512×512, etc.) when d is smaller than twice the side of the pixel.

If the PSF is a Gaussian function with standard deviation σ, then the distance of resolution is $d = 2.35\sigma$.

A perfect 2D imaging system would present a point for each object point. Degradation occurs in real devices, and images of points appear as an area whose intensity decreases from the central region to the periphery. Considering that each activity distribution can be represented by a set of points, the loss of sharpness for the images of large sources is caused by the overlapping of the images of the numerous points of the object.

The function that represents the 2D response of an imaging system to a point object of unit intensity is the function of the degradation of a point or the normalized point spread (PSF), $h(x, y)$.

Since the PSF can vary with direction, the line spread function (LSF) is often used along an axis of an orthogonal coordinate system.

The LSF $hL(x)$ can be calculated from $h(x, y)$, if $h(x, y)$ is symmetric and independent of the direction.

In general, for gamma cameras the PSF is approximately constant in planes perpendicular to the collimator axis. Therefore, for a 2D image at a distance z in front of the collimator, the image $g(x, y)$ is obtained by 2D convolution of the object function $f(x, y)$ with $h(x, y)$, that is,

$$g(x,y) = \int_{-\infty}^{+\infty} \int_{-\infty}^{+\infty} f(x - x', y - y') h(x', y') \, dx' \, dy'. \tag{5.21}$$

The output is therefore, an overlap of displaced versions of the PSF weighted for each position to the input amplitude in that point.

Equation 5.21 can be applied if the image is linear and invariant to displacements. For NM systems, linearity and invariance in space are only approximately verified owing to the effects of dead time and nonuniformity of response.

Equation 5.21 can also be presented as

$$g(x,y) = f(x,y) * h(x,y), \tag{5.22}$$

where * means convolution.

Equation 5.21 can be applied to sections of homogeneous objects provided that the PSF, corrected for attenuation, is known in a number of planes perpendicular to the center line of the collimator.

Equation 5.21 takes into account the degradation factors of the imaging system, such as those that depend on the device characteristics, from the acquisition to the interpolation in the screen display. However, it does not consider noise. A more objective approach to modeling the process of images would be

$$g(x,y) = f(x,y) * h(x,y) + \eta, \tag{5.23}$$

where η represents the noise.

The increase of the detector thickness in scintillation cameras causes degradation of the intrinsic spatial resolution. This can be improved, however, by increasing the energy of the photons and increasing the number of PMTs.

5.2.2 Modulation Transfer Function

The information contained in the PSF, defined above, is rarely used directly; another quantity is used instead, the modulation transfer function (MTF), which is more convenient for some purposes. The MTF is based on the concept that all radioactivity distribution in a plane normal to the axis of the collimator can be expressed by a series of sinusoidal components, relative to two rectangular coordinate axes, and having their own spatial frequencies, amplitudes, and phases. The frequencies of the various components are integer multiples of a frequency usually termed *fundamental*.

Activity distributions with high variations or activity that show fine detail have a large contribution from high-frequency components in their spectral content, whereas images that involve only low spatial variation essentially have only low-frequency components [35].

Based on this interpretation and using the Fourier analysis, it is assumed that the images obtained by a particular device are approximate descriptions of the objects in terms of sinusoidal components with features that depend on the device.

Hence, images of objects with a simple sinusoidal structure can be employed to assess the response of imaging systems.

The MTF expresses the response of the system, the system components, or the modulated sinusoidal distributions of the object parameter when the frequency varies.

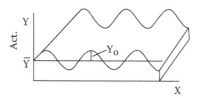

FIGURE 5.15
Object consisting of the sum of a constant value of activity, with activity sinusoidally modulated in the xx direction.

An instrument to study the response of an NM imaging system could be a set of activity distributions consisting of sums of sinusoidal modulations in the direction of the xx axis, with different spatial frequencies summed to a constant value $(y;^{-})$ as shown in Figure 5.15.

For each spatial frequency the MTF is the ratio of the contrast in the image to the contrast in the object and represents the fraction of the information of the object that is retained in the image. An ideal system has a unit value for MTF at all frequencies.

For a sinusoidal activity variation, the modulation or contrast is the ratio between the AC component (half of the peak-to-peak value of activity) and the average activity, or DC component.

The image of a sinusoidally modulated object is supposed to be a sinusoidally modulated image with the same frequency, although the amplitude and phase can be different.

A general expression for MTF can be obtained from Equation 5.24 by applying the 2D Fourier transform to both members of the equation and performing some transformations.

Since the Fourier transform of the convolution between two functions is equivalent to the product of the Fourier transform of the two functions, we get

$$G(\omega_x, \omega_y) = F(\omega_x, \omega_y)H(\omega_x, \omega_y), \tag{5.24}$$

where the functions G and g, F and f, and H and h are pairs of Fourier transforms. G represents the output spectrum, F the input spectrum, and H a characteristic function of the system defined in the frequency domain.

The angular frequencies ω_x and ω_y are related to the spatial frequencies v_x and v_y in directions x and y by

$$\omega_x = 2\pi v x \quad \text{and} \quad \omega_y = 2\pi v y. \tag{5.25}$$

The DC contribution to the frequency spectrum of f and g are the values F and G when $\omega_x = \omega_y = 0$, that is, $F(0,0)$ and $G(0,0)$. Then

$$G(0,0) = F(0,0)H(0,0). \tag{5.26}$$

The Fourier transform of the function of response to a point $h(x, y)$ is by definition

$$H(\omega_x, \omega_y) = \int\limits_{-\infty}^{+\infty} \int\limits_{-\infty}^{+\infty} h(x, y) e^{-j(\omega_x x + \omega_y y)} dx \, dy. \tag{5.27}$$

The Fourier transform of $h(x, y)$, $H(\omega_x, \omega_y)$, is usually called the transfer function of the imaging device, which is a quantitative measure of the ability of the system to preserve the contrast and sharpness of the image.

The Fourier transform of the PSF when $\omega_x = \omega_y = 0$ is given by Equation 5.28.

$$H(0,0) = \int\limits_{-\infty}^{+\infty} \int\limits_{-\infty}^{+\infty} h(x, y) \, dx \, dy = \xi \tag{5.28}$$

$H(0,0)$ is consequently the total information transferred in 1 s to the plane of the image when the object is a source point with a disintegration speed of one disintegration per second, that is, the camera efficiency ξ.

The MTF value (Equation 5.29) is the ratio between the image and object modulations, considering that they are sinusoidal functions (second term of equation). The third member of Equation 5.29 is given by applying Equation 5.24:

$$\text{MTF} = \frac{G(\omega_x, \omega_y)/G(0,0)}{F(\omega_x, \omega_y)/F(0,0)} = \frac{H(\omega_x, \omega_y)}{H(0,0)} \tag{5.29}$$

The final result for the MTF can be obtained as the quotient between the $H(\omega_x, \omega_y)$ and $H(0,0)$, that is, the transfer function divided by the system efficiency.

The MTF is defined for all frequencies as the ratio of the modulations or contrasts in the image and in the object (Figure 5.16).

We can also write

$$\text{MTF} = \frac{\text{Image Modulation}}{\text{Object Modulation}}. \tag{5.30}$$

The MTF is a decreasing function of frequency. For spatial frequencies higher than c, called the *cutoff frequency*, the contrast of the object is not transmitted to the image, and the MTF is zero.

Cutoff frequencies for some imaging diagnostic methods are

- Film/screen: 10–20 line pairs/mm
- CT scanner: 1–2 line pairs/mm
- Gamma camera: 0.3 line pairs/mm
- Image intensifier tube with CSI screen: 4–5 line pairs/mm

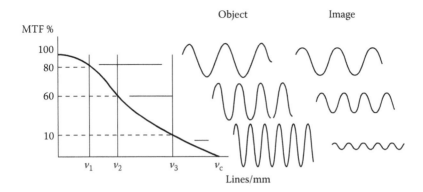

FIGURE 5.16
Modulation transfer function vs. spatial frequency.

5.2.3 Detection Efficiency or Sensitivity

The efficiency of detection or sensitivity applies to the specific energy of the radiation used and measures the fraction of energy emitted by the source that is actually employed by the detector to produce the image. In general, the efficiency of a detector can be divided into three parts: geometric efficiency, quantum (or intrinsic) efficiency, and conversion (or extrinsic) efficiency.

The first is related to the geometric scheme exploited in detecting the radiation. The second is the fraction of incident energy that is absorbed in the detector. The third measures the yield from the transformation of the absorbed energy by the detector into useful signals.

5.2.3.1 Geometric Efficiency

The geometric efficiency, E_g, is the relationship between the solid angle seen by the detector and solid angle of emission. Essentially, it depends on the size of the detector and on the source–detector distance. The presence of dead regions in detectors must often be considered: at the edges of the detectors or between detectors that have been combined. The last effect is very important in detectors whose large area of detection is achieved by joining up small detector units. In these cases, the fill factor is defined as the quotient between the effective area of detection and total area.

5.2.3.2 Quantum Efficiency

The quantum efficiency, E_Q, measures the probability of interaction, that is, the fraction of energy of incident photons that is absorbed by the detector:

$$E_Q = 1 - e^{-\mu x_d}, \tag{5.31}$$

where x_d is the thickness of the detector and μ is the linear attenuation coefficient of the detector material.

Therefore, the quantum efficiency depends on the μ of the detector material, the thickness of the detector, and the photons' energy. Quantum efficiency is modified by absorbent materials that are placed in the path of incident rays, such as the protection that encloses the detectors such as windows (aluminum, glass, titanium, etc.), grids, and collimators.

Equation 5.31 presupposes that any energy transferred by incident photons to the detector produces a usable signal, despite a fraction being discarded when there is energy selection.

The human eye operating in the visible range of the electromagnetic spectrum (λ from 400 to 700 nm) has an E_Q of nearly 1%. For the same energy range, the E_Q of the film is typically 5–20%; and for CCDs, it is 50–90%.

5.2.3.3 Conversion Efficiency

The efficiency of conversion, E_c, is the fraction of the photon energy that is absorbed by the detector and that is converted into a measurable signal, either electrical or luminous.

5.2.3.4 Total Efficiency

The total efficiency (or sensitivity) of a detector is the product of the geometric, quantum, and conversion efficiency, $E_t = E_g \times E_Q \times E_c$.

In addition, the total efficiency of the detector depends on the dead time of the system. This is the period after detection during which the detector is unable to carry out a new detection.

A system with high sensitivity includes more information in the images in the same time than a system with lower sensitivity can.

5.2.4 Noise

The LSF and MTF of the imaging systems previously described were considered under noise-free conditions.

Due to the statistical nature of the production of gamma photons, random fluctuations in the radiation intensity emitted by radionuclides are expected. These are recognized as Poisson noise and cause degradation of the contrast.

The probability distribution for p photons in a time of T seconds, when the average intensity is i photons per second, is

$$P(p; i, T) = \frac{(iT)^p e^{-iT}}{p!}. \tag{5.32}$$

The interaction with the detector can be considered binomial with an E_Q success probability.

The distribution of the photons that interact is the Poisson distribution with standard deviation equal to

$$\sigma = (iTE_Q)^{1/2}, \tag{5.33}$$

where iT is the number of incident photons. The SNR is

$$\text{SNR} = 10\, \log_{10}(N_0E_Q)\ \text{dB}. \tag{5.34}$$

If the detection process is followed by a process with gain g, the average signal amplitude will be

$$S = N_0E_Qg. \tag{5.35}$$

The standard deviation of this quantity should include the noise and the noise gain in amplification process, g. Considering the two sources, the total noise is

$$\sigma_s = \left\{ N_0E_Q \left[g^2 + \sigma g^2(1 + N_0E_Q) \right] \right\}^{1/2}. \tag{5.36}$$

In modern systems, the noise introduced by the amplifiers is usually negligible. In order to completely define the noise, it must be known how the signal and noise are affected by spatial frequency.

The power spectrum or Wiener spectrum, which describes the noise as a function of the spatial frequency, allows a complete description of the noise.

It tells us what MTF frequencies contain the most contribution to noise.
White noise corresponds to equal power at all frequencies.
Image processing such as filtering or reconstruction may change the noise level.

5.3 Methods of Image Production in Nuclear Medicine

The two ways to obtain images in NM are considered in general in Figure 5.17.

a. Includes all types of conventional NM imaging except for SPECT, that is, static, whole body, and dynamic scintigraphy and scintigraphy with synchronized acquisition
b. Includes the SPECT and PET techniques

The selective retention of compounds marked with γ emitters enables images of the distribution of the tracer to be obtained that not only include morphology aspects but also contain data on the metabolic capabilities of

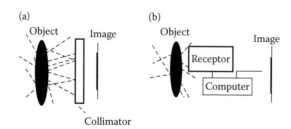

FIGURE 5.17
(a) Images obtained by direct emission. (b) Images obtained by indirect techniques.

the organs, with local and frequently quantitative information. The NM can depict, through images, the spatiotemporal distribution variations of the biomolecules that compose the human body.

These images are plane orthogonal projections of the local concentration of a radiopharmaceutical that is present in a partial volume of the body, and they are usually called planar images, or simply scintigraphies. The images are similar to 2D maps where one dimension has been removed from the original 3D distribution.

Dynamic studies can also be achieved by acquiring rapid sequences of planar images. This capability is important, because a static image of a biological system obtained at a certain instant provides little information, as dynamic processes are the essence of physiology.

The specific functional content of the NM techniques was a decisive step forward in the study of the metabolic dynamics of numerous organs.

The radial acquisition around an object followed by reconstruction enabled the acquisition of tomographic maps of radionuclide concentration. This methodology, when used with the detection of a single photon, is known as SPECT and offers a better contrast than planar scintigraphy. Moreover, the acquisition of several transverse sections of the activity distribution allows a reasonable reconstruction of its space location. The availability of rotating gamma cameras at a reasonable cost has enabled SPECT to become a standard technique in conventional NM.

The most important factors to be considered when choosing the scintillation detectors to detect gamma radiation for the purpose of obtaining images are the field of view, the stopping power of the detector material, the efficiency, the response time, and the energy resolution.

Scintillation cameras have steadily improved and they are now almost at the top of their performance relative to cost. This performance has been achieved even though the NaI (Tl) scintillator, universally used in gamma cameras, does not offer good energy resolution or good time response, besides having other undesirable physical properties such as high hygroscopicity. The main strengths of the NaI (Tl) are its high attenuation coefficient for medium energy photons and high luminosity.

5.3.1 Gamma Camera and SPECT

The scintillation camera, the imaging device restricted to conventional NM, is a position sensitive gamma radiation detector; and it is used in more than 90% of routine studies using γ rays of 140 keV from technetium 99m (^{99m}Tc). This is the radionuclide used in over 90% of diagnostic studies in NM, because it emits virtually no particles; it has an almost optimal gamma energy for detection by the NaI (Tl), because it has excellent properties with regard to the chemical labeling of molecules; it has a short period; and, finally, because it can be produced through generators.

The gamma camera was developed by Hal Anger [23] and has not undergone major changes since its creation in 1958. For this reason, the gamma camera is also known as the Anger camera. The detection portion of the gamma camera, the head (Figure 5.18), consists mainly of a crystal of sodium iodide activated with thallium, NaI (Tl), in the form of a disk or rectangle 30–50 cm wide and 0.9–1.2 cm thick. This is enclosed in a hermetic aluminum cylinder, coated with a layer of a light diffuser and equipped with a transparent base (Lucite), a set of (60–90) photomultiplier tubes optically coupled to the transparent window of the crystal, a lead collimator, and some electronics. The thickness of the crystal is a compromise between intrinsic spatial resolution and efficiency of detection.

The main function of the collimator is to limit the photons that interact with the crystal to only those that have a particular path relative to the detector surface, which allows 2D projections of the activity distribution of the tracer molecule to be obtained. The selection of photons makes the detection less efficient. Typically, 99% of the emitted photons are stopped at the collimator. Nevertheless, it is essential to the formation of a projection that is spatially correlated with the activity distribution of the radiolabeled molecule. The collimator significantly reduces the sensitivity of the camera while, at the same time, it affects the spatial resolution of the gamma camera and makes it dependent on the distance of the object to the detector and on the characteristics of the collimator.

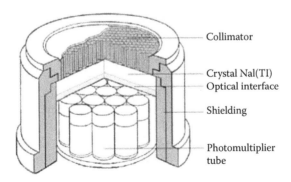

FIGURE 5.18
Diagram representing the components of the gamma camera.

The energy of an incident photon in the scintillator is partially converted into visible or near UV light (scintillation) and then transformed into an amplified electrical signal by a PMT. The intrinsic energy resolution expresses the ability of the detector (without collimator) to differentiate between two photons of different energies. Of particular importance is the differentiation between photopeak and scattered photons. The value is given by the energy range that corresponds to the photopeak width at half height expressed as a percentage of the energy of the photopeak.

Impulses of the total absorption peak correspond to the total energy absorption of the incident photon and can result from (a) interaction in the crystal by photoelectric effect, with both the photoelectron energy and the X-radiation energy due to electron transition or Auger electrons being dissipated in the crystal and (b) interaction by Compton effect followed by absorption of the Compton electron and the scattered photon by the crystal.

The PMT collects the light generated in the crystal by γ rays, by ionization and excitation, giving rise to electrical signals that contain information about the energy and position of the interactions taking place in the crystal. The interactions that occur in the crystal are processed individually, which require the use of fast electronics and low dead time in the detector to allow high counting rates [36].

The proportionality between the energy transferred and the electrical impulse produced allows the scintillation detector to be used for gamma radiation spectrometry.

The gamma spectrum, that is, the distribution of the number of pulses for each amplitude value, from zero to a maximum value (for each energy value dissipated in the crystal), can be obtained with a pulse height analyzer.

These signals are sent to charge amplifiers where they are amplified to be the input of position circuits, which then computes the individual coordinates, x and y, of the detections. These steps should be performed by digital circuits.

The performance of the scintillation camera is compromised by an intrinsic spatial resolution close to 3 mm, aggravated by the response of the collimator and by deterioration with the distance to the detector. These factors lead to poor spatial resolution compared with the morphological techniques. Resolution distances expected for the scintillation camera are of the order of 5 mm for most exams of deep organs.

Poor spatial resolution leads to the partial volume effect that prevents the collection of quantitative information about the intensity of objects whose size is less than twice the system resolution distance.

Planar scintigraphy produces 2D images of 3D objects. The planar images are impaired by the overlap of active and nonactive regions that limit the contrast and the functional accuracy of the information.

Radial acquisition around the object followed by reconstruction allows the acquisition of tomographic maps in NM [37].

Currently, most studies show only stationary distribution of activity, and the analysis of the resulting images is mainly qualitative.

The problem of quantification in SPECT is complex, because it is necessary to determine the distribution and concentration of an unknown tracer in an unknown set of absorbent tissues.

Quantification through the use of dynamic and quantitative functional imaging can, however, improve the quality of diagnosis with SPECT; but this objective is only approximately achieved in some studies.

Although conceptually the gamma camera has not undergone major changes over time, the same does not hold for the design, the detectors, and the materials used to make it. One of the most easily recognized changes is the number of heads that the current gamma cameras have. The gamma camera quickly evolved to increase the number of heads from one to two and then to three. This development meant an improvement in sensitivity, which, in turn, led to a reduction of the dose of radioactivity administered and/or the time taken for the examination (Figure 5.19).

Moreover, in recent years, new devices have emerged:

1. Cameras with multiple small crystals. PMTs are used in a new design.
2. Cameras with photodiodes. Silicon diodes are used instead of PMTs. CsI scintillators are used, as their emission spectrum fits the photodiodes better.
3. Compton effect cameras.
4. Cameras with CZT detectors; this is a semiconductor detector capable of operating at room temperature.

5.3.1.1 Collimators

The collimator is a device that is placed between the crystal and the object of study and consists of a gamma radiation absorbing material with thousands of holes distributed in a network [38]. This makes it possible to select certain directions of the incident photons and to filter scattered rays that are not suitable for the formation of the image. Despite making the technique less

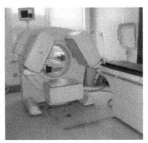

FIGURE 5.19
Gamma cameras.

efficient by reducing the counting rate, collimators are essential for the formation of the image, as the gamma photons cannot be refracted; and, therefore, lenses cannot be used, as they can with photons in the visible range.

The walls that confine the holes are called the *collimator septa*. The thickness of the collimator, the diameter of holes, and the septa thickness relate directly to the properties of the image. The dimensions of the collimator are the same as the crystal and it can be square or circular in shape. The material generally used for collimators is lead, because it combines a high density with a high atomic number. A high-density material is required, because it is directly related to the attenuation coefficient, which must be high to ensure gamma photon absorption. Further, the atomic number should be high to ensure that the fraction of interactions by photoelectric effect is greater than the fraction of interactions by Compton and Rayleigh–Thomson scattering. We recall here that in the NM energy range (80–511 keV), the fraction of the photoelectric effect increases with atomic number. This condition is essential to image quality since the secondary photons, produced by the Compton effect or by elastic scattering, will degrade the image.

Other materials such as tungsten, gold, or tantalum exhibit the required characteristics for their use in collimators; however, the high unit and/or manufacturing cost of the collimator make the use of these materials impracticable.

Collimators differ mainly in their geometry. The usual distinctions are between parallel-hole, convergent, divergent, and pinhole collimators. The holes in the collimators may also differ in shape with the three basic ones being square, triangular, or hexagonal (Figure 5.20).

The parallel-hole collimator is the most widely used, in which is the arrangement of holes is perpendicular to the surface of the crystal. In this case, the image has the same size as the object, regardless of distance from the collimator to the subject. This characteristic is intrinsic to the orthogonal projection that is provided (Figure 5.21).

When the organ under study is smaller than the crystal of the gamma camera, a converging collimator can be used, which can use all the surface of the crystal to build the image. In this case, the geometry is not orthogonal,

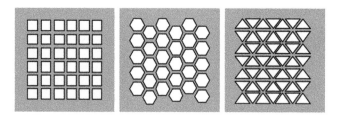

FIGURE 5.20
Various types of hole arrays in collimators.

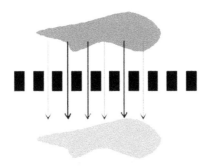

FIGURE 5.21
Schematic of a parallel-hole collimator. The size of the object is preserved.

causing a magnification that depends on the distance of the collimator to the object (Figure 5.22).

The diverging collimator is the inverted version of the converging collimator and is used when attempting to increase the camera field of view. The image obtained is smaller than the real object (Figure 5.23).

The pinhole collimator is based on the camera obscura in which light rays pass through a small opening and are projected to form an inverted image.

Since the rays have to pass through a small hole, only a small percentage of the emitted photons can reach the crystal; and, consequently, one of the major disadvantages of the pinhole is its reduced sensitivity. Another drawback is the distortion that occurs as a result of the finite size of the hole. However, the possibility of magnifying and improving spatial resolution makes the pinhole the preferred choice for studies of small organs (Figure 5.24).

The use of collimators has direct implications for both the resolution and sensitivity of the system. These factors can be analyzed based on geometric arguments. *Resolution* is defined as the ability to discriminate two points close together and is given by the FWHM of the response PSF (Figure 5.25).

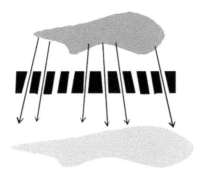

FIGURE 5.22
Schematic of a converging collimator.

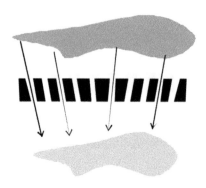

FIGURE 5.23
Schematic of a diverging collimator.

Consider a point source at a distance, f, from a gamma camera with a parallel-hole collimator with septa thickness s and hole diameter h. Let ℓ represent the height of the septa and c the distance of the collimator to the scintillator crystal. The angle α is defined as the angle between a ray perpendicular to the crystal and the oblique ray of maximum inclination that still reaches the crystal. Taking into account the geometrical aspects considered and using the properties of similar triangles, it is deduced that

$$d = \frac{h}{\ell}(c + \ell + f). \tag{5.37}$$

The expression shows that the characteristics of the collimator influence the spatial resolution. Given that, generally, the distance from the source to the collimator f is greater than the height of the septa, ℓ, and the distance

FIGURE 5.24
Schematic of a pinhole collimator.

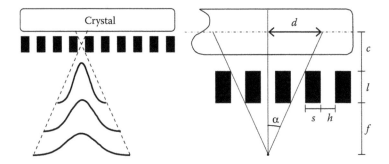

FIGURE 5.25
Variation of spatial resolution with depth. Geometric relations associated with the resolution of a collimator.

from the collimator to the crystal c the factors that determine resolution are the distance from the source to the collimator and the relation between the septa thickness and the hole diameter ($d \approx hf/\ell$). Thus, the smaller the holes (h) relative to the height of the septa (ℓ), the lower the resolution distance d. That is, the greater will be the power to discriminate two point sources that are very close. Another conclusion to be drawn from Equation 5.36, which is also shown in Figure 5.25, is that the spatial resolution varies with the distance from the source to the collimator. The further the collimator is from the sources, the worse the resolution will be.

Another aspect that must be taken into account is the geometric efficiency of the collimator, which is defined as the fraction of the number of photons isotropically emitted and the number of photons effectively collimated [39]. Assuming an infinite plane of activity in the air, the detector response is invariant with regard to the distance to the plane; so the geometric efficiency, Ω_{ef}, is generally defined, assuming that the plane is adjacent to the collimator. Given the definition of geometric efficiency, we have

$$\Omega_{ef} = \frac{\text{Number of collimated photons}}{\text{Number of emitted photons}} = \frac{\text{Total hole area}}{\text{Total area}} \times \frac{\Omega}{4\pi}, \qquad (5.38)$$

where Ω is the solid angle in which a hole is seen from the center of the section in a plane in contact with the collimator. Thus, the geometric efficiency, Ω_{ef} is best described by

$$\Omega_{ef} = \bar{k}\frac{h^2}{(d+s)^2} \times \frac{h^2}{4\pi\ell_e} = k\left[\frac{h^2}{\ell_e(h+s)}\right]^2, \qquad (5.39)$$

where k is a factor that reflects the properties associated with the type of hole (hexagonal, triangular, or quadrangular), h is the size of the hole, s is the septum thickness, and ℓ_e is the effective thickness of the collimator.

5.3.1.2   Data Acquisition

In gamma scintigraphy, a simple 2D projection is formed perpendicular to a given direction, keeping both the gamma camera and the patient at rest during data acquisition (Figure 5.26a). The result is an image similar to a radiographic projection.

In SPECT, one or more gamma cameras are rotated around the patient to acquire a large number of projections along different directions (Figure 5.26b). These projections are combined using standard tomographic reconstruction algorithms to build a 3D volume representing the radiotracer's activity. This volume may then be digitally manipulated to produce 2D images of the activity distribution along any chosen direction. Current SPECT scanner technology uses two or three gamma cameras, known as detector heads, which rotate around the patient in circular or elliptical trajectories, known as orbits [40]. Decay events are accumulated at fixed points of the orbit, usually 3° or 6° apart. As a rule, the heads complete a full 360° rotation to minimize geometric distortions in the reconstructed images; such distortions may arise from the dependence of the spatial resolution of each head on the distance from the radioactive source to the head's collimator (Figure 5.26c).

5.3.1.3   Data Storage

5.3.1.3.1   Data Formats: List-Mode and Histograms

The mapping of each gamma photon detected in the active volume of the scanner allows the raw data in these nuclear techniques to be compiled in a sequential list of detection events, where each entry stores a certain set of

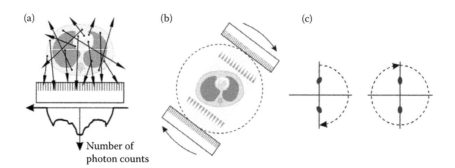

FIGURE 5.26
(a) Production of a 2D projection in gamma scintigraphy along a given direction. The gamma camera and the patient remain still during data acquisition. (b) Production of multiple 2D projections in a SPECT scan with a dual-head scanner. The projections are taken at fixed positions along the orbit of the heads as these rotate around the patient. (c) Typical geometric distortions generated in the reconstructed images of two point-like sources placed symmetrically relative to the detector axis, for a 180° orbit (left) and a 360° orbit (right) in a single-head SPECT scanner.

relevant information about the event. This information includes the position at which the photon interacted with the scintillation detector or the energy deposited by the photon. This data format, known as list mode [41], is, however, generally inappropriate for image reconstruction; and the histogram data format is usually preferred. The histogram, a 2D array of integers, groups the recorded events according to the location where the photon was detected [42]. The active volume of the scanner is therefore, discretized into channels or bins, and the total number of photons detected in each bin is counted during the exam. For gamma scintigraphy scans, the correspondence between array entries and scanner bins is chosen so that a unique histogram coincides with the 2D projection of the activity distribution along the direction of the collimator channels [40].

In SPECT imaging, the detection events are stored in several histograms, known as *sinograms*. Contrary to what one would assume, sinograms are not the 2D projections of the activity distribution along the acquisition positions of the orbit. Instead, they are assembled in a way that turns out to be more suitable for the image reconstruction algorithms [43].

Most SPECT scanners record the raw event data in list-mode format, converting it later to histogram format. This conversion is easily performed once the discretization of the scanner's active volume is known relative to the location of the heads. Although the histogram is better suited to image reconstruction, list-mode data have a larger information content which is often important for the correction of physical effects that decrease image quality, such as scattered radiation [44]. Further, it is the most flexible format, as, unlike histogram data, list-mode data do not depend on the particular discretization resulting from dividing up the scanner's active area into channels.

5.3.1.3.2 SPECT Sinograms

The histogram data structure used in SPECT is the sinogram, which consists of a 2D array with the photon counts along all possible directions in a plane perpendicular to the rotation axis of the scanner [43]. The two indexes of each sinogram entry identify the spatial orientation of the detected photons stored in that entry, according to the radial coordinate x_r and the azimuthal angle, ϕ, shown in Figure 5.27. These two coordinates are defined using a reference frame OX_rY_r linked to one of the scanner's heads, obtained from the negative rotation of the stationary frame OXY fixed to the scanner by an angle $\phi \in [0; 2\pi]$. In the stationary reference frame OXY, the origin O is located in the scanner's axis, and OX and OY are, respectively, the horizontal and vertical directions (Figure 5.27a and b). An additional axis OZ allows an axial coordinate, z, to be defined for each sinogram, locating it along the scanner's axis (Figure 5.27b). The set of sinograms acquired for a given object in all the planes is compiled in a 3D array $s(x_r, \phi, z)$. It should be noted that the number of elements in each dimension of the sinogram depends on the discretization of the scanner's heads (dimension corresponding to x_r) as well as on the number of acquisition angles used (dimension corresponding to ϕ).

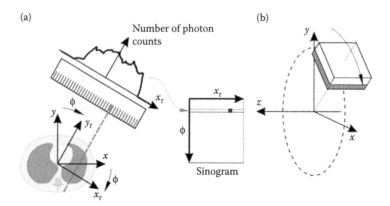

FIGURE 5.27
(a) Definition of the x_r and ϕ coordinates for a photon emission direction (dashed line) in the OX_rY_r reference frame linked to one of the heads, and corresponding location in the sinogram. (b) Orientation of the stationary reference frame XYZ. The XY, YZ, and XZ planes are usually referred to as the transaxial, sagittal, and coronal planes.

If the scanner has more than one head, the number of counts along a direction labeled by a set of x_r, ϕ, and z values is the sum of all counts detected in each head for that direction.

According to its definition, each row of the sinogram groups photon emission directions that are all parallel and defines an angle ϕ with the horizontal; whereas each column groups different directions, but all with the same radial coordinate x_r, that is, which are tangent to a circle of radius x_r centered in O (Figure 5.28). Mathematically, a sinogram is the 2D radon transform of an object [45] and is so called, because a point-like source of photons always gives rise to a sinusoidal pattern in its sinogram [43] (Figure 5.28).

The way photon counts are organized in a sinogram is particularly appropriate for image reconstruction algorithms, as each entry stores the number of photons emitted in a well-defined direction, which, in the absence of effects such as Compton scattering or self-absorption, is proportional to the total activity of the radiotracer along that direction. Each row of the sinogram is, thus, the set of parallel projections of the activity (or, more simply, the projection of the activity) of the imaged object along the direction ϕ in the plane z, $p(x_r, \phi, z)$ [43]:

$$p(x_r, \phi, z) = s(x_r, \phi, z). \tag{5.40}$$

The set of activity projections along all possible orientations in a given z plane is the information used to reconstruct the image in that plane. The final result of typical image reconstruction algorithms is an image volume consisting of all reconstructed planes stacked together.

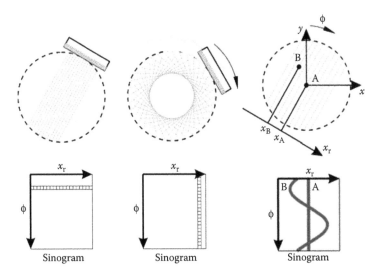

FIGURE 5.28
(a) Representation of all photon emission directions involved in the definition of a sinogram row (left) and column (right). (b) Sinogram originated from a point-like source located in the scanner's axis (A) and away from that axis (B). As a function of ϕ, the sinogram describes a complete sine wave for a rotation of 360°, with an amplitude equal to the source-axis distance.

5.3.1.4 Preprocessing of Data

The reconstruction of a SPECT image assumes that the number of photon counts in each emission direction is proportional to the integrated activity of the object over that direction. There are, however, several factors that disturb the counting and change that proportionality. These are mainly physical effects, such as the occurrence of Compton scattering in the patient, which leads to disperse radiation or photon flux attenuation due to interaction between the emitted photons and the patient's body. There are different strategies for the correction of those effects, most of which pre-process the acquired data before image reconstruction. This preprocessing, known as data correction, is a key factor in ensuring that the reconstructed image is a reliable estimate of the activity distribution of the object being studied. The main effects corrected and methods used are addressed in Chapter 6.

5.3.1.5 Image Reconstruction

The SPECT imaging is a tomographic technique that generates slices of the object, showing the radiotracer distribution in the patient's volume. Reconstructing the image from the number of photon counts along different spatial directions starts presuming that, in the absence of attenuation and Compton scattering, the number of photons detected in a given direction is, as previously mentioned, proportional to the integral of the activity distribution

$f(x, y, z)$ along that direction, that is,

$$\sum_{\text{direction } i} (\gamma \text{ counts}) \propto \int_{\text{direction } i} f(x, y, z) \, dr_{\text{direction } i}. \qquad (5.41)$$

Using the notation introduced in the previous section (see, e.g., Figure 5.27), this line-integral approximation [42] can be expressed in terms of the parallel projections of the object's activity,

$$p(x_r, \phi, z) = c \int f(x, y, z) \, dy_r, \qquad (5.42)$$

where the constant c is usually ignored.

The image reconstruction algorithms assume that the data are in a parallel-projection format given by sinograms, the creation of which is, therefore, mandatory for the reconstruction step. Image reconstruction approaches can be either analytical or iterative [41], depending on whether they construct an estimate of the activity distribution by analytically inverting Equation 5.41 or by trying to find the activity distribution that best matches the measured projections according to pre-established criteria, using methods based on successive approximations. Some of the main reconstruction algorithms are discussed in Chapter 6.

5.3.2   PET

5.3.2.1   Physical Principles and Limits

In a PET scanner, the photons produced in annihilation events occurring within the patient leave its body and are individually detected by a set of discrete detectors, usually arranged in a circular ring around the patient. If two of those photons are detected within a very short space of time, the system's time window typically being a few nanoseconds, it is assumed that the pair has been detected in coincidence, that is, that both photons originated from the same annihilation event. Knowing the location of the triggered detectors and taking into account that each annihilation gives rise to two photons emitted in opposite directions, it is possible to trace the direction of emission of the two detected photons with the line joining the detectors and to define the LOR of the coincidence event (Figure 5.29).

If the photon detectors are particularly fast and the signal they produce is short enough, of the order of hundreds of picoseconds, a portion of the LOR where the annihilation took place can be identified by measuring the time difference between the instants when the two photons are detected. This technique is known as TOF-PET. Only a few scanners currently in operation have TOF capabilities, although it is likely that TOF-PET scanners will be dominant very soon. This will increase sensitivity, which is greater the

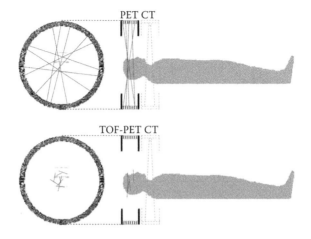

FIGURE 5.29
Typical acquisition geometry of a whole-body PET exam. Top: conventional PET, in which lines of response join pairs of detectors. Below: TOF-PET, where the annihilation region is reduced to a portion of the triggered LOR. The faster the detectors used, the smaller is the portion, and the annihilation site becomes better defined. The location of the annihilation is assumed to be described by a Gaussian probability distribution along that portion, contrary to conventional PET, where it is assumed that the annihilation event has equal probability of having been generated over the full extent of the LOR.

shorter the LOR portion within which the annihilation event can be located. It should be noted that no increased spatial resolution of TOF-PET scanners is foreseen, as detector development does not indicate that we shall be having very fast detectors in the near future, which would be capable of pinpointing the annihilation site within the LOR with the necessary accuracy.

The PET scanner, thus, counts the number of 511 keV photon pairs in coincidence for each LOR defined in the system (Figure 5.30); since the direction of the emitted pair of photons in a given annihilation event is random, it is possible to create parallel projections of the activity distribution in the patient by organizing the counts in sets of parallel LORs. We, thus, get position information about the annihilation location without having to use physical collimation, contrary to what happens with SPECT, where a collimator is essential to the formation of parallel projections, as noted in Section 5.1. This feature of PET is known as *electronic collimation*, and it gives PET an important advantage over other NM techniques, because it dispenses the need for a physical collimator. Hence, PET has the potential to accept pairs of photons emitted in any direction, using all the decay events occurring in the patient and not just those that lead to photons being emitted along the narrow solid angle accepted by a physical collimator. The PET is, therefore, considerably more sensitive compared with all the other NM techniques; it requires the administration of a smaller amount of radiotracer to the patient for the same image quality, which reduces the risk of harming the patient's health and the cost of the exam.

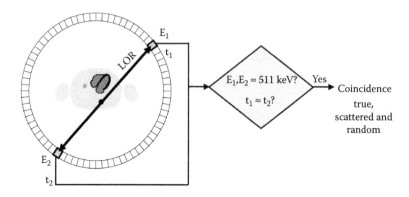

FIGURE 5.30
If two photons are detected in coincidence, it is assumed that they originated in the same anni-hilation event. The line that joins the two detectors triggered in the coincidence is known as line-of-response (LOR) and corresponds to the direction of the pair of photons emitted in the annihilation event.

Although the spatial resolution of PET scanners is usually better than that achieved in SPECT, there are certain physical limits imposed on the resolution in PET that do not exist in SPECT. These limits arise from two independent aspects: the range of the positron before the annihilation and the fact that the emission directions of the two annihilation photons are not perfectly collinear. The combination of these two effects leads to spatial resolutions that, in prac-tice, are no better than about a tenth of a millimeter to several millimeters, depending on the radioisotope employed, the radius of the detector ring, and whether there are intense magnetic fields.

The average energy with which the positron is emitted in the beta decay depends on the particular radioisotope being used and determines the aver-age length of the path it will travel before being annihilated. The location of the annihilation event, thus, slightly differs from the decay site, leading to an intrinsic uncertainty about the position of the radiotracer; and this imposes a fundamental limit on the spatial resolution attained in a given PET scanner, with the chosen radioisotope.

The other physical limit on the spatial resolution arises from the fact that the positron–electron pair may have nonzero momentum at the time of anni-hilation. In this case, the angle between the directions of the two annihilation photons is slightly deviated from 180° (about ±0.25°) [2], which results in a small error while locating the position of the radiomarker. This degrades the spatial resolution, with degradation worsening the further the radiation detectors are from the patient.

The spatial resolution of an imaging system is a very important parameter, as it corresponds to the minimum distance between two points in the object that allows them to be observed as two distinct points in the image. This feature is parameterized by the FWHM of the system's PSF, obtained by the imaging of a point-like activity source. The full width at tenth maximum (FWTM) is often also used, if the PSF has a non-Gaussian shape.

The near future foresees these two physical effects as the main factors limiting the spatial resolution in a PET scanner, although at present spatial resolution is still dominated by aspects related to technical implementation that limit the resolution in commercially available systems to about 1 mm [46] for animal PET, about 2 mm in state-of-the-art whole-body PET scanners [47], and above 4 mm in most ordinary scanners. Today, the main factors governing the spatial resolution of a PET system are the size of the detectors, the preponderant factor until recently, and the image reconstruction process, which implies the use of filters to increase the SNR but that blur the image and decrease the ability of the system to distinguish between fast spatial changes of the activity. This parameter can be estimated in systems with detector blocks by an empirical equation proposed in [48], obtained from a comprehensive analysis of a large number of scanner models. The equation reads

$$\Gamma = 1.25\sqrt{\left(\frac{d}{2}\right)^2 + (0.0022D)^2 + r^2 + b^2}, \qquad (5.43)$$

where Γ is the spatial resolution of the reconstructed image, d is the width of a single crystal detector, D is the diameter of the detection ring (included to take into account noncollinear emission), r is the effective dimension of the point-like source (including the positron's path), and b is the uncertainty in the position of a photon detection event inside one detector block (which is larger when the scintillation light is shared among crystals and smaller when the light readout is performed by individual detectors). The constant factor of 1.25 affecting the whole value is due to the image reconstruction process. This equation supposes that there is no under-sampling, a factor which may also degrade resolution.

Spatial resolution also depends on the position of the activity source (although considerably less than in SPECT); it is generally worse on the periphery of the field-of-view (FOV) of the scanner, due to the curvature of the camera and the depth-of-interaction (DOI) effect. This has been the subject of recent technical developments aimed at its correction. Patient motion is yet another element that degrades image resolution; some current systems have optical setups that measure the external motion and use that information to correct the image. Internal motion such as breathing or heart movement is harder to correct, but several methods are being currently developed to address that problem.

Finally, there is the possibility of bypassing the resolution limit imposed by the positron's path before annihilation using high-intensity magnetic fields in the range of 5–10 T [49–51]. Nevertheless, the limit due to noncollinear photon emission still holds; so this application will be mostly important in PET scanners with small detector ring radii, that is, those used in animal imaging, where record-breaking values for spatial resolution will certainly be attained.

5.3.2.2 Data Acquisition

5.3.2.2.1 Types of Events

Although coincidence detection in a PET scanner always assumes that each detected coincidence adds useful information to the radiotracer's activity map, in reality not all of them have valid positional information regarding annihilation events. When an individual photon is detected, it is referred to as the detection of a "single." In spite of not being used in image formation, singles are easily the most basic events handled by the PET scanner, and they may be over two orders of magnitude more numerous than the number of coincidences. If the system detects three or more singles in the same time window, the event is called a *multiple coincidence* and is rejected, as it becomes impossible to ascribe an LOR to that event. It is only when the system detects two singles in the same time window that a coincidence is said to have happened; coincidences are classed as true (or nonscattered), scattered, and random (Figure 5.31).

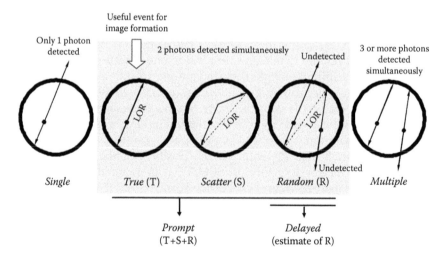

FIGURE 5.31

Types of events detected in PET. For true coincidences, the detection event is counted in the correct position (LOR); true coincidences are the only type of event of interest to noiseless image formation. For scattered coincidences and random coincidences, at least one of the two photons detected is not emitted along the LOR defined by the two detectors triggered (dashed line), and so incorrect spatial information is conveyed to the data, which is expressed as image noise. Besides being named according to the type of event, coincidences can be further classed as "prompts" and "delayed," respectively, for the total of two-photon coincidences detected by the system (=T+S+R) and for the estimate of random coincidences using the delay line method.

True coincidences are the only ones having valid position information, because they correspond to the detection of two photons that were actually emitted along the LOR of that coincidence. Scattered and random coincidences, on the other hand, impart counts to LORs with photons that were not originally emitted along the LOR in which they are added, and they solely contribute to the addition of noise in the final image.

In a scattered coincidence, at least one of the two emitted photons in the annihilation undergoes Compton scattering, an event that lowers the energy of the scattered photon and changes its direction. Using the triggered LOR leads to a wrong location for the annihilation site. In certain acquisition planes of a typical whole-body exam, the number of scattered coincidences can easily surpass the number of true coincidences.

For a random coincidence, the two photons detected arise from two distinct, but almost simultaneous, annihilations; and the corresponding photon pairs are not detected by the system. The number of these events can also be larger than the number of true coincidences, especially if the activity in the FOV is high and the detectors are slow, which forces the use of a wider time window. The probability of detecting a random coincidence in a given LOR increases with the singles count rates R_{s1} and R_{s2} in the two individual detectors defining that LOR and also with the length of the system's time window. It is easy to show that the random coincidences rate in that LOR is given by

$$R = 2\Delta t R_{s1} R_{s2}, \tag{5.44}$$

as long as the singles count rates are much larger than the coincidences count rate, and any dead time effects may be ignored. If one assumes that $R_{s1} \cong R_{s2} \cong R_s$ (individual detectors as similar as possible with approximately the same flux of coincidence photons in each detector), Equation 5.44 becomes

$$R = 2\Delta t R_s^2. \tag{5.45}$$

5.3.2.2.2 Acquisition Geometry

Most PET scanners are composed of thousands of small scintillation crystals arranged around the region where the patient lies, the FOV. The layout of the crystals, each of which is an individual photon detector, is usually cylinder shaped, which allows coverage of a large solid angle of detection, because the pairs of annihilation photons are emitted in just about all spatial directions. In whole-body PET systems (Figure 5.29), the gantry can move the patient inside the FOV along the axial direction and perform sequential acquisitions to build up the full image of the patient's body.

The individual crystals are normally grouped in detector blocks, that is, of 8×8 crystals coupled to four photomultipliers. The crystals belonging to a detector block are usually produced from one large single crystal, which is segmented by longitudinal cuts of different depths that are subsequently filled with a reflective material. Detector blocks can also be grouped in larger structures called modules (typically with 4 blocks/module, although this number varies considerably), and the whole set constitutes the full detector. This organization is very efficient from the point of view of the electronics needed to detect photons and determine coincidences, but it is not useful when it comes to organizing the data in sinograms and image reconstruction; for that purpose, it is better to consider that the individual scintillation crystals are grouped in rings lined up along the scanner's axis.

In the early twenty-first century, commercial PET systems began to include a CT subsystem with the intention of combining anatomic and functional information. This had a considerable impact in the world of clinical diagnosis and led to the fast spreading of PET or CT scanners. Nowadays, almost all commercially available systems are PET or CT scanners; not only do these combine the anatomic information offered by CT with the functional information conveyed by PET, but they also offer better quality attenuation correction (less noise and better spatial resolution) by constructing attenuation maps from CT information, duly converted to PET photon energies. However, this change in the attenuation correction has also raised new problems, namely, artifacts due to misalignments between the PET and CT images. Multimodal systems such as PET or CT are further described in Section 5.3.3.

5.3.2.2.3   *Acquisition Modes*

The principle of image formation in PET is based on counting the number of true coincidences detected in each LOR. This LOR by LOR reckoning, together with the extremely large number of crystals in current PET systems, generates several tens of millions of distinct LORs and leads to volumes of data gathered in a PET exam, which can easily reach several hundreds of MB of information. In PET, data can be collected using different acquisition philosophies, corresponding to the 2D and 3D modes described below, and organized (or not) in structures that may or may not be directly used in image reconstruction methods.

Until the end of the 1980s, PET systems were solely operated in an acquisition mode known as 2D mode, in which the scanner's rings are separated by physical septa of a lead–tungsten alloy, projected by about 10 cm into the camera [52]. This mode is similar to data acquisition in SPECT, as the septa play the same role in providing physical collimation, similar to what the collimators of gamma cameras do. In PET, the septa, besides limiting data acquisition and the corresponding image reconstruction to sets of 2D planes stacked along the scanner axis, also protect each ring from photons scattered off-ring and eliminate the need to correct the data from scattered coincidences. In 1988, pioneering work developed at London's Hammersmith Hospital showed that the sensitivity could be substantially increased if the septa were removed (see [53,54], and references therein). Since then, all commercial PET systems started offering the possibility of executing exams with no septa, in a new acquisition mode, termed 3D mode.

The greatest sensitivity to true coincidences of the 3D mode, typically 5 times better than that attained in the 2D mode [53,54] (Figure 5.32a), is due to the fact that acquisition is performed over a larger solid angle, and information from more LORs is used (Figure 5.32b). In fact, the 2D mode does not use the full potential of electronic collimation, because the septa introduce physical collimation and limit the acquisition to LORs between crystals residing in the same or adjacent rings, leaving out steeper LORs.

By being more sensitive to true coincidences, the 3D mode has the advantage of allowing the amount of radiotracer given to the patient to be reduced (or, using the same amount, the acquisition time can be reduced, or a better image quality is provided). Nevertheless, detecting coincidences over a wider solid angle by removing the septa also improves the sensitivity to scattered and random coincidences [53] (Figure 5.33). The enhanced sensitivity to scattered coincidences means that the data from this type of background have to be corrected, which often involves sophisticated models of correction. Additionally, a greater number of random coincidences drives the system nearer to saturation, requiring the use of detectors and electronics with a very short dead time.

5.3.2.3 Data Storage

5.3.2.3.1 Data Formats

As happens in SPECT, events detected in a PET system may be stored using either the list-mode format or a histogram format. In the first format, the data are recorded as a sequential list of coincidence events, where in each entry a

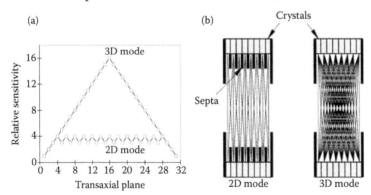

FIGURE 5.32
(a) Estimate of the sensitivity to true coincidences in the 2D and 3D modes calculated from the number of LORs that intersect each of the transaxial planes possible to define in a 16-ring scanner. The sensitivity in the 3D mode changes linearly along the camera axis, being maximal in the centre of the axial FOV; for the 2D mode, this quantity is approximately constant over the axial FOV. (b) Axial view of the LORs used in a PET scanner in 2D and 3D modes. The absence of septa in the 3D mode makes it possible to use considerably more LORs, with a subsequent increase in sensitivity.

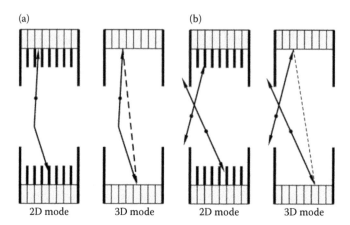

FIGURE 5.33
(a) Effect of removing the septa on the sensitivity to (a) scattered and (b) random coincidences. The angular acceptance of the crystals is larger in the 3D mode, which increases the sensitivity to true coincidences and scattered and random coincidences.

set of items of information about the coincidence is written. This information includes the indexes of the two crystals triggered (in the numbering scheme defined for the detector), the energies of the two detected photons, or the instant when the coincidence was detected. In the second format, the data are collected in a multidimensional array which maps all the LORs that can be defined in the system, one per entry of the array; the integer number stored in each element of the array coincides with the total number of coincidences detected in the corresponding LOR.

Data acquisition in PET systems, just as in modern SPECT scanners, is normally performed in list-mode format and then converted to a histogram format. The conversion tools use knowledge about the physical arrangement of the detector, that is, the positions of the crystals. The list-mode format also has greater potential for data correction in PET and depends neither on the particular choice of "possible" LORs done for a particular system nor on the spatial discretization imposed by those LORs. In addition, it is, in many situations, the most compact type of format for the storage of 3D mode data.

5.3.2.3.2   2D Sinograms

As happens in SPECT, the basic histogram structure used for data acquired in the 2D mode is typically the sinogram, which is now a 2D array of the coincidences recorded in LORs belonging to a specific plane. The plane is simultaneously perpendicular to the scanner axis and parallel to the detector rings. The two indexes of a sinogram's entry define the spatial orientation in that plane of the LOR corresponding to the entry, according to the LOR's radial and azimuthal coordinates x_r and ϕ (Figure 5.34a). In addition the definition of these coordinates underlies the existence of a OX_rY_r reference frame rotated

(a)   (b)

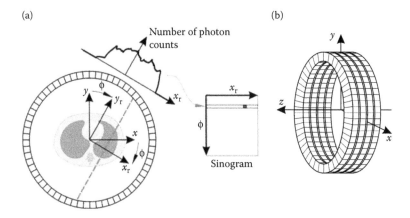

FIGURE 5.34
(a) Definition of the coordinates x_r and ϕ of an LOR (dashed line) relative to the OX_rY_r reference frame (obtained by rotating the OXY frame fixed to the scanner by an angle ϕ), and corresponding location on the 2D sinogram. (b) Orientation of the XYZ reference frame fixed to the scanner. The XY, YZ, and XZ planes are again the transaxial, sagittal, and coronal planes, respectively.

relatively to the scanner by a clockwise angle $\phi \in [0; \pi]$. The axial coordinate z of each sinogram is defined by the position of the plane along the OZ axis coinciding with the scanner axis (Figure 5.34b). In PET, the 2D sinograms of an object relative to all 2D acquisition planes are also grouped in a 3D array $s_{2D}(x_r, \phi, z)$. It should be noted that in PET the azimuthal coordinate varies in the range 0° to 180° due to the symmetry in the emission of two annihilation photons along opposing directions; whereas in SPECT the rotation angles are in the range 0° to 360°, because there is only one photon emitted per radiomarker decay.

Each row in a 2D sinogram groups parallel LORs that form an angle, ϕ, with the horizontal direction, and each column collects LORs with the same radial coordinate x_r. The layout of rows and columns in a 2D sinogram is identical to that shown in Figure 5.28 for SPECT, with the difference that the azimuthal coordinate runs in the interval $[0; \pi]$. The range of radial positions for each row of the sinogram covers the FOV's diameter (twice the FOV's radius, i.e., $2 \times R_{FOV}$), which is normally less than the size of the crystal rings, as the LORs farther from the FOV's centre, which have a low information content and need specific processing, are usually ignored. The 2D sinogram in PET is also well suited to the usual image reconstruction algorithms, as each element represents the number of coincidences detected in the corresponding LOR, and the proportionality between that number and the integrated activity of the radiomarker over the LOR's direction still holds. Each row of the sinogram is again the projection of the object's activity along the direction ϕ in the plane z, $p(x_r, \phi, z)$, that is,

$$p(x_r, \phi, z) = s_{2D}(x_r, \phi, z), \tag{5.46}$$

a projection that is employed to reconstruct the image in that plane. As we shall see later, in PET the total number of reconstructed planes in the 2D mode is $2N - 1$, where N is the total number of rings in the scanner.

5.3.2.3.3 3D Sinograms

The 3D mode also uses sinograms for storing data in histogram format but which include the counting of coincidences between crystals in distant rings. Thus, besides all the planes defined in the 2D mode parallel to the rings, the 3D mode also includes sinograms corresponding to planes oblique to the rings. Each of these oblique sinograms stores the coincidences in LORs defined between two different rings (Figure 5.35a); so, in a system of N rings, a maximum of N^2 sinograms can be defined for the 3D mode, whereas only $2N - 1$ exist in the 2D mode. In the 3D mode, it is, therefore, necessary to specify for a given plane not only its position along the detector's OZ axis but also its inclination relative to that axis. The latter is usually given in terms of the ring difference between the two rings in coincidence, $\Delta r = r_2 - r_1$, where r is an integer that positions each ring along the detector axis (Figure 5.35a). The axial coordinate z of the plane is the average of the positions of the two rings,

$$z = \frac{r_1 + r_2}{2} \Delta d_Z, \tag{5.47}$$

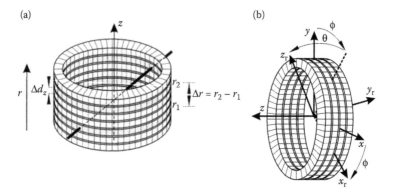

FIGURE 5.35
(a) Sketch of the rings involved in the definition of an oblique sinogram, and corresponding definition of the values for Δr and z of that sinogram. (b) Definition of the $OX_r Y_r Z_r$ reference frame obtained by the rotation of the $OXYZ$ frame fixed to the detector by the azimuthal and co-polar angles ϕ and θ. It should be noted that in the 2D mode the direction of each LOR with azimuthal coordinate ϕ is parallel to the axis OY_r (see Figure 5.34a), whereas the presence of the co-polar coordinate θ in the 3D mode forces the direction of an LOR with angular coordinates (ϕ, θ) to be parallel to the OZ_r axis.

where Δd_Z is the width of each ring. As in the 2D mode, the 3D sinograms of an object are grouped in an array $s_{3D}(x_r, \phi, z, \Delta r)$, now four-dimensional, which collects all the oblique and nonoblique sinograms. The organization of this array as a function of Δr and z will be addressed later.

In 3D sinograms, the definition of the coordinates x_r and ϕ is now performed using a reference frame $OX_rY_rZ_r$ rotated relative to the $OXYZ$ frame by the azimuthal angle, $\phi \in [0, \pi]$, and the copolar angle, $\theta \in [0, \pi]$ (Figure 5.35b). The copolar angle θ is a way of representing the inclination of the plane equivalent to using Δr, as the LORs located in the central radial positions for all the azimuthal angles (which constitute the central column of the sinogram) make an angle of $\pi/2 - \theta$ with the axis of the detector. For a sinogram with inclination Δr and axial coordinate z, each row groups LORs that make exactly the same angle, ϕ, with the vertical axis OY, measured in the OXY plane (Figure 5.35b). Those LORs are not all strictly parallel, and, except for the central LOR, they form an angle with the detector axis that is slightly lower than $\pi/2 - \theta$ (i.e., are more slanted than the central LOR).

Since each sinogram acquired in the 3D mode does not directly define a plane, owing to this slight nonparallelism, a problem that does not arise in the 2D mode, 3D mode data are often immediately transformed into parallel projections of the activity of the object being observed [42]. In this mode, parallel projections are 2D, generated in the plane OX_rY_r defined by the angular coordinates ϕ and θ (Figure 5.36), and are a function of (x_r, y_r). The mapping between all the 3D sinograms $s_{3D}(x_r, \phi, z, \Delta r)$ and the parallel projections

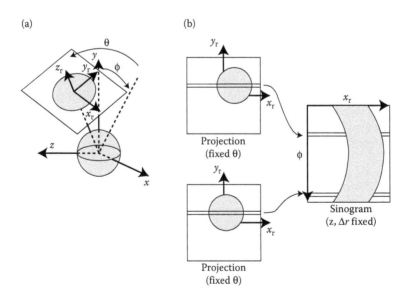

FIGURE 5.36
(a) Parallel projection of an object in the plane OX_rY_r along the angles (ϕ, θ). (b) Relation between parallel projections and the rows of an oblique sinogram.

$p_{3D}(x_r, y_r, \phi, \theta)$ is performed by

$$p_{3D}(x_r, y_r, \phi, \theta) = s_{3D}(x_r, \phi, z, \Delta r), \tag{5.48}$$

where the coordinates (x_r, y_r, ϕ, θ) of the projections are given in terms of the sinogram coordinates $(x_r, \phi, z, \Delta r)$ by

$$\begin{cases} x_r = x_r \\ y_r = -z \cos \theta \\ \phi = \phi \\ \tan \theta = \dfrac{\Delta r \times \Delta d_Z}{2\sqrt{R_D^2 - x_r^2}} \end{cases}, \tag{5.49}$$

with R_D being the radius of the detector rings.

If the radius, R_{FOV}, of the FOV and the length of the detector axis are small when compared with R_D, then it is possible to ignore the nonparallelism in an oblique sinogram and assume that the LORs grouped in a sinogram's row corresponding to an azimuthal angle, ϕ, are all parallel and make the same angle $\theta = \arctan(\Delta r \times \Delta d_Z / 2R_D)$ with the OXY plane. With this approximation, there is a direct correspondence between that sinogram row and the row with coordinate $y_r = -z \cos \theta$ in the projection array along the direction (ϕ, θ); the projection along (ϕ, θ) corresponds, thus, to the rows of coordinate ϕ of all the sinograms with different z coordinates that define an angle $\pi/2 - \theta$ with the detector axis. If that approximation is not valid, for each value of the radial coordinate x_r it becomes necessary to interpolate the count values stored in the corresponding columns in sinograms with different inclinations to obtain the parallel projections. The reconstruction of the final image for the whole volume of the patient in the 3D mode uses all the projections $p_{3D}(x_r, y_r, \phi, \theta)$ simultaneously, which means that the reconstruction algorithms in the 3D mode operate in a four-dimensional space and involve a considerably larger degree of numerical sophistication than the algorithms operating in the 2D space used for the 2D mode, as will be seen later.

5.3.2.4 Preprocessing of Data

As in the case of SPECT, image reconstruction in PET assumes that the coincidence counts in each LOR are proportional to the integrated activity of the object along the direction of the LOR, and similarly there are several physical effects which modify that proportionality relation. The perturbation introduced by those effects is different from LOR to LOR, and obtaining a reconstructed image that is as near as possible to the true distribution of the activity in the patient means there must be a preprocessing step which corrects the data from those effects. Attenuation of the annihilation photons and scattered radiation (which leads to scattered coincidences) also involves physical effects that are found in PET, and they considerably affect the image. There are still more effects specific to PET, such as the existence of random coincidences or instrument-related effects which arise from the nonuniformity of the detection efficiency between distinct crystals, or the fact that the system has LORs with slightly different geometrical conditions. All the physical effects which alter a PET exam can, in principle, be corrected, allowing a quantitative use of this technique, that is, that images showing absolute values for the activity concentration, in, for example, Bq/cm^3, are produced. In SPECT, quantitative measurements are much harder to obtain, because it is also considerably more difficult to properly correct attenuation effects.

5.3.2.4.1 Data Reduction

Besides data correction, image reconstruction procedures in PET are usually preceded by another preprocessing task, one that aims at reducing the number of LORs used in the reconstruction. This task consists of summing coincidence counts of adjacent LORs, a process that is known as *mashing* [42,55]. Mashing techniques are used in most current scanners due to the huge number of LORs that these systems possess, a number which results in low counting statistics in individual LORs and creates difficulties in data processing because of the large size of the data structures that have to be handled. The sampling of the radial, azimuthal, copolar, and axial coordinates and the way LOR mashing is accomplished are described next.

5.3.2.4.2 Radial Sampling: Interleaving

The detection of coincidence in a PET system, in which each LOR samples a prism-like volume of the FOV, implies the discretization of the whole FOV's volume. The discretization process determines a maximum spatial resolution for the system, which depends on the properties of the detector; for instance, the sampling of the radial coordinate x_r depends directly on the size of the crystals, whereas the sampling of the azimuthal angle, ϕ, depends on the number of crystals per ring.

The circular geometry of the detection system in PET entails a slight spatial oversampling of the transaxial FOV's central region relative to its peripheral regions. Spatial sampling is normally further improved by a strategy known as interleaving [42], which increases radial sampling by reducing the angular sampling in a plane. In this process, the LORs of two consecutive ϕ angles are alternately combined in a single row of the sinogram, as shown in Figure 5.37. Supposing that the number of crystals per ring is even and equal to a value N_c, in the 2D mode, those LORs of consecutive angles define two sets of $N_c/2$ LORs that are shifted apart by a half-radial position. By combining these two sets in a single row of the sinogram, the sampling frequency of the radial coordinate is doubled, at the cost of halving the azimuthal sampling frequency and introducing a minor error in the angular position, ϕ, of the coincidence counts recorded in each radial position, x_r.

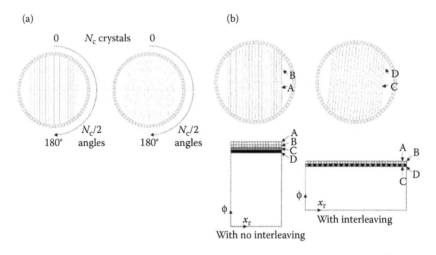

FIGURE 5.37
(a) Diagram of the two sets of LORs shifted by a half radial position combined in the interleaving process (N_c even, 2D mode). (b) The way the interleaving process combines those shifted sets to build the interleaved sinogram rows. The spatial sampling frequency doubles at the cost of halving the angular sampling.

In addition to the interleaving, the radial sampling is also usually restricted to an interval smaller than the transverse dimension of the detector. The FOV is, therefore, smaller than the volume defined by the rings, but it avoids practical problems related to the nonuniform spatial sampling of peripheral LORs and to the low count rates they exhibit. There is rarely any activity at the periphery, because the patient is quite a distance from the surface of the rings, and even if there were any activity the marked inclination of the peripheral LORs relative to the crystal surfaces would reduce the solid angle of acquisition.

5.3.2.4.3 *Angular Sampling: Angular Compression*

The frequency of the angular sampling in a PET system, which depends on the total number of crystals in each ring, also influences the maximum spatial resolution of the system. In most of today's PET scanners, far more crystals are needed to ensure good spatial resolution along the radial coordinate than are necessary to define good spatial resolution along the azimuthal coordinate, especially in the center of the transaxial FOV. Therefore, the reduction of the angular sampling frequency which results from the interleaving process does not compromise spatial resolution, and a further reduction is even achieved by what is called angular compression (or angular mashing) [42], that is, the postinterleaving sum of coincidence counts of LORs and consecutive ϕ values. This procedure is equivalent to adding sinogram rows in groups of two-by-two, four-by-four, etc. (sets of powers of 2), dividing the number of sampled angles by a factor of two, four, etc. (Figure 5.38). Angular compression has

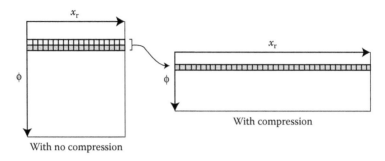

FIGURE 5.38
Diagram of angular compression procedure. In the sinogram, rows are added according to the angular compression factor, and a new sinogram is constructed with a single row per added set corresponding to the average angular position of the set.

the advantage of lowering both storage and data processing needs without significantly compromising the spatial resolution.

5.3.2.4.4 *Axial Sampling in the 2D Mode: 2D Mashing*

In the 2D mode, the mashing of LORs along the axial direction is accomplished by summing the coincidences between crystals located in the same ring k, with those reported between the next ring $(k + 1)$ and the preceding ring $(k - 1)$, the rings in the next succeeding position $(k + 2)$, second preceding position $(k - 2)$, etc., up to a maximum even value for the ring difference Δr_{max} between the rings in coincidence, as shown in Figure 5.39a. The summing is carried out for each pair of x_r and ϕ coordinates, which is the same as adding together the elements in the 3D sinograms of the planes defined by those ring combinations, that is,

$$s_{2D}(x_r, \phi, z = k \times \Delta d_Z) = s_{3D}(x_r, \phi, z = k \times \Delta d_Z, \Delta r = 0)$$
$$+ s_{3D}(x_r, \phi, z = k \times \Delta d_Z, \Delta r = \pm 2)$$
$$\vdots$$
$$+ s_{3D}(x_r, \phi, z = k \times \Delta d_Z, \Delta r = \pm \Delta r_{max}) \tag{5.50}$$

(a)

$\Delta r = +2$
$\Delta r = 0$
$\Delta r = -2$

$z = k \times \Delta d_z$

Direct planes

(b)

$\Delta r = +3$
$\Delta r = +1$
$\Delta r = -1$
$\Delta r = -3$

$z = (k+1/2) \times \Delta d_z$

Crossed planes

FIGURE 5.39
Representation of the axial mashing performed in the 2D mode, which gives rise to (a) direct and (b) crossed planes. In a system with N rings, there are N direct planes and $N - 1$ crossed planes.

from which a single plane is created, called a direct plane, coinciding with the axial position of ring k. Additionally, crossed planes are created by summing the coincidences between the rings k and $(k + 1)$, $(k − 1)$, $(k + 2)$, etc., (Figure 5.39b), also up to a maximum ring difference, but now an odd number. The crossed plane produced with this mashing is ascribed to the middle position between the k and $(k + 1)$ rings, which has an axial coordinate given by $z = (k + 0.5) \times \Delta d_Z$:

$$s_{2D}(x_r, \phi, z = (k + 0.5) \times \Delta d_Z) = s_{3D}(x_r, \phi, z = (k + 0.5) \times \Delta d_Z, \Delta r = \pm 1)$$
$$+ s_{3D}(x_r, \phi, z = (k + 0.5) \times \Delta d_Z, \Delta r = \pm 3)$$
$$\vdots$$
$$+ s_{3D}(x_r, \phi, z = (k + 0.5) \times \Delta d_Z, \Delta r = \pm \Delta r_{max}) \tag{5.51}$$

In this way, it is possible to define N direct planes and $N − 1$ crossed planes in a detector with N rings, in a total of $2N − 1$ planes for the 2D mode.

5.3.2.4.5 Copolar and Axial Sampling in the 3D Mode: 3D Mashing and the Michelogram

For the 3D mode, the axial mashing is combined with the mashing of the parameter that determines the inclination of the sinograms, Δr. This process involves summing 3D sinograms with the same axial coordinate z and adjacent Δr inclinations, yielding a total sinogram with that same z coordinate and an inclination which is the average of the inclinations of the planes figuring in the sum. This method creates sets of total sinograms that have the same inclination but different axial coordinates z [42], called segments (Figure 5.40).

To create the summed sinogram with axial coordinate $z = q \times \Delta d_Z$ (where q is an integer or half-integer, depending on whether the plane is direct or crossed) belonging to the segment with inclination $\Delta r = m$, we add all the original 3D sinograms with that axial coordinate $z = q \times \Delta d_Z$, inclinations Δr within a range centered in m, and with a width known as span:

$$s_{3Dseg}(x_r, \phi, z = q \times \Delta d_Z, \Delta r = m)$$
$$= s_{3D}(x_r, \phi, z = q \times \Delta d_Z, \Delta r = m − (span − 1)/2)$$
$$+ s_{3D}(x_r, \phi, z = q \times \Delta d_Z, \Delta r = m − (span − 1)/2 + 1) \tag{5.52}$$
$$\vdots$$
$$+ s_{3D}(x_r, \phi, z = q \times \Delta d_Z, \Delta r = m + (span − 1)/2)$$

The span is the maximum number of direct and crossed planes used to obtain the total-sinogram and is necessarily an odd quantity. It is easy to check that the sum of Equation 5.52 is conducted over all sinograms that

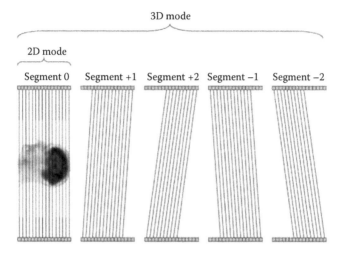

FIGURE 5.40
Diagram of the 3D mashing of sinograms. In a given segment, the total sinograms obtained from the mashing process cross all the possible integer (direct planes) and half-integer (crossed planes) z positions. Segment zero corresponds to all the direct and crossed planes defined in the 2D mode.

collect coincidences between the rings r_1 and r_2 which simultaneously satisfy the conditions $r_2 - r_1 \in [m - \text{int}(\text{span}/2), m + \text{int}(\text{span}/2)]$ and $(r_2 + r_1)/2 = q$. Each segment is labeled by its inclination m, which may take the values of $0, \pm\text{span}, \pm 2 \times \text{span}, \pm 3 \times \text{span}$, etc., in a total of N_seg number (always odd) of segments. The number of segments is determined not only by the span but also by the maximum ring difference (MRD), Δr_max, which expresses the maximum inclination of combined rings that it is allowed to use, according to

$$\Delta r_\text{max} = \frac{(\text{span} - 1)}{2} + \frac{(N_\text{seg} - 1)}{2} \times \text{span}. \tag{5.53}$$

It should be noted that once Δr_max is chosen, only certain span values remain possible (and vice versa) so that the relation of Equation 5.53 holds, with the number of segments being an integer quantity.

The 3D mashing process is very aptly described by a diagram called the Michelogram, developed by Christian Michel. The Michelogram is a 2D plot in which all the possible combinations of ring pairs are represented by dots along the horizontal (first ring index r_1) and vertical (second ring index r_2) axis (Figure 5.41). Each dot represents a sinogram, and the sum of sinograms is indicated by lines joining the corresponding dots. Segments are delimited by dashed lines, which differentiate the separate regions in the plot. This type of plot is highly intuitive, and its careful analysis clarifies all the quantitative relations shown in this section. In addition, it makes it possible to find the parameters of the 3D mashing very quickly, because the maximum ring difference is read directly from the number of dots in the first row or column,

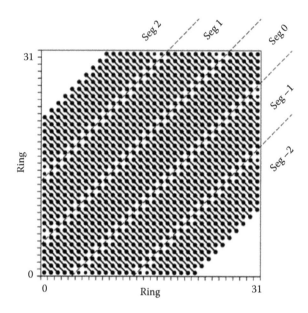

FIGURE 5.41
Example of a Michelogram for a PET system with 32 rings, with mashing span parameters $= 9$ and $\Delta r_{max} = 22$ (which defines $N_{seg} = 5$ distinct segments). Each point represents the sinograma for the pair of rings $(r1, r2)$, the lines connecting points represent the sum of sinogramas for these points.

whereas the number of dots in a given row or column inside the same segment is the span.

5.3.2.5 Image Reconstruction

Image reconstruction in PET, as in SPECT, is based on the line-integral approximation in both the 2D and 3D modes:

$$p_{2D}(x_r, \phi, z) = c \int_{-R_D}^{+R_D} f(x, y, z) \, dy_r \quad \text{(2D mode)}$$

$$p_{3D}(x_r, y_r, z, \theta) = c \int_{-R_D}^{+R_D} f(x, y, z) \, dz_r \quad \text{(3D mode)}$$

$$(5.54)$$

For the purpose of image reconstruction, the constant c is ignored as well, although in PET this constant can be experimentally measured. Its value is routinely determined during PET scanner calibration procedures, and it is a scale factor that enables PET to measure the absolute concentration of the radioactive tracer throughout the patient's body.

The image reconstruction algorithms for PET normally assume that the data are organized in the format of parallel projections and follow analytical or iterative methods, depending on whether they invert the equations (Equation 5.54) or numerically compute the activity distribution that best reproduces the 2D or 3D projections acquired. Some of the main image reconstruction algorithms are discussed in Chapter 6.

5.3.3 Multimodal and Dedicated Systems

After some years of discussion 10 or 20 years ago on the relative advantages and disadvantages of each imaging modality and after much speculation about which modality would survive this fight, the dominant perspective changed and discussion began to focus on the possibility of merging different modalities to optimally achieve a certain clinical or research goal. Thus, for each specific application the objective is to develop or use a system, multimodal or not, that has the best set of performance characteristics, while optimizing the resources to achieve this purpose. In some cases, a system based on a single modality and optimized for a given purpose can be the simplest and cheapest system and can perform perfectly adequately. In other cases, the synergy achieved by integrating two or more methods may lead to the sharing of components and functions, to a better overall performance, and to an optimization of resources that makes a system better than the sum of the parts, although more complex. This section will give examples of systems optimized for specific applications, unimodal and multimodal, existing or under development.

5.3.3.1 Systems for Specific Applications

With the gradual exploitation of the potential of a technique, often we see a diversification of the design of systems and the introduction of variants of existing systems to meet the demands of the market and the emergence of new applications. In the case of PET and SPECT, the development of systems optimized for the study of a particular organ is especially attractive, because placing the camera close to the organ increases the solid angle, and this can significantly improve the sensitivity, while at the same time lowering the cost of the system, as it may require less detector volume and fewer electronic devices. As an additional benefit, the spatial resolution may increase in both SPECT and PET: in SPECT, the resolution improves with a shorter distance from the source to the detector (due to the effect of the collimator, as mentioned in Section 5.3.1.1); and in PET the smaller distances between detectors reduces the noncolinearity effect of the two annihilation photons. There may be more gains, too, for example, with regard to noise and detected event rates, resulting from the specific geometries used (e.g., radiation scattered from outside the field of view decreasing with the reduction of the radius of the PET detector).

5.3.3.1.1   Brain PET

The development of scanners dedicated to brain studies, with better spatial resolution, good sensitivity, and lower cost, has had a market in PET for many years, especially in research. The ECAT 953B camera [56] appeared in 1990; and, more recently, the HRRT (high-resolution research tomograph) system has been developed, which has the best intrinsic spatial resolution for studies in humans, about 2 mm [57]. Other systems have been developed but not necessarily marketed and are used primarily in research. In PET, these systems aim at maximizing spatial resolution by reducing the size of crystals, while seeking to improve sensitivity by gaining in solid angle of detection. One desirable feature is the ability to measure the DOI in PET, so that the spatial resolution is more uniform in the field of vision. The HRRT camera was the first commercial PET tomograph capable of measuring the depth of interaction, through the joint use of two crystals in each detector block, LSO and LYSO, in a technique called Phoswich. The different signal produced by each scintillator can distinguish whether the interaction was on the front or back of the detector, enabling improvement of the resolution at the periphery of the field of vision. The Phoswich arrangement using two crystals can discriminate the depth information at two levels. The use of three or more crystals increases complexity but would identify more levels of DOI. A viable alternative, giving a greater number of levels of DOI, involves reading the scintillation light at both ends of the crystal, for example, by using APDs (see Section 5.1.3.2). The relationship between the readings of the two APDs can distinguish the depth of interaction in the crystal with an accuracy of the order of millimeters [58].

5.3.3.1.2   Animal PET

The growing importance of animal experimentation to help our understanding of humans makes the development of systems dedicated to the study of animals quite attractive. It offers testing methods for the better analysis of both diseases and the behavior of normal tissues and organs, in addition to trying out new therapies and techniques, such as *in vivo* observation of gene expression. The first commercial cameras dedicated to small animals were produced in the late 1990s [59] and quickly became popular due to the large number of research centers interested in their images. Nowadays, there are dedicated animal testing cameras for all the main imaging techniques: microPET, microSPECT, microCT, and microRMN. In these small cameras, too, the smaller the area of detection the lower the price of components, and increasing proximity to the subject enhances the sensitivity of the camera, while it is possible to optimize the system in other ways, keeping the cost relatively low. In this application, spatial resolution is essential and dominates the other specifications (with the possible exception of sensitivity, which should increase with the improvement of spatial resolution in order to maintain a sufficient number of decays detected by a voxel and, thus, an adequate signal-to-noise ratio). This factor leads to the development of systems with more radical solutions, such as the HIDAC system, which achieves better spatial resolution, 0.95 mm [46], through nonconventional detectors in PET (multiwire proportional chambers, which are not based on crystal scintillators). Indeed, the most successful attempts at approaching the physical limit of spatial resolution are found in animal PET, where technical solutions exist that can reach spatial resolutions close to a few tenths of millimeters [60,61].

5.3.3.1.3 *Other Dedicated Systems*

Several systems have been developed to observe specific organs, including systems dedicated to mammography (PEM), where several groups are actively researching this area; and there are several prototypes in development [62,63]. Systems specifically designed for other organs are also being developed [64], with the following characteristics in common: an important clinical application, with a large number of patients able to justify a dedicated and optimized system for the purpose; detection systems smaller than the traditional PET whole body systems, which can be placed closer to the patient, in order to increase the solid angle and thus ensure good sensitivity at a low cost; and high spatial resolution obtained with smaller individual crystals, preferably able to measure the DOI. The best performance in sensitivity and spatial resolution, for a lower cost, makes these systems particularly interesting, whereas the fact that they are dedicated to a specific application improves the functional aspects considered important. For example, a PEM system can be optimally adapted to stereotactic systems to guide the biopsy, to conventional mammography systems to aid the diagnosis with image fusion, and to the geometric constraints of data acquisition (including the possibility of inspecting the axillary region), among other possibilities.

5.3.3.2 *PET/CT*

The development of the first PET or CT system with a configuration similar to current commercial systems began in 1995 with a prototype installed at the University of Pittsburgh Medical Center that entered into clinical trials in 1998. This prototype had a PET camera and a CT camera in a tandem or in-line configuration, where the patient enters a CT tunnel followed by a PET tunnel (Figure 5.42). This design was adopted, because it was simpler and more modular than another option originally designed and tested in two prototypes [65]. In these prototypes, the two modalities were integrated in a single ring, consisting of a partial ring of two PET detectors in coincidence, and the CT component, with an x-ray source and detector, using the remaining space of the ring [65]. This geometry with a unique rotating partial ring required the development of electronics, sensors, and mechanics for a new type of camera with integrated PET and CT. This forced the resolution of many technological problems whose complexity was less advantageous than the option of simply adding two existing and tested cameras and performing as few adjustments

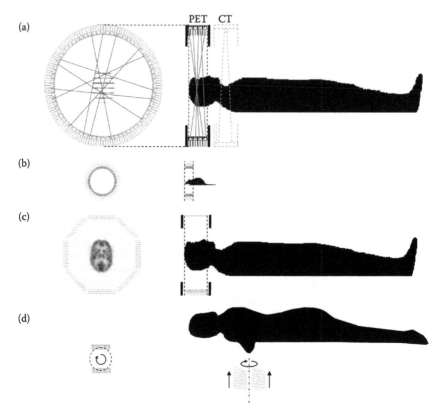

FIGURE 5.42
Some proposed PET camera geometries for systems designed for specific applications. (a) Traditional whole-body system (PET/CT), with indication of some coincidences detected. (b) Animal PET. (c) Brain PET. (d) Positron emission mammography (PEM).

as necessary, especially mechanical ones, to allow the patient's bed to enter the two tunnels. Thus, PET or CT was born, combining two techniques with the minimum of adaptation and, given its success, there is now a trend in opposite direction, with the development of dedicated new systems that optimize the integration of two techniques, in terms of both software and hardware.

The first commercial PET or CT system was introduced in 2000, and the first systems were installed in 2001 [66]. The success of PET or CT was huge; and within a few years, the market for PET cameras had been replaced and more than a dozen models had quickly emerged from various manufacturers. The advantages of joining the morphological information of CT with the functional information of PET in clinical practice were central to this success. Additionally, there was the benefit from using CT information to correct for attenuation in PET with greater speed, lower noise, and better contrast. From the perspective of the patient, PET or CT had the advantage of being able to acquire all data in a single and shorter study, thanks to the faster attenuation correction. From the standpoint of the medical imaging service, reducing the acquisition time of PET or CT both increased the number of studies and facilitated the management of patients by halving the number of times they have to come for examinations and the number of reports.

The rapid evolution of PET or CT has not yet been matched by an effective and complete clinical validation of the advantages of PET or CT compared with conventional PET alone or to PET coregistrated by software with CT data acquired in another scanner [67]. Although many studies have been carried out for different clinical applications, the information for an overall assessment is still limited, partly for methodological reasons: it would be difficult to justify the systematic exposure of patients to multiple tests and a higher dose of radiation for these studies. Apart from this, it is nevertheless agreed that the observation of the function together with anatomy is a good practice, improving the confidence of physicians in assessing the status of the patient. Whether this improvement is expressed in an actual clinical benefit of PET or CT relative to other options, thus offsetting the disadvantage of using a higher dose of radiation than in PET alone or surpassing the advantage of using existing cameras and software for coregistration and fusion, depends on the characteristics of the specific clinical application; and more time is needed for analysis. According to [66], a review of publications on PET or CT suggested an incremental benefit of PET or CT over PET in 10% of patients analyzed. Some of the additional difficulties of such conclusions should be noted: technology is rapidly changing, not only in terms of scanners and software, which change the conditions of the comparison, but also with the expected introduction of new and more specific radiopharmaceuticals, for example, the clinical information available mainly refers to studies with FDG.

Although the integration of PET and CT in the first PET or CT systems attempted to minimize changes in the systems, several aspects had to be substantially modified. The patient table had to be redesigned, as it would have to move more than usual because the tunnel was now wider. One of the problems found was related to the vertical deflection of the table due to the weight of the patient and the larger extension outside the support as it entered the tunnel, which caused coregistration errors between PET and CT data. One solution was to move the table forward on rails instead of having a fixed base for it on the floor, so that the progression of the patient and, therefore, the deflection was not significant. Another concern was the use of CT data for attenuation correction in PET. The CT data relate to the linear attenuation coefficient of tissues for photons emitted by the CT tube, with an energy of about 70–140 keV, and not to the photons detected in PET, with 511 keV (assuming there is no Compton interaction). Ways to convert the attenuation coefficient from CT to PET are discussed in Chapter 6. The difference between the acquisition time for PET, typically several minutes, and for CT, of the order of seconds, also generates artifacts: The movements of organs (heart, lungs) cause greater blurring of the PET image relative to CT. This changes the shape of organs and leads to badly overlapped images and errors in attenuation correction. The resulting effect is the appearance of areas with wrong attenuation correction factors, for example, in the diaphragm. Other artifacts may arise due to metallic implants or other objects of higher density, causing beam hardening in CT. This effect causes an increase of local values of the attenuation coefficient, leading to an excessive attenuation correction in the region around the implant, which then appears to have greater activity than it really does. Other artifacts include those due to the presence of contrast in CT and the truncation of the patient, in some cameras that have a CT with a smaller field of view radius than that of PET [68]. These artifacts led to the development of scanners with a larger field of view radius for CT, with 70 cm instead of the usual one of about 50 cm, to reduce the effects of truncation of the CT image of the patient and the corresponding attenuation correction errors (over-correction of attenuation within the patient, leading to overestimation of the activity; the opposite situation in the region where there are no attenuation coefficients measured by CT [69]).

5.3.3.3 PET/MRI

The integration of PET and MRI (magnetic resonance imaging) is underway, with several prototypes of cameras in development [70–73]. The marketing of some of these systems is starting, at the moment only for animals, although commercial clinical systems for the whole body are expected soon. A multimodal PET or MRI system has essentially the same advantages as a PET or CT system; it integrates the functional and morphological information, ensuring a good coregistration of the two types of information, with the benefits that result for the patient, clinic and service from doing a single study, as noted in the previous section. However, there are also some notable differences: the MRI has better contrast for different tissue types and a different flexibility in terms of other parameters measured, with the variants of magnetic resonance spectroscopy (MRS) and functional nuclear magnetic resonance (fMRI). The morphological information of MRI is, thus, of particular interest, for example, not only in brain studies where it can easily distinguish between white and gray matter (unlike CT) but also in other parts of the body, such as studies of the pelvic and locomotor system. The MRS and fMRI can also provide additional functional information, complementary to PET. These modalities have been the subject of intense research, especially in brain studies; and a wide range of applications has already been developed, which would benefit if they could be simultaneously obtained with the information from PET. Although there are software tools that facilitate the accurate fusion and coregistration of the images, particularly in brain studies, they use data acquired at different times and in different conditions. The flexibility of MRI, especially if simultaneous with PET, can also, in principle, address a wider range of issues than PET or CT, which is sequential in time in current systems. Basically, it is the possibility of developing new types of metabolic studies, correlating a variety of physiological variables that can be measured by fMRI and MRS, and combining the excellent spatiotemporal resolution of MRI with the excellent sensitivity and specificity of PET that enables us to predict great opportunities for the study of biological systems *in vivo*, with the possibility of conducting studies that have not been possible until now and of developing new clinical applications in the short or medium term.

As with PET or CT, where the integration of modalities brought additional benefits such as CT attenuation correction, in PET or MRI, too, it is possible to correct for partial volume effects more easily and accurately. This correction, discussed in Chapter 6, is of great interest in PET to quantify the activity in small structures in absolute terms and with good precision. This is useful in brain studies, for instance, where one intends to study the direct action of drugs in the brain or to diagnose and study the development of pathologies. However, this correction is rarely performed owing to the lack of good quality anatomical information, inadequate acquisition or coregistration, or simply because of the greater complexity of procedures involved. These limitations disappear with PET or MRI systems, which could well serve to improve the accuracy of the quantitative information provided by PET in small structures.

Another advantage that may result from the simultaneous use of PET and MRI is the fact that the magnetic field can reduce the positron range, in principle, reducing this limitation of PET physics and thereby improving the spatial resolution of PET in the plane perpendicular to the magnetic field. This possibility has been mentioned several times [74,75], but it would require very strong magnetic fields of the order of 9 T or more to achieve a significant effect with the most common radionuclides. This seems to be possible in the medium term, with the first signs that the studies on patients with high fields (9.4 T) are safe and with the development of MRI scanners for patients with fields above 11 T. With such high fields, the specification of the instrumentation becomes more difficult to meet, so there is still a long way to go.

Even though these advantages were known long ago, only now the first PET or MRI systems are appearing, mainly because their development poses more important technical difficulties than PET or CT. The high magnetic fields of MRI, its fast gradients, and high intensity of radio-frequency pulses interfere with the electronic devices used in PET; and these, in turn, may influence aspects of the MRI system such as the uniformity of the magnetic field. The PMTs used in PET detector blocks do not work in the fields of MRI and must be replaced by other devices such as APDs (Avalanche PhotoDiodes). This replacement of PMTs is beneficial for PET, as APDs are smaller and can help produce systems with better spatial resolution, while reducing the size of the detection system, which can then enter the MRI ring in a device usually known as PET insert (a system that is inserted when necessary to acquire simultaneous PET or MRI, generally used for small animals). The shielding of the detectors and electronics is another concern of the groups that are developing these systems, and it has been possible to develop scanners where the interference between the PET electronics and MRI field is reduced to negligible values.

The first PET or MRI systems for animals [70–73] and the brain [76,77] are emerging and new and interesting applications are expected. It is not clear how this technology will evolve in an environment dominated by PET or CT [66], but it seems certain that the integration of imaging modalities with important synergies between them that compensate for the additional complexity of the systems will open up new paths that may lead to exciting developments and discoveries in the coming years.

References

1. Y. Charon, P. Lanièce, and H. Tricoire. Radio-imaging for quantitative autoradiography in biology. *Nucl. Med. Biol.*, 25, 699–704, 1998.
2. M. E. Phelps, E. J. Hoffman, N. A. Mullani, and M. M. Ter-Pogossian. Application of annihilation coincidence detection to transaxial reconstruction tomography. *J. Nucl. Med.*, 16, 210–224, 1975.
3. S. R. Cherry. *In vivo* molecular genomic imaging: New challenges for imaging physics. *Phys. Med. Biol.*, 49, 13–48, 2004.
4. G. F. Knoll. *Radiation Detection and Measurement.* John Wiley and Sons, New York, NY, 1979.
5. M. A. Omar. *Elementary Solid State Physics: Principles and Applications.* Addison-Wesley, Reading, MA, 1975.
6. C. Scheiber and G. C. Giakos. Medical applications of CdTe and CdZnTe detectors. *Nucl. Instrum. Meth. A*, 458, 12–25, 2001.
7. W. Moses, V. Gayshan, and A. Gektin. The evolution of SPECT from Anger to today and beyond, in *Radiation, Detectors for Medical Applications*, S. Tavernier, et al. (eds.), 37–80, Springer, 2006.
8. C. W. E. van Eijk. Inorganic scintillators in medical imaging. *Phys. Med. Biol.*, 47, 85–106, 2002.
9. J. Pawelke. Methodische Untersuchungen zum Einsatz der Positronen-Emissions-Tomographie in der Leichtionen-Tumortherapie. PhD Thesis—Technische Universität Dresden, Dresden, 1995.
10. D. E. Groom. Particle detectors. *The European Physical Journal C*, 3, 1–4, 154–162, 1998.
11. Crystal Clear Collaboration, available at http://crystalclear.web.cern.ch/crystal clear/, April 2007.
12. A. I. Santos, P. Almeida, M. V. Martins, N. Matela, N. Oliveira, N. C. Ferreira, J. D. Aguiar, et al. Design and evaluation of the clear-PEM detector for positron emission mammography. *Proc. IEEE 2004 Med. Ima. Conf.*, Vol. 6, pp. 3805–3809, Roma, Itália, 2004.
13. J. L. Humm, A. Rosenfeld, and A. DelGuerra. From PET detectors to PET scanners. *Eur. J. Nucl. Med. Mol.*, I. 30, 1574–1597, 2003.

14. G. Muehllehner, J. S. Karp, and S. Surti. Design considerations for PET scanners. *Q. J. Nucl Med.*, 46, 1, 16–23, 2002.
15. M. E. Casey and R. Nutt. A multicrystal two dimensional BGO detector for positron emission tomography. *IEEE Trans. Nucl. Sci.*, 33, 460–463, 1986.
16. T. R. De Grado, T. G. Turkington, J. J. Williams, C. W. Stearns, J. M. Hoffman, R. E. Coleman. Performance characteristics of a whole-body PET scanner. *J. Nucl. Med.*, 35, 1398–1406, 1994.
17. K. Wienhard, M. Dahlbom, L. Eriksson, C. Michel, T. Bruckbauer, U. Pietrzyk, and W. D. Heiss. The ECAT EXACT HR: Performance of a new high resolution positron camera. *J. Comput. Assist. Tomo.*, 18, 110–118, 1994.
18. L. E. Adam, J. S. Karp, M. E. Daube-Witherspoon, and R. J. Smith. Performance of a whole-body PET scanner using curve-plate NaI(Tl) detectors. *J. Nucl. Med.*, 42, 1821–1830, 2001.
19. M. Conti, B. Bendriem, M. Casey, M. Chen, F. Kehren, C. Michel, and V. Panin. Implementation of time-of-flight on CPS HiRez PET scanner. *Proc. IEEE 2003 Med. Ima. Conf.*, Vol. 5, 2796–2800, Portland, 2003.
20. W. W. Moses, S. E. Derenzo, and T. F. Budinger. PET detector modules based on novel detector technologies. *Nucl. Instrum. Meth. A*, 353, 189–194, 1994.
21. V. Yu Chepel. A new liquid xenon scintillation detector for positron emission tomography. *Nucl. Tracks Radiat. Meas.*, 21(1), 47–51, 1983.
22. J. Colloti, S. Jan, and E. Tournefier. A liquid xenon PET camera for neuro-science. In B. Aubert and J. Colas, editors, *CALOR2000 Proc. 9th Int. Conf. on Calorimetry in High Energy Physics*, Frascati Phys. Ser., 305–313, 2000.
23. H.O. Anger. Scintillation camera. *Rev. Sci. Instrum.*, 29, 27–33, 1958.
24. W. W. Moses and S. E. Derenzo. Empirical observation of resolution degradation in positron emission tomographs using block detectors. *J. Nucl. Med.*, 34, 101–102, 1993.
25. S. Suzuki, T. Nakaya, A. Suzuki, H. Suzuki, K. Yoshioka, and Y. Yoshizawa. PMTs of superior time resolution, wide dynamic range, and low cross-talk multi-anode PMTs. *IEEE Trans. Nucl. Sci.*, 40(4), 431–433, 1993.
26. H. Kyushima, Y. Hasegawa, A. Atsumi, K. Nagura, H. Yokota, M. Ito, J. Takeuchi, K. Oba, H. Matsuura, and S. Suzuki. Photomultiplier tube of new dynode configuration. *IEEE Trans. Nucl. Sci.*, 41(4), 725–729, 1994.
27. C. P. Datema, I. P. Pleasents, and D. Ramsden. Hybrid photodiodes in scintillation counter applications. *Nucl. Instrum. Meth. A*, 387, 100–103, 1997.
28. C. Field, T. Hadig, M. Jain, D. W. G. S. Leith, G. Mazaheri, B. N. Ratcliff, J. Schwiening, and J. Va'vra. Timing and detection efficiency properties of multi-anode PMTs for a focusing DIRC. *Proc. IEEE 2003 Nucl. Sci. Symp.*, N38, Portland, 2003.
29. M. D. Petroff and M. G. Stapelbroek. Photon-counting solid-state photomultiplier. *IEEE Trans. Nucl. Sci.*, 36(1), 158–162, 1989.
30. P. Buhzan, B. Dolgoshein, L. Filatov, A. Ilyin, V. Kantzerov, V. Kaplin, A. Karakash, et al. Silicon photomultiplier and its possible applications. *Nucl. Instr. Meth. A*, 504, 48–52, 2003.
31. N. T. Ranger. The AAPM/RSNA Physics Tutorial for Residents: Radiation detectors in nuclear medicine. *Radiographics*, 19(2), 481–502, 1999.
32. P. W. Nicholson. Pulse processing and shaping. *Nuclear Electronics*. Wiley, New York, 1974.
33. J. L. Prince and J. M. Links. Planar scintigraphy. *Medical Imaging Signals and Systems*. Pearson Prentice-Hall Bioengineering, New Jersey, 2006.

34. S. Webb. *The Physics of Medical Imaging*. Medical Science Series, New York, 66–79, 1988.

35. G. K. Von Schulthess and J. Hennig. Eds. *Functional Imaging*. Lippincott-Raven, Philadelphia, 1997.

36. V. Fidler. Current trends in nuclear instrumentation in diagnostic nuclear medicine. *Raiol. Oncol.*, 34(4), 381–385, 2000.

37. P. Zanzonico. Technical requirements for SPECT. In Kramer, E. L. and J. J. Sanger, *Clinical Spect Imaging*; Raven Press, New York, 1994.

38. M. N. Wernick and J. N. Aarsvold. *Emission Tomography—The Fundamentals of PET and SPECT*. Elsevier Academic Press, San Diego, 2004.

39. S. C. Moore, K. Kouris, and I. Cullum. Collimator design for single photon emission tomography. *Eur. J. Nucl. Med.*, 19(2), 138–150, 1992.

40. G. L. Zeng, J. R. Galt, M. N. Wernick, R. A. Mintzer, and J. N. Aarsvold. Single-photon emission computed tomography. In Wernick, M. N. and Aarsvold, J. N. eds., *Emission Tomography: The Fundamentals of SPECT and PET*, San Diego, CA: Elsevier, 2004.

41. M. N. Wernick and J. N. Aarsvold. Introduction to emission tomography. In Wernick, M. N. and J. N. Aarsvold, eds., *Emission Tomography: The Fundamentals of SPECT and PET*. Elsevier, San Diego, 2004.

42. M. Defrise, P. Kinahan. Data acquisition and image reconstruction for 3D PET. In B. Bendriem and D. W. Townsend, eds., *The Theory and Practice of 3D PET*. Kluwer Academic, Dordrecht, 1998.

43. P. E. Kinahan, M. Defrise, and R. Clackdoyle. Analytic image reconstruction methods. In M. N. Wernick and J. N. Aarsvold, eds., *Emission Tomography: The Fundamentals of SPECT and PET*. Elsevier, San Diego, CA, 2004.

44. A. J. Reader, K. Erlandsson, and R. J. Ott. Attenuation and scatter correction of list-mode data driven iterative and analytic image reconstruction algorithms for rotating 3D PET systems. *IEEE Transactions on Nuclear Science*, 46, 2218–2226, 1999.

45. S. R. Deans. *The Radon Transform and Some of Its Applications*. Wiley, New York, 1983.

46. A. P. Jeavons, R. A. Chandler, and C. A. R. Dettmar. A 3D HIDAC-PET camera with sub-millimetre resolution for imaging small animals. *IEEE Trans. Nucl. Sci.*, 46 No. 3 Parte 2, 468–473.

47. Brochura Are you HD ready? Siemens Medical, disponível em http://www.medical.siemens.com/siemens/en_GB/gg_nm_FBAs/files/brochures/Biograph/HD-Ready_Brochure.pdf, acedido em Abril de 2007.

48. S. E. Derenzo, W. W. Moses, R. H. Huesman, and T. F. Budinger. *Critical Instrumentation Issues for <2 mm Resolution, High Sensitivity Brain PET. Quantification of Brain Function*. Elsevier Science Publishers, Amsterdam, 25–37, 1993.

49. N. L. Christensen, B. E. Hammer, B. G. Heil, and K. Fetterly. Positron emission tomography within a magnetic field using photomultiplier tubes and lightguides. *Phys. Med. Biol.*, 40 (4), 691–697, 1995.

50. B. E. Hammer, N. L. Christensen, and B. G. Heil. Use of a magnetic field to increase the spatial resolution of positron emission tomography. *Med. Phys.*, 21(12), 1917–1920, 1994.

51. R. R. Raylman, E. Hammer, and N. L. Christensen. Combined MRI-PET scanner: A Monte Carlo evaluation of the improvements in PET resolution due to the

effects of a static homogeneous magnetic field. *IEEE Trans. Nucl. Sci.*, 43(4, pt. 2), 2406–2412, 1996.

52. D. W. Townsend, and B. Bendriem. Introduction to 3D PET. In B. Bendriem and Townsend, D.W., eds., *The Theory and Practice of 3D PET*. Kluwer Academic, Dordrecht, 1998.

53. S. R. Cherry, M. Dahlbom, and E. J. Hoffman. Three-dimensional positron emission tomography using a conventional multislice tomograph without septa. *J. Comput. Assist. Tomogr.*, 15, 655–668, 1991.

54. D. W. Townsend, T. J. Spinks, T. Jones, A. Geissbuhler, M. Defrise, M. C. Gilardi, and J. Heater. Three-dimensional reconstruction of PET data from a multi-ring camera. *IEEE Trans. Nucl. Sci.*, 36, 1056–1065, 1989.

55. G. Brix, J. Zaers, L. E. Adam, M. E. Bellemann, H. Ostertag, H. Trojan, U. Haberkorn, J. Doll, F. Oberdorfer, and W. J. Lorenz. Performance evaluation of a whole-body PET scanner using the NEMA protocol. *J. Nucl. Med.*, 38(10), 1614–1623, 1997.

56. B. Mazoyer, R. Trebossen, R. Deutch, M. Casey, and K. Blohm. Physical characteristics of the ECAT 953B/31: A new high resolution brain positron tomography. *IEEE Trans. Med. Imag.*, 10(4), 499–504, 1991.

57. K. Wienhard, M. Schmand, M. E. Casey, K. Baker, J. Bao, L. Eriksson, W. F. Jones, et al. The ECAT HRRT: Performance and first clinical application of the new high resolution research tomography. *Trans. Nucl. Sci.*, 49(1), Part 1, 104–110, 2002.

58. M. C. Abreu, J. D. Aguiar, E. Albuquerque, F. G. Almeida, P. Almeida, P. Amaral, P. Bento, et al. First experimental results with the Clear-PEM detector. *IEEE Nucl. Sci. Symp. Conf. Rec.*, 3, 1082–3654, 2005.

59. S. R. Cherry, Y. Shao, R. W. Silverman, K. Meadors, S. Siegel, A. Chatziioannou, J. W. Young, et al. MicroPET: A high resolution PET scanner for imaging small animals. *IEEE Trans. Nucl. Sci.*, 44(3), Parte 2, 1161–1166, 1997.

60. A. Blanco, N. Carolino, C. M. B. A. Correia, R. Ferreira Marques, P. Fonte, D. González-Díaz, A. Lindote, M.I. Lopes, M.P. Macedo, and A. Policarpo. An RPC-PET prototype with high spatial resolution. *Nucl. Instrum. Methods Phys. Res. A*, 533(1–2), 139–143, 2004.

61. J. R. Stickel, J. Qi, and S. R. Cherry. Fabrication and characterization of a 0.5-mm lutetium oxyorthosilicate detector array for high-resolution PET applications. *J. Nuc. Med.*, 48(1), 115–121, 2007.

62. K. Murthy, M. Aznar, C. J. Thompson, A. Loutfi, R. Lisbona, and J. H. Gagnon. Results of preliminary clinical trials of the positron emission mammography system PEM-I: A dedicated breast imaging system producing glucose metabolic images using FDG. *J. Nucl. Med.*, 41(11), 1851–1858, 2000.

63. I. N. Weinberg, P. Y. Stepanov, D. Beylin, V. Zavarzin, E. Anashkin, K. Lauckner, S. Yarnall, M. Doss, R. Pani, and L. P. Adler. PEM-2400—A biopsy-ready PEM scanner with real-time x-ray correlation capability. *IEEE NSS Conf. Rec.*, 2, 1128–1130, 2002.

64. R. R. Raylman, S. Majewski, A. G. Weinberger, V. Popov, R. Wojcik, B. Kross, J. S. Schreiman, and H. A. Bishop. Positron emission mammography-guided breast biopsy. *J. Nucl. Med.*, 42(6), 960–966, 2001.

65. M. F. Smith, R. R. Raylman, S. Majewski, and A. G. Weisenberger. Positron emission mammography with tomographic acquisition using dual-planar detectors: Initial evaluations. *Phys. Med. Biol.*, 49(11), 2437–2452, 2004.

66. N. K. Doshi, R. W. Silverman, Y. Shao, and S. R. Cherry. A dedicated mammary and axillary region PET imaging system for breast cancer. *IEEE Trans. Nucl. Sci.*, 48(3), 811–815, 2001.
67. G. C. Wang, J. S. Huber, W. W. Moses, J. Qi, and W. S. Choong. Characterization of the LBNL PEM camera. *IEEE Trans. Nucl. Sci.*, 53(3), 1129–1135, 2006.
68. J. S. Huber, W. S. Choong, W. W. Moses, J. Qi, J. Hu, G. C. Wang, D. Wilson, S. Oh, R. H. Huesman, and S. E. Derenzo. Characterization of a PET camera optimized for prostate imaging. *Conf. Rec. Nucl. Sci. Symp.*, 3, 23–29, 2005.
69. D. W. Townsend. From 3-D positron emission tomography to 3-D positron emission tomography computed tomography: What did we learn. *Mol. Imaging Biol.*, 6(5), 275–290, 2004.
70. T. Beyer and D. W. Townsend. Putting "clear" into nuclear medicine: A decade of PET/CT development. *Eur. J. Nucl. Med. Mol. Imaging*, 33, 857–861, 2006.
71. C. Pettinato, C. Nanni, M. Farsad, P. Castellucci, A. Sarnelli, S. Civollani, R. Franchi, S. Fanti, M. Marengo, and C. Bergamini. Artefacts of PET/CT images. *Biomed. Imaging Interv. J.*, 2(4), e60, 2006.
72. O. Mawlawi, J. J. Erasmus, T. Pan, D. D. Cody, R. Campbell, A. H. Lonn, S. Kohlmyer, H. A. Macapinlac, and D. A. Podoloff. Truncation artifact on PET/CT: Impact on measurements of activity concentration and assessment of a correction algorithm. *AJR*, 186, 1458–1467, 2006.
73. C. Catana, Y. Wu, M. S. Judenhofer, J. Qi, B. J. Pichler, and S. R. Cherry. Simultaneous acquisition of multislice PET and MR images: Initial results with a MR-compatible PET scanner. *J. Nucl. Med.*, 47, 1968, 2006.
74. B. J. Pichler, M. S. Judenhofer, C. Catana, J. H. Walton, M. Kneilling, R. E. Nutt, S. B. Siegel, C. D. Claussen, and S. R. Cherry. Performance test of an LSO-APD detector in a 7-T MRI scanner for simultaneous PET/MRI. *J. Nucl. Med.*, 47, 639, 2006.
75. R. R. Raylman, S. Majewski, S. K. Lemieux, S. S. Velan, B. Kross, V. Popov, M. F. Smith, A. G. Weisenberger, C. Zorn, and G. D. Marano. Simultaneous PET and MRI imaging of a rat brain. *Phys. Med. Biol.*, 51, 6371, 2006.
76. Y. Shao, S. R. Cherry, K. Farahmi, R. Slates, R. W. Silverman, K. Meadors, A. Bowery, and S. Siegel. Development of a PET detector system compatible with MRI/NMR systems. *IEEE Trans. Nucl. Sci.*, 44, 1167, 1997.
77. R. R. Raylman, B. E. Hammer, and N. L. Christensen. Combined MRI-PET scanner: A Monte Carlo evaluation of the improvements in PET resolution due to the effects of a static homogeneous magnetic field. *IEEE Trans. Nucl. Sci.*, 43(4), Part 2, 2406–2412, 1996.
78. N. Blanco, C. M. Carolino, B. A. Correia, L. Fazendeiro, Nuno C. Ferreira, M. F. Ferreira Marques, R. Ferreira Marques, P. Fonte, C. Gild, and M. P. Macedo. Spatial resolution on a small animal RPC-PET prototype operating under magnetic field. *Nucl. Phys. B—Proc. Suppl.*, 158, 157–160, 2006.
79. University of Illinois at Chicago (2008, January 2). World's most powerful MRI ready to scan human brain. *ScienceDaily*. Acedido em 9 April, 2008 em http://www.sciencedaily.com/releases/2007/12/071204163237.htm.
80. H. Schlemmer, P. J. Pichler, K. Wienhard, M. Schmand, C. Nahmias, D. Townsend, W. Heiss, and C. D. Claussen. Simultaneous MR/PET for brain imaging: First patient scans. *J. Nucl. Med.*, 48 (Suppl. 2), 45P, 2007.

6

Imaging Methodologies

Nuno C. Ferreira, Francisco J. Caramelo, J. J. Pedroso de Lima,
Carina Guerreiro, M. Filomena Botelho, Durval C. Costa,
Hélder Araújo, and Paulo Crespo

CONTENTS

6.1 Physical Aspects of Nuclear Medicine Functional Imaging

A medical image is a planar representation of the local values of a parameter in a region or organ under study that is accessible to the eye. It is obtained by projection or emission techniques or by indirect evaluation.

Advances in physics and technology since the end of the nineteenth century have brought imaging methods such as conventional radiology, scintigraphy, ultrasound, computed tomography (CT), and magnetic resonance imaging (MRI) to medicine.

There has been a steady stream of remarkable improvements in these techniques in recent years, providing powerful diagnostic tools and opening up increasingly sophisticated and promising perspectives.

In very general terms, the techniques of medical imaging can be divided into two categories: passive and active.

The passive techniques use only endogenous signals spontaneously generated in the human body, for example, thermography and the visualization of electrical activity of the brain.

The active techniques make use of the properties of various forms of radiant energy from exogenous sources that can propagate through the matter and provide information, either anatomical or physiological. Examples are radiology and ultrasound.

In addition, based on the properties or parameters that are acquired, the medical images can be divided into morphological (anatomical) images and functional (physiological) images. The morphological images provide information about physical structures, form, and some properties of the mass of patients. They tend to have good resolution (≤ 1 mm). Examples are conventional radiography, MRI, CT, and ultrasound.

Functional images portray the movement of materials associated with the physiological processes that occur in patients. They are images containing information, in some cases quantitative, on metabolism and other aspects of biological dynamics [1]. Examples of this category are images provided by radioisotopic techniques: single-photon emission computed tomography (SPECT), positron emission tomography (PET), and emerging techniques such as functional magnetic resonance imaging (fMRI).

The functional images of nuclear medicine (NM) generally have poor spatial resolution (3–5 mm or more).

The medical images directly obtained from the detectors are often represented by analog functions with two variables of position x and y in the image plane or, in some techniques, with three variables of position x, y, and z in the domain of 3D image space.

When it comes to image sequences, a new variable, time t, is added, giving rise to functions of three or four variables, with dynamic information $I(x, y, t)$ and $I(x, y, z, t)$, respectively.

In NM, the range of existence of the image function I corresponds to values of activity (or concentration of the tracer) below the limits imposed by radioprotection criteria, in the image points (x, y) or (x, y, z), at a particular time (t).

When it comes to handling the results for a list of parameters, it is often convenient to use the images in vector or matrix notation.

A matrix is a rectangular table of values that can represent an image. For a matrix I with n rows and m columns, Iij is the value of the matrix element corresponding to the line i and column j, where i ranges from 1 to n and j varies from 1 to m.

A vector can be represented as a list of values and is usually written as a column. Some operators in image processing are represented by vectors. For a vector $\mathbf{w}$ with n components, $\mathbf{w}_i$ refers to component i, which would vary from 1 to n. A vector of intensity w is equivalent to a column matrix of n by 1.

Scalar functions are used to describe monochrome images, but for colour images it is necessary to use vector functions that represent the three components of color:

$$I_R(x, y, t), \quad I_G(x, y, t), \quad I_B(x, y, t), \qquad (6.1)$$

where R is for Red, G is for Green, and B is for Blue.

Other vector functions with multiparametric components can be found in some medical images such as MRI.

The use of vector and matrix operators can simplify the resolution of complex systems of equations and facilitate the execution of many tasks in imaging sciences [2].

Analog image functions $I(x, y, t)$ are not suitable for processing in digital computers, and this justifies the systematic transition from conventional analog images (using film) to digital mapping, which has been occurring in medical imaging since the 1960s.

Digitization converts analog functions into digital functions.

In planar digital images, x, y, and t are discrete finite quantities. In NM images, the digitization of images involves either the values of activity or these and the parameters of position.

In the first case, the data are sampled for discrete values of spatial coordinates, and small image elements with dimensions that depend on the detectors used are defined. We can, thus, represent a sampled image as a function

$$f(m, n) = f(m\Delta x, n\Delta y) \qquad (6.2)$$

with $0 \leq m \leq M, 0 \leq n \leq N$, and where Δx and Δy are the sampling intervals along two normal directions x and y, respectively. The elementary area $(\Delta y \Delta x)$ is called a pixel (image element).

The values of activity in the interval between the minimum and maximum that can be attributed to pixels of a particular image are quantified in a discrete form as binary numbers in a set of intervals that depends on the number of bits (depth) available. Current options for the medical image data are 8 bits (256 intervals), 12 bits (4096 intervals), or 16 bits (65,536 intervals).

In NM, the selective uptake in tissues or organs or spatiotemporal variations of distributions of biomolecules labeled with gamma or positron emitters yield images which contain functional data that are often quantitative, in addition to morphology.

In NM, the values of the pixels that make up a digital image represent values of the activity in volume elements of the object, called voxels, which, in the example of planar scintigraphy, are projected on the image plane as pixels.

Thus, a digital 2D image of NM is a mapping of binary integer numbers representing, on a plane, the quantified activity of discrete volume samples of the object. The dimensions of the voxel and pixel depend on the technique used. In tomography studies, the volume of the voxel is the product of the pixel area and the slice thickness.

The accuracy associated with each pixel value depends on the number of bits assigned to it. The error associated with each pixel value depends on the number of events accumulated.

In a 3D image, voxels with dimensions (Δx, Δy, Δz) have average values of activity $I(x, y, z)$ expressed as binary numbers and with the identification of the spatial position of the element (x, y, z).

NM has several special features that strongly influence the type of information which is made available in NM imaging studies.

The emission of radiation by the organs or tissues under study is the basis of imaging methods in NM.

The detection in NM refers to only the radiation emitted by radioactive atoms that exist in the object, which is typically gamma radiation, that is, electromagnetic radiation with a very short wavelength ($<1.24 \times 10^{-10}$ m).

Substances with marked radioisotopes (tracers) are administered to patients, and these are chemically the same (or functionally identical) as the native molecules to be studied; but they are physically different, as they are identifiable by the emission of radiation. NM shows spatiotemporal changes of the distribution of biological molecules in the human body by identifying and tracking the labeled replicas. It is now understood that the static images of a biological system, in a particular moment, may be entirely inappropriate, because dynamics is the essence of physiology and life.

The information that is transmitted by NM is essentially functional and inherently different from that provided by other imaging methods such as CT (computed tomography with x-rays) or ultrasound, which are essentially structural.

Physiological or *functional information* is a general term that encompasses metabolism, secretion and excretion (kidney, liver), and movements of organs (heart, lungs, and blood) [1].

This information is important, especially in early detection and diagnosis, because metabolic disturbances precede the structural changes in the evolution of pathological processes and also as the dimensions of functional lesions are often different from the sizes of corresponding anatomical lesions.

It is also remarkable that all these processes may be viewed using NM techniques without any interference to the biological system, because the amounts of tracer used are extremely small. For example, the mass of $^{99m}TcO_4^-$ injected in a study with 20 mCi (740 MBq) of this ion is about 3.9×10^{-9} g.

The NM images represent mere distributions of appropriate radioactive agents within the body. Nonradioactive matter, which may or not be related to the amount of radioactivity present, is ignored during detection or considered only for the purposes of correction of attenuation. This may also suggest that good signal or noise ratios are to be expected in NM images. This is true in many cases but not a rule, because often you want to see the contrast between slightly different concentrations of radioactive material and not between radioactive and nonradioactive tissue.

In NM images, the geometric limits of the matter that did not uptake radioactive material are undetectable and are often assumed or ignored. Only the limits of the distribution of radioactive atoms are known, and even these not very clear, due to limitations in the spatial resolution of NM images. Thinking in terms of waves alone, this poor spatial resolution appears to be in disagreement with the small wavelength of the γ rays. The nature of physical interactions for the energies of gamma radiation justifies this apparent anomaly, impeding the occurrence of phenomena such as refraction and reflection. Strong technical limitations and the often unfavourable statistics of photons in NM techniques also contribute to the poor definition of images in NM.

One of the biggest problems that conventional single-photon NM has had to face since its inception is related to the fact that the vast majority of biological molecules consist of elements with low atomic numbers. This simple fact is a real disaster for conventional NM. In fact, there are no emitters of gamma radiation, with usable lifetimes and that enable external detection, among the artificial radioisotopes of biological elements of low atomic number (C, O, H, N, S). The artificial radioisotopes of these elements, with useful periods, are pure beta emitters (β^- or positrons).

This led to the development of molecules labeled with nonbiological gamma emitting radioisotopes of elements of high Z that are supposed to behave similar to the native molecules, at least in the initial metabolic steps. With the exception of some radioisotopes of elements such as iodine, iron, and sodium, which have important roles in human metabolism, and some biological molecules containing heavy elements in their constitution (vitamin B12, hemoglobin, etc.), the techniques of conventional NM use radiopharmaceuticals that are intrinsically different from the molecules that are to be studied. The history of conventional NM is also the history of the steps that were taken to overcome this and other difficulties and ultimately to use this technology with the original concept of using isotopic tracers.

Biological systems are defined by multiple variables and are poorly represented, in general, by a single parameter, as happens in medical images.

In fact, the information provided by most imaging techniques refers to only one or a few parameters, probably having minimal biological involvement with the pathological condition under study and, even so, suffering constraints of various types (intrinsic, technical, protection of patients, etc.). Despite such constraints, the use of a particular technique of choice in specific situations can provide important information.

CT reports on the local average density of electrons (or the average atomic number Z) in voxels, elements that are basically imaginary slices of the patient and whose average properties are transcribed in pixels. This information is quantitative. The differences in the parameters for conventional radiography and CT make it possible to distinguish bone from soft tissue, fat tissue, etc. MRI reports on the proton density and chemical bonds or interactions. The differences can distinguish between gray matter and white, soft tissue of nerves, etc. Ultrasound signals detect the changes in acoustic impedance in the media they pass through. These variations identify contours, the presence of masses, changes in structure, etc [3,4].

We have stated that the information provided by most medical imaging modalities was limited to a single parameter; but we have not drawn attention to the great exception, that is, NM, whose images are not confined to reporting on a single parameter, as dozens of different molecules are used in this speciality, with specific information. The NM images provide functional information associated with the labeled molecules that we use. This sets NM imaging apart from other techniques that can only report on one or a few properties. Thus, PET and SPECT are as many techniques as the number of molecules that we are able to label. These new medical imaging techniques have become windows for the noninvasive observation of the anatomy, physiology, and human pathophysiology and are essential to the practice of modern medicine.

Overall, with medical imaging, one tries to detect changes in complex and multiparametric systems, measuring essentially a limited number of parameters, and in the case of projection techniques, previously changing the geometry.

NM imaging is an exception in terms of the number of parameters evaluated, because each labeled molecule administered provides specific functional information.

6.2 Sources of Degradation in NM Imaging Methods

6.2.1 Introduction

An NM image is only a rough representation of what we wish to see in living organisms. This simple observation is equally valid for planar images obtained by scintigraphy in a gamma camera; a set of raw data, such as pro-jection data acquired in a SPECT or PET study; a series of axial, sagittal, or transaxial slices obtained by tomographic reconstruction; or calculated parametric images of a physiological parameter of interest, such as the local cerebral blood flow. Not only do these images have an error associated with each pixel or voxel value, resulting from various physical and technical fac-tors affecting the measurement, but also if not properly taken into account this error can spread to any calculation or inference we make from these images, whether to obtain a simple curve of activity over time, a clinical diagnosis report, or the choice of a particular treatment.

When we analyze an NM image, we must be aware that each pixel value has a statistical error and that typically this error is, in relative terms, far greater than the errors occurring on a photograph or radiograph. If we had the tools to indicate the exact extent of this error, we could eventually correct the measured value and thus obtain a perfect image. However, the error affecting each pixel is only known as an uncertainty range. We must be aware of this uncertainty, because it ultimately tells us to what extent we can trust NM images.

There are several factors affecting the quality of NM images: they are not very clear, with pixels of several millimeters that do not correspond to the actual detail of the body; they are discrete versions of a continuous reality. The counts in pixels or voxels are affected by statistical noise, which itself gives rise to fluctuations that add to the effects of blurring caused by the response of the camera, the distortion due to the attenuation of radiation, the loss of contrast caused by Compton effect, the patterns of spatial vari-ations of the efficiency of detectors, the effects of filters, and the simplistic assumptions made during image reconstruction. When some of these effects are compensated for, using appropriate methods, the corrections themselves often introduce new random and systematic errors, which are expected to be smaller than the systematic errors they are supposed to correct.

The presence of such errors or artifacts is unavoidable. Awareness of their origin, amplitude, shape, and how they are distributed in the image gives us an opportunity both to better understand the virtues and vices of the image formation processes, leading to the development of better systems, and to improve the perception and interpretation of information in order to draw the appropriate conclusions based on the actual content of the images.

6.2.2 Correction of Image Degradation Effects

6.2.2.1 SPECT

6.2.2.1.1 Attenuation Correction

After having been emitted in the decay process of molecular markers, the gamma photons in SPECT travel a given distance inside the body of the patient before hitting the detector. Along that path all photons have a finite probability of interacting with the tissues of the patient, which leads to the attenuation of the photon flux emitted by the activity distribution of the marker. Since the attenuation of the photons emitted in a particular point of the activity distribution depends on the length traveled by the photons inside the patient, the reconstruction of the activity distribution directly from the raw measured projections usually ascribes less activity to the regions of the patient that are thicker and less to those that are thinner, as shown in Figure 6.1.

The correction of attenuation implies knowing the relation between the number of photons emitted by the activity distribution along each direction in space and the number of photons which actually hit the detector heads of the camera in that direction. This relation depends on the probability that each photon emitted in each point of that direction has to interact along its path, which, in turn, depends on both the nature of the different structures that the photons pass through and the distance traveled in each of those structures. The situation is quite complex, as it involves knowing the attenuation map of the patient, that is, the value of the linear attenuation coefficient in all points imaged by the scanner [5].

Determining the attenuation map is therefore crucial to correcting attenuation in SPECT. When there is no independent way of obtaining it other than from the projections acquired with the SPECT scanner, the attenuation map has to be simultaneously determined with the activity distribution. This necessarily implies using an iterative method to perform the image reconstruction and the computation of a map precise enough [6–8]. When it is possible to obtain the attenuation map by other methods, that information can be used

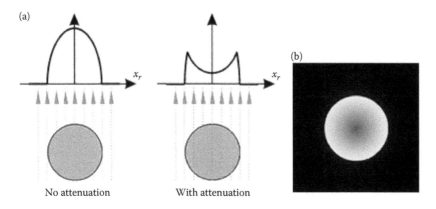

FIGURE 6.1
Attenuation effect in SPECT. (a) Diagram of a uniform activity distribution and its nonattenuated and attenuated theoretical projections along a chosen direction (supposing the inexistence of statistical noise and scattered radiation). (b) Reconstructed image of the activity distribution using the attenuated projections. The central region of the reconstruction exhibits lower activity than the original distribution.

to correct the SPECT data. This can be done either when the activity distribution is analytically reconstructed, in which case the correction is implemented in a preprocessing procedure before the reconstruction [9–12], or when the reconstruction is iterative, for which the correction is incorporated in the iterative process [13,14]. The attenuation map can not only be estimated with a transmission measurement using the SPECT scanner itself [5], as noted later, but it can also be based on pure morphologic images obtained with other methods (CT, MRI), in which attenuation coefficients are attributed to the different anatomical regions according to literature values for the different types of tissue [15,16], or even by anatomical information gathered using scattered radiation [17,18].

6.2.2.1.2 Transmission Measurement

A transmission measurement consists of determining the fraction of photons that is transmitted throughout the patient in all the directions defined by the projection planes. Some SPECT cameras incorporate a CT scanner that performs this task, but the transmission measurement can also be obtained using the gamma heads of the SPECT system and an external collimated γ emitting source, which is rotated around the patient (Figure 6.2).

The γ source may be point-like, linear or planar, but the collimator must be chosen to suit the type of source used; the most popular combination is a sweeping linear source with a parallel collimator. The radioactive isotope for the source has to be long-lived to be of practical use and produce γ photons with energy near to that emitted by the SPECT marker, so that the attenuation coefficients are approximately the same. When the marker is ^{99m}Tc, which emits 140 keV photons, it is customary to use a ^{153}Gd source for the transmission measurement, as it produces 100 keV photons and has a half-life of 242 days. The attenuation measurement can be carried out at the same time as or after the SPECT emission measurement; the first mode requires a coincidence

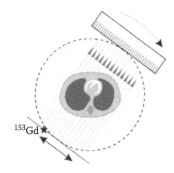

FIGURE 6.2
Layout of a transmission measurement, performed with the SPECT system's detector heads and an external collimated radioactive source, moved along the projection plane and around the patient.

system between the transmission source and the illuminated collimators of the detector head at each instant, but it offers faster exams, thereby reducing the amount of motion artifacts in the data.

The transmission measurement builds parallel projections of the attenuation map, as the quotient between the number N of transmitted photons and the number N_0 emitted by the source in a given direction is related to the line integral along that direction of the linear attenuation coefficient by the exponential attenuation law:

$$\ln \left. \frac{N_0}{N} \right|_{\text{direction } i} = \int_{\text{direction } i} \mu(x,y) dr_{\text{direction } i}. \tag{6.3}$$

The nonattenuated measurements N_0 are obtained in a blank scan, that is, a measurement in the same acquisition conditions as the transmission measurement but with no patient. The attenuation map $\mu(x,y)$ for each plane is eventually constructed from the parallel projections of Equation 6.3 using the filtered backprojection method or any other image reconstruction method discussed in this chapter.

The use of attenuation correction in the clinical applications of SPECT, formerly dismissed due to its complexity, cost, and quality control demands, is nowadays becoming more widespread owing to the increasing number of SPECT or CT cameras. The clinical impact of attenuation correction has been enjoying growing acceptance and is recommended, for example, in cardiology for myocardial perfusion exams using gated-SPECT [19].

6.2.2.1.3   Correction Methods

Once the attenuation map $\mu(x,y)$ is known, the attenuation correction can be carried out using methods that directly act on the reconstructed image. These may be analytical methods that correct the SPECT projections before image reconstruction or iterative methods that correct the projections during image reconstruction.

The direct method of image correction most often employed in everyday clinical practice is that proposed by Chang [9], according to which the estimate of the activity distribution $f_{\text{est}}(x,y)$ in each plane is corrected by

$$f_{\text{corrected}}(x,y) = c(x,y) \times f_{\text{est}}(x,y), \tag{6.4}$$

where the correction coefficients $c(x,y)$ are the inverse of the average attenuation calculated over all the emission directions that contain the point (x,y), that is,

$$c(x,y) = \frac{1}{\dfrac{1}{2\pi} \displaystyle\int_0^{2\pi} e^{-\bar{\mu}_\phi(x,y) l_\phi(x,y)} d\phi}, \tag{6.5}$$

with $\bar{\mu}_\phi(x,y)$ being the average linear attenuation coefficient computed from the point (x,y) up to the border of the patient along direction ϕ and $l_\phi(x,y)$ being the thickness of that path. This method, although very easy to implement, yields poor results when there is a broad activity distribution; In those cases, it is preferable to apply this type of correction iteratively, successively improving the correction coefficients by comparing the projections of the corrected activity distribution, calculated by Equation 6.4, with the original projections.

Analytical correction methods operate on the acquired projections before image reconstruction. This type of method is less popular in the clinical environment, but its outcome is usually better than the method of Chang. At present, the main analytical correction method is based on the mathematical inversion of the attenuated Radon transform proposed by Markoe [10] from the inversion of the exponential Radon transform previously obtained by Tretiak and Metz [20]. For each z plane, the SPECT measurements are the projections of the activity distribution modulated by the attenuation map according to the exponential attenuation law,

$$p_{at}(x_r, \phi) = \int_{-\infty}^{+\infty} f(x,y)e^{-\int_{y_r}^{+\infty} \mu(x',y')\,dy_r}\,dy_r, \tag{6.6}$$

where $p_{at}(x_r, \phi)$ are the parallel projections of $f(x,y)$ attenuated by $\mu(x,y)$. If $\mu(x,y)$ is constant, with a value of μ_0 over the region where $f(x,y)$ is nonzero, then the attenuated projections can be converted into the exponential projections of $f(x,y)$ with a μ_0 factor, $p_{\mu_0}(x_r, \phi)$ by

$$p_{\mu_0}(x_r, \phi) = \int_{-\infty}^{+\infty} f(x,y)e^{\mu_0 y_r}\,dy_r = e^{d(x_r,\phi)}p_{at}(x_r, \phi). \tag{6.7}$$

In this equation, the function $d(x_r, \phi)$ is given by

$$d(x_r, \phi) = \mu_0 y_r^{min}(x_r, \phi) + \int_{y_r^{min}(x_r,\phi)}^{+\infty} \mu(x,y)\,dy_r, \tag{6.8}$$

whose value can be computed if we know the attenuation map $\mu(x,y)$ and the lower boundary $y_r^{min}(x_r, \phi)$ of the region where $f(x,y)$ is nonzero, expressed in the OX_rY_r reference frame fixed to the detection heads (see Section 5.3.1.3, Figure 5.27). The value of μ_0 is also obtained from the attenuation map, corresponding to the average value of $\mu(x,y)$ where $f(x,y)$ is nonzero. It should be noted that assuming that the linear attenuation coefficient is constant where the activity is nonzero is generally a good approximation, because the radiotracer usually concentrates in a single type of tissue. The exponential projections of $f(x,y)$ are finally converted into nonattenuated projections $p(x_r, \phi)$ by a translation in Fourier space given by the relation

$$\mathcal{F}_1\{p\}(v_{x_r}, \phi) = \mathcal{F}_1\{p_{\mu_0}\}\left(\sqrt{v_{x_r}^2 + \left(\frac{\mu_0}{2\phi}\right)^2}, \phi + i\sinh^{-1}\frac{\mu_0}{v_{x_r}}\right), \quad (6.9)$$

where $\mathcal{F}_1$ is the one-dimensional Fourier transform with regard to x_r of the operand function. With this equation, and executing a Fourier series expansion to compute the complex angles which are the arguments of the Fourier transform of $p_{\mu_0}(x_r, \phi)$, we obtain the nonattenuated projections that will be used in the analytical image reconstruction process.

For the iterative correction methods, the attenuation correction is included in the iterative image reconstruction process itself. This is currently the most widely used correction method in clinical practice today [21,22], operating at the level of constructing the activity projections too. In each iteration, the attenuation map is employed via Equation 6.6 to calculate the attenuated projections from the activity distribution estimate proposed in that iteration. Those attenuated projections are compared with the measured projections, and the estimated activity distribution is updated according to the result of that comparison. Image reconstruction, thus, follows the usual course of iterative methods, with the single difference being in the model of formation of the parallel projections from the activity distribution estimates.

6.2.2.1.4 Correction of Scattered Radiation

Correcting scattered radiation in SPECT involves determining the number of γ photons detected along each direction in space that has not suffered any Compton scattering interaction. The products of this type of interaction, as referred to in Chapter 5 (see Figure 5.6), are one electron and one γ photon that share the energy of the incident photon. The scattered photon is emitted along a direction θ relative to the incident photon and carries an energy given by Equation 5.5.

The energy E' carried by the scattered photon is always smaller than the energy of the incident photon, and this fact can be used to distinguish photons that have undergone Compton scattering from those that have not, by simply measuring the energy of the detected photons. The correction of scattered radiation must be performed before the attenuation correction, so that the latter will solely operate on the nonscattered photons, which are the only ones whose direction bears a causality relation with their emission points.

Executing the energy discrimination of the detected photons is, therefore, the simplest way of estimating the number S of scattered photons, a value that can then be subtracted from the total number of detected photons C to obtain the number of nonscattered photons T in each direction of the parallel projections:

$$T = C - S. \quad (6.10)$$

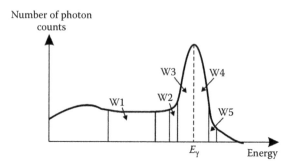

FIGURE 6.3
Typical energy histogram of the photons detected in a channel of a SPECT scanner and corresponding division in the energy windows normally used to perform the correction of scattered radiation (W1–W5). The number of windows actually employed depends on the method chosen for the correction.

There are several empirical methods to estimate the value of S in a given direction from the energy histogram of the detected photons, based on counting the number of photons inside different energy windows around the photopeak (Figure 6.3) [23–25]. The actual estimate of S is calculated by taking into account the total number of counts in each energy window, the width of the windows, and other simple parameters obtained from empirical studies of scattered radiation in SPECT.

Scattered radiation can alternatively be corrected using methods that model the existence of Compton scattering in the body of the patient. One of the simplest consists of using attenuation coefficients that are lower than those obtained in the transmission measurement [26]. The underassessment of the attenuation map takes into account the increase of the number of detected photons due to the presence of scattered radiation. The degree of underassessment that should be used depends, however, on the nature of each tissue and has to be known beforehand for all the structures present in the activity distribution. One other, rather more sophisticated, correction model is based on computing the scattered radiation map of the patient from his or her anatomical features [27,28]. This map is constructed using the attenuation map in conjunction with the knowledge of the type of tissue existing in each point of that map, to estimate the probability of Compton scattering occurring in that point.

6.2.2.1.5 Spatial Resolution Correction

In addition to the physical effects that affect SPECT measurements, particularly attenuation and scattered radiation, certain instrumental factors also influence image formation. The strong dependence of the spatial resolution of a SPECT scanner on the distance to the detector heads, due to the use of physical collimators, is the most important of those factors. Associated with that dependence, moreover, is the fact that the point spread function, that is, the function which quantifies the spatial resolution of the tomograph [29], has a finite width, giving rise to what is known as the partial volume effect. This effect mostly influences the observation of small structures, with dimensions of less than 2 to 3 times the spatial resolution of the camera, and it is expressed in an undervaluation of the activity in high-activity structures relative to the surrounding tissues as well as in an overvaluation of those which have low activity [30]. The finite width of the point spread function produces a smoothing of the boundaries between adjacent tissues with different activity concentrations in the final image, blurring all the abrupt transitions of activity (Figure 6.4). In other words, the high spatial frequencies found in the activity distribution are filtered by the point spread function of the system. The spatial resolution correction is only carried out after the scattered radiation and attenuation corrections, although it is seldom used in clinical practice.

The correction methods for the partial volume effect in SPECT, which include the dependence of the spatial resolution on the distance to the detector heads, can be grouped in three categories: deconvolution methods [31,32], methods that model the effect of spatial resolution during iterative image reconstruction [33–35], and methods that use morphological information independently obtained from the SPECT measurement [36].

In the first category, the measured projections are corrected just before image reconstruction. The correction is based on the fact that the image produced in a tomograph is the convolution of the activity distribution with the point spread function of the tomograph. For each reconstruction plane, this function

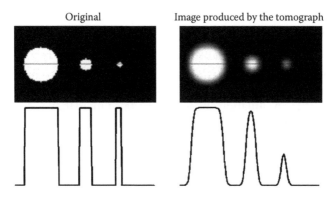

FIGURE 6.4
The partial volume effect. (left) Activity distribution in an ideal tomograph capable of exactly reproducing the object in the field of view and (right) corresponding image in a tomograph with finite spatial resolution, equal to the dimension of the smallest structure in the image. The other structures have dimensions of 2, 3, and 4 times the spatial resolution. Below the two top images, the activity measured along the dashed line is also shown. In the right image, the abrupt activity transitions have been smoothed by the tomograph; the image produced is the convolution of the activity distribution with the point spread function of the system. The error in the activity recorded in small structures can be quite significant, as observed in the rightmost structure.

is given to a good approximation by a Gaussian function of the type

$$PSF(x,y) = e^{-\frac{(x-x_0)^2+(y-y_0)^2}{\sigma^2(x_0,y_0)}} \tag{6.11}$$

in each point (x_0, y_0) of that plane, with a width $\sigma(x_0, y_0)$ that changes from point-to-point and which may be experimentally determined during the characterization routines of the tomograph. Knowing the point spread function, it becomes possible to correct the measured projections using the mathematical deconvolution of the data, which reduces simply to a division in the Fourier space.

In the second group, the point spread function is also used to convolute the estimate of the activity distribution in each iteration of the image reconstruction process, before computing the parallel projections corresponding to that estimate. The comparison of the convoluted estimated projections with the acquired projections takes into account the fact that the acquired projections are subject to the partial volume effect.

Finally, the third category includes recently proposed correction methods which adjust the activity distribution measured by the SPECT camera to the morphology of the tissues underlying that distribution. The adjustment is performed by forcing the activity in a given region to be confined to the tissues where the radiotracer is located; in particular, the activity in supposedly nonactive tissues is transferred to the nearest tissues capable of fixing the radiotracer. These methods imply the use of images separately obtained with morphological techniques (CT, MRI) from the SPECT measurement.

6.2.2.2 PET

6.2.2.2.1 Attenuation Correction in PET

Just as in SPECT, PET measurements are liable to be affected by attenuation, which occurs when the two photons of an annihilation pair traverse the path leading from the annihilation site to the crystals where they will be detected. This means that for a given LOR, the number of photon pairs hitting the two detectors is smaller than the number of photon pairs emitted along that LOR.

There are, however, two major differences between PET and SPECT with regard to attenuation. First, attenuation is more pronounced in PET, as both photons have to escape the patient's body without interacting for a coincidence to be detected. The fraction of decays in the patient that gives rise to a counting event (a coincidence in the case of PET and a single photon in SPECT) along a given direction is, therefore, considerably smaller in PET than in SPECT.

The second difference lies in the fact that attenuation in PET does not depend on the position of the emission point, in marked contrast to what happens in SPECT. This is easily seen if we take a simple situation consisting of a point source located inside a uniform body with a constant linear attenuation coefficient; for that object, the coincidence count rate recorded in an line of response (LOR) passing through the source, A_{at}, is given as a function of the activity of the source in that direction by

$$A_{at}(x_r, \phi) = A \left(e^{-\mu y_{r_1}} \times e^{-\mu y_{r_2}}\right) = Ae^{-\mu(y_{r_1}+y_{r_2})} = Ae^{-\mu l(x_r, \phi)}, \qquad (6.12)$$

as the sum of the distances y_{r_1} and y_{r_2} traveled by the two photons of one annihilation pair inside the body is the total thickness $l(x_r, \phi)$ of that body along the direction of the LOR. The factor $e^{-\mu l(x_r, \phi)}$ depends on the full thickness traveled by the photon pair and not on the specific location of the source along the LOR. This is why estimating the attenuation factors is very simple and precise in PET compared with the attenuation correction methods of SPECT.

The distribution of attenuation factors in the object is obtained using transmission measurements that can be carried out with multienergy x-ray sources in PET or CT systems (described in the next section) or in PET systems with single-energy sources, which can be either positron or gamma photon sources; Figure 6.5 shows these three possibilities. For the latter, the most common sources used are [68]Ge, a positron emitter, or ^{137}Cs, a gamma emitter that produces single gamma photons with an energy of 662 keV [37,38]. The number of photons transmitted in each LOR, N, is given as a function of the number of photon pairs emitted in that LOR, N_0, by the exponential attenuation law

$$N = N_0 e^{-\int \mu(y_t)\,dy_t}, \qquad (6.13)$$

where the nonattenuated measurements are collected in a blank scan carried out in exactly the same conditions as the transmission measurement but having no patient. To compute the correction sinograms having (the inverse of) the attenuation factors $e^{-\int \mu(y_r)\,dy_r}$, we divide the blank scan sinograms by the sinograms of the transmission measurement. The attenuation correction

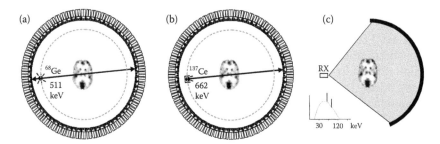

FIGURE 6.5
Diagram showing the three ways of measuring attenuation factors of a patient, using (a) a positron emitter; (b) a gamma emitter; (c) and an x-ray source. The measured attenuation factors depend on the layout of organs and tissues in the patient and on the energy of the radiation produced by the source.

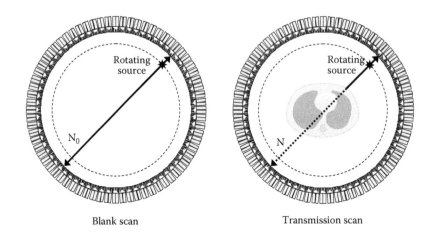

Blank scan   Transmission scan

FIGURE 6.6
Representation of the attenuation correction in PET using a transmission measurement with a rotating radioactive source. (Left) blank scan, that is, an acquisition without having the patient in the scanner's FOV; (right) transmission measurement with the patient in the FOV.

is eventually performed by multiplying point by point the sinograms of the PET acquisition with the sinograms of the attenuation factors [37–39].

The comparison of the transmission measurement data collected with and without the patient in the FOV of the camera, thus, provides a direct estimate of the attenuation in each LOR. The quotient sinograms obtained are enough to perform the correction, although there are situations when it is necessary to actually know the attenuation map, implying the reconstruction of the transmission data. Once this image is computed, the data can be segmented to suppress noise, for example, and only afterward it should be used to generate the attenuation correction factors (Figure 6.6).

6.2.2.2.2 Attenuation Correction in PET or CT

6.2.2.2.2.1 Attenuation Based on Polyenergetic Transmission (x-rays) In computed axial tomography (CAT or CT) scans, the values proportional to the linear attenuation in each point of the slice under consideration are stored in digital form in the computer memory, starting from the smaller to larger values, addressed in well-determined positions. The wide variety of attenuation coefficient values that can occur in CT images can be estimated from Figure 6.7 [40].

The figure shows the attenuation coefficient values from media that absorb x-rays in the human body, with the values given as percentages. Water is attributed the value 0, and air corresponds to the value −100. It can be seen that fat tissue absorbs 10% less than water, soft tissue can absorb up to 4% more than water, and bone can exceed 100% absorption also relative to water. The attenuation coefficients obtained with CT scans are not usually expressed

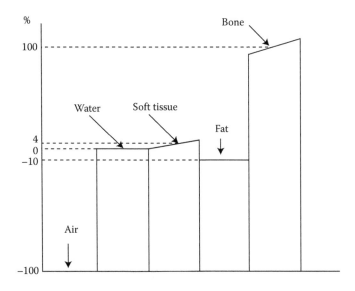

FIGURE 6.7
Attenuation coefficients for several human components of biological relevance, given in percentage with water and air having values of 0 and −100%, respectively. (From De Lima, J. J. P. *Eur. J. Phys.*, 19:485–497, 1998.)

in percentages but in Hounsfield units (HU). On this scale, named after its inventor, air corresponds to −1000 HU and water to 0 HU.

Table 6.1 gives various HU values and their corresponding linear attenuation coefficients, obtained for two peak voltages commonly used in clinical practice.

In modern PET or CT scanners, the attenuation maps used to compute the attenuation of 511 keV radiation detected with the PET system are generated from the CT scan of the patient, performed either before or straight after the PET scan. The CAT scan data are given in HU, as mentioned. These HU values cannot be directly used to correct the attenuation of the emission data,

TABLE 6.1

Hounsfield Units (HU) and Linear Attenuation Coefficients for Two Peak Voltages and Different Materials and Tissues

		84 kVp μ (cm⁻¹)	12 kVp μ (cm⁻¹)
Air	−1000	0.0003	0.0002
Water	0	0.180	0.160
Fat	−100	0.162	0.144
Blood	40	0.182	0.163
Gray matter	43	0.184	0.163
White matter	6	0.187	0.166

Source: From De Lima, J. J. P. *Eur. J. Phys.*, 19:485–497, 1998. With permission.

but there are three methods that can perform this conversion: segmentation, scaling, and CT scans with two different x-ray energies.

6.2.2.2.2 Segmentation   Segmentation methods can be used to separate the CT image in regions that correspond to different types of organic tissue (e.g., soft tissue, lung, and bone). The HUs obtained with the CT scan are replaced by the attenuation coefficients for a photon energy of 511 keV. A significant problem arises from the fact that the density of some regions of tissue continuously varies, and so it cannot be represented by a discrete ensemble of segmented values. For example, some lung regions exhibit density variations amounting to a 30% difference [41].

6.2.2.2.3 Scaling   The image values produced by the CT scan are generally linearly correlated with the attenuation coefficient of the corresponding tissue. This makes it possible to estimate the patient attenuation map for 511 keV radiation. All that is needed is to multiply the entire CT image by the ratio of the attenuation coefficients of water (representing the soft tissues) for the energies of the CT and PET scans. Since the spectrum of energy of the CT scan is a continuum, just one value is usually estimated (typically 70 keV), which then represents the whole spectrum. La Croix et al. [42] carried out simulation studies to investigate different scaling techniques for SPECT scan attenuation coefficients for an energy of 140 keV. The results show that the mentioned rescaling technique yields correct attenuation coefficients only for low atomic number (Z) materials (e.g., air, water, and soft tissue). For bone, with a higher atomic number, the study reveals that the results of this linear scaling method are not satisfactory, because the photoelectric contribution

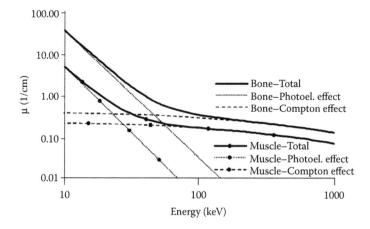

FIGURE 6.8
Linear attenuation coefficients for bone and muscle between 10 and 1000 keV. The components corresponding to photoelectric absorption and Compton scattering are also shown. The total absorption coefficient shown considers all types of interactions possible.

is dominant for the lower energies in CT scans (Figure 6.8). In other words, different factors have to be used in bone and soft tissue scaling to transform CT images acquired at an effective energy of 70 keV into an attenuation map calibrated for 511 keV.

One method to compensate for the presence of high Z material arises from noting that HU values such that $-1000 < HU < 0$ correspond mostly to regions that contain mixtures of lung and soft tissue; whereas regions with $HU > 0$ contain soft tissue and bone. Blankespoor et al. [43] have suggested that a bilinear scale could be used to convert CT images into attenuation maps for 140-keV SPECT. In the method suggested, two factors are used for the rescaling: One factor is used for water and air $(-1000 < HU < 0)$, and the other is used for water and bone $(HU > 0)$. The attenuation coefficients resulting from this method, but corresponding to PET radiation at 511 keV, are displayed in Figure 6.9. This method was proposed for PET by Burger et al. [44] and Bai et al. [45].

An alternative method that could be used to convert CT images into attenuation maps is the so-called *hybrid* method. This combines the segmentation and scaling methods. The attenuation map for 511-keV radiation is estimated by using an initial pedestal that separates the bone component in CT, followed by two other factors that rescale bone from nonbone components. This method is motivated by the information displayed in Figure 6.10, which shows that the ratio of the mass attenuation coefficients for radiation with 70 keV and 511 keV is quite similar for all organic materials except bone. This is due to the larger photoelectric fraction in bone caused by the presence of calcium (amounting to about 22.5% in cortical bone). The behavior of the hybrid method when converting HU to linear attenuation coefficients at 511 keV contrasts with the

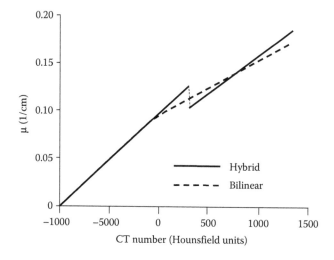

FIGURE 6.9
Conversion of HU into linear attenuation coefficients at 511 keV. Note the change in slope at $HU = 0$ for the bilinear method and the discontinuity at $HU = 300$ in the hybrid method.

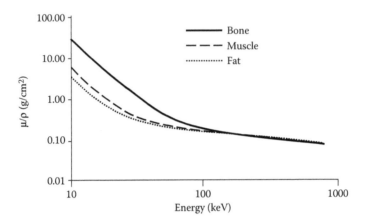

FIGURE 6.10
Mass attenuation coefficients for three biological materials.

bilinear rescaling method, as shown in Figure 6.9, where, for heuristic reasons, the pedestal chosen to separate bone from nonbone regions was HU = 300 [46]. This means that the hybrid method assumes regions of $-1000 <$ HU < 0 to be mixtures of air or water and regions of HU > 300 to be mixtures of air or bone (Figure 6.10).

It has been shown that both the bilinear scaling method and the hybrid method yield reasonable results for biological material [45,47,48]. Recent studies compare patient transmission images based on both positron sources and x-ray sources [49,50]. The impact of these transmission images on the reconstructed images in studies with the biotracer FDG show slight, even negligible, effects. This is no longer true if the patient is injected with a contrast agent or when the effects contain metal objects.

6.2.2.2.2.4 Dual-Energy X-ray Imaging A precise solution of the problem of converting HU into linear attenuation coefficients for 511 keV radiation can be obtained if two x-ray images with two different spectra are obtained. The attenuation coefficient is now a weighted combination of the photoelectric absorption and Compton scattering, that is, in essence it is a system with two components. If the individual photoelectric and Compton components could be determined, they could be separately converted into any desired energy and then added together to give the total attenuation coefficient. Alvarez and Macovski [51] developed one such quantitative CT imaging method. It was used by Hagesawa et al. [52] to create a monoenergetic attenuation map at 140 keV for a prototype SPECT or CT detector. One potential disadvantage of the method arises from the fact that it obtains the attenuation map by computing the difference between the two CT scans obtained with different energies, which means that the noise of both images is added to the quadrature. Consequently, even though double-energy techniques potentially offer the highest degree of precision, they also result in degraded noise characteristics compared with methods obtained either from a single CT scan or from transmission scans with positron emitting radionuclides.

6.2.2.2.3 Random Coincidences Correction

As mentioned in Chapter 5, coincidences can be true (T), random (R), or scattered (S). The number of coincidence events, C, detected per unit time in an LOR is the sum of these three types of coincidence,

$$C = T + R + S, \tag{6.14}$$

but only the true coincidences contain valid information about the location of the source that generated the photon pair detected. The correction of random and scattered coincidences consists of obtaining an independent estimate of the random coincidences rate $\hat{R}$ and scattered coincidences rate $\hat{S}$ for each LOR in order to estimate the true coincidences rate by

$$\hat{T} = C - \hat{R} - \hat{S}. \tag{6.15}$$

The random coincidences rate can be estimated in two ways: by using Equation 5.43 directly, if the singles count rates R_{s1} and R_{s2} are known for each LOR and the coincidence window of the system Δt is also known [53]; or by using a hardware-implemented delay line, which delays the time signals of each detector by a time lag considerably larger than Δt [54,55]. Since the large time lag prevents there being any correlation between the delayed line of one detector and the nondelayed, or prompt, line of another, the coincidences existing between the delayed and prompt lines of any two crystals are random and provide a way to estimate the random coincidences rate $\hat{R}$ for the corresponding LOR (Figure 6.11). Using a delay line is the commonest way of obtaining $\hat{R}$, because it can be done online during the acquisition, and it does not need any type of input from the operator of the system.

6.2.2.2.4 Scattered Coincidences Correction

The correction of scattered coincidences involves determining whether in each recorded coincidence any of the two annihilation photons detected interacted with an electron inside the patient's body via Compton scattering. Just as in SPECT, the simplest way of reducing the effect of Compton scattering consists of measuring the energy of the photons hitting the detectors and rejecting all the coincidences in which one of the photons exhibits energy lower than a preset value [56]. This discriminating value is typically around 350 keV; but it must be chosen carefully, because if it is too low then the rejection of scattered coincidences will not be efficient, whereas if it is too high then there is the risk of rejecting true coincidences for which one of the annihilation photons did not deposit all its energy in the scintillator that detected it.

To supplement the simple energy window discrimination, a more precise way of correcting the data from scattered coincidences is often used. One

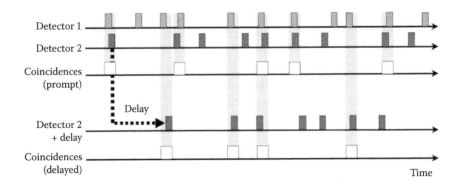

FIGURE 6.11
Diagram of the random coincidence correction using a hardware delay line. Besides the prompt line of timing signals, each crystal has a delay line whose coincidences with the prompt lines of other crystals enable the estimation of the random coincidence rates of the corresponding LORs.

method involves using the attenuation map determined with the transmission measurement together with a model of the geometry for the scanner or patient set to compute the fraction of scattered photons hitting each crystal. The computation takes into account the probability of each annihilation photon interacting with the patient (given by the attenuation map); and, should it interact, the probability of the scattered photon being emitted along a direction θ is given by the Klein–Nishina equation for the differential cross-section $d\sigma/d\Omega$ as a function of θ,

$$
\frac{d\sigma}{d\Omega}(\theta) = \frac{Zr_0^2}{2} \frac{1 + \cos^2\theta}{\left(1 + (E_\gamma/m_ec^2)(1 - \cos\theta)\right)^2}
$$
$$
\left(1 + \frac{\left((E_\gamma/m_ec^2)(1 - \cos\theta)\right)^2}{(1 + \cos^2\theta)\left(1 + (E_\gamma/m_ec^2)(1 - \cos\theta)\right)}\right), \tag{6.16}
$$

where Z is the effective atomic number of the medium, m_e is the electron's mass, c is the speed of light, and r_0 is a quantity known as the classical electron radius. One of the most precise and efficient methods uses a simplified simulation, assuming that there is only one Compton scattering event in the path of the photon pair giving rise to a scattered coincidence [57,58], a situation that corresponds to the large majority of Compton coincidences in PET [59].

There are many other techniques for the correction of scattered coincidences, as described in detail in [38]. They are all based, in one way or another, on the physics of Compton scattering and in the geometric and technical specificities of PET. Since scatter in PET can be counted in projection lines external to the object (contrary to SPECT where it is unlikely that this could happen), there are several methods that exploit this fact and estimate its spatial distribution by fitting the activity tail observed on the outside of the object's frontiers with, for example, a Gaussian function in one [60] or two dimensions [61]. An assumption often used as a starting point is that the distribution of Compton scattering events over space slowly changes and mainly comprises low spatial frequencies, which allow modeling of the distribution inside the object with Gaussian or polynomial functions, or even by curves obtained by convolution [62] or deconvolution [63] from the total distribution of coincidences. Another approach uses methods based on data acquisition with one [64,65] or more [66,67] additional energy windows covering energy ranges below [64] or above [65] the usual energy window centered around 511 keV. Finally, there are yet other methods that compute the scattered radiation distribution with sophisticated Monte Carlo simulation techniques [68] and methods that explore the complementarity of all these techniques in an effort to optimize correction [69,70].

Scattered radiation correction is usually ignored in data acquired with the 2D mode, because the physical collimation provided by the septa drastically reduces the amount of scattered coincidences. In the 3D mode, however, correcting the data from scattered radiation is mandatory, as the sensitivity to scattered coincidences is much greater. Figure 6.12 illustrates that requirement, as the impact of the scattered coincidences correction is clearly seen in 3D mode data, but it is almost imperceptible in the 2D mode.

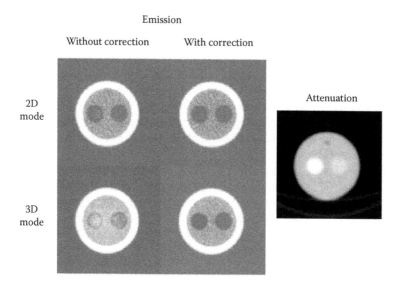

FIGURE 6.12

Impact of the scattered coincidence correction in the 2D and 3D modes. On the right, the attenuation map of the object is shown, with the lighter regions indicating areas of higher linear attenuation coefficient. Comparing this map with the noncorrected 3D mode image, the existence of more scattering in the more attenuating media is apparent, as expected.

6.2.2.2.5 Normalization Correction

The coincidence detection efficiency of a PET system is not uniform for all LORs, that is, the probability of a coincidence being detected varies from LOR to LOR. This is due both to variations in the intrinsic efficiency of each crystal and its coupled photomultiplier tube, variations that may vary over time, and to systematic effects related to the geometrical arrangement of crystals. The correction of the nonuniformity of response for the LORs in a PET system is called normalization correction, in which dead-time effects are also included.

The normalization correction commences by carrying out a measurement, the normalization measurement, in which all LORs are illuminated equally by a uniform activity source [71]. In these conditions, the number of counts obtained in a given LOR, properly corrected from the existence of attenuation and random and scattered coincidences, is directly proportional to the probability of detecting a coincidence in that LOR. The inverse of that number, $1/C_{uniform}$, called the LOR's normalization coefficient, NC, can then be used to correct the coincidences, C, of regular measurements using the equation

$$C_{norm} = NC \times C. \qquad (6.17)$$

This normalization method, known as the direct method [38,71], is often used to normalize data acquired in the 2D mode, but it is frequently impractical for the normalization of 3D mode data due to the overwhelming number of 3D LORs that are found in modern PET systems (several tens of millions), which implies the acquisition of a massive number of coincidence events in the normalization measurement in order to obtain a reasonable statistical precision for the NC factors and renders the normalization procedures incompatible with everyday clinical use. Further, the larger solid angle covered by the 3D mode subjects the coincidence processing system to an event rate considerably higher than in the 2D mode, forcing the use of low activity sources in the normalization procedure, which makes it even more time consuming. Additionally, storing and handling such a large number of NC coefficients can also be a problem.

The normalization correction of 3D data, therefore, employs other methods called indirect or component methods to model the individual response of each detector instead of directly measuring the response of each LOR to coincidence events [72–74]. The normalization coefficients are separated into the product of several components, one for each effect modeled:

$$NC = NC(\text{intrinsic efficiency})$$
$$\times NC(\text{geometric efficiency})$$
$$\times NC(\text{dead time})$$
$$\vdots \qquad\qquad (6.18)$$

The components relative to the intrinsic and geometric efficiencies of the system are normally broken down into several other factors, depending on the specific features of the tomograph, especially its geometry [75,76].

With this method, it is possible to reduce the number of parameters needed for the 3D normalization from several million (the number of LORs in a PET scanner) to just a few thousand (of the order of the total number of crystals), making it practical to implement [71]. In addition, the number of coincidence events that has to be accumulated during the normalization measurement is substantially smaller than that needed for the direct method, although it is still quite substantial, partly because the methodologies used to estimate the different components from the normalization measurement are not fully independent in the presence of noise.

6.2.2.2.5.1 Intrinsic Efficiency The correction of intrinsic efficiency aims at offsetting the variations in the intrinsic efficiency of the crystals in the PET scanner. The simplest way of doing this is by estimating the individual efficiency ε_i of each crystal from the normalization measurement using the fan-sum method [72,73]. This method estimates the relative efficiency of each crystal i with regard to the average efficiency of all crystals in the ring where it is located, by dividing the sum C_i^{fan} of all counts recorded in the fan facing that crystal, that is, all the LORs it establishes with the other crystals in the ring (Figure 6.13) by the average value of the fan sums of all the crystals of the ring:

$$\varepsilon_i = \frac{C_i^{\text{fan}}}{\left(C^{\text{fan}}\right)_{\text{ring}}}. \tag{6.19}$$

Since the measurement of the efficiency for a crystal i is not totally independent of the efficiencies of the other crystals in its fan, this method may

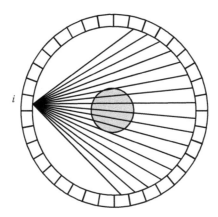

FIGURE 6.13
Diagram of all the LORs defined between a crystal i and the other crystals of its ring; this set of LORs is called the fan of crystal i.

produce inaccurate values if, for any reason, one of the crystals in the ring has a very different efficiency from the others [77].

The normalization coefficient expressing the intrinsic efficiency of the LOR defined by the crystals i and j reads

$$NC(\text{intrinsec effeciency}) = \frac{1}{\varepsilon_i \times \varepsilon_j}, \tag{6.20}$$

assuming that all the rings exhibit identical overall efficiencies. Should that not be true, the relative efficiency of each crystal given by Equation 6.19 needs to be multiplied by the efficiency of the ring where it is located.

6.2.2.2.5.2 Geometric Efficiency Geometric efficiency correction consists of compensating the variations arising from the geometric specificities of the camera's construction that affect the detection efficiency of each LOR. Since these variations are purely geometrical and do not change over time, they are typically determined only once for each PET system.

The main geometrical factor responsible for the differences in the response of LORs belonging to the same parallel projection is the decrease of the incidence angle of the photons on the crystal surface as LORs become more peripheral (Figure 6.14). The change of angle reduces the solid angle of detection, which leads to poorer LOR efficiency. The systematic variation of the detection efficiency with the radial coordinate is called the geometrical profile [75]; its inverse is the normalization coefficient for the geometric efficiency of the camera.

In addition to the geometrical profile, it is also usual to carry out a second-order geometric correction in systems where the crystals are grouped in detector blocks. The systematic variation of the crystal positions inside a block, which is reflected in slightly different acquisition geometries due to

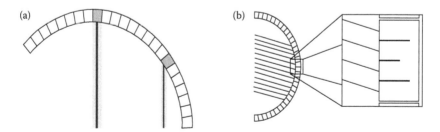

FIGURE 6.14
(a) Variation of the incidence angle of the LORs on the crystal surfaces in a ring; since the incidence is further from perpendicularity, the cross-section of the LORs diminishes, decreasing also their detection efficiency. (b) Geometrical interference between crystals in a system assembled from detection blocks. Each distinct position inside a block corresponds to a slightly different acquisition geometry of the crystal.

features such as the depth of the cuts between crystals or the coupling to the photomultiplier tube, modulates the geometrical profile; this modulation is known as crystal interference [76].

6.2.2.2.5.3 *Dead Time* As mentioned in Chapter 5, the sensitivity of PET is higher than that of SPECT, because it dispenses with the use of physical collimators to build parallel projections of the activity. This leads to a count rate in PET that is much higher than in SPECT; and, therefore, it becomes important to consider the dead time that exists after a crystal detects a photon, time during which the crystal is incapable of detecting new incoming photons. This dead time decreases the number of coincidences actually recorded relative to the number of coincidences whose photons hit the detector and interacted with the crystals. In the simplest model to describe the dead time effect in a PET scanner, which ignores crossed interference between crystals in a block, the number of singles counts in a crystal i, n_{single}, is given as a function of the number of incident photons n_{inc} that interact with the crystal and of the dead time parameter τ_i of that crystal by

$$n_{single} = (1 - \tau_i) \times n_{inc}, \tag{6.21}$$

where τ_i is the dead time of the crystal divided by the total acquisition time.

Since each coincidence involves the detection of two photons in different crystals, the normalization coefficient for the dead time component in an LOR defined between the crystals i and j that corrects the incident coincidences rate from the detected coincidences rate can be given by

$$NC(\text{dead time}) = \frac{1}{(1 - \tau_i) \times (1 - \tau_j)}. \tag{6.22}$$

6.2.2.2.6 *Other Corrections*

In addition to the corrections just mentioned, there are others that, in certain circumstances, should be performed when quantitative PET measurements are wanted.

Partial volume correction can also take place in PET; it is based on the same principles and uses correction strategies similar to those applied in SPECT (cf. Section 6.2.2.1.5) [78]. This correction is mostly performed in PET, especially in research studies that need a precise absolute quantification in small structures, with dimensions up to 3 times the camera's spatial resolution. Still, with the increasing use of PET systems for animals, where the need for imaging small structures is obvious and multimodal hybrid systems PET or CT and PET or MRI, where the existence of high quality anatomical information and good coregistration techniques means this correction can be performed easily, partial volume correction is being employed more often, although it is not yet routinely included in commercial cameras.

Other auxiliary corrections include the decay and the branching ratio corrections. The first is important when very short-lived markers are used, and it corrects for the decline in activity that occurs from the instant when it is administered to the patient until the start of the acquisition (and, if the exam is long, the change in activity during the acquisition) by directly applying the radioactive decay law. The second corrects the fraction of radioactive markers that decay by processes other than β^+ decay, a correction which is important in quantitative procedures if that fraction is not negligible.

From the geometric point of view, there are two more corrections that may prove to be necessary in PET. One of them stems from the curvature of the rings and consists of correcting the nonuniformity of spacing between consecutive LORs along a sinogram row. The spacing narrows as the LORs become more peripheral, being given by

$$\Delta x_r = \Delta d_t \sqrt{1 - \left(\frac{x_r}{R_D}\right)^2}, \tag{6.23}$$

where Δd_t is the width of each crystal and R_D is the camera's radius. This correction must be taken into account during image reconstruction whenever the region being observed in a patient lies near to the limit of the FOV. The usual practice involves interpolating new sinograms with a constant spacing between radial coordinates, based on the measured ones.

The second correction of a geometrical nature that may be important is related to the DOI effect, sometimes also referred to as the *parallax error*. This effect arises from the possibility that photons hitting a given crystal obliquely might penetrate beyond that crystal and be detected in neighboring crystals, leading to an error in determining the direction of the LOR of the photon (Figure 6.15). The primary result of this effect is the reduction of image resolution in the periphery of the camera. It is only possible to correct or minimize this effect using detectors providing DOI information, that is, detectors capable of measuring the distance from the crystal surface at which the photons interact [79].

6.3 Image Processing

The analysis capabilities of medical images produced by different techniques have improved considerably over time. These analysis procedures have been the object of constant optimization so that they can better reveal structural anomalies or functional behavior. Segmentation algorithms (i.e., for anomaly delimitation) have been developed in this field, along with procedures to determine image characteristics (by defining parameters that quantify visual details), and their coping with previously obtained limits. The

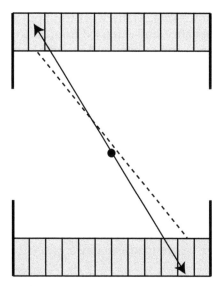

FIGURE 6.15
Depth of interaction (DOI) effect. If a photon pair hits the crystals obliquely, it is possible that the two photons penetrate neighboring crystals and produce a coincidence count in an LOR with a wrong direction (dashed line).

limits are established by observing a very large number of normal individuals, preferably belonging to the population one intends to study.

The practical objective of anomaly detection in medical imaging is to provide a simple yes or no answer on the existence of a given anomaly. This answer is, in principle, substantiated by a set of elements based on certain image features; but when the detection is doubtful or it is not possible to identify the relevant features, then we have to look for contextual information. This is a procedure of image interpretation, somewhat more complex than simple detection. Different image modalities can help to reach a conclusion and they provide an important way of clarifying the presence of pathologies.

Image processing consists of applying a broad set of algorithms and tools to repair, analyze, develop, and visualize images. For instance, it is possible to improve noisy or degraded images to enhance certain image details in order to better detect lesions, to extract a range of parameters from the image, to characterize shapes and textures, and to perform the coregistration of images obtained with different modalities. There are basically three levels of information processing in image techniques:

- Low-level processing is characterized by both the input and the output data being images. It involves basic initial operations, such as preprocessing for noise reduction, contrast enhancement, and image detail improvement.

- Medium-level processing: At this level, the input is an image, but the outputs are image attributes extracted from the input image, such as peripheral regions, contour lines, and even the identification of image objects. Medium-level processing involves tasks such as segmentation (partitioning an image into regions or objects), the description of objects and corresponding data reduction for digital processing, and the classification (and recognition) of individual objects.
- High-level processing: This type of processing involves searching for a set of characteristic objects based on the analysis of images for the performance of cognitive functions usually associated with vision.

In general terms, the most important steps in image processing are the acquisition, transformation (preprocessing), segmentation, selection of characteristics, and classification. The processing operations that can be performed on medical images can be grouped as follows:

- Image enhancement and repair, which includes image addition and subtraction, linear and nonlinear filtering, filter projections, deblurring, and automatic contrast enhancement.
- Image analysis, which includes texture analysis, detection of contour lines, detection of morphology, segmentation (partition into features), region-of-interest (ROI) processing, measurement of image characteristics, and the extraction of relevant image information.
- Image recognition.
- Image compression.
- Visualization and processing of multidimensional images (3D and 4D).
- Colour image processing, which includes color format conversions and the import and export of machine-independent ICC (International Color Consortium) profiles; an example of this type of processing is the color transformation from the RGB (red, green, blue) format used in the visualization to the CMY (cyan, magenta, yellow) format used for hard copy printing.
- Image transformation to alternate versions that are more efficient, where the FFT (fast-Fourier transform), DCT (discrete cosine transform), Radon, and fan-beam projection transforms are included.
- Multimodal images processing and analysis.
- Reconstruction from projections.
- DICOM (digital imaging and communications in medicine) import and export.
- Interactive visualization of images and modular tools to generate GUI (graphical user interface) images.

In digital NM imaging systems, a computer is involved in the acquisition process, as the measured signals become image data only after signal processing. In these systems, it is usual to make a distinction between the processes that operate on the raw data and lead to the formation of an image directly observable on a screen (preprocessing) and the computer-assisted methods for subsequent analysis and modification of the image content. The latter are often called digital image processing, or postprocessing, methods, and they target the extraction and restoring of implicit information that is not directly accessible via visual inspection.

The preprocessing of data has a number of purposes, including attenuation and scattered radiation correction, normalization correction, for example, in intrinsic efficiency, geometric, dead time, decay, and partial volume, or the scaling, sampling, time realignment of data, etc., all of which are intended to produce corrected digital images optimized for input in the subsequent processing tasks.

As a first step, the camera or acquisition device generates an image with supposedly appropriate statistics. This is followed by a second phase, where the raw data are subjected to preprocessing. The next step consists of adding context to the image, which is done by a specialist aided by computer-assisted analysis. The image context takes into account the probable pathology of the patient, its position, the type of study, the working parameters of the equipment, etc., and converts the recorded data into new images, possibly with quantitative information that will support the task of the clinician.

The nuclear imaging software should be able to examine extensive regions of the patient (up to his whole body) in a short time and make use of sophisticated correction methods (e.g., the quality control of the camera) based on efficient algorithms. As such, it must be able to handle very large sets of data and allow the processing and analysis of complex 3D data. It is also necessary to have fast CPU processing speeds for the direct acquisition of 3D or 4D images in real time. The computing industry seems to be reaching a point where it may no longer respond to the highly demanding conditions of real-time image processing, in which case image segmentation techniques will prove to be an important support.

It is also usually necessary to perform several corrections for scintillation cameras to prevent anomalies such as spectral shifts (energy correction), wrong positioning of regions (spatial linearity correction), local counting differences (uniformity correction), loss of counts due to pile-up (pile-up correction), and stabilization of the photomultiplier tubes (high-voltage adjustment).

The radiopharmaceuticals administered to the patients to image physiological functions and identify diseases are becoming increasingly specialized (i.e., organ and disease specific). Those employed in PET and SPECT are widely used at present, but only PET is able to quantify certain important metabolic measurements in absolute physiological conditions and offer the most cost-effective means of staging several types of cancer. The high cost of PET studies and, in particular, of the facilities needed for those studies have restricted its application to a limited number of centers. Nevertheless, PET has provided much information that helped the development of radio-pharmaceuticals marked with gamma emitters for SPECT, which, in many cases, became a good alternative to PET, for imaging the same function. New perspectives were introduced with coincidence detection using dual-head gamma cameras. Comparing SPECT and PET imaging devices in terms of sensitivity, spatial resolution, SNR, performance limits, etc., it seems unlikely that there will be any major changes in the current state of the two techniques in the near future.

Although there have been many advances in radiochemistry and radio-pharmacology, ^{99m}Tc is the most widely used radioisotope in conventional NM diagnostics, and it has been so for over 30 years. It will probably retain its popularity due to its availability, low cost, low radiation dose, and highly convenient chemical properties. Research is currently going on in many labo-ratories all over the world to develop ligands marked with ^{99m}Tc and ^{123}I for SPECT, and with ^{18}F or ^{11}C for PET, designed to bond to chosen receptors in neurology, oncology, and cardiology.

The radiopharmaceuticals used in NM imaging diagnostics should provide useful clinical information and expose patients to minimal radiation doses. For this, the radiation emitted by the tracer should not contain particles, the half-life of the nuclide has to be of the order of the exam duration, the energy of the gamma photons needs to be appropriate for efficient detection with gamma cameras, and the specific activity of the radiopharmaceutical has to be high enough not to affect the biochemistry of the patient.

The accurate reconstruction of 3D and 4D images is difficult, mainly owing to noise; the lack of efficient analytical reconstruction methods, especially for certain acquisition geometries; image degradation factors; and the mis-alignment of imaging systems, particularly in devices used for the imaging of small animals. Further, with the increasing interest in minimal invasive surgery and in intensity modulated radiotherapy (IMRT), the precise locating of anatomical targets has become more and more important.

6.3.1   Image Reconstruction

As noted in Chapter 5, tomographic images obtained in PET and SPECT result from the application of reconstruction algorithms that can be classed as ana-lytical and iterative algorithms and which, in general, take some model with regard to how the activity distribution of the radioactive tracer in the volume of the patient is translated into measurements, normally in the form of parallel projections of the distribution. The inversion of the image formation model allows recovery of the original tracer distribution as a volume of images cor-responding to sequential slices of the object along a chosen direction. The model most often used to describe the formation of the parallel projections of the object presumes that, in the absence of attenuation or Compton scattering, the number of photons detected in a projection line which passes through the object is proportional to the integrated activity of the distribution $f(x, y, z)$ along that direction, that is,

$$\sum_{\text{direction } i} (\gamma \text{ counts}) \propto \int_{\text{direction } i} f(x, y, z) \, dr_{\text{direction } i}. \tag{6.24}$$

This notion corresponds to the notion of a Radon transform of an object, a mathematical operation that is central to all tomographic techniques.

Taking the notation introduced in Chapter 5 (see Figure 5.27), this line-integral approximation [80] can be expressed in terms of the parallel projections of the object's activity, in two or three dimensions, by

$$
\begin{aligned}
p_{2D}(x_r, \phi, z) &= c \int_{-R_D}^{+R_D} f(x, y, z) dy_r \\[2em]
p_{3D}(x_r, y_r, z, \theta) &= c \int_{-R_D}^{+R_D} f(x, y, z) dz_r
\end{aligned}
\tag{6.25}
$$

For image reconstruction purposes, the constant c can be ignored, although it is usually experimentally determined in PET when we want to have an absolute calibration of the image values. This constant is a scale factor, which endows the imaging technique with the ability to measure the absolute concentration of the radioactive tracer throughout the body of the patient and whose value can be found with the aid of phantoms during camera calibration.

The reconstruction algorithms normally assume that the data are organized in parallel projections and follow analytical or iterative strategies, depending on whether they invert Equations 6.25 or numerically compute the activity distribution that better reproduces the 2D or 3D projections acquired.

6.3.1.1 2D vs. 3D Reconstruction

The term *2D reconstruction* refers to the determination of the activity distribution for a given plane or slice of the object, using for that purpose the projections of the object measured along that plane. This term also means the process of computing the 3D distribution of the activity of the object by independently reconstructing the activity in each 2D plane, using a 2D reconstruction algorithm. For the latter, the 3D volume obtained has the same dimensions as those obtained by the 3D reconstruction of the object. The 3D reconstruction in PET differs from 2D reconstruction in that it uses

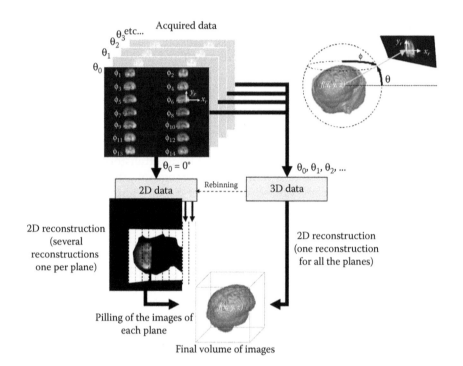

FIGURE 6.16
Outline of the PET image reconstruction process in 2D and 3D.

information collected in oblique planes and by the fact that it uses the information from all the projections together (rather than independently, plane by plane) to construct the volume of the object (Figures 6.16 and 6.17).

6.3.1.2   2D Reconstruction

6.3.1.2.1   Analytical Methods

In a 2D reconstruction algorithm, image reconstruction is carried out in planes perpendicular to the axis of the system, which means that the algorithms handle projection arrays corresponding to fixed z planes, $p(x_r, \phi, z)|_{z=z_0}$. These arrays can be looked at as the compilation of one-dimensional parallel projections along the different acquisition angles ϕ (Figure 6.18a), which enables the simplest analytical reconstruction procedure to be defined, the backprojection [81,82]. In this procedure, the number of photon counts for a given pair of (x_r, ϕ) coordinates is added to all the image points located along the corresponding projection line, as depicted in Figure 6.18b. By performing the backprojection of all the one-dimensional projections over all the ϕ angles, one obtains an estimate $f_{est}(x, y, z)|_{z=z_0}$ of the activity distribution $f(x, y, z)|_{z=z_0}$, which gave rise to those projections.

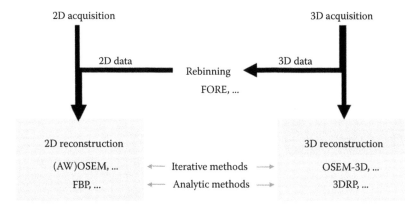

FIGURE 6.17
Summary of the reconstruction strategies in PET, indicating some of the most popular algorithms.

However, the finite sampling of projection angles, together with the uniform distribution of the photon count values resulting from backprojection along the projection line, generates artifacts, that is, there are points in the reconstructed image to which some counts are attributed but that in reality have no activity (Figure 6.19). In the limit when the number of projections used in the reconstruction becomes infinite, those artifacts give rise to a blurring effect [81], which is the spreading out of counts from the point where the activity source is located; the magnitude of the spread is proportional to $1/r$, with r being the distance from the source (Figure 6.18c). This effect is directly due to the oversampling that the spatial point where the source lies is subjected to, as it concentrates a density of projection lines, which, in the limit when the variation of ϕ is continuous, tends to infinity [80]. Mathematically, this effect corresponds to a convolution in each point (x, y) of the true image

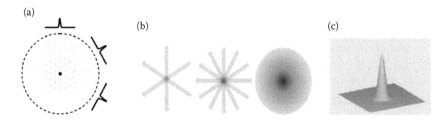

FIGURE 6.18
(a) One-dimensional projections in three different Π angles generated by a point source located in the detector axis. (b) Image of the point source shown in (a) reconstructed by the backprojection of 3, 6, and a very large number of projections. When the image is reconstructed using few projections, star-shaped artifacts are created, which become blurred when the number of projections is very large. (c) The blurring effect affecting the backprojection.

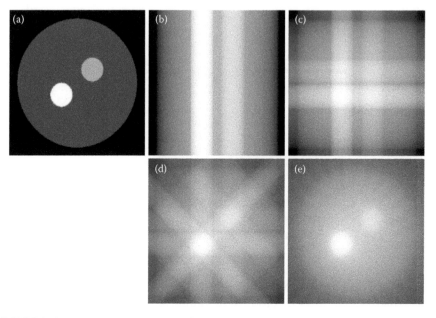

FIGURE 6.19
Comparison between the true activity distribution (a) with images reconstructed using the backprojection of 1, 2, 4, and 16 parallel projections of a (b–e).

with the function $1/r$ [82], that is,

$$f_{est}(x,y,z)\big|_{z=z_0} = f(x,y,z)\big|_{z=z_0} {}^*1/r, \tag{6.26}$$

where $*$ denotes the convolution operation, and r is the distance to the point (x,y) for the plane $z = z_0$.

The solution to the blurring effect is based on not allowing the addition of counts along the projection lines to be uniform, but instead weighting it so that fewer counts are added to points near the activity source. A simple way of implementing this unequal distribution is by convoluting, or filtering, the backprojected image with a given filter [81], an operation that is performed by multiplying in the Fourier space the Fourier transforms of the filter, $w(\nu)$, and of the image data. In the method known as filtered-backprojection (FBP) [81,82], the filter is directly applied to the one-dimensional projections $p(x_r, \phi, z)\big|_{z=z_0}$ for each ϕ angle, and the result, a set of filtered projections $p(x_r, \phi, z)\big|_{z=z_0} {}^*w(x_r)$, is then backprojected. If $\Re\{\}$ denotes the backprojection operation[*],

$$\Re\left\{ p(x_r, \phi, z)\big|_{z=z_0} \right\} = \int_0^{2\pi} p(x_r, \phi, z)\big|_{z=z_0}\, d\phi = f_{est}(x,y,z)\big|_{z=z_0}, \tag{6.27}$$

---

[*] It should be noted that with PET, the upper limit of the integral in Equation 6.27 is π, because there are two detected photons and not just one, as in SPECT.

then one has for the FBP method

$$f_{estFBP}(x, y, z)\big|_{z=z_0} = \Re\left\{p(x_r, \phi, z)\big|_{z=z_0} {}^*w(x_r)\right\}. \tag{6.28}$$

An alternative method consists not of filtering the one-dimensional projections but of first backprojecting them and only then applying the filter; this is the backprojection-filtering method (BPF) [81,82]:

$$f_{estBPF}(x, y, z)\big|_{z=z_0} = \Re\left\{p(x_r, \phi, z)\big|_{z=z_0}\right\} {}^*w(x, y). \tag{6.29}$$

This method is seldom used, as it produces images of a quality comparable to those obtained using the FBP, with the aggravation that it involves the computation of a 2D Fourier transform, which makes it slower to execute.

The filtering tries to correct the effect of convolution with a function of the type $1/r$; as the Fourier transform of $1/r$ is $1/v$; the function that naturally appears as the filter to use is the ramp function in the frequency domain, $w(v) = v$. In practice, however, the finite spatial sampling of the activity distribution f forces the filter to be truncated at the Nyquist frequency, $v_N = 1/2\Delta x_r$, and prevents the exact shape of f from being restored [80]. Further, the statistical noise existing in the data is amplified by the ramp filter in the high frequencies, which diminishes the contrast of the reconstructed image. To minimize this problem, the ramp filter is usually apodized with decreasing functions of v, such as the Hann and Hamming windows [83] (Figure 6.20). Apodization causes the smoothing of the image by suppressing abrupt changes but decreases image resolution [81]; for that reason, the apodization window parameters must be chosen while taking into account that one is trading less noise for less resolution and that the level of compromise between the two must be governed by each specific situation.

6.3.1.2.2   Iterative Methods

The analytical reconstruction methods, based on Fourier transforms and backprojection operations, are very important in clinical practice, as they are fast, stable, and easy to implement. However, these methods are rather inflexible, because the possibility of incorporating information from the process of image formation in the reconstruction process is very limited, as is the imposition of certain known physical conditions, such as the degree of smoothness of the imaged object or the shape and volume occupied by it. In particular, the fact that they employ filters truncated at or near the Nyquist frequency often generates artifacts by underestimation at the frontiers between regions with very different activities, which sometimes yields negative (i.e., nonphysical) values for the activity in certain areas of the image [80].

Iterative reconstruction methods, on the other hand, are very flexible and allow the incorporation of image formation models that correct the effects of image deterioration, such as Compton scattering or attenuation, during the reconstruction process itself [80]. This is a major advantage over the analytical

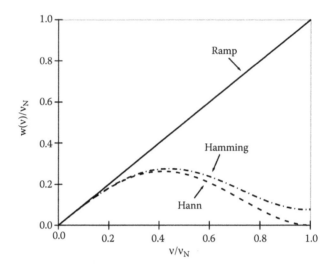

FIGURE 6.20
The ramp filter, truncated in the (spatial) Nyquist frequency $vN = 1/2\Delta xr$, and the two most common apodization windows used, the Hann and the Hamming windows.

methods, which, by assuming that the value of a projection is proportional to the integral of the activity along the corresponding projection line, force any data correction procedure to be performed before the reconstruction. The main disadvantage of the iterative methods compared with the analytical ones is their slowness, because they are far more intensive from the point of view of numerical processing (typically they take up to 2 orders of magnitude times longer than analytical methods to reconstruct the same image). However, advances in the processing power of modern computers observed in the last few years have made these methods commonplace in many circumstances of clinical practice, and they are nowadays the predominant reconstruction methods used in commercial cameras. Still, the nonlinear character of iterative methods may, on rare occasions, give rise to erratic behavior capable of seriously affecting the reconstructed image.

The working principle of iterative methods is simple: Plane by plane, one generates an estimate of the activity distribution in the patient, $f_{est}(x, y, z)\big|_{z=z_0}$, which is iteratively improved by comparing the parallel projections $p_{est}(x_r, \phi, z)\big|_{z=z_0}$ generated by the estimate with the projections $p(x_r, \phi, z)\big|_{z=z_0}$ actually measured. If the two projections are significantly different, the estimate of the activity distribution is corrected by an amount that is a function of the differences found between the projections (Figure 6.21) [84]. This cyclic procedure is normally initiated with an empty (or sometimes unitary) activity distribution, and it is repeated until the projections generated by the estimate and the measured projections are equal (within a preset tolerance), or a certain condition on the activity distribution is met.

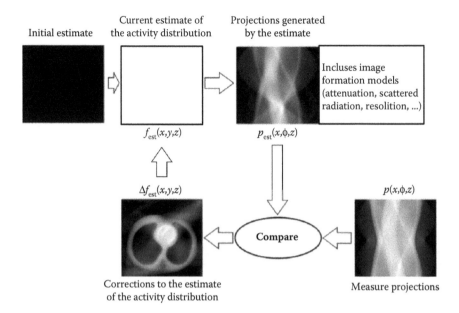

Initial estimate | Current estimate of the activity distribution | Projections generated by the estimate

Incluses image formation models (attenuation, scattered radiation, resolution, ...)

$f_{est}(x,y,z)$

$p_{est}(x,\phi,z)$

$\Delta f_{est}(x,y,z)$

$p(x,\phi,z)$

Compare

Corrections to the estimate of the activity distribution

Measure projections

FIGURE 6.21

Outline of an iterative reconstruction algorithm. The cycle shown is processed plane by plane, until a stopping condition is met.

Obtaining the projections $p_{est}(x_r, \phi, z)|_{z=z_0}$ from the estimate of the activity distribution $f_{est}(x, y, z)|_{z=z_0}$ is a fundamental component of the iterative process. This task is carried out using a system matrix A, which establishes a relation between the values measured along the different projection lines, often organized in the shape of a one-dimensional data vector, y, with the volume elements that constitute the object, usually discretized in voxels whose values are stored in a vector x:

$$y = Ax + n. \tag{6.30}$$

The one-dimensional vector n contains the noise present in the measurement, which can be modeled in many different ways although it is customary to assume that it obeys a Poisson or Gaussian probability function.

The existing iterative methods mainly differ in how the projections generated by the activity estimate and the measured projections are compared and also in how the corrections to the estimate in each iteration are computed from the differences between those two (this also implies different processing times for different methods). One of the simplest iterative methods is the Algebraic Reconstruction Technique (ART), which is based on a purely algebraic iterative algorithm that improves the activity estimate angle by angle, comparing each measured projection with the calculated projection using a simple subtraction [85]. This method, although simple, is slow compared with most other iterative algorithms for the number of pixels commonly used in SPECT and PET, and it is, therefore, rarely used.

6.3.1.2.2.1 The MLEM Algorithm The Maximum Likelihood Expectation Maximization (MLEM) algorithm describes the acquisition of counts in the parallel projections, y, as

$$y_j = \sum_i a_{ij} x_i, \qquad (6.31)$$

where x_i is the activity in the voxel i, contributing with weight a_{ij} for the projection channel j. The inversion of the matrix $A = \{a_{ij}\}$ in Equation 6.31, which would lead to the direct solution of x to yield the activity distribution, is impractical for several reasons: (1) The elements of the system matrix a_{ij} are not known with enough accuracy, as they depend on the activity x_i due to Compton scattering; (2) There might not be enough measured projections to render Equation 6.31 a fully determined system of equations; and (3) The statistical nature of radioactive decay and its detection introduces ambiguities in the system of Equation 6.31. For these reasons, the inversion of Equation 6.31 has to be done under further assumptions, which leads to a solution given by an iterative procedure.

The quantitative aspect of that procedure may be deduced by considering that the probability of observing a given distribution of projected counts, $y = y_{obs}$, given a hypothetical voxel activity distribution, x_{hyp}, can be maximized if maximum likelihood arguments are applied to those variables. The probability of success, that is, of detecting the activity of the voxel x_i in a given projection channel y_j, is much smaller than the total number of possibilities, leading to the conclusion that the detection process obeys Poisson statistics. Further, since a pair of photons interacting in a given pair of crystals cannot be detected in any other pair (ignoring the existence of Compton scattering in the crystals), one concludes that, besides obeying Poisson statistics, the variables are independent as well. It is, therefore, possible to use the Poisson probability distribution to write the conditional probability, $P(y_{obs}|x_{hyp})$, and maximize it in the sense of having maximum likelihood. The manipulation of that function [86] shows that it is concave and that the iterative convergence to the maximum follows the recurrence relation

$$x_i^n = x_i^{n-1} \sum_j a_{ij} \frac{y_j}{\sum_k a_{kj} x_k^{n-1}}, \qquad (6.32)$$

where n is the iteration number, $i = 1, \ldots$ and $k = 1, \ldots$ refer to the voxels in the image, and $j = 1, \ldots$ refers to the projection channels. The sum in k in the denominator of the second factor expresses the projection of the activity distribution x^{n-1}, that is, the measured projection counts if x^{n-1} is the true image. The sum in j in the numerator corresponds to the multiplication of the measured projections by the transpose of the system matrix A, and it represents the backprojection of the ratio between the measured data and the estimated projections in the iteration n. The vector x^1 is usually a uniform distribution ($x_i^1 = 1$).

This algorithm can be easily implemented in series, processing the data directly from list-mode format, as it works only with the projection channels in the vector y, which contribute to the backprojection, that is, have nonzero counts. In terms of its stability, the algorithm converges to the activity estimate that best approaches the data in terms of a Poisson likelihood. However, this approximation assuming Poisson noise induces high-frequency instabilities whose amplitude sharply increase after a certain number of iterations. This is expressed in high-frequency artifacts in the reconstructed image, which can be remedied by several techniques. For instance, it is possible to apply a 3D Gaussian filter after the image is reconstructed, in which the intrinsic spatial resolution of the filter must correspond to the one that we want to obtain in the image for the chosen SNR. Possibilities consist of filtering the data before applying the algorithm or even stopping the reconstruction in a previously chosen iteration. There are several methods that estimate the most appropriate number of iterations to run [87,88], although all clinical applications use a number of iterations empirically determined [89].

6.3.1.2.2.2 Other Methods The Ordered Subset Expectation Maximization (OSEM) method [90] became very popular, because it significantly accelerates the convergence to a final solution relative to the MLEM algorithm, although it is based on exactly the same principles. The fast speed of the OSEM algorithm is achieved by dividing the data into subsets: in each iteration, the estimate of the activity distribution obtained from one subset is used as the initial estimate in the next one. The convergence process of OSEM is swifter than MLEM; but, on the other hand, it has an increased numerical noise, a fact that is borne out by the different solutions found by the two methods. An improvement of the OSEM method often used today is the Attenuation Weighted OSEM (AWOSEM) [91], which is based on a better modeling of the image noise to achieve better image quality without significant reductions in the computation speed.

6.3.1.3 3D Reconstruction

In the 3D mode of PET, image reconstruction using analytical methods resorts to the generalization to three dimensions of the methods developed for the 2D mode [89,93]. There is, however, an important difference that raises several problems from the practical point of view: In the 2D mode, one has 2D arrays corresponding to the projections $p_{2D}(x_r, \phi, z)|_{z=z_0}$ of each plane separately; whereas in the 3D mode all projections have to be simultaneously used, which forces reconstruction to deal with the projection arrays $p_{3D}(x_r, y_r, \phi, \theta)$ in the totality of their four dimensions. In this way, 3D reconstruction exhibits a new aspect relative to 2D reconstruction: there is information redundancy in the measured data, so it is possible to reconstruct the imaged object using only a part of the 3D mode data (e.g., it is possible to reconstruct the object using only the nonoblique projections, i.e., those whose co-polar angle is 0°, which corresponds exactly to the 2D mode data).

Additionally, 3D reconstruction implies a drastic step-up in the amount of data, which now includes oblique projections and makes computational processing quite a lot more complex and lengthy. The result, however, is images of better quality than the 2D ones using the same amount of injected radiotracer, because one has a larger number of detected coincidences in the 3D mode. This fact justifies the increasing interest in the development and implementation of 3D image reconstruction methods in many PET systems, in spite of their slowness.

6.3.1.3.1 Analytic Methods

Adapting the analytical methods of the 2D mode to the 3D mode is straightforward, as the backprojection operation of parallel projections, now two-dimensional in the coordinates (x_r, y_r), is still possible for each direction of space specified by the angular coordinates (ϕ, θ). Thus, to reconstruct the activity distribution, one just needs to backproject all the projections $p_{3D}(x_r, y_r, \phi, \theta)$ along all the directions in space (ϕ, θ). Just as for the 2D mode, filtering is still necessary to prevent the oversampling leading to blurring effects. However, the process of backprojection in the 3D mode has a fundamental problem that does not exist in the 2D mode: depending on the acquisition parameters and on the geometrical relation between the imaged object and the detector, there may be parallel projections that are incomplete, meaning that the sampling of the object volume is not uniform (Figure 6.22) [80].

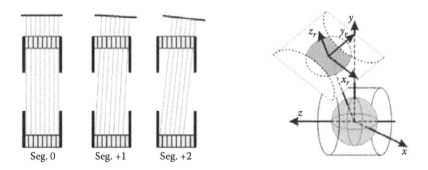

Seg. 0   Seg. +1   Seg. +2

FIGURE 6.22
(a) Axial view of planes defined on a PET system in the 3D mode. The cross-section of the parallel projections decreases as the inclination θ of the projection lines increases. (b) Drawing of an incomplete parallel projection due to the limited FOV of the PET system.

The problem of the incomplete projections can be solved by introducing a previous step of 2D mode reconstruction, which enables an initial estimate of the activity distribution $f^i_{est}(x, y, z)$ to be obtained. This estimate is then projected to fill the regions in the (incomplete) oblique sinograms, from which a set of complete parallel projections $p_{3Dcompl}(x_r, y_r, \phi, \theta)$ is drawn, and to which it is possible to apply a 3D FBP reconstruction operation with a chosen filter. The whole process is known as 3D-Reprojection (3DRP) [89], and it is depicted in Figure 6.23.

6.3.1.3.2   *Iterative Methods*

The philosophy of image reconstruction in PET using iterative methods is the same as that depicted in Figure 6.21 for SPECT and 2D mode PET. It is a cyclic process that involves the gradual improvement of an estimate for the activity distribution by comparing, in each iteration, the projections generated by the estimate with the measured projections and the computation of a correction to the activity distribution estimate from that comparison. The implementation of this principle is identical for the PET 2D and 3D modes, differing only in the dimension of the projections used (two-dimensional in the 2D mode, four-dimensional in the 3D mode) and of the system matrix employed to obtain the projections corresponding to the estimate of the activity distribution in each iteration. The different iterative methods available for 3D PET include ART, MLEM, OSEM, and others.

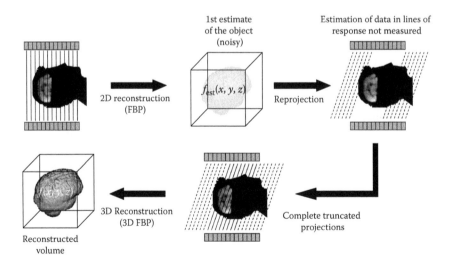

FIGURE 6.23
Outline of the analytical reconstruction algorithm 3DRP [92]. For the standard analytical reconstruction algorithm 3D FBP to be correctly applied, the measured projections have to be complete, though, in general, the oblique projections are truncated and that does not happen. Hence, the 3DRP method aims at completing the oblique projections by projecting the volume obtained from the reconstruction of the nonoblique projections.

As in SPECT and 2D-mode PET, these methods allow the inclusion of image formation models which enable the correction of several physical factors that degrade the image as well as incorporating a priori information about the image one intends to reconstruct. Some examples of those effects are the camera finite point spread function (PSF), the existence of scattered radiation, or the noise embedded in the attenuation correction; whereas *a priori* information may include the setting of mathematical conditions for the formed image, such as the positivity of the activity in all points of space, or the confinement of the activity within the frontiers of the object, etc.

Although iterative methods in PET also exhibit nonlinear behavior that is difficult to foresee and interpret, their advantages over the analytical methods largely overcome the disadvantages, which is why they are today the most common approach for image reconstruction in the 2D mode [94–96]. In the 3D mode, however, it is rare in clinical practice to perform image reconstruction using pure 3D data, and either the analytical or iterative methods described in the next section are used. This is due to matters of practical order related to the large amount of data contained in the parallel projections $p_{3D}(x_r, y_r, \phi, \theta)$ and to the fact that they have to be simultaneously dealt with. This makes the system matrix so big that it is difficult for computers to cope with, both from the point of view of storage in memory and from the point of view of processing speed [97]. The iterative reconstruction of 3D data ends up being mostly employed in research, almost always with the purpose of creating new computational methodologies capable of making the reconstruction of 3D data feasible in practice.

6.3.1.4 Rebinning Methods

The analytical or iterative reconstruction of PET data acquired in 3D mode is significantly more time consuming than reconstruction in 2D by the same methods, so it is almost completely excluded from clinical practice unless dedicated hardware is used. However, the fact that 3D mode data offer a substantially better sensitivity has prompted the search for methods that, using all the information contained in the 3D data, would carry out the image reconstruction in the 2D mode. The solutions found have a simple philosophy that consists of performing the regrouping, or rebinning, of 3D data into 2D data (Figure 6.24), adding the counts in oblique planes to the 2D planes. Assuming that the data are organized in the sinogram format, this operation corresponds to the transformation

$$s_{3D}(x_r, \phi, z, \Delta r) \rightarrow s_{2D}(x_r, \phi, z), \tag{6.33}$$

where each specific rebinning method establishes a rule to attribute counts from oblique directions to the transaxial planes used in the 2D mode.

For the simplest rebinning method, the Single-Slice Rebinning (SSRB) [98], the counts in an oblique LOR with axial coordinate z are added to the transaxial plane with the same coordinate z independently of the LOR inclination Δr.

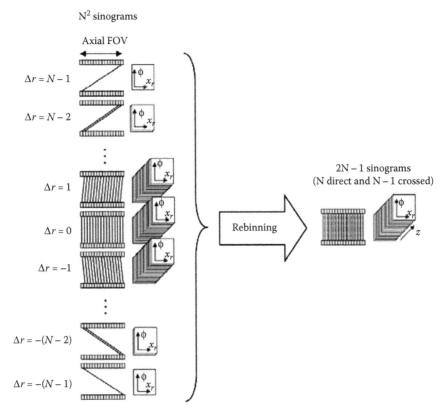

FIGURE 6.24
Outline of the rebinning process. The 3D sinograms are grouped in a smaller number of 2D sinograms corresponding only to transaxial planes.

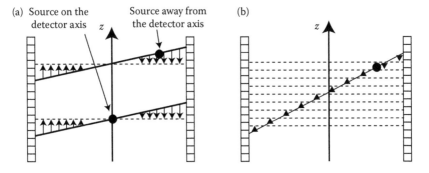

FIGURE 6.25
Attribution of counts of oblique LORs to transaxial planes in the (a) SSRB and (b) MSRB methods. The error locating the axial coordinate of activity sources that are outside the detector axis in SSRB is substantially reduced in the MSRB method, as the counts of oblique LORs are distributed over all the planes crossed by the LORs.

The reduction of the sinograms follows

$$s_{3D}(x_r, \phi, z) = \sum_{\Delta r} s_{3D}(x_r, \phi, z, \Delta r)|_{(x_r, \phi, z)\text{ fixed}}, \qquad (6.34)$$

implying that the counts of each oblique LOR are added to the middle plane of the two detectors defining that LOR (Figure 6.25a). The simplicity of this rebinning rule does, however, have the downside of introducing errors in the determination of the axial coordinates of activity sources placed away from the detector axis [80]. The Multi-Slice Rebinning (MSRB) [99] method is an alternative which addresses that problem and equally distributes the number of counts in an oblique LOR by all the planes traversed by that LOR (Figure 6.25b). The MRSB method is fast to process numerically, but it is not very robust to the presence of noise in the data, as it displaces the activity of very oblique LORs (which are those with fewer counts) by several planes.

These two methods are often ignored in favor of a third rebinning method, known as Fourier Rebinning (FORE) [100]. This method is very stable to noise and involves the numerical computation of the 2D Fourier transform $S_{3D}(v_x, v_\phi, z, \Delta r)$ of each oblique sinogram $s_{3D}(x_r, \phi, z, \Delta r)$ collected in the 3D mode. The FORE method is based on an approximation, known as frequency–distance relationship [80], according to which $S_{3D}(v_x, v_\phi, z, \Delta r)$ depends only on the activity distribution $f(x, y, z)$ in a set of discrete points along each LOR. The transform $S_{3D}(v_x, v_\phi, z, \Delta r)$ of each oblique sinogram (z and Δr fixed) is the data structure suffering the rebinning process, with the value of each coefficient of that transform being added to a sum-sinogram $S_{2D}(v_x, v_\phi, z)$ in the z plane coordinate that intersects the LOR in question in the point where the frequency–distance relationship is satisfied. The regrouped 2D sinograms $s_{2D}(x_r, \phi, z)$ are finally obtained by calculating the (two-dimensional) inverse Fourier transform of $S_{2D}(v_x, v_\phi, z)$ for each plane individually. A more detailed description of this method, in particular about the frequency-distance relationship and the region where it is valid, can be found in [80].

Any of these rebinning methods adds numerical noise to the data, leading to a performance after the 2D reconstruction that is worse than if pure 3D reconstruction had been performed. However, rebinning makes the image reconstruction task much faster; and the noise problem can be largely compensated when the associated 2D reconstruction method belongs to the class of iterative methods, because they are capable of modeling and suppressing noise; the FORE+OSEM combination, one of the most popular methods of reconstruction today, is a good example of this type of approach [80].

The combination FORE+AWOSEM has also become very popular, and it is an improvement over FORE+OSEM, because it allows the reconstruction algorithm to be applied in conditions nearer to the hypotheses on which they are based. In fact, the FORE algorithm assumes that the data from which it starts are corrected for the different physical effects that affect the measurement. However, the carrying out of that correction destroys the Poisson character of the data, a hypothesis which underpins the OSEM algorithm. So, to apply the FORE algorithm in its best conditions one would have to subsequently apply the OSEM algorithm in bad conditions, that is, knowing that the data output from the FORE procedure is not Poisson distributed, having a variance that does not equal the average. To avoid this problem, the solution found with the FORE+AWOSEM procedure consists of first applying the FORE algorithm and afterward correcting the data with a weighting factor that tries to restore the Poisson character to the data before applying the OSEM method. After applying OSEM, the correction is reversed, and a final result is eventually produced. As the name indicates, the correction in the AWOSEM method is based on the attenuation correction coefficients. The attenuation correction is the one that affects the amplitude of the measured values the most and due to that, it has a large influence on removing the Poisson-like character of the data; it also has the advantage of being easy to apply, as the attenuation factors are available, being based on the transmission measurement. Other weighting methods have been proposed, including other corrections, such as the normalization correction [91], as well as other criteria directly related to noise [101,102].

There are two other specific rebinning alternatives, named FOREX [103] and FOREJ [104], which are more precise than the FORE method, which is only valid up to θ angles of about 20°. These methods are referred to in the literature as exact methods, in contrast to the approximated methods mentioned earlier (SSRB, MSRB, and FORE). As the name indicates, they are based on the frequency-distance relationship as well, but they employ additional correction terms that allow their validity to be extended to larger θ angles, although at the cost of a longer computation time and larger implementation complexity. Finally, rebinning methods appropriate for TOF-PET have been recently put forward [105].

6.3.2 Concepts of Digital Image Processing

6.3.2.1 *Image Preprocessing*

Image preprocessing concerns image operations aiming at correcting or improving the image. Point-based processing is a special type of image preprocessing. Point-based operations are applied pixelwise without taking into account the values of the pixels in the neighborhood. As a result of these operations, the histogram of the gray image is changed.

6.3.2.2 *Point-Based Preprocessing*

In this type of preprocessing, each gray level value is changed into another value depending on the type of preprocessing operation. Let the set of gray level values be $[0, K - 1]$. Given a specific gray level value $u \in [0, K - 1]$, it is changed into another gray level value $v \in [0, K - 1]$ depending on a function

f such that

$$v = f(u)$$

Several types of image transformation can be implemented depending on the function f. In general, f is monotonic and increasing. As a result, the ordering of the gray level values between black and white is preserved [106]. This mapping can be defined by a look-up table (LUT), as the gray level values are discrete and integer (the function is discrete and, therefore, it defines a mapping between discrete values). This mapping can be represented by Figure 6.26.

The use of the LUT enables the contrast and image intensity to be changed. If the function is linear or piecewise linear, then changing the slope of the line permits changes in image contrast. Figure 6.27 shows an example of a piecewise linear change of contrast. This transformation can be defined by

$$v = f(u) = \begin{cases} au & 0 \le u < I \\ b(u - I) + I' & I \le u < J \\ c(u - J) + J' & J \le u < K \end{cases} \qquad (6.35)$$

If the line slope is greater than 1, then the mapping increases the contrast. Image histogram analysis enables the values of the parameters to be chosen, namely, a, b, c and I and J. The coordinates of points (I, I') and (J, J') define the type of transformation. If $I = I'$ and $J = J'$, the transformation is a linear function that does not change the gray levels. If $I = J, I' = 0$, and $J' = K - 1$, then the mapping performs an image binarization. In this case, the image has only two gray level values, in general 0 and 1.

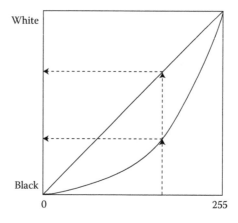

FIGURE 6.26
Gray level mapping (Adapted from A. K. Jain. *Fundamentals of Digital Image Processing*, 2nd Edition, Prentice-Hall, Englewood Cliffs, NJ.).

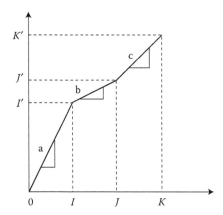

FIGURE 6.27
Changing the image contrast using piecewise linear transformations. (From A. K. Jain. *Fundamentals of Digital Image Processing,* Prentice-Hall, Englewood Cliffs, NJ, 1989. With permission.)

6.3.2.3 Gamma Correction

In general, images with high contrast are easier to visually analyze. In many cases, the nonlinear adjustment of the contrast offers better results. One of the commonest nonlinear adjustments of the contrast is known as "gamma correction." In this case, the function $f(u)$ is defined as follows:

$$v = f(u) = cu^{\gamma} \tag{6.36}$$

Values of $\gamma > 1$ increase the contrast whereas values of $\gamma < 1$ decrease the contrast. However, saturation does not occur for gray level values within the variation range [107]. These transformations are also known as mappings according to the power law. Many devices used for printing and visualizing images operate according to the power law (Figure 6.28).

It is, in fact, the process by which these devices correct their response obeying the power law that is called gamma correction. For example, cathode ray tubes have a relationship between light intensity and voltage that follows the power law. Gamma correction is important for the correct visualization of the images. Another type of correction that can be used is based on a sigmoid function:

$$v = f(u) = \frac{1}{1 + e^{-\alpha u + \beta}} \tag{6.37}$$

This nonlinearity has two degrees of freedom and can, therefore, produce a more balanced contrast improvement.

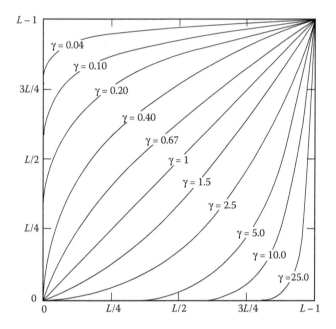

FIGURE 6.28
Gamma correction for several gamma values. (From R. C. Gonzalez, R. W. Woods, *Digital Image Processing*, 2nd Edition, Prentice-Hall, Englewood Cliffs, NJ. With permission.)

6.3.2.4 Logarithmic Transformations

Logarithmic transformations are usually employed to increase the range of gray level values corresponding to dark pixels while simultaneously compressing the range of gray level values corresponding to the whitest pixels [107]. The function $f(u)$ specifying the transformation is generally defined as follows:

$$v = f(u) = c \log(1 + u) \tag{6.38}$$

Transformations defined according to the power law tend to be more flexible than this one, but logarithmic transformation provides a more effective compression of the dynamic range of the gray level values when the image contains considerable variations in gray level values. A transformation defined by the inverse of the logarithm has the opposite effect.

6.3.2.5 Histogram Processing

The histogram image represents the relative frequency of occurrence of gray levels in one image. Histogram modeling methods change the image so that the resulting histogram matches a predefined type. This modification improves the image quality. One of the most widely used methods is histogram equalization [106]. The goal is to change the histogram so that the final histogram is uniform, that is, a histogram in which all the gray levels occur with the same frequency. Let m_k be the number of pixels in an image with gray level k. If M is the total number of pixels in one image, the normalized histogram is given by $p(k) = m_k/M$. In the normalized histogram, the sum of all its elements is 1. To obtain an equalized histogram, the following value is computed for each gray level k:

$$r = T(k) = \sum_{i=0}^{k} p(i) = \sum_{i=0}^{k} \frac{m_i}{M} \qquad (6.39)$$

The transformed image is obtained by replacing each gray level k by gray level r. The image transformation has an inverse and is monotonic, which means that it will preserve the gray level ordering from black to white. This transformation ensures that one gray level will never be replaced by a darker one. This type of histogram modification is usually employed to increase the image contrast. The procedure can be generalized to obtain other types of histogram.

6.3.2.6 Correction of the Geometric Distortion

Geometric distortions often occur in image acquisition processes, leading to changes in the spatial relationships between the pixels. Spatial transformations are used to suppress geometric distortions. The spatial transformations change the relative positions between pixels. Once the geometric changes have been suppressed, the new gray level values for the pixels have to be estimated. These new values are estimated by means of interpolation. Geometric transformations are usually expressed by means of a matrix. The commonest geometric transformations include translation and rotation (which are usually called Euclidean because they preserve distances and angles). A slightly more general set of transformations includes rotations, translations, scaling, and reflection (these are called affine transformations because they preserve parallelism). Given a pixel of coordinates (x, y), the coordinates of the transformed pixel (x', y') can be calculated by

$$\begin{bmatrix} x' \\ y' \end{bmatrix} = \begin{bmatrix} a & b \\ c & d \end{bmatrix} \begin{bmatrix} x \\ y \end{bmatrix} + \begin{bmatrix} s \\ t \end{bmatrix} \qquad (6.40)$$

In the general case, the coordinates (x', y') may not be integers. It is, therefore, necessary to estimate the corresponding gray level (Figure 6.29). The usual procedure consists of considering that the coordinates (x, y) correspond to the pixel in the image with distortion and then estimate the coordinates of the corresponding pixel (x', y') in the image without distortion. Since there is an analytic model for the geometric distortion which implies that, in general, the corresponding gray level value has to be estimated based on the

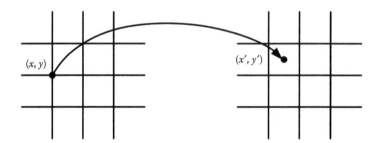

FIGURE 6.29
Correction of geometric distortion and interpolation.

gray level values of the neighboring pixels. The simplest solution consists of interpolating the value based on the gray level value of the nearest neighbor. This kind of interpolation is usually called zero-order interpolation. It involves choosing for the gray level value of the pixel in the image without distortion (with coordinates $[x, y]$) the gray level value of the pixel closest to (x', y'). The coordinates for this pixel can be obtained by rounding off the coordinates (x', y'). A widely used solution is bilinear interpolation. Consider four neighboring pixels whose coordinates are (x_i, y_i), (x_i, y_{i+1}), (x_{i+1}, y_i), and (x_{i+1}, y_{i+1}). Let $g(x, y)$ be the gray level value of pixel (x, y). Then the gray level value $g(x', y')$ estimated using bilinear interpolation is given by

$$g(x', y') = (1 - t)(1 - s)g(x_i, y_i) + t(1 - s)g(x_{i+1}, y_j)$$
$$+ (1 - t)sg(x_i, y_{j+1}) + tsg(x_{i+1}, y_{j+1})$$

where $x_i < x' < x_{i+1}, y_i > y' < y_{i+1}$ and

$$t = \frac{x' - x_i}{x_{i+1} - x_i} \qquad s = \frac{y' - y_i}{y_{i+1} - y_i} \qquad (6.41)$$

Other types of interpolation are possible, in particular, cubic interpolation using splines. However, results obtained with bilinear interpolation are generally good enough for most applications.

6.3.2.7 Color Preprocessing

Human perception of color is a complex phenomenon that is not yet fully understood. It has been experimentally determined that for most viewers only three colors (usually called primary colors) are required to obtain (by combination) any other color. However, for this to be possible, it has to be admitted that the combination can be subtractive and that the primary colors are independent (i.e., mixing two of the primaries cannot produce the third). This principle is usually known as the trichromatic principle. It is explained by admitting that there are three types of photoreceptors in the retina. In general, given the same primary colors and a test color, most viewers choose the same mixture of primaries to obtain the test color. Based on these facts, one of the ways most often used to represent a specific color is to define a set of primary colors and express any visible color by means of three multiplying weights corresponding to those that an average observer would employ to get that specific color. The retina photoreceptors responsible for color perception are designated by cones. Experimental evidence shows that the 6 to 7 million retina cones can be divided into three types roughly corresponding to red, green, and blue. About 65% of the cones are sensitive to red, 33% are sensitive to green, and only about 2% are sensitive to blue (but they are the most sensitive) [107]. The spectral sensitivities of the three types of cones are represented in Figure 6.30. These are the colors known as additive primaries.

It is important to distinguish between light primary colors and pigment primary colors (also known as subtractive primary colors). Pigment primary colors are defined as those colors that subtract or absorb a light primary color and reflect or transmit the other two [107]. Therefore, the pigment primary colors are cyan, magenta, and yellow. Additive primary colors are employed in cathode ray tubes; and they are red, green, and blue. Pigments, however, combine based on the subtraction of colors.

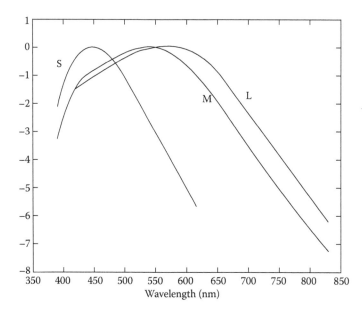

FIGURE 6.30
Cone responses to visible radiation: logarithms of the spectral sensitivities of the three types of cones in the human retina. The S curve corresponds to blue cones, the M curve corresponds to green cones, and the L curve corresponds to red cones. (From R. C. Gonzalez, R. W. Woods, *Digital Image Processing*, 2nd Edition, Prentice-Hall, Englewood Cliffs, NJ. With permission.)

Colors can be represented in several types of spaces. One of the most common linear spaces for representing colors is RGB. This is a linear space using primary colors. Each of the primary colors in this space corresponds to single wavelength radiation (645.16 nm for red, 526.32 nm for green, and 444.44 nm for blue). In this space, colors are represented in a cube, the RGB cube, whose edges represent the weights of the primary colors (Figure 6.31). The cube is represented in a Cartesian coordinate system. Black is located at the origin, and white is located at the vertex farthest from the origin. The vertices on the axes of the coordinate system correspond to the primary colors red, green, and blue. In this model, different colors are points inside the cube and are defined by vectors that connect them with the origin [107]. If the pixels in each of the images are represented by 8 bits, then each color pixel is represented by 24 bits.

The spaces used to represent the subtractive colors are different. The simplest color space for subtractive colors is CMY made up of cyan, magenta, and yellow. Cyan is obtained by subtracting red from white, magenta is obtained by subtracting green from white, and yellow is obtained by subtracting blue from white. However, in practice, printing systems require at least four colors (cyan, magenta, yellow, and black), because the mixture of the three subtractive primary colors does not generally produce good quality black colors.

Additive colours can be represented in other, nonlinear spaces. In these spaces, the goal is to represent color properties and their perceptual

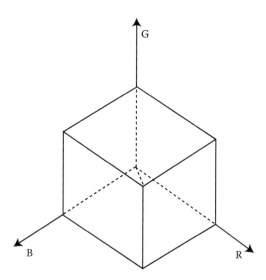

FIGURE 6.31
The RGB cube. The cube has unit dimensions, and in this space all the colors can be obtained combining the primary colors red, green, and blue, corresponding to the three axes. (From D. Forsyth, J. Ponce, *Computer Vision: A Modern Approach*, Prentice-Hall, Engle-wood Cliffs, NJ. With permission.)

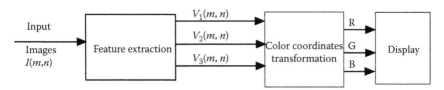

FIGURE 6.32
Image transformation using pseudo-color.

relationships. One of these spaces is the HSV (Hue, Saturation, Value). Hue is the pure color or the dominant wavelength. Saturation is the whiteness of the color or the relative bandwidth of the color. A high saturation color is a color with low bandwidth. Value is the brightness or intensity of the color. This space can be readily obtained from RGB.

The use of pseudo-colour for image processing aims at increasing the perception of some aspects or details of the information available in an image. Visualization and interpretation of images become easier. This is because the human visual system is able to discriminate many more colours than gray levels. Pseudo-color processing deals also with mapping a set of images $I_i(m, n)$ with $i = 1, \ldots, k$ into a single color image. This mapping is usually defined so as to ensure that the different elements or structures in the dataset can be distinguished by means of the colors [106]. This makes it easier to visualize complex datasets (Figure 6.32).

The usual procedure to determine pseudo-colour mappings consists of mapping the set of input images into three feature images (by means of the functions $V_1(m, n)$, $V_2(m, n)$, and $V_3(m, n)$) and then mapping those three images into the primary colors. If the goal is to transform a gray level image into a pseudo-color image, however, a mapping must first be defined between the gray levels and the color space. This establishes a correspondence between each gray level and a color. If the number of initial images is bigger than three, then it will be necessary to generate three intermediate images. These images have to be generated by processing them so that the relevant goals can be achieved. After the three images have been generated, the color mapping can then be applied. In general, pseudo-color mappings are defined by trial and error so that colors are selected to best ensure the visualization of the features.

6.3.3 Image Registration

Image registration is the process of placing different images in the same frame of reference with the goal of integrating and comparing their information. It consists of aligning images by finding correspondences between common characteristics [109,110]. For this, it is necessary to determine the transformation that maps the points in an image in the corresponding points in the other image.

Image registration aims at benefiting from integrating information from several images. Usually, it is possible to extract more useful information from many images than can be obtained by separately analyzing each of these images. This is the main reason for the interest in attempting to solve the difficulties posed by the problem of image registration.

Registration can be performed for images of the same type or for images of different types. The term *multimodal registration* is used in the second case. Different types of images add extra information to the final result, thus increasing the amount of useful information. For instance, CT images provide anatomical information, and PET images provide functional information. When these images are registered, in addition to the existing information, we gain from learning the precise location of the functional information. In many situations, it is advantageous to align images of the same type, either to analyze the variations in time or to perform studies between different subjects. In the case of NM, we should note that although one may use images of the same modality and of the same individual, the information can vary if different tracers are used.

The importance of registration in the medical field is essentially based on two circumstances: on one hand, diagnosis often implies the acquisition of data from many image modalities; and on the other hand, an individual goes through several exams in his or her lifetime. It is often up to the physician to combine the information from the different acquired images in his or her mind or to compare studies acquired at different times, searching for differences between them. If the task of merging only two exams is already complex enough, trying to combine information from a sample of subjects with the goal of determining a pattern is almost impossible to be mentally performed. Image registration creates a map between different images, allowing the accumulation of point-to-point information or the search of small differences, either in time or in a population.

Besides allowing the creation of better images, registration also helped new technologies such as guided surgery to emerge. The images acquired before intervention are usually registered (in real time) with the surgical device and used as a guide by the surgeon.

The work in this field is vast, and the algorithms found in the literature are very diverse. The diversity is easily understood if we take into account the different paths that can be taken to reach the final goal, which is to establish a correspondence between the points of the images to be coregistered.

The problem can be generically seen as the search for the optimal transformation, τ, between two images, where by transformation we consider a way to map points between the two images. Thus, we have

$$\arg\min_{T} S(A - \tau(B)). \tag{6.42}$$

where S represents a measure of the similarity between points in image A and points in the transformed image B. If the two images were exactly the same, τ

would represent a unitary operation; and the measure of similarity, S, applied to all points in the image would give zero.

Although the previous equation suggests that registration is always seen as an optimization problem to which the usual optimization methods and techniques are applied, there are examples where the transformation τ is obtained by other means, for instance, by solving a differential equation. However, even in these cases, it is common to use iterative algorithms with a stopping criterion defined by a function of similarity. Except for the due differences, we can say that the problem is still that of finding adequate mapping between the two images.

Equation 6.42 allows us to check the various questions that are posed in solving this problem. Summarizing, we can have different types of initial data, different forms of considering the transformation that is inherent to the correspondence, several ways of measuring the similarity or the proximity associated with the correspondence, and, finally, different approaches to minimization. Apart from what is generically indicated by Equation 6.42, it is necessary to take particular aspects into account that are proper to each specific application, for example, an increase or decrease of the area in a zone of tracer uptake does not usually have the same anatomical meaning.

One of the first classification criterion for registration methods was proposed by Elsen [111], which was later returned to by Maintz [112,113]. The criteria adopted in this work are found in subsequent literature, with small variations [114–116]. Essentially, the following classifiable aspects of the method are distinguished:

1. Dimension
2. Nature of the basis of registration
3. Nature and domain of the transformation
4. Interaction
5. Optimization
6. Modalities
7. Subject
6. Object of registration

These aspects will now be further described.

6.3.3.1 Dimension

The initial data can originate from 2D or 3D images. Registration can be achieved through either 2D or 3D images and, in some cases, between 2D and 3D images. The type of image used at the start has direct implications for the mapping of the data being searched for; at least, there is an implication regarding the number of dimensions that need to be taken into account. However, the large subdivision that is made at the dimension level is to whether to consider only spatial dimensions or to take into account the temporal dimension. If only spatial dimensions are considered, then it is common to discriminate the registration methods into 2D/2D, 2D/3D, and 3D/3D.

6.3.3.2 Nature of the Basis of Registration

A distinction is often drawn between approaches that use only a few points in the images and those that use images as a whole, working on the intensity of pixels or voxels. We can also distinguish between the extrinsic methods, which make use of artificial objects that are placed in the individual(s), and the intrinsic methods, which are solely based on information contained in the images. The latter can be further subdivided, depending on what the data are, into (1) intensity of pixels or voxels, (2) characteristics of images obtained by segmentation, and (3) characteristic marks that can be seen in the image, for example, the extremity of the bifurcation of a blood vessel.

6.3.3.3 Nature and Domain of the Transformation

One of the aspects that determines the type of registration algorithm, with direct implications on its complexity and on the number of variables to process, is the kind of transformation that is applied and the domain (global or local) of its application. The specialized literature contains some splits between algorithms dedicated to rigid transformations and algorithms, known as nonrigid, that introduce deformations. It is consensual that nonrigid methods are used to improve the registration obtained by rigid methods.

In rigid transformations, the correspondence between points in the images can be described only by translations and rotations, which means that the relative distance between points is constant. In 2D or 3D registration (the simplest), only three parameters are needed to characterize the transformation: two for the translation and one for the rotation. In the registration of two 3D images, three translations and three rotations are necessary.

Rotation and translation are operations that conserve colinearity, which means that a straight line in one image is transformed into a straight line in another image. Operations that maintain colinearity and the ratio of distances (Equation 6.43) between three collinear points, p_1, p_2, and p_3, are called affine operations.

$$\frac{\|p_2 - p_1\|}{\|p_3 - p_2\|}.$$
(6.43)

Besides rotation and translation, which are special kinds of affine transformations, there are also scaling and skewing transformations. An affine transformation is often defined as a linear transformation followed by a translation, that is, an affine transformation transforms a vector x into another vector x' using the following rule:

$$x' = Mx + t, \tag{6.44}$$

where M is a combination of rotation, scaling, and skewing and t is a translation. The elements $m_{i,j}$ of the matrix M can have any value. The number of degrees of freedom increases to 12 in this case, where 6 are due to the rigid-body transformation, 3 to scaling, and 3 to skewing. Figure 6.33 shows the effect of applying each of these transformations to a 2D image.

In a rigid transformation, the elements m_{ij} of matrix M have to be such that the determinant is equal to one, which reveals the rotation character of the matrix. Matrix M is then defined by

$$M = \begin{bmatrix} 1 & 0 & 0 \\ 0 & \cos\alpha_1 & -\text{sen}\,\alpha_1 \\ 0 & \text{sen}\,\alpha_1 & \cos\alpha_1 \end{bmatrix} \begin{bmatrix} \cos\alpha_2 & 0 & -\text{sen}\,\alpha_2 \\ 0 & 1 & 0 \\ \text{sen}\,\alpha_2 & 0 & \cos\alpha_2 \end{bmatrix}$$
$$\times \begin{bmatrix} \cos\alpha_3 & -\text{sen}\,\alpha_3 & 0 \\ \text{sen}\,\alpha_3 & \cos\alpha_3 & 0 \\ 0 & 0 & 1 \end{bmatrix}. \tag{6.45}$$

where the angle α_i characterizes the rotation performed around the axis i.

As previously mentioned, a given transformation can be locally or globally applied, meaning that it can be applied to the whole image or to a part of the image. The application of local transformations requires special care, as it may cause violations to local continuity. A single local transformation is, therefore, rarely applied: Instead, one generally uses several local transformations in different parts of the image. Figure 6.34 shows some examples of the transformations described for 2D images.

In a nonrigid transformation, the relative positions of the points are not invariant, and the number of parameters that characterizes this transformation is larger than that for an affine transformation. They can, theoretically, be more than the number of points in the images; in an extreme case, any point

(a)   (b)   (c)

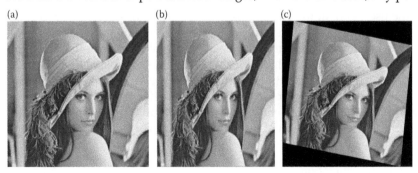

FIGURE 6.33
(a) Original image; (b) change of scale along xx; (c) skewing.

Global   Local

Affine

Nonrigid

FIGURE 6.34
Examples of transformations applied on to 2D images. Local transformations can lead to the appearance of discontinuities in the image, in case adequate restrictions are not adopted. (Adapted from P. A. van den Elsen, E.-J. D. Pol, M. A. Viergever. Medical image matching—a review with classification. *IEEE Eng. Med. Biol. Mag.*, 12, 26–39, 1993.)

can move in a particular way. So, these algorithms usually have different approaches to simplify the problem.

6.3.3.4 Interaction

We can basically distinguish three levels of interaction of the algorithms with the user. At one end, there are algorithms that do not require any intervention by the user; and, at the opposite end, there are algorithms that require total intervention from the user in order to work. The former are known as automatic whereas the latter are known as interactive. Between these extremes, there are semiautomatic algorithms where the user has an active role in processing the registration but does not completely execute it. In these methods, the user usually participates in the process, whether in the initialization of the algorithm, in image segmentation, or even by guiding the method through the acceptance or rejection of hypotheses that are being assumed.

6.3.3.5 Optimization

Optimization involves three different aspects: the measure of similarity that is used, that is, the objective function that is to be minimized or maximized;

the method for searching for the extremity; and, finally, the technique for resampling or interpolation used to translate the image transformation.

6.3.3.5.1 Objective Functions

An exhaustive enumeration of the cost functions that have been used to register images is beyond the scope of this book. However, among the functions we can highlight are the distance between points, difference in intensity, correlation and mutual information [117].

Markers (intrinsic or extrinsic) are widely used in methods that use the distance between points. In this case, the problem is mathematically described by

$$\arg\min_{T} \|P_A - \tau(P_B)\|, \tag{6.46}$$

where P_A and P_B represent a set of points to be coregistered in the two images and τ is the transformation to be determined. Usually $\|\cdot\|$ expresses the Euclidean norm, but other distances can be defined.

In the particular case where the distance between points is used for the objective function and for a rigid transformation, the algorithm is direct. The problem, thus, defined is known as "Procrustes" [118] or as "absolute orientation" in the context of photogrammetry [119–122]. It is known that the solution is unique if and only if the existence of at least three noncolinear points is guaranteed. In 1966, Schönemann [123] proposed a closed-form solution for this problem that involves two steps [124]: first, the translation is determined as a vector between the centroids of the points in the two images; second, the rotation between the two data sets is calculated (already corrected for translation) by decomposition of the covariance matrix in singular values. Let $\bar{\mu}_{P_A}$ and $\bar{\mu}_{P_B}$ be the centroids of the sets of points considered in the two images. Defining vectors $\bar{P}_A$ and $\bar{P}_B$ as the position vectors corrected by the corresponding centroids, we have

$$\bar{P}_A = P_A - \mu_{P_A} \quad \text{and} \quad \bar{P}_B = P_B - \mu_{P_B} \tag{6.47}$$

Thus, the rotation matrix R is such that it verifies the equation

$$\arg\min_{R} \|\bar{P}_A - R(\bar{P}_B)\|, \tag{6.48}$$

which is subjected to the orthogonality constraint

$$RR^T = 1. \tag{6.49}$$

The rotation matrix is then found by

$$R = UV^*, \tag{6.50}$$

where matrices U and V^* are obtained by decomposition in singular values of the matrix $S = \bar{P}_A(\bar{P}_B)^{-1}$:

$$S = U\Sigma V^*. \tag{6.51}$$

The methods that are based on the intensity of image values exhibit different characteristics from those just described. The difference between intensities [125,126] or, more precisely, the sum of the quadratic pixel-to-pixel differences (SDQ) is one measure of similarity that can be easily established. Considering images A and B and a subset Ω of voxels where there is correspondence between the two images, the sum of quadratic differences can be defined by

$$\text{SDQ} = \sum_{X\in\Omega} [A(X) - \tau(B(X))]^2. \tag{6.52}$$

This is a good objective function for cases where the images differ from one another only by Gaussian noise [127], but this is a rare situation even for intramodal registration. Still, the application of this function in magnetic resonance series is common [128,129].

Correlation is a measure of similarity between two images which assumes that there is a linear relationship between the intensities of the two coregistered images. The correlation coefficient (CC) is then defined by

$$\text{CC} = \frac{\sum_{X\in\Omega}(A(X) - \bar{A})(\tau(B(X)) - \bar{B})}{\sqrt{\sum_{X\in\Omega}(A(X) - \bar{A})^2 \cdot \sum_{X\in\Omega}(\tau(B(X)) - \bar{B})^2}}. \tag{6.53}$$

where $\bar{A}$ and $\bar{B}$ represent the average intensity of images A and B in the same domain Ω.

Since correlation is not too restrictive relative to the nature of the images, it can be used between different modalities, as long as there is a linear relationship between voxel intensities, as assumed. Another interesting characteristic of correlation is the possibility of applying it in the frequency domain, taking advantage of the use of the Fourier transform [130]. For a rigid transformation, therefore, a phase difference in the frequency domain will correspond to a translation in the spatial domain. Rotation can be determined by polar representation in the frequency domain [131–133].

Mutual information is based on the idea that registration corresponds to the maximization of the information that is common to the images being registered. It assumes that the alignment of two images implies the superposition of the same structures contained in the images and that in the opposite case there is duplication of the structures in the final result. This enables a measure of information to be used in the registration. The commonest measure for the information contained in a signal was suggested by Shannon [134] in 1940 and is known as a measure of entropy. The average information, H, given by a series of symbols j whose probabilities are given by p_j, is described by

$$H = -\sum_j p_j \log p_j. \tag{6.54}$$

Entropy, H, is maximal if the probabilities p_j are all equal, that is, if all the symbols j occur with the same probability. The entropy is minimal if only one symbol occurs: In that case, the probability is one for that symbol and zero for the remaining ones.

Registration implies the existence of two images, so there are two symbols per voxel (each relative to the intensity of its own image). Consequently, the concept of joint entropy is defined, which quantifies information in the two combined images. The more similar the two images, the larger the joint entropy. The mathematical definition of the joint entropy between two images A and B is

$$H(A,B) = -\sum_i \sum_j p_{AB}(i,j) \log p_{II'}(i,j). \tag{6.55}$$

where $p_{AB}(i,j)$ represents the joint probability that is usually determined from the normalized joint histogram. As previously seen, for any superposed voxel we have a pair of intensities that can be represented on the axes of a chart. The histogram is then obtained by calculating the relative frequency of each pair. Figure 6.35 presents some examples for PET images, aligned and misaligned.

The joint histogram is increasingly scattered as the images become more misaligned, and entropy is used as a measure of this dispersion [135,136]. Mutual information is another measure of the same type used in multimodal registration and was proposed by both Collignon et al. [137] and Viola and Wells [127,138]. Pluim [139] reviews three similar definitions for mutual information, of which we only discuss one, as they are all much the same. Mutual information is directly related to both the entropy of the images and the joint entropy. Mutual information, $I(A, B)$ of two images A and B is given by

$$I(A,B) = H(A) + H(B) - H(A,B). \tag{6.56}$$

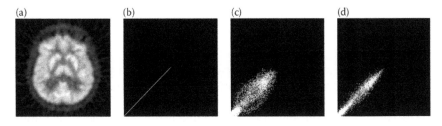

FIGURE 6.35
Joint histogram between the same PET image (slice) (a); (b) aligned; (c) translated by 2 pixels; (d) rotated by $2°$.

The subtractive term points to an equivalence between the maximization of the mutual information and the minimization of entropy. However, mutual information presents an advantage over entropy, which is that it also includes the entropy of the images. This means that, in the optimization process, we do not obtain a superposition zone containing only the background of the images, which may occur if only joint entropy is used. Mutual information can also be thought of as a measure of how well one image explains the other.

6.3.3.5.2 Search for the Minimum

The techniques involved in the search for the minimum of the objective function directly depend on the characteristics of the function itself. For smooth and convex functions, Newton's method and its derivations are usually preferred, as they are both fast and easy to implement [140–142]. Newton's method is an iterative technique to search for zeroes in a function. It relies on the approximation of a function using the Taylor series expansion [143]. A function, $f(x)$, infinitely differentiable in the neighborhood of a point a, can be approximated around this point by a polynomial obtained using the expression

$$f(x) \approx f(a) + f'(a)(x - a) + \frac{f''(a)}{2}(x - a)^2 + \cdots + \frac{f^n(a)}{n!}(x - a)^n + \cdots . \quad (6.57)$$

Considering a linear approximation and assuming that a is a point close to a zero of function $f(x)$, we deduce the relation

$$x_{n+1} = x_n - \frac{f(x_n)}{f'(x_x)}, \quad (6.58)$$

which forms the basis of Newton's method. From a first estimate of a zero of the function, an increasingly better approximation is obtained. This is a method that is used for local minima search by searching for the zeroes of the first derivative. Since the objective functions usually have several variables, Equation 6.58 is adapted to

$$X_{n+1} = X_n - [Hf(X_n)]^{-1} \nabla f(X_n), \quad (6.59)$$

where $X_n = \left[x_n^1 \; x_n^2 \; x_n^3 \ldots x_n^m\right]^T$ is a vector that specifies a point in the space of function f, ∇ represents the gradient of the function, and H is the Hessian matrix that contains the second derivatives of the function. Newton's method rapidly converges to the solution; but the first estimate is crucial, as it must be close to the solution. Another essential aspect for this method's success, mentioned before, is the type of function; for smooth and convex functions, the method is excellent.

One of the easiest ways of modifying Newton's method is to include a variable, $\gamma > 0$, to set the step

$$X_{n+1} = X_n - \gamma \left[Hf(X_n) \right]^{-1} \nabla f(X_n). \tag{6.60}$$

The inversion of the Hessian matrix is frequently complex or even impossible. Alternative forms are used in these circumstances, such as the gradient method (or maximum descent) or conjugated gradient, which do not use the second derivatives. In the gradient method, the search direction of the minimum is given by the gradient, so a new point is obtained according to

$$X_{n+1} = X_n - \gamma \nabla f(X_n). \tag{6.61}$$

In the case of conjugated gradients, the descending direction is obtained to speed up the search, using for that purpose a set of directions that are conjugated between themselves and closest to the gradient [144].

For least squares problems, the Gauss–Newton (GN) algorithm or Levenberg–Marquardt algorithm [145–147] is also often used, and it combines the advantage of the GN method with that of the gradient. The Gaussian method can be rapidly derived from Newton's method, considering its application to the mean squares. In the case of the mean squares, the gradient is given by

$$\nabla f(X) = 2J_f(X)^{\mathsf{T}} f(X). \tag{6.62}$$

where J_f represents the Jacobian of the function. Consequently, the Hessian is approximately equal to

$$Hf(p) = 2J_f(X)^{\mathsf{T}} J_f(X). \tag{6.63}$$

So, finally, we obtain

$$X_{n+1} = X_n - \left[J_f(X_n)^{\mathsf{T}} J_f(X_n) \right]^{-1} J_f(X_n)^{\mathsf{T}} f(X_n). \tag{6.64}$$

Powell's method is a frequently used optimization technique [148–151]. It is characterized by considering a set of mutually conjugated directions and by searching the minimum in one direction at a time. The appeal of this method is the absence of the use of gradients. The minimum along a line is usually determined by means of Brent's method [152], which combines a slower but more robust search with a faster search when the function is smoother. When the function is more difficult, the search for the extreme involves bisecting the interval. In this case, a triplet of points is considered a, b, and c so that $f(b) < f(a)$ and $f(b) < f(c)$; and if the function is continuous in the interval (a, c), then there is a minimum within the interval. This minimum can, therefore, be isolated by the bisection of intervals (b, c) or (a, b). In fact, there is no true bisection: given the previous triplet (a, b, c), the new point, x, tested will be at a fraction of 0.38197 of the largest segment measured from the central point. This procedure guarantees a golden ratio* between the points of the triplet, even if it was initiated without its existence; and so, it assures a linear convergence in the search for the minimum. In situations where the function is smoother, a faster technique to search for the minimum can be used, based on a parabolic interpolation. In this case, for the triplet of abscissae (a, b, c) and ordinates $(f(a), f(b), f(c))$, the parabola that contains them is calculated; and from this, the value of x that is the minimum of the parabola's abscissa is found. For noncolinear points we, thus, have

$$x = b - \frac{1}{2} \frac{(b-a)^2 \left[f(b) - f(c)\right] - (b-c)^2 \left[f(b) - f(a)\right]}{(b-a)\left[f(b) - f(c)\right] - (b-c)\left[f(b) - f(a)\right]}. \qquad (6.65)$$

After finding the point x, which is the minimum in a given direction, the algorithm tests a new direction searching for a new minimum, and so on.

Other methods besides those described above are commonly used in image registration, in particular genetic algorithms and simulated annealing.

Genetic algorithms are implemented as a computational simulation that emulates biological evolution to solve a given problem [153–155]. It is, thus, not uncommon to find characteristics such as heredity, mutation, selection, and recombination in genetic algorithms. For a given problem, a set of possible solutions (population) is given to the algorithm, along with an objective function that is used to evaluate each particular solution (candidate). The initial solutions are usually randomly obtained, and it is hoped that the algorithm will improve them with time. The candidates differ by a set of abstract characteristics that have meaning for the problem being solved. The candidates are scored and depending on the score are selected to be reproduced, thereby generating a new population. The candidates of the new population will be different from the previous ones, despite heredity, due to recombination and eventual mutation of their characteristics. The stopping criterion of the algorithm can be the number of generations produced or the level established for the objective function.

Simulated annealing is inspired, as the name implies, from the tempering of a metal by annealing. This technique involves heating a metal to its melting point and then cooling it slowly. The heat causes a change in the positions of atoms, which are in a state of minimum energy, promoting random rearrangements of higher energy. The slow cooling facilitates the establishment of settings with less energy than the original one. The simulated annealing algorithm was independently proposed by Kirkpatrick et al. [156] and Cerny [157], using the Metropolis algorithm [158]. Similar to the metallurgical process, the simulated annealing algorithm replaces the current solution by another that

---

* A golden ratio is algebraically defined as a triplet of points (a, b, c) on a straight line, similar to those in the relation $\overline{ac}/\overline{ab} = \overline{ab}/\overline{bc}$.

is chosen based on a distribution of probabilities which is a function of a parameter of temperature. This solution is then accepted or rejected by the Metropolis condition, which takes into account not only the improvement of the values of the objective function but also the level of temperature. The process is repeated until a balance is reached, followed by a new lowering of temperature and repetition of the process. The fundamental idea of the algorithm is that at the beginning (high temperature) almost all the states (solutions) proposed are accepted, including some that worsen the value of objective function, in order to escape from local minima. As the temperature falls, the number of possible transitions also decreases until, at the points when the temperature tends to zero, only the transitions that reduce the value of the objective function are accepted. At this point, the solution obtained is the minimum of the objective function.

6.3.3.5.3   Resampling and Interpolation

Since the registration methods are iterative, at each iteration it is necessary to "move" one of the images, which involves performing a resampling. That is, it is essential to determine new values of the intensity of pixels or voxels. This calculation uses interpolation, an approach that, according to Thèvenaz et al. [159], is the method of recovering a continuous model from discrete values within a known abscissae range. Therefore, the value, $f(X)$, at any point X of the space of dimension q is given by linear interpolation, such that

$$f(X) = \sum_{K \in \mathbb{Z}^q} f_K \varphi_{\text{int}}(X - K) \quad \forall_{X = [x_1, x_2, \dots, x_q]^T \in \mathfrak{M}^q}. \tag{6.66}$$

The value so determined is, therefore, a linear combination of known values, f_k, sampled in positions $K = [k_1, k_2, \dots, k_q]$ and weighted by the values given by the function $\varphi_{\text{int}}(X - K)$. The interpolation is, therefore, usually carried out by convolution with an appropriate kernel. The sync function is an optimal interpolator; however, in practice, it is difficult to use, because it is spatially unlimited. Other interpolators are, therefore, preferred, including nearest neighbor, bilinear, bicubic, and cubic B-splines.

Over time, several methods of interpolation have been proposed that differ fundamentally and, in practice, in their accuracy and computational complexity [160]. The method to be used must be carefully chosen for each specific situation, but bilinear interpolation is one of the methods that gives the best compromise between accuracy and complexity [115].

6.3.3.6   Modalities

Two kinds of tasks are usually distinguished in relation to modality. Registration can be done on the same modality, in which case it is monomodal, or on different modalities, in which case it is multimodal. A typical example of monomodal registration is what occurs in myocardial SPECT images in which two tests are performed: one at rest and another in stress conditions. The diagnosis in this case benefits from the comparison (through registration) of the two examinations. As for the multimodal case, it is common to find registration between PET brain images and MRI. Attenuation correction in PET needs transmission images that can be obtained from external sources and, more recently, from images of CT (see Section 6.2.2.2) that have to be registered. There are many applications of multimodal registration, and it is common to categorize registration into anatomical and functional modalities, that is, anatomical/anatomical, anatomical/functional, and functional/functional.

Besides these two categories, Maintz et al. [112] proposed two other types: registration modality-model and registration patient-modality. An example of the first type is the registration of a brain study for a compartmental model involving large structures, and of the second type, the alignment performed in radiotherapy. Note that the radiotherapy planning is done beforehand and that in that case the registration is based on acquired images; but the task of positioning the patient for the radiotherapy table is itself a type of registration.

6.3.3.7 Subject

Registration serves several different purposes that in this chapter can be divided into (1) registration on images of the same individual acquired at different times and in different modalities, (2) registration on different individuals (same modality), and (3) between the image of an individual and an atlas. The shorter names of these three cases are (i) intrasubject, (ii) intersubject, and (iii) atlas. An atlas is usually taken to be an image that is built from a database of images of different subjects. An image thus built is the pattern of a given population.

6.3.3.8 Object of Registration

There are several objects that require the use of registration, including the brain, eye, heart, the breast, kidney, and liver. This list is far from exhaustive, and even if it were, it is not thought to be closed, because we are seeing the implementation of registration in an increasing number of objects and organs.

6.3.3.9 Nonrigid Methods

There are situations for which rigid transformations are enough to perform the registration. That is the case of intrasubject anatomical brain images. The movements are, in these circumstances, constrained by the rigid characteristics of the skull; hence the adoption of more complex solutions is not justified. However, for other applications nonrigid transformations are required to map the points between the two images in question. This happens with intersubject registration, where the natural anatomical variations are adjusted by means of the transformation. Many different approaches have been tested. The deformation map obtained by thin-plate splines tends to be used when registration is based on fiducial marks. When registration is based on intensity values, the transformation can be described by a linear combination of polynomials or by basis functions. Another approach that has been used is based on physical models such as elastic deformation or on the flow of fluids.

6.3.3.9.1 Thin-Plate Splines

The term *thin-plate spline* refers to a physical analogy with a long blade made of metal or wood used to shape boat hulls. These layers were folded by placing weights on certain points. Similarly, we can think of a way to translate a nonrigid spatial transformation to the mapping between two images. Therefore, when the deformation of the plate is considered, it is as if there were movement in the x and y measures on the original plane. Taking into account the corresponding fiducial marks in the two images, the problem is reduced to determining the transformation that best aligns these points, which, in this context, are usually called control points.

Given a set of control points, $\{p_1, p_2, \ldots, p_n\}$, let us say that a radial base function* $\phi(r)$ defines a spatial map that corresponds to each position x a new position $f(x)$ and is given by

$$f(x) = \sum_{i=1}^{n} c_i \phi \left(\|x - p_i\| \right), \tag{6.67}$$

where c_i are coefficients of the map. One of the possible choices for the radial base function is the thin-plate spline, independently introduced by Duchon [161] and Meinguet [162] and defined by

$$\phi(r) = r^2 \log(r) \quad \beta \in N. \tag{6.68}$$

Gaussian and multiquadratic functions have also been used. The choice of the base function has implications for the space of possible deformations, that is, the ability to produce large deformations in a given region with minimum effect on other areas of the image.

The thin-plate splines are particularly used in interpolation of surfaces and have been proposed as an approach to image registration by Goshtasby [163]. Taking into account a set of control points in the two images, $\{[(x_1, y_1), (X_1, Y_1)], [(x_2, y_2), (X_2, Y_2)], \ldots, [(x_n, y_n), (X_n, Y_n)]\}$, whose correspondence is known, the problem of registration of the two images can be reduced to determining two smooth surfaces: one that contains the points $\{(x_1, y_1, X_1), (x_2, y_2, X_2), \ldots, (x_n, y_n, X_n)\}$ and the other that contains the points

---

* A radial base function is a real function whose value depends only on the distance to a given point, a, such that $\phi(x, a) = \phi(\|x - a\|)$, or more simply $\phi(r, a) = \phi(r)$, where r is the distance, hence $r = \|x - a\|$.

$\{(x_1, y_1, Y_1), (x_2, y_2, Y_2), \ldots, (x_n, y_n, Y_n)\}$. It is as if at each point (x_i, y_i) of one of the images is associated with a load that causes a deflection of the surface of equal value X_i in the first case and of value Y_i in the second case. Therefore, given any point (x, y) of the image, the surfaces determined by thin-plate splines can be used to calculate the corresponding point in the other image. The intuitive physical interpretation and the associated algebraic simplicity are the great advantages of this method that continue to attract researchers [164].

6.3.3.9.2 Elastic Registration

Elastic registration [165] is based on the idea that the field of deformation can be modeled as a physical process of deformation of an elastic material which is described by the Navier–Stokes equation:

$$\mu \vec{\nabla}^2 \vec{u} + (\lambda + \mu) \vec{\nabla} \left(\vec{\nabla} \cdot \vec{u} \right) + \vec{F} = \vec{0}, \tag{6.69}$$

where $\vec{u} \equiv \vec{u}(x, y, z)$ represents the field of displacements and $\vec{F} \equiv \vec{F}(x, y, z)$ is the external force that acts on each point of the elastic body. The parameters μ and λ are the Lamé elasticity constants that are characteristics of the elastic body.

The differential equation 6.69 is solved in order to the field of displacement, $\vec{u}$, which is determined from the external force $\vec{F}$ for which different ways of calculation have been proposed. The gradient of a measure of similarity applied to the values of intensity of the images is one of the commonest ways of calculating the external force.

Finite differences are used on the whole, and a scheme of successive relaxation [152] for solving the equation is used that results in a discrete displacement field defined on each voxel. Another scheme is the determination of the displacement field only for those voxels that correspond to nodes of a physical model for which the external force is known. The displacements for the remaining voxels are then obtained afterward by interpolation.

6.3.3.9.3 Fluid Registration

One of the disadvantages of elastic registration is the fact that intense localized deformations are not possible due to the balance of internal forces (elastic) and external forces. This aspect is overcome in fluid registration whereby large local deformations can be modeled. The model that reflects the flow of a fluid that is used in the registration is again described by the Navier–Stokes equations:

$$\mu \vec{\nabla}^2 \vec{v} + (\lambda + \mu) \vec{\nabla} \left(\vec{\nabla} \cdot \vec{v} \right) + \vec{F} = \vec{0}, \tag{6.70}$$

which is similar to Equation 6.69, with the exception that now the field of velocities is involved. However, there is a relationship between velocity and displacement given by

$$\vec{v} = \frac{\partial \vec{u}}{\partial t} + \vec{v} \cdot \vec{\nabla} \vec{u}. \tag{6.71}$$

Equation 6.70 can also be solved using a scheme of successive overrelaxation [166], which unfortunately results in a rather slow algorithm that requires a much longer computation time. One way to speed up the algorithm is through the use of a convolution filter [167].

6.4 Advanced Methods in Nuclear Oncology

NM provides functional information related to the biodistribution of the radiopharmaceuticals used. Further, advances made in recent years in understanding the molecular mechanisms underlying the disease have enabled diagnoses to be made at increasingly early stages, sometimes several months before the onset of the molecular changes underlying the disease. These new approaches are especially important in oncology. NM allows you to map and quantify the local activity related to the metabolic pathways involved in malignant transformation or tumor proliferation.

This section provides a longitudinal overview of the development of malignant transformation; and it shows how it is possible to evaluate each step, using NM to get information not only about the metabolic situation of tumor tissue but also about the therapeutic response.

6.4.1 Introduction

The advances achieved in recent years in understanding the molecular mechanisms underlying disease have led to major changes in methods of diagnosis and therapeutic approaches. Based on molecular and biochemical changes in the cells, it is possible to arrive at a diagnosis ever earlier, even several months before the onset of morphological changes. These new approaches apply both to new treatments and to imaging methods, particularly those that provide functional information.

Molecular imaging techniques aim at quantifying the molecular changes associated with the disease, such as cell differentiation, the changes associated with various intracellular metabolic pathways, and even with the pathways of cell death, functioning as an alternative to the image that provides information about morphological changes resulting from functional alterations.

These new perspectives are especially important in oncology. NM is able to map and quantify local activity related to specific characteristics of some metabolic pathways involved in malignant transformation or in tumor proliferation. Some of these metabolic pathways are responsible for malignant transformation, such as protein synthesis and apoptosis, or for malignant proliferation, as occurs with angiogenesis or tumor hypoxia [168].

Any of these pathways is evaluated in NM by means of radiotracers, which are molecules that allow us to obtain functional or molecular images. The right choice of radiopharmaceutical is very important, because the biochemical pathways involved have to be considered in order to obtain a good and efficient tracer. As a general principle, the labeling procedure should not introduce structural changes in the molecule chosen. However, sometimes small changes are acceptable if the molecule maintains its biological behavior and, thus, its purpose.

This is easily accomplished if the labeling is done through a covalent connection, where displacement or addition reactions are used to put an isotope in the molecule chosen. Examples of this type of labeling occur when fluorine-18 (^{18}F), iodine-123 (^{123}I), iodine-131 (^{131}I), bromine-75 (^{75}Br), bromine-77 (^{77}Br), or carbon-11 (^{11}C) are used. This type of labeling is normally demanding and is associated with both a low yield and a high cost [169].

However, this is not the most common labeling technique, because it usually requires a chelating agent to bind metal isotopes. This requirement not only makes the chemical procedures much more complex but also the chelator can change the chemical properties of the molecule. Examples of isotopes used for this type of labeling procedure are technetium-99m (^{99m}Tc), rhenium-188 (^{188}Re), and gallium-68 (^{68}Ga), all of them obtained from generators. By contrast, despite involving the addition of a chelating agent, this type of labeling is normally simple and is associated with a high yield and a low cost.

The radioisotope most often used to label radiopharmaceuticals is ^{99m}Tc. This preference is due to favorable energy emitted (140 keV) by the isotope, easy chemistry, and low cost. In terms of tracers of interest to oncology, several chelators have been proposed for labeling with ^{99m}Tc. Examples are N_4, N_3S, N_2S_2, NS_3, S_4, diethylenetriamine pentaacetic acid (DTPA), O_2S_2, and hydrazine nicotinamide (HYNIC). Of these chelators, those containing atoms of nitrogen and sulfur are stable chelators for ^{99m}Tc-bis-aminoethanethiol tetradentate ligands, also known as diaminodithiol compounds, which form very stable Tc(V)-O complexes that can bind to two thiosulfur and two amine nitrogen atoms [170].

The complexes DTPA forms with ^{99m}Tc are less stable; whereas HYNIC requires two chemical intermediaries, thiphenylphosphine and tricine, in order to be labeled with ^{99m}Tc. L, L-ethylenedicysteine (EC), which uses an N_2S_2 chelate, may be labeled with both ^{68}Ga and ^{99m}Tc, with high efficiency, high radiochemical purity, and high stability, as the preparation remains stable for several hours [170].

An important feature of the ^{99m}Tc-labeled tracers is that they may indicate a potential therapeutic target to be the target of another radionuclide, such as rhenium-188 (^{188}Re). Similar to ^{99m}Tc, ^{188}Re is a generator-produced radionuclide that has a short physical half-life (16.9 h), with a favorable dosimetry and good characteristics for imaging, depending on its γ-ray emission of 155 keV. It also has high potential for therapeutic use, based on its beta emission of 2.1 MeV.

A generator is a device that contains a parent–daughter nuclide pair in which a relatively long-lived parent isotope decays to a short-lived daughter isotope, capable of being used for imaging and/or for metabolic therapy. With the generators, the parent isotope can be produced in a cyclotron and sent to where it will be put to clinical use, where the daughter isotope is obtained by elution with an isotonic saline solution.

There are also some positron emitters that are provided by generators. This is the case of gallium-68 (^{68}Ga) that has a half-life of 1.13 h, which is produced from germanium-68 that has a half-life of 275 days. The long parent half-life makes this generator a good alternative to the isotopes that are most often used in PET, such as ^{18}F and iodine-124 (^{124}I).

6.4.2 Tumor Proliferation

One of the characteristics of the malignant transformation of a tissue is uncontrolled cellular proliferation that is associated with high mitotic and metabolic activities. These characteristics are directly correlated with increased cellular anaplasia and tumor aggressiveness. The development of an imaging agent able to visualize this proliferative activity would have high specificity for the detection of malignant tumors and could be used to distinguish between high-grade and benign or low-grade tumors. Additionally, the tracer should be able to detect the transformation of a low-grade tumor into a high-grade one, and it could be used to plan the best approach for biopsy, surgical resection, or even for radiation therapy.

6.4.3 Tumor Metabolism

There are several pathways involved in malignant transformation that can be used in NM to obtain functional information about the state of a disease or about the monitoring of therapy.

6.4.4 Glucose Metabolism

When a malignant transformation occurs, changes take place at different levels of the cell related to glucose transport and metabolism. There is increased activity of the glucose transporters, particularly GLUT-1; biosynthesis of a new class of transporters; increased glucose uptake through the membrane; increased glucose metabolism; and also increased normal glycolytic enzymes such as hexokinase, phosphofructokinase, and pyruvate dehydrogenase [171].

These changes are the basis of the use of fluorodeoxyglucose (FDG), a glucose analog that has similar first metabolic steps in the glycolytic pathway. Once inside the cell, it undergoes phosphorylation by hexokinase and gives origin to fluorodeoxyglucose-6-phosphate (FDG-6-phosphate). However, the similarity with the glycolytic pathway ends here, as the glucose-6-phosphate isomerase, which acts in the next step, does not recognize the FDG-6-phosphate as a substrate and does not metabolize it. This metabolite has a very slow transmembrane clearance, which allows greater retention in tumor cells than in normal cells. This intracellular accumulation reflects energy consumption in the tissues and is an index of cellular activity. In fact, several studies have shown that in a wide variety of tumors the FDG uptake is related to the degree of tumor invasiveness and proliferation status. Overall, FDG uptake is lower when invasiveness is low and in tumors with a low proliferation rate than in poorly differentiated tumors with a high proliferation rate [172,173].

If the FDG is labeled with 18F (18F-FDG), the increased uptake correlates with a high-grade of malignancy or biological aggressiveness and also appears to be strongly related with the number of viable cells in the tumor mass. The biochemical mechanism involved in the use of this tracer helps differentiate tumors from postradiotherapy necrosis, because necrotic regions have low metabolism. Moreover, changes in the FDG uptake during treatment may be an indicator of therapeutic response [173].

However, as increase in glycolysis is associated with increased metabolic activity, cell growth rate, and malignancy of the tumor, higher FDG uptake is not specific for malignancies.

6.4.5 Nucleoside Metabolism

The S phase of the cell cycle, the step when deoxyribonucleic acid (DNA) is synthesized, is fundamental to cell growth. At this step, nucleosides are incorporated into the DNA whose pyrimidine nitrogen bases are thymine and cytosine and whose puric nitrogen bases are adenine and guanine. Of the four nitrogen bases, thymine is only incorporated into DNA, whereas the other three are also incorporated into ribonucleic acid (RNA).

The nitrogen bases associated with a pentose form the nucleosides. If there is a tumor, the nucleoside incorporation is directly dependent on how fast it grows. This means that the accumulation of nucleosides by the tumor tissue and organs with a high proliferative rate reflects the relative degree of synthesis of DNA, which is an index of cell division.

This information is very important, because the growth rate of the tumor is inversely proportional to the timing of the appearance of radiotherapy effects. This means that quantitative information about the tumor's proliferative activity can be obtained by noninvasive imaging methods, which is crucial for suitable monitoring of the response to radiotherapy [173].

We can also consider the purine analogs that have been labeled using chelating agents for the labeling reaction. An example of this is the labeling of guanine with ^{99m}Tc or ^{68}Ga using ethylenedicysteine (EC) as chelator. Two tracers were thus obtained, ^{99m}Tc-ethylenedicysteine-guanine (^{99m}Tc-EC-Guan) and ^{68}Ga-ethylenedicysteine-guanine (^{68}Ga-EC-Guan), to assess cell proliferation as well as to differentiate it from inflammation [170,173,174].

6.4.5.1 Thymidine

Thymidine, as noted earlier, is one of the nucleosides that are incorporated into the DNA. It can either come from the bloodstream or originate from intracellular synthesis. Since the rate of DNA synthesis mainly depends on the proliferative status of the tissue, the incorporation of nucleosides and the respective nucleotides in DNA reflects the rate of cell synthesis.

When thymidine is used for medical imaging, the molecule must be labeled with an appropriate radiation emitter. The most widely used option is labeling with ^{11}C, which can be done in two positions, that is, in the 5-methyl position, giving rise to ^{11}C-methylthymidine, or in position 2 of the ring that originates from the ^{11}C-thymidine [170,173].

After intravenous injection, both molecules are rapidly metabolized, but their final metabolites are different. For ^{11}C-methylthymidine the metabolites are the β-ureidoisobutyric and β-aminoisobutyric acids, whereas for ^{11}C-thymidine it is CO_2 labeled with carbon-11 ($^{11}CO_2$). These metabolisms have different kinetic models, which must be taken into account for the correct quantification of tumor uptake [173].

^{11}C-thymidine and ^{11}C-methylthymidine are both taken up by a great variety of human cancers such as lymphoma; brain tumors; sarcomas; and tumors of lung, kidney, and head and neck.

6.4.5.2 Thymidine Derivatives

The rapid metabolism and slow kinetics of incorporation of ^{11}C-thymidine into the DNA as well as the need to have a cyclotron nearby due to the short half-life of ^{11}C make it of little use in clinical applications. These limitations led to the development of several analogs capable of being labeled with emitters that have longer half-lives and are more resistant to *in vivo* metabolic breakdown. These features can be obtained by labeling with the ^{18}F, but the fluorine itself induces changes in the *in vivo* behavior of the molecule. The changes induced make the molecule more resistant to metabolic breakdown, while maintaining the ability to visualize the tumor proliferative capacity *in vivo*.

Several analogs have been developed, and the most interesting from the point of view of functional imaging is 3'-deoxy-3-^{18}F-fluorothymidine (^{18}F-FLT). In both animal models and patients, this radiopharmaceutical exhibits better performance than FDG and is also more useful for monitoring therapy. This improved performance is due to several factors. The most important are that FLT uptake due to inflammatory response is lower than FDG uptake, and that cytostatic drugs have a greater impact on cell division than on the metabolism of glucose. Additionally, the uptake of FLT in normal brain regions is very low compared with that in brain tumors, due to reduced cell division of neuronal cells [170,173,175].

These characteristics mean that ^{18}F-FLT is a radiopharmaceutical of special interest in brain tumors, for restaging after treatment.

6.4.5.3 Other Nucleosides and Analogs

In addition to FLT, other nucleoside derivatives have been proposed, also aiming at overcoming both the rapid degradation of thymidine and the short half-life of ^{11}C. One of the derivatives chosen was deoxyuridine. This molecule was labeled with several radioisotopes such as iodine-131 (^{131}I), iodine-124 (^{124}I), bromine-76 (^{76}Br), or bromine-77 (^{77}Br). ^{131}I and ^{77}Br are isotopes that can be used in NM by single photon emission, whereas ^{76}Br and ^{124}I are used in PET [169].

The four different possibilities of labeling deoxyuridine resulted in four complexes: ^{76}Br-deoxyuridine, ^{77}Br-deoxyuridine, ^{131}I-deoxyuridine, and ^{124}I-deoxyuridine, all of which can be taken up by the tumor tissue. However, bromodeoxyuridine (BrdU) has a higher tumor uptake due to its smaller size and hydrophilicity than thymidine. Nevertheless, *in vivo* dehalogenation leads to greater background activity, with a consequent fall in SNR. Later studies showed that a significant part of the tissue signal is a result of the presence of free Br (^{76}Br or ^{77}Br), that is, the tracer available is not totally incorporated into the DNA. In fact, the rate of incorporation into DNA is relatively low compared with the total activity in the tissue, with no significant improvements in SNR when the diuresis is forced. Given the characteristics described, BrdU is a thymidine analog that is used cold to find the labeling index or fraction of cells in mitosis by immunohistochemistry; and when labelled with ^{76}Br or ^{77}Br, it can be used as a tracer in PET or in NM using single photon emission, respectively [173].

Another thymidine analog developed for labeling with the ^{76}Br is 1-(2'-deoxy-2'-fluoro-β-D-arabinofuranosyl)-5-[^{76}Br]bromouracil (^{76}Br-BFU); it is stable to metabolic degradation and is incorporated in greater amounts into DNA with a bigger uptake in proliferative tissues than in nonproliferative ones, as has been shown in animal models.

Other nucleoside analogs currently under study include ^{18}F-1'-fluoro-5-(C-methyl)-1-β-D-arabinofuranosyluracil (FMAU), ^{124}I-iododeoxyuridine, and ^{124}I-5-iodo-1-(2-fluoro-2-deoxy-D-β-arabinofuranosyl)-uracil (FIAU) [170,173].

6.4.6 Amino Acids Metabolism

After nucleosides, the amino acids are the next step if we want to study tumor proliferation, as they too can be used as markers of protein synthesis. Tumors need amino acids for protein synthesis as metabolic fuel or to create secretory products. These requirements show significant variations both for different amino acids and for different types of tumor.

When labeled amino acids are used, the bigger part is taken by the cell, and only a small amount is used for protein synthesis. However, if amino acids are used as tracers to obtain an image all fractions are evaluated, and their total concentration is related to the metabolic activity of viable tumor cells.

One important issue related to the use of amino acids as tumor imaging tracers is that their uptake by the cells involved in inflammatory processes is poor. This is extremely important, because patients with cancer frequently suffer inflammatory processes associated with chemotherapy or radiotherapy. Given this, the imaging evaluation obtained with amino acids of a patient with cancer can give functional information with the objective of achieving better staging with fewer false positives [172].

Several amino acids have been labeled with either gamma or positron emitters with a view to them being *in vivo* tumor imaging agents, after intra venous injection. Examples are methylmethionine labeled with ^{11}C (^{11}C-MET), thyrosine labeled with ^{11}C (^{11}C-TYR), and a phenylalanine derivative, fluoro-phenylalanine, labeled with ^{18}F (^{18}F-Phe). However, the amino acids with the most potential to obtain functional imaging in nuclear oncology are an artificial amino acid, L-3-iodo-α-methyl thyrosine (IMT), and some fluorinated derivatives such as O-(2-fluoroethyl)-L-thyrosine (FET) and α-fluoromethyl-thyrosine (FMT). The amino acid IMT can be labeled with iodine-123 (^{123}I-IMT), whereas FET and FMT can be labeled with the fluorine-18, to give the ^{18}F-FET and ^{18}F-FMT complexes, respectively [170,172,173].

Regarding thyrosine, since it is an amino acid precursor of melanin, the thyrosine derivatives may have great importance for imaging patients with malignant melanomas.

6.4.7 Enzymes

Due to their involvement in very specific stages of a particular metabolic pathway, some enzymes may be especially important, as they are good image targets for assessing cellular functional information. This assessment may be related to the choice of patients to be included in a specific therapy or to the information about the therapeutic response.

6.4.7.1 Thymidine Phosphorylase

Thymidine phosphorylase is an enzyme that catalyzes the hydrolysis of thymidine to thymine and deoxyribose-1-phosphate.

The overexpression of thymidine phosphorylase by cells of colorectal, head and neck, bladder, and cervix cancer tumors is associated with increased tumor angiogenic activity and a decrease in patient survival. In this context, data from animal models show that 5-chloro-6-(2-iminopyrrolidin-1-yl)methyl-2,4(1*H*,3*H*)-pyrimidinedione, a specific inhibitor of thymidine phosphorylase, is able to reduce tumor size [170].

Additionally, since thymidine phosphorylase also catalyzes the reverse reaction, that is, the conversion of thymine to thymidine, it may also help to retain within the cell therapeutic thymidine analogs such as capecitabine, which is converted into fluorouracil.

Analogs radiolabeled with astatina-211 (^{211}At), iodine-125 (^{125}I), or iodine-131 (^{131}I) can also be used both as therapeutic agents to destroy tumor cells and as tracers to identify tumors with high levels of thymidine phosphorylase that can be treated and/or to monitor the therapeutic response and tumor angiogenesis.

6.4.7.2 Tyrosine Kinase

Tyrosine kinase inhibitors such as gefitinib (Iressa®) or erlotinib (Tarceva®) showed that they are capable of preventing tumor growth. The effectiveness of this therapy can be assessed using a derivative of tyrosine labeled with ^{99m}Tc, using cyclam (1,4,8,11-Tetraazacyclotetradecane) as a chelating agent.

Studies performed *in vitro* showed a good correlation between the cellular uptake of ^{99m}Tc-cyclam-tyrosine and the expression of the protein quantified by Western blot. Consequently, tumors with a high expression of tyrosine kinase also show a good correlation with data obtained from cell cultures [170].

Studies with ^{99m}Tc-cyclam-tyrosine indicated that the tracer can be useful in the selection of patients undergoing therapy with tyrosine kinase inhibitors.

6.4.8 Tumor Hypoxia

Tissue hypoxia is characterized by a decrease in the partial pressure of oxygen in the tissues; it underlies the pathogenesis of several diseases, and it has also been considered essential for the growth of malignant solid tumors. In cancer, tumor hypoxia is one of the most important factors implicated in resistance to radiotherapy and conventional chemotherapy, resulting in higher local tumor recurrence. Moreover, it is known that hypoxia also induces angiogenesis, which contributes to cancer invasion and metastization, which is also associated with poor prognosis.

Tumor resistance related to hypoxia can be overcome through the use of hypoxic cell radiosensitizers or hyperbaric oxygen, as low levels of hemoglobin and low tumor partial pressure of oxygen (pO_2) are associated with higher resistance to treatment.

Getting information about tumor hypoxia status, so as to predict the outcome of radiotherapy or the possible use of radiosensitizers, is now a reality, through either biochemical analysis or noninvasive functional imaging [173].

There is a nuclear protein called hypoxia-inducible factor (HIF), which provides information on intracellular hypoxia status. This protein is composed of two subunits, alpha and beta. The alpha subunit is responsible for a specific hypoxia response, but the beta subunit has no such specificity. Under normoxia conditions, depending on the von Hippel–Lindau tumor suppressor protein (pVHL), the subunits are rapidly degraded by the ubiquitin–proteasome pathway. Under hypoxia conditions, the subunits are stabilized and translocated to the nucleus, where the β-subunit can be dimerized with a DNA molecule.

In terms of detection, the HIF alpha subunits are not present in most normal human tissues, but they are expressed in many malignant tumor cells, particularly in areas adjacent to necrosis. This pattern is seen in prostate, breast, lung, gastrointestinal tract, brain, ovary, melanoma, and mesothelioma tumors. In clear-cell renal carcinoma and hemangioblastoma, it is mainly seen throughout the tumor mass. In brain tumors, since necrosis and hypoxia are histological invasiveness hallmarks, the degree of HIF α-subunit expression is correlated with the tumor type.

NM is important, as it is possible to radiolabel some molecules involved either in the intracellular pathways of hypoxia or in the impact on other metabolic routes. This labeling means that an image can provide qualitative and quantitative information about whether there is tumor hypoxia.

It is known that hypoxic tissues can take up bioreductive molecules that contain an imidazole group, such as misonidazole. These molecules have the capacity to accept an electron, with the consequent production of a free radical anion that, after reduction, is incorporated in the constituents of the cell if it is under hypoxic conditions. Consequently, the quantity of bioreductive molecules taken up by hypoxic cells can be seen as a direct sensor of oxygen partial pressure in tissues.

The first molecules developed with the objective of being used as molecular imaging agents were fluoromisonidazole (FMISO) and fluoroerythronitroimidazole (FETNIM), both having a fluorine atom. This fluorine atom in molecules means that they can be labeled with ^{18}F, to form the ^{18}F-FMISO and ^{18}F-FETNIM complexes, respectively [168,174,176].

With the same goal, other molecules that do not contain an imidazole group were developed. They include iodoazomycin arabinoside (IAZA) and iodovinylmisonidazole (IVM). These molecules have an iodine atom, enabling them to be labeled with different iodine isotopes, including ^{123}I and ^{124}I. This labeling enables them to be used in NM by single photon emission (when labeled with ^{123}I) or PET, if ^{124}I is used.

There are other compounds, not derived from nitroimidazole, that can be used for the same purpose. The ones giving the best results were 4,9-diaza-3,3,10,10-tetramethyldodecan-2,11-dione dioxime, known as HL-91, which can de labeled with ^{99m}Tc, and Cu(II)-diacetyl-bis(N-4-methylthiosemicarbazone) (ATSM), which can be labeled with several copper isotopes [176].

More recently, two compounds have been developed, ^{18}F-fluoro-PR-170 for use in PET and SR 4554 (*N*(2-hydroxy-3,3,3-trifluoropropyl)-2-(2-nitro-1-imidazolyl)acetamide) for use in nuclear MRI. This compound also appears to have potential to be labeled with positron emitters and thus, can be used as a PET tracer [168].

6.4.9   Tumor Angiogenesis

Angiogenesis is the proliferation of endothelial and smooth muscle cells to form new blood vessels. This neovessel proliferation is a crucial factor in the metastatic process and has an impact at two levels. First, the new vessels provide the main route through which tumor cells leave the primary tumor site and spread to other parts of the body, with the possibility of distant secondary locations. Second, the new vessels supply tumor tissue with the oxygen that enables the primary tumor growth more.

The growth of these new blood vessels is mediated and controlled by several angiogenic growth factors, particularly the vascular endothelial growth factor (VEGF), some cellular receptors such as estrogen receptors, and some adhesion molecules such as integrin αvβ3. This integrin is expressed in vascular endothelial cells only during angiogenesis and vascular remodeling, particularly in the metabolic pathways stimulated by VEGF; it is not expressed in mature vessels or in the nonneoplastic epithelium. Integrin vβα3 also binds to several ligands if they contain the pattern-Arg-Glu-Asp-(RGD) in the extracellular matrix [177].

With regard to this bond, it was shown that disruption of this ligand interaction by competitive antibody binding blocks the proliferation of new blood vessels. This was clinically used with the development of a new therapeutic monoclonal antibody, called LM609, which prevented the proliferation of new vessels.

An *in situ* tumor can grow up to 1–2 mm without additional nourishment from the blood supply. However, the oxygen needs increase as it grows, and this leads to the appearance of a more acidic microenvironment that induces tumor cells to produce angiogenic factors which stimulate the development of new vessels. The neovascularization allows for tumor expansion and also provides an escape route for tumor cells to travel to other areas of the body, a long way from the primary tumor location, to produce new tumor foci: these are metastases. This process is particularly true for tumors with high vascular density that show a higher incidence of metastasis than poorly vascularized tumors.

This type of tumor expansion is the basis of the development of angiogenesis inhibitors that are a highly promising new approach for anticancer therapy. Four types of molecules can be considered as antiangiogenic therapeutic agents. They are antibodies, protein fragments, modulation of the fibroblast growth factor (FGF), and synthetic small molecules. In the case of antibodies, we have anti-integrins, anti-EGFR (epidermal growth factor receptor), anti-VEGF monoclonal antibody, and antiendoglin glycoprotein. In the case of protein fragments, plasminogen and collagen can be used. In relation to FGF modulation, interferons are very important molecules. As far as small synthetic molecules are concerned, protease inhibitors, urokinase inhibitors, cyclooxygenase inhibitors, and tyrosine kinase inhibitors are new approaches that must be considered [177].

Based on these new approaches, several studies using cyclic peptides with the RGD sequence labeled with ^{18}F, ^{99m}Tc, and ^{111}In are now in progress to assess angiogenic status through a functional image.

In fact, if the results obtained from animal models confirm the high tumor uptake of such molecules, they could be used as image tracers for diagnosis and also to give information about the therapeutic response to $\alpha v\beta 3$ integrin antagonists. One of these molecules is ethylenedicysteine endostatin (EC-endostatin) labeled with ^{99m}Tc (^{99m}Tc-EC-endostatin). This complex is under study as a potential noninvasive image tracer, able to give qualitative and quantitative information about the tumour's response to antiangiogenic therapy. This is because *in vitro* cell viability and TUNEL (terminal deoxynucleotidyl transferase biotin-dUTP nick end labelling) assays indicate no clear difference between EC-endostatin and endostatin. Moreover, biodistribution of ^{99m}Tc-EC-endostatin in tumor-bearing rats showed that the tumor to normal tissue count ratio increased with time, which was confirmed by planar images. These images reveal that the tumors were well visualized two hours after ^{99m}Tc-EC-endostatin administration. In addition, this uptake by the tumor can be used to assess the effectiveness of antiangiogenesis therapy.

Cyclooxygenase-2 (COX-2) is another molecule with an important role in angiogenesis and, indirectly, in cancer progression. Since many tumours express COX-2, functional images obtained with a COX-2 inhibitor such as Celebrex (CBX) labeled with ^{99m}Tc (^{99m}Tc-EC-CBX) may give noninvasive information about tumor COX-2 expression as well as information about clinical responses to anti-COX-2 therapy or even about patient selection for treatment with these agents [170].

Another important factor that plays an important role in cell division, tumor progression, angiogenesis, and metastasis is EGFR. Since many tumors express EGFR on their surface, the functional images using monoclonal antibodies that target the EGFR, such as the chimeric monoclonal antibody C225 labeled with ^{99m}Tc (^{99m}Tc-EC-C225), may give noninvasive information about EGFR expression. This information can also be useful for the evaluation of clinical responses to anti-EGFR therapy or even for patient selection for treatment with this kind of molecule [170].

Results already obtained in nude mice bearing xenografts showed that C225 in combination with cytotoxic drugs or with radiotherapy is effective in the eradication of well-differentiated EGFR-expressing tumors.

6.4.10   Apoptosis

Apoptosis, which is programmed cell death or cell suicide, is an active energy-dependent mechanism for the removal of injured, infected, or immuno-logically recognized as harmful or superfluous cells. Although a large number of stimuli can initiate the apoptotic process, they all culminate in caspase cascade activation. The activation of these proteases leads to irreversible changes in cell components, such as cytoskeletal disruption, chromatin clumping, internucleosomal DNA cleavage, and, eventually, disintegration of the cell into small membrane-bound leftovers targeted for quick removal by macrophages. The apoptotic process is fast and is not usually accompanied by an acute inflammatory response.

The role of apoptosis has been demonstrated in a great variety of phys-iological processes, including during embryogenesis, in postnatal brain remodeling, in the development of immune tolerance through clonal deletion of T cells, and in the turnover of senescent cells in the intestinal mucosa.

In oncology, apoptosis is also important in a wide variety of malignant tumors, particularly in hypoxic regions adjacent to areas of necrosis, which is currently the endpoint of most forms of anticancer therapy.

With regard to anticancer therapy, some authors see apoptosis as the most critical factor influencing tumor sensitivity or resistance to chemotherapy. In some cancers, therefore, the apoptotic index of the tumor can be considered as a prognostic factor of the clinical response to chemotherapy and as overall survival.

It is known that apoptosis is the major mechanism in cell death after radio-therapy. In fact, doses of 1 to 5 Gy induce apoptosis, which appears about 1 to 2 h after irradiation and reaches a maximum almost 3–6 h after irradiation.

The intensity of apoptosis can be quantified by using a fluorescent TUNEL stain and Bcl-2 (B-cell leukemia/lymphoma 2) and Bax (Bcl-2-associated-X-protein) by semiquantitative immunohistochemical assays [170].

One of the characteristics of normal membranes is that phosphatidylserine, a native phospholipid membrane anion, is generally limited to the inner leaflet of the plasma membrane lipid bilayer. However, in cells that are in apoptosis, phosphatidylserine is selectively and quickly translocated to the outer leaflet and is, thus, exposed on the membrane surface.

Annexin-V is an endogenous human protein that is able to bind with very high affinity to exposed phosphatidylserine. This characteristic makes it an apoptosis marker which has been used after being labeled with radioac-tive isotopes to obtain functional images that show whether apoptosis is occurring at the site under study. For this purpose, annexin-V was labeled with iodine and technetium using ethylenedicysteine (^{99m}Tc-EC-annexin-V) or hydrazinonicotinamide (^{99m}Tc-HYNIC-annexin-V) as chelating agents for labeling with ^{99m}Tc. This type of labeling yields functional information related to apoptosis for monitoring the dynamics of cell death induced by chemotherapy and/or radiotherapy, and it also assesses the effectiveness of therapies [170,178].

Results obtained in animal models where annexin-V labeled with ^{99m}Tc were used showed that it is taken up by heart, lung, and liver allografts; by arteriosclerotic plaques; and neonatal rabbit brain after unilateral carotid artery ligation to induce cerebral hypoxia.

Similarly, studies in humans show increased annexin-V uptake in a cardiac allograft recipient when there is histological evidence of transplant rejection as well as in some patients with acute myocardial infarction.

Also, *in vitro* studies confirm the uptake of annexin-V labeled with ^{99m}Tc by breast cancer cell lines when they are exposed to paclitaxel and 10 to 30 Gy of radiation. Similar results were obtained in breast tumor bearing rats when they were treated with paclitaxel.

The results obtained so far indicate that annexin-V uptake correlates well with the extent of cell death but is not specific to apoptosis, because it can also occur in cell necrosis.

6.4.11 Multiple Drug Resistance

Multiple drug resistance (MDR) is the development of cross-resistance to several cytotoxic drugs, not related either structurally or functionally after tumor exposure to an individual cytotoxic drug [179–181].

In clinical practice, chemotherapeutic protocols almost always involve a combination of drugs that act on different cellular targets to maximize the therapeutic effect. This means that the drugs must be different in order to operate in different intracellular metabolic pathways. If, after a few therapeutic cycles, a malignant tumor starts to express MDR, this indicates that its response to chemotherapy is going to be reduced, which indicates a worse prognosis.

There are several groups of transmembrane transport proteins related to multidrug resistance, four of which are P-glycoprotein (Pgp), multidrug resistance-related protein (MRP), lung resistance-related protein (LRP), and breast cancer related protein (BCRP), all of which are ATP-binding cassette multidrug transporters.

P-glycoprotein is a transmembrane glycoprotein with a molecular weight of 170 kDa, encoded by the MDR1 gene, located on chromosome 7 (7q21.1), which has 12 transmembrane domains and two ATP binding sites. P-glycoprotein acts as an ATP-dependent membrane efflux transporter that pumps cationic and lipophilic drugs out of cells and so dramatically reduces their intracellular accumulation. This extrusion mechanism confers resistance on a large number of cytotoxic drugs, such as anthracyclines, vinca alkaloids, epipodophyllotoxins, and taxanes, which are not related either structurally or functionally. In addition to Pgp being present in oncology tissues after cytotoxic drug exposure, it is also found in a variety of normal tissues including the adrenal cortex, the intestinal mucosa, the gastrointestinal epithelium, the biliary canalicular surface of hepatocytes, pancreatic duct cells, proximal tubule cells in kidney, myocytes, capillary endothelial cells of the brain and testes, the uterus during pregnancy, and stem cells from bone marrow positive for CD34 [182,183].

The presence of Pgp in normal cells acts as a protection mechanism, as lipophilic and cationic molecules, which are toxins, are extruded to outside the cells. Some malignant tumors overexpress Pgp from the time of initial diagnosis, even before exposure to any cytotoxic drug. This overexpression limits the effectiveness of a wide variety of chemotherapeutic agents such as daunorubicin, vincristine, ectoposide, and adriamycin, which are substrates for their activity.

Another transmembrane transport protein, also belonging to the superfamily of ABC transporters, is MRP. It is encoded in humans by several closely related genes located on chromosome 16 (16p13.1). This protein has a molecular weight of 190 kDa, has 17 transmembrane domains with two ATP binding sites, transports lipophilic and anionic drugs, and is expressed in almost all normal epithelial cells. The MRP transports several noncytotoxic drugs such as anionic compounds and leukotrienes (LT) (in particular LTC4, LTD4, and LTE4) and cytotoxic drugs such as vinca alkaloids, epipodophyllotoxins, anthracyclines, and camptothecins, most of which are also substrates for Pgp. Therefore, both the Pgp and the MRP may be overexpressed at the same time in drug-resistant cells, which also limits the effectiveness of a number of cytotoxic drugs such as doxorubicin, epirubicine, ectoposide, methotrexate, cisplatin, vincristine, vinorelbine, and mitoxanthrone, because they function as their substrates [180,182].

Owing to this behavior, both the Pgp and the MRP are now considered as major targets for pharmacological intervention. As a result, chemosensitizer molecules able to interact with the referred transporters were developed which can make the tumors that express Pgp or MRP sensitive to several cytotoxic drugs.

The action of these modulating molecules or chemosensitizers of MDR is to block the transmembrane efflux proteins, especially Pgp, and so to increase the intratumor concentrations of cytotoxic drugs.

Over the last few years, several molecules have been used as MDR modulators. The first generation of modulators includes molecules such as verapamil, azidopin, cyclosporin A, and FK506, which act by binding to Pgp, competing with cytotoxic drugs, and are themselves transported by membrane transporters. The clinical utility of such molecules is, however, very limited due to their adverse effects, as to have an MDR modulation effect their concentrations must be very high.

The second generation of modulators includes molecules such as dexverapamil, tamoxifen, progesterone, GF120918, and PSC833. These molecules compete for binding sites, which inhibits the transport of cytotoxic agents, but they are not transported. These modulators are more potent and less toxic than the first-generation ones. One of the most promising modulators of the second generation is PSC833. This modulator is a derivative of cyclosporine but lacks its immunosuppressive effects and its nephrotoxicity. Results obtained in preclinical tests and clinical trials showed that it is a powerful MDR modulator. However, these tests also showed that it was toxic to the central nervous system (CNS), that it was a substrate for Pg, and it is currently considered as a partial antagonist [182,184].

The third-generation modulators include molecules such as tariquidar and zozuquidar and were developed to overcome the limitations of the second-generation modulators. Studies have shown that they are able to specifically inhibit Pgp.

With regard to the assessment of MDR, NM has some tracers that were originally developed for studies on myocardial perfusion. These tracers are lipophilic and monocationic molecules, labeled with ^{99m}Tc, as ^{99m}Tc-hexakis(2-methoxyisobutylisonitrile) (^{99m}Tc-MIBI), ^{99m}Tc-tetrofosmin (^{99m}Tc-TF), and ^{99m}Tc-furifosmin (^{99m}Tc-FUR), which are currently in use to assess the *in vivo* expression of MDR by tumors.

The MIBI is an isonitrile derivative, whereas tetrofosmin is a diphosphine derivative. These two molecules are substrates for both Pgp and MRP at picomolar concentration, and either can be used as a tracer to evaluate resistance to cytotoxic drugs or to give information about resistance reversal if the patient has been previously treated with an MDR modulator.

The initial uptake and concentrations of the tracers referred to earlier are due to nonspecific factors such as perfusion and electrostatic interactions in the mitochondrial membrane; whereas in the tumor, efflux kinetics reflect the level of expression of MDR1 or other similar genes that confer resistance to multiple drugs. An assessment of the Pgp activity level can predict the response to chemotherapy or indicate the need of adjuvant therapy with Pgp inhibitors such as verapamil or cyclosporine, or even with the newer agents such as PSC833, GF120918, VX-710, tariquidar, or zozuquidar [181,184].

The usefulness of chemosensitizers to reversing Pgp function and, thereby, increasing the intracellular accumulation of cationic tracers labeled with ^{99m}Tc has not been well studied. In fact, most studies about the capacity of the two lipophilic and cationic agents to recognize MDR expression, or its reversal through the use of modulators, were performed *in vitro* [184].

In addition to these tracers used in NM by single photon emitters, several molecules have been proposed to be labeled with positron emitters. These include the colchicine labeled with ^{11}C (^{11}C-colchicine), verapamil labeled with ^{11}C (^{11}C-verapamil), daunorubicin labeled with ^{11}C (^{11}C-daunorubicin), and N-acetyl-leukotriene E4 labeled with ^{11}C (^{11}C-LTE4) [180].

6.4.12 Tumor Receptors

Among the characteristics of malignant transformation are the expression of a large number of membrane receptors and the emergence of new ones.

6.4.12.1 Folic Acid Receptors

Folic acid (FA) receptors of the membrane mediate the intracellular accumulation of FA and its analogs, such as methotrexate. Their expression is limited in normal tissues, but receptors are overexpressed in several types of tumor cells. Using ethylenedicystein as a chelator, the FA can be labeled with ^{99m}Tc (^{99m}Tc-EC-FA). This tracer can be used to obtain images that reflect the cellular uptake and consequent intracellular accumulation [185].

6.4.12.2 Sigma Receptors

Although the biology of sigma opioid receptors is not yet fully understood, its expression has been detected outside the CNS in a large variety of tissues including the heart, kidneys, adrenal glands, gonads, gastrointestinal tract, liver, and spleen.

In terms of location, these receptors can be found in the plasma membrane of some cellular organelles, such as the Golgi complex, endoplasmic reticulum, and nucleus, where they may be also associated with G-proteins. Two subtypes are described, the sigma-1 and sigma-2, both expressed in some types of tumor. The sigma-1 receptors are mainly expressed in prostate tumor, whereas the sigma-2 receptors are overexpressed in melanomas, breast cancer, prostate cancer, and small cell lung cancer. With regard to these receptors, the higher the sigma-2 receptors' expression by a tumor, the greater the degree of proliferation of tumor cells. In some cases of breast cancer, their overexpression can be so intense that each cell can express more than a million copies of the receptor.

According to studies performed in cells of breast adenocarcinoma, a specific complex of ^{99m}Tc for the sigma-2 receptors has been recently synthesized, [N-[2-((3′-N′-propyl-[3,3,1]azabicyclononan-α3-yl)(2-methoxy-5-methyl-phenylcarbamate)(2-mercaptoethyl)amino) acetyl]-2-aminoethane-thiolato]Tc(V)oxide) [186–188].

6.4.12.3 Breast Cancer Receptors

Hormonal therapy in breast tumors has been well established for a number of years, because more than a half of primary breast tumors express receptors for estrogen and progesterone. For tumors that express receptors, the stimulation with estrogen has a pro-proliferative effect.

Tamoxifen and raloxifene are selective estrogen receptor modulators that act as agonists in some tissues such as cardiovascular tissue and as antagonists in other tissues such as the breast and breast cancers [189].

With this type of distribution in mind, estrogen blocking action is associated with the slowdown of tumor growth, increasing the survival time, and lower incidence in patients who undergo prophylactic therapy.

This type of correlation is in the basis of the use of estrogen analogs labeled with radioactive isotopes to assess the expression of estrogen receptors and to monitor the antiestrogen therapy response in patients with breast cancers.

The HER-2/neu is a tyrosine kinase transmembrane receptor that is overexpressed in many of the primary breast cancers, and it is associated with a poor prognosis. However, its expression is also a predictive sign of response to adjuvant treatment with doxorubicin. Nevertheless, if the patient has undergone neoadjuvant therapy with tamoxifen or anthracyclines, the overexpression of HER-2/neu receptor predicts a poor response to therapy [189].

When overexpressed, this receptor is an important therapeutic target in breast cancers. With this goal, therefore, a monoclonal antibody that binds to the extracellular portion of HER-2/neu, trastuzumab (Herceptin®) was developed. The use of this molecule improves the therapeutic response by acting as an adjuvant to first-line chemotherapy, particularly in tumors that overexpress HER-2/neu receptor.

This receptor was achieved by developing monoclonal antibodies anti-HER-2/neu labeled with ^{131}I and ^{111}In. The uptake of these labeled antibodies by the tumor and its visualization are a sign that can predict the response to immunotherapy with trastuzumab.

6.4.12.4 Cholecystokinin-B/Gastrin Receptors

The receptor of cholecystokinin-B (CCK-B) or gastrin is expressed in almost all the telencephalon and in the stomach and is not found in any organ in normal conditions. It may be overexpressed, however, in human pancreatic, gastric, and colorectal carcinomas; small cell lung cancer; ovarian stromal tumors; and astrocytomas. In the thyroid, more than 90% of medullary carcinomas express this receptor, and the injection of cold pentagastrin has been widely used as a provocative test for detection of primary, recurrent, or metastatic medullary thyroid carcinoma [190].

In terms of NM, gastrin heptadecapeptide labeled with ^{131}I is taken up by medullary thyroid carcinomas, thus allowing its use in the field of metabolic radiotherapy and in imaging. The uptake of this complex by medullary thyroid carcinomas reflects the receptor's overexpression [191].

6.4.12.5 Other Receptors

Besides the receptors already mentioned, tumors may overexpress receptors for other peptides and other small molecules, such as G-protein receptors, tyrosine kinase receptors, and nontyrosine kinase receptors.

Gastrin releasing peptide receptors, vasoactive intestinal peptide (VIP) receptors, and type-2 somatostatin receptors are examples of protein-G receptors. Regarding tyrosine kinase receptors, we can refer to the platelet-derived growth factor-B receptor (PDGF-R), the vasoendothelial growth factor receptor (VEGF-R), the insulin receptor, the insulin-like growth factor type-1 receptor, the EGF-R, and the fibroblast growth factor receptors 1 and 4. With regard to the nontyrosine kinase receptors, we have the leukemia inhibitory factor receptor and the β-subunit of interleukin-2 receptor [192].

Similar to trastuzumab for the treatment of breast cancer, other antibodies and peptide antagonists have been developed to explore receptors as therapeutic targets. As a consequence, several receptors are considered as targets, particularly the gastrin-releasing peptide receptor, the EGF-R, the PDGF, and VEGF receptors.

As in the trastuzumab example, too, the use of ligands labeled with radioactive isotopes can give information about the correct choice of therapy as well as allow the monitoring of anticancer therapy [193].

Several ligands have been tested in this context. Examples are the ligands for the endothelin receptors such as PD156707 labeled with ^{11}C (^{11}C-PD156707) and FBQ3020 labeled with ^{18}F (^{18}FBQ3020), the oxytocin analog ligands such as DOTA-lys8-vasotocin labeled with ^{111}In (^{111}In-DOTA-lis8-vasotocin), and a specific ligand for melanoma, N-(2-diethylaminoethyl)-2-iodobenzamide labeled with ^{123}I (^{123}I-N-(2-diethylaminoethyl)-2-iodobenzamide).

6.4.13 Molecular Image as Target

After reviewing the several types of approaches to the use of NM, we can say that molecular imaging provides information about many metabolic steps of angiogenesis, hypoxia, and the different transformations and genetic changes of cells.

In oncology, the added value of molecular imaging is the possibility of obtaining functional information related to cell differentiation and therapeutic response. This means that discriminating between an inflammatory reaction and a tumor recurrence, or knowing in advance whether the tumor will respond to a specific therapy, gives the clinician the ability to wisely choose which patients to include in a particular therapeutic protocol.

After analyzing the available possibilities for evaluating the various metabolic pathways through functional imaging, when a malignant transformation occurs, we can say that the agents that can differentiate between inflammation and tumor recurrence are apoptosis and DNA markers, because they are able to accumulate in cell nuclei. Agents that are able to provide information in terms of predicting therapeutic response are usually enzymatic markers or ligands for membrane receptors. The most commonly used enzymatic markers are those of glycolysis, hypoxia, and tyrosinase; whereas the most important receptors are the estrogenic and the androgenic receptors.

6.5 CNS: Physiological Models and Clinical Applications

6.5.1 Introduction

Now that you have reached this section, we may say that your CNS has performed important and relatively essential tasks for the reading, apprehension, and understanding of the book's subjects, matters, and goals. You have probably not noticed the significantly high amount of substrates you had to use nor even the cellular interaction and coordination mechanisms you triggered. We are convinced that other readings have already given you the opportunity to understand that the use of these substrates and fairly complex mechanisms of the functioning of many millions (approximately 100×10^9) of SNC cells can nowadays be investigated by means of medical imaging. However, you may have also noticed that new medical imaging techniques enable us to build functional maps [194] of the CNS, which, to some extent, refute some of the functional anatomy knowledge so rigorously and artistically developed by basic research [195] and clinical [196–198] scientists. Very recently, we were stunned by some remarks about the need to change concepts that we thought were well established: Contrary to the number of neuronal cells decreasing from birth on, as was previously thought, it is now considered as certain that these cells can be re-created during our life [199] to offset functional failure due to disease or trauma—neuronal plasticity. Further, the concept of an anatomical and functional map known as "homunculus" seems to be under reconsideration [200] in view of the results of medical imaging research, especially functional magnetic resonance and optical imaging [201].

The objective of this section is to make you think a little more deeply about the anatomical and functional bases of CNS medical images, especially those using radioactive molecules as radiopharmaceuticals. More than just informing you, our intention is rather to stimulate your curiosity and provide you with some clues to develop your potential interest in functionally investigating the CNS by means of medical images obtained with radionuclide-labeled molecules, that is, radiopharmaceuticals.

So, come along with us through a short summary of the anatomy, physiology, pharmacology, and neuronal interaction mechanisms relevant to the most important clinical applications in this field of knowledge.

6.5.2 Anatomical Basis

The anatomy of the CNS describes the morphology and macroscopic and microscopic components of the brain and spinal cord. Here, we will only remind you of some important aspects to help understand medical images. The morphological study of the spinal cord and cranial pairs is almost exclusively covered by magnetic resonance radiology, which plays the leading role in diagnosing disease and evolutive changes. Radiopharmaceuticals are much more often associated with the study and investigation of brain physiology and pathophysiology, either supra- or infra-tentorial. Therefore, we will focus on the brain, including the cerebellum and midbrain, with the medulla oblongata and the pons, where we can find a reticular substance that is functionally implicated in some neurological diseases currently under extensive research.

6.5.2.1 Brain

Some authors argue that several anatomical features of the brain can be used as an index of mental capacity [202], including brain weight, predominance of left hemisphere, and the complexity of surface brain gyri of the frontal and parietal lobes. All these have been reported in brain autopsies by famous scientists. Even you if find it hard to fully believe in such morphological assumptions of human intelligence, you will have to consider their importance in the assessment of qualitative and quantitative studies of brain functions and diseases [203].

The average weight of an adult brain is approximately 1300 to 1400 g, and the average volume is about 1400 mL. These figures from postmortem studies are not error free, especially due to water loss. Structural studies with very precise spatial and volumetric resolution are possible with MRI. They enable us to establish such *in vivo* values by means of automated and semi-automated algorithms for the extraction of the skull and extracranial structures [204]. This methodology gives us an *in vivo* brain volume of 1286.4 ± 133 mL and 1137.8 ± 109 mL in normal male and female volunteers, respectively. The weight is then calculated from these volumes using the equation

$$p_c = V_T \times 1.0365 \, \text{g mL}^{-1} + V_L \times 1.00 \, \text{g mL}^{-1} \qquad (6.72)$$

in which V_T is the total volume in milliliters and V_L is the cerebrospinal fluid volume in milliliters [205,206].

Brain volume is age-related that is, it increases exponentially during childhood and adolescence to achieve a peak between 12 and 25 years of age. From then until the age of 80, the brain volume slowly decreases, with a reduction of about 26% of the peak value between 71 and 80 years of age. At that point, brain volume is lower than that of healthy 2- or 3-year-old children. These findings are very similar to those reported in research using data from postmortem studies [207].

The brain surface is composed of gray matter (cell bodies of neuronal cells) gyri with gaps between them—the sulci—filled in by meninges. This surface is divided into four lobes, as depicted in Figure 6.36: frontal (8), parietal (1), temporal (7), and occipital (2) in each of the hemispheres. There are well-defined geographical boundaries between the frontal and the parietal lobe—central sulcus or the fissure of Rolando—between the temporal and frontal

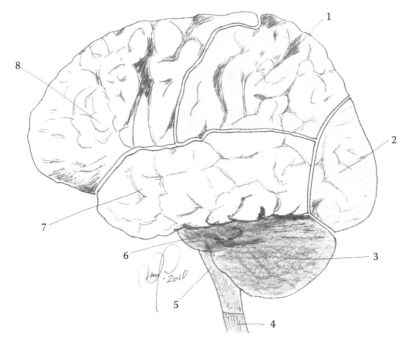

FIGURE 6.36
Brain surface anatomy as an intracranial component of the central nervous system. For reference, the cerebellum (3), the medulla oblongata (5), the pons or pons Varolii (6), and the upper end of the spinal cord (4) are shown.

lobe, including the most anterior and inferior regions of the parietal lobe—Sylvian or lateral fissure. However, the division between the rest of the parietal lobe and the temporal lobe, between the parietal and the occipital lobes and between the temporal and the occipital lobes, is simply functional and has no well-defined geographical representation.

The parietal-temporal-occipital association cortex is located at the most posterior and inferior region of the cortex (brain surface composed of the outermost brain tissue made of cell bodies—gray matter). As its name implies, it extends from the parietal cortex to the most posterior area of the temporal cortex and invades the occipital cortex. Small sulci separate cortical areas with different cytoarchitecture and functions (52 Brodmann areas; 15 were originally described) in each lobe. However, there are no geographical boundaries in the above mentioned association cortex that make an accurate distinction between parietal, temporal, and occipital lobes possible. Some other functions are referenced to cortical areas in two other brain lobes called subsidiary lobes; one is the insular cortex, which forms the medial wall of the Sylvian fissure or sulcus, and the other is the limbic cortex or lobe, located above the most anterior (cephalic) or rostral region of the midbrain next to the corpus callosum. The Brodmann functional areas (Figure 6.37) are a very important working medium for imaging experts and neurosurgeons conducting brain functional provocation studies during presurgery preparation in neuro-oncology and epilepsy.

Within the brain, besides lateral ventricles and most of the white substance made of axons (some of which are very long) of neuronal cells with soma (cell body) in the cortex, there are also gray matter nuclei with very important regulation and transition functions. These nuclei accommodate synapses (transitions from presynaptic neurons to postsynaptic neurons) in which neuronal information is transmitted either electrically or chemically. Several types of neurotransmission belonging to different physiological–pharmacological systems, with such different actions as stimulation or inhibition, may coexist in the same nuclei. These nuclei can be identified or visualized only by means of tomography techniques. This chapter does not set out to describe tomographic anatomy in detail, but in the next section we give an overall idea of it. We briefly describe some of the problems we face when comparing images from the same brain section, in the same subject, showing different functions in the same structures.

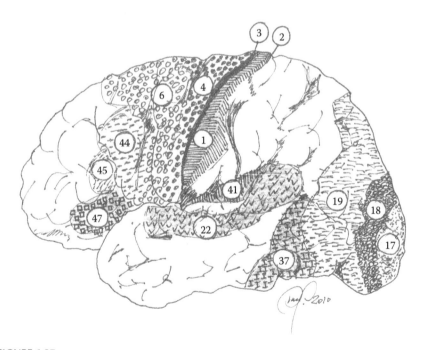

FIGURE 6.37
Brodmann areas on the outer surface of the left hemisphere. Some are worth mentioning as an example: 1, 2, and 3 (sensory); 4 and 6 (motor and extrapyramidal); 17, 18, and 19 (visual); 22, 37, 41, and 47 (auditory); 44 and 45 (Broca's speech production).

6.5.2.1.1 In Vivo Comparative Anatomy

The knowledge about CNS morphology and structure, especially the brain, was fundamentally based on postmortem dissection studies. Even the microscopic structure, described so clearly and accurately by Ramón y Cajal, was characterized at the expense of painstaking cadaver studies. Today, computerized tomography and MRI studies detect important changes in brain morphology and architectural structure with a spatial resolution of less than one millimeter in some cases. Compared with autopsy-derived data, the CNS structural design above and below the brain tentorium is obtained in an elegant manner and can be shown in concordance and with the same orientations. The advent of new sequences used by the most recent MRI instruments and software brought functionality to brain images. Each image obtained with a different sequence also has a different meaning (Figure 6.38).

Even this ability of MRI to offer us high resolution, fine images of the brain structures with functional characteristics does not depict "life" that is, cognitive function, blood flow, perfusion, or substrate use. To do that, we need *in vivo* functional markers, such as radiopharmaceuticals. Figure 6.39 shows slices taken at the same level as those in Figure 6.38, but this time each slice has its functional and chemical meaning in vivo. On the left, the distribution of a radioligand (^{99m}Tc-exametazime) whose brain uptake is proportional to the regional cerebral blood flow (perfusion) is shown. On the right, the distribution in the same subject (normal volunteer), at the same level, of another radiopharmaceutical (^{123}I-ioflupane), whose image shows the uptake distribution in the basal ganglia, which is directly proportional to the distribution of dopamine transport systems located on the membrane of the presynaptic neuronal end, is shown.

All these images need suitable coordinates for intermodality and interfunction referencing. By defining planar representation grids (such as those in Figure 6.39) and, more importantly, volumetric representation grids, such as those defined by Talairach [209], it is possible to superimpose images using

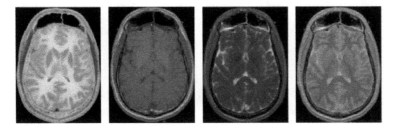

FIGURE 6.38
Images of the same slice (at the level of basal ganglia, with caudate nucleus, lentiform nucleus, and thalamus in both hemispheres) of one cadaver. The leftmost image is a picture (gray shades) of the cadaver slice, followed by sequence MR images to demonstrate, from left to right, T1, T2, and proton density.

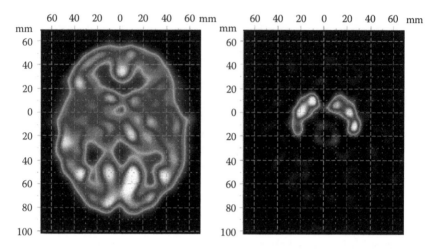

FIGURE 6.39
Comparative functional anatomy obtained from the same normal volunteer, at the level of basal ganglia, in a transversal slice parallel to the anterior commissure-posterior commissure reference line, as is usual in neuroimaging. The two slices obtained on different days are referenced to a grid (1 cm² units) similar to the one used by Talairach in his *Co-Planar Stereotaxic Atlas of the Human Brain*. (Adapted from J. Talairach, P. Tournoux. *Translated by Mark Rayport*. Thieme Medical Publishers, Inc., Stuttgart, New York, NY, 1988.)

software and relatively complex algorithms. These images are the coregistration of anatomical and functional information. Further, these images allow us to obtain composite maps of the correlation between different functions. Coregistration techniques are already part of routine clinical application, for instance, in PET studies with ¹⁸F-DG and CT. Figure 6.40 illustrates an example of this. The image on the right is the coregistration of the voxel-by-voxel information from the two images on the left. The leftmost of these

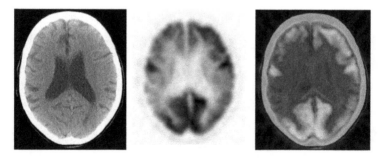

FIGURE 6.40
The images are of transversal slices at the same level (lateral ventricles). From left to right: x-ray attenuation map (computerized axial tomography, CAT, or CT), distribution map of brain glucose metabolism (¹⁸F-DG), and voxel-by-voxel coregistration of both images, that is, glucose metabolism distribution superimposed on the attenuation map (CT).

represents the attenuation map, and the middle image represents a distribution map of glucose metabolism in the brain. In this specific case, a marked reduction of brain metabolism in the parietal cortices (more to the right than to the left) can be seen, and in the frontal cortices, too (lower than in the parietal cortices), with the metabolism conserved in the motor cortices of both hemispheres and in the occipital visual cortex. The leftmost CT image only shows slight ventricular dilatation without further significant abnormalities. This pattern of cortical hypometabolism is typical of a CNS degenerative disease evolving into memory loss and other cognitive disturbances typical of dementia. This is a case of asymmetric Alzheimer's dementia.

This example shows the simplest method of coregistering images, as the object (brain) is encased in a static shell (as long as the subject does not move their head), and the positioning of each voxel containing different information is the same in the two acquisitions. In this case, there is no need for external fiducial markers to act as a reference while the superimposition program is run. These fiducial markers are contained in the images themselves. The two instruments that acquire the images are contiguous, and the object just moves automatically and with (computer determined) precision between the CT (left image) and PET (middle image). In other cases, the physical proximity of data acquisition instruments is not compatible with object immobilization (the subject), so coregistration has to be digitally performed, based on more elaborate techniques (see Section 6.3.3). Most of these techniques require the use of mathematical algorithms to manipulate the data in several ways and directions; consequently, the final product is not error free. To complicate the coregistration system even more, images have to be often rescaled, because, during acquisition, voxels of different sizes are used and this immediately introduces a scale-correction problem. This can be solved, but artifacts may arise. This can be seen in a recent work comparing the coordinates from several methods used for brain imaging functional data analysis [210].

6.5.2.2 Cerebellum

The cerebellum, formed by two lateral lobes and a median lobe, the vermis, is located under the occipital lobes and is separated from them by a membrane called the tentorium cerebelli. That is why it is considered infratentorial, similar to the midbrain, the medulla oblongata, and the pons. The main cerebellar function is motor coordination, which is very important for maintaining body posture and balance. For this reason, it receives axons from the brain cortex after they have crossed the middle line. The right cerebellar hemisphere receives axons originating in the left brain hemisphere and vice versa; the left cerebellar hemisphere receives axons from the right brain hemisphere. The cerebellum and periphery (muscular structures) are connected through the superior, middle, and inferior cerebellar peduncles on the right and left sides.

6.5.2.3 Midbrain (Mesencephalon)

References to the brain stem are common in the literature published in English. This includes the cerebellum, the uppermost part of the spinal cord, the medulla oblongata, and pons. The latter two together form what is called the midbrain pons in the Anglo-Saxon literature. In Portugal, we consider the mesencephalon as the pons or pons Varolii, located between the medulla oblongata below and the cerebral base above, to which it is connected by the cerebral peduncles. All the axons communicating between the brain and the spinal cord and passing through the cerebellum have to pass through the midbrain. For the purpose of this book, this section will refer both to the actual midbrain itself and to the Anglo-Saxon brain stem, because we want to make it clear that the 12 pairs of cranial nerves that mediate the senses, that is, sight, hearing, taste, and smell, arise from the lateral sides of the brain stem. Also, all of them have specific representation in the brain through connections between their source nuclei and the cortical areas directly related to their functions. However, it is even more important to mention the *formatio reticularis* or reticular formation or matter; although this is chiefly a mesencephalic structure, it also extends to the brain stem from the upper end of the spinal cord to the base of the cerebral peduncles. The reticular formation or matter is a mixture of interneuronal arrays forming an intricate network of functional regulation and modulation between sensitive and sensory neuronal input and peripheral motor region output of what we call the reflexes. As far as we know, it is in this mesencephalic area that some inputs are enhanced and maximized, whereas others are minimized or even suppressed. The main functions of the reticular formation or matter are based on four types of connections: (1) upward to the thalamus and cortex, which act as an activation or warning system. This is especially important in pain perception and its unpleasant connotation; (2) downward to the spinal cord and periphery, which modulate skeletal muscle tension; (3) input, using cranial pairs, allowing it to modulate heart rate and blood pressure originating from the carotid sinus (neurovegetative reflex); and (4) downward through the reticulospinal bundle, ending at the dorsal horn of the spinal cord and attenuating pain by reducing the effects of pain-mediating sensitive nerves. In brief, it serves as a high-level controller between the spinal cord and the brain, which helps responses to undesirable events, always on the alert and ready to choose the best response under normal functional conditions. Due to its micro-structural complexity, with many axon branches and neuronal interconnections, reticular formation or matter dysfunction causes poorly identified syndromes, which are sometimes medically disparaged [211,212].

6.5.3 Physiological and Pharmacological Basis

Substrate input to the CNS obeys pharmacokinetic laws. If we exclude absorption, that is, the route from the administration site to blood circulation, this includes a distribution that is directly proportional to local blood flow and, consequently, to the percentage of blood volume per tissue mass unit (for the purpose of our case the CNS, chiefly the brain). The substrate may be present in blood, more specifically in plasma or serum, in its free form or bound to plasma proteins or blood corpuscles, that is, red blood cells, white blood cells, and platelets. The substrate concentration available to appear in brain tissue cells is mainly the substrate concentration in its free form. However, according to pharmacokinetic principles, a substrate or a drug may be metabolized (hepatically or by other means) or excreted (hepatically and/or renally), which tends to reduce plasma-free concentration. This means that substrates (for the purpose of this book on NM, radiopharmaceuticals) are, after intravenous administration, available to brain tissue depending on blood flow distribution. They will have to cross the blood–brain barrier and find their more-or-less specific binding sites; they may be metabolized in either the CNS or other organs, and finally a part of the molecules will be excreted, either via the liver or the kidneys.

The brain and the rest of the intracranial CNS are irrigated by the carotid (carotid arteries run along the neck's anterolateral side) and vertebral (vertebral arteries run beside the cervical spine) arterial systems. These arterial systems communicate with each other both extra- and intracranially. Intracranial communication is better organized. The Circle of Willis is well known: it is formed by the anterior and posterior communicating vessels connecting anterior and posterior cerebral arteries, respectively. If not permanently patent, these communications become so in pathological states, so that the right system may easily supply blood to the left system and vice versa, and the posterior system may supply blood to the anterior system and vice versa. Besides these communications, there is also a capillary distribution network that allows communication among all neighboring vascular territories in the brain, especially at their boundaries, also called watershed boundaries. It is essentially here that the development of reactive hyperemia after vascular insult can be observed, along with the formation of new collateral capillary vessels to supply ischemic territories in areas of ischemic penumbra, that is, the halo surrounding the territory without blood supply.

The vascular and capillary network of the brain is not a rigid system of branching tubes. Quite on the contrary, as noted earlier, it is an intricate, flexible, and dynamic system of multidirectional vascular routes connecting terminal arterioles to venules. The main driving force for regional cerebral blood flow is the so-called perfusion pressure. This is the difference between the arterial influx pressure and the outflow pressure in the veins. Cerebral blood flow has self-regulation mechanisms that maintain a relatively constant blood flow even under the influence of a number of factors. Under physiological conditions, any change in brain metabolism evokes a blood flow change of similar amplitude and direction. The brain vascular network reacts to changes in arterial CO_2 concentration, in ionic (Ca^{2+}, K^+) concentration, and in the interstitial concentration of drugs such as adenosine and neuropeptides. All these can be used to trigger vascular stress and thus, neuronal metabolism. Under pathological conditions, in particular during the development of brain ischemia, flow-metabolism dissociation occurs. Rather than showing the normal behavior described above, blood flow changes may not produce recognizable metabolism changes or trigger opposite metabolic responses.

Once they arrive at the capillaries, the substrates or drugs then face cell barriers separating blood from brain tissue. These cell barriers comprise the capillary endothelium, a basement membrane, and astrocyte endfeet (podocytes) amid a variable amount of interstitial liquid. These barriers are explained next.

6.5.3.1 Barriers

The components of the CNS are demarcated by three types of barriers that separate blood from the brain parenchyma with extracellular fluid (blood–brain barrier), blood from cerebrospinal fluid (blood–cerebrospinal fluid barrier), and cerebrospinal fluid from the brain parenchyma (brain–cerebrospinal fluid barrier). These barriers play an important role in the formation of images of the CNS. Intrathecal administration (in the cerebrospinal fluid of the spinal cord subarachnoid space) of a myelography solute shows cerebrospinal fluid distribution and, hence, the state of the brain–cerebrospinal fluid barrier. This imaging method has been superseded by the tissue definition characteristics of MRI, which enable an evaluation of the brain–cerebrospinal fluid interface with very good spatial resolution. However, MRI does not provide information on cerebrospinal fluid kinetics, especially on its production and absorption. These can be better defined by intravenous administration of sodium pertechnetate (^{99m}Tc), which diffuses from blood serum into intraventricular space through choroid plexi (blood–cerebrospinal fluid barrier, i.e., production) and by intrathecal administration of, for instance, indium-111- or technetium-99m-labeled DTPA, which, similar to cerebrospinal fluid, will be absorbed into blood through the arachnoid villi or arachnoid granulations and through the CNS and pia mater capillary walls. The other barrier—the blood–brain barrier—is highly selective and regulates substrate uptake. It therefore regulates the uptake of potentially useful radiopharmaceutical molecules for the investigation of physiopharmacological functions and processes of the brain.

6.5.3.1.1 Blood–Brain Barrier

In the early twentieth century, Ehrlich [213,214] found out during his research studies on chemotherapeutical agents that intravenous dye injection did not cause any coloring in the brain, unlike all the other organs. Later, when studying how various substances moved from blood into the brain and what effects these substances had on the CNS, Stern and Gautier [215] realized that (1) there was a barrier between the blood and the brain that excluded some substances from the CNS and they called it the blood–brain barrier—*barrière hématoencéphalique*; (2) substances unable to cross this barrier could not be found in the cerebrospinal fluid; and (3) some of the substances unable to cross the barrier moved from the cerebrospinal fluid to the CNS, for example, iodine. Although not completely right, Stern and Gautier's hypothesis established for the first time the selectivity feature of the blood–brain barrier [217]. It still holds good today and is in many ways better understood. The other very important feature of the blood–brain barrier, liposolubility, was suspected by Becker and Quadbeck [218] and determined with the use of quantitative analysis algorithms with defined parameters for membrane crossing (rate constants) [219,220].

The concept of blood–brain barrier was thus based on two physical-chemical characteristics: selectivity and liposolubility. Although liposolubility is easy to understand and locate in cell membranes of capillary wall cells (endothelial cells) and their adjacent cells (astrocytes), the barrier selectivity is far more difficult to anatomically ascribe. Some authors [221,222] concluded that the endothelium was responsible for selectivity, whereas others [223] noted that the endothelium was covered by a protoplasmic layer—the podocytes—of astrocytes closely knitted by a basement membrane and that there were no apparent gaps in these membranes [224]. The contact points between two endothelial cells without gaps are called tight junctions and are thereafter regarded as the structural location of the blood–brain barrier according to Brightman and Reese [225]. These researchers showed that after intraventricular injection, horseradish peroxidise diffuses through the interstitial cell space around brain capillaries but cannot reach the bloodstream and stops next to the tight junctions.

The blood–brain barrier [226] is permeable to liposoluble substances and is apparently built of brain capillary endothelial cells, with their tight junctions. Another of its properties is selectivity. Together, these two features explain the results of many studies showing the absence of pharmacological effects of nonlipophilic drugs on the CNS after intravenous administration. However, the same nonlipophilic, hydrosoluble drugs can cross the cerebrospinal fluid–brain barrier when given by intrathecal route and can cause measurable pharmacological effects.

6.5.3.1.2 *Structural Components of the Blood–Brain Barrier*

The blood–brain barrier concept has been evolving a long time, and it is still fundamentally based on the CNS capillary endothelial cells. Their functional and morphological features are modulated by the astrocytes with which they are in very close contact [227,228]. The endothelial cell layer has no gaps as such. On the contrary, due to the intervention of astrocytes, there are tight junctions, which build up a barrier that can only be crossed by means of transport mechanisms in the cell barriers, without any type of passage through pores. The modulation of the endothelial layer by glial cells was shown by cross-experiments in which a nonvascular cell tissue transplanted into bird embryonic coeloms originated a blood–brain-type barrier [229]. On the other hand, when coelomic tissue grafts were transplanted into embryonic brain tissue, capillaries were seen to have the same characteristics as peripheral capillaries, that is, they had pores in the endothelial membranes and there were no tight junctions. This was later confirmed by Janzer and Raff [230] when they showed that astrocytes are responsible for forming tight junctions in nonneuronal tissue endothelial cells. To supplement these modulating and inducing capacities of astrocytes on endothelial cells, it was also shown that these glial cells increase the frequency, size, and complexity of tight junctions that are formed in cell growth media. Cultured endothelial cells develop fragmented tight junctions of about 13 μm, whereas the size of endothelial cells grown together with astroglia increases to approximately 165 μm.

Endothelial cells and astrocytes share yet another structure—the perivascular basement membrane (BM)—which is composed of a fibrous basement sheet with high electron density inserted between two much less dense electronic layers (one in contact with the endothelial cell membrane and the other in contact with the astrocyte cell membrane) The basement sheet contained glycoproteins, laminin, fibronectin, variable amounts of different types of collagen, and glycosaminoglycans [232].

This means that the blood–brain barrier should be thought of as a physiological barrier composed of three structures: endothelial cells with tight junctions, a perivascular basement membrane, and astroglia (astrocytes).

These three components of the blood–brain barrier make it selective and, according to Crone [233], show well-defined physiological features, which are as follows:

1. It is a cell layer ensheathed by a continuous basement membrane
2. Endothelial cells are connected to one another by tight junctions
3. Permeability to hydrophilic substances (nonelectrolytes) is very low
4. Ionic permeability is very low
5. Water conduction is very low
6. Passive permeability to solutes is mainly through intercellular junctions
7. It has facilitated transport mechanisms for some organic solutions
8. It shows stereospecificity is saturable and allows competitive interactions
9. It has induction mechanisms
10. It has high electric resistance ($\sim$2000 cm^2) [234]
11. Its permeability is increased by high osmolarity levels
12. Na$^+$–K$^+$ pumps are located on the abluminal membrane of endothelial cells

All these characteristics help one way or the other to define the most suitable transport mechanism for each substance or substrate present at the blood–brain barrier during regional blood flow distribution [235]. For instance, it is increasingly accepted that endothelial cells are active in both the regulation of brain extracellular fluid components and the brain's amino-acid content [236]. Active amino-acid transport occurs on the abluminal membrane, the facilitated transport of other amino acids is regulated on the luminal side, and the neutral and long amino acids cross the barrier via mechanisms located on the two sides of the endothelium [236].

6.5.3.1.3 Transport Mechanisms

Intracerebral radiopharmaceutical uptake is governed by the same properties as solute uptake, which is primarily determined by the nonionized and, therefore, lipophilic fraction of each of the components in question. The product uptake and its relation to the circulating portion not taken up can be quantified by the indicator diffusion technique. According to Crone [237], permeability across the blood–brain barrier is given by the following formula:

$$P = -\frac{F}{S} \ln(1 - E), \tag{6.73}$$

in which P is the permeability or partition coefficient, F is the blood or plasma flow, S is the capillary surface area, and E is the initial extraction fraction of the substance in question.

6.5.3.1.3.1 Influence of Liposolubility or Lipophilicity Although the first contact and limitation to the passage of solutes from the blood to the CNS is the endothelial cell membrane, due to its properties of lipophilicity and selectivity, the blood–brain barrier is not fully impervious to water and polar solutions. Apparently, water and some low molecular weight polar solutions can pass through hydrocarbon chains in cell membrane lipid bilayers. Further, the tight junction dynamics is such that in some cases a small percentage of them may open and allow the nonselective crossing of small ions [238]. According to Equation 6.73, during the first passage through the blood–brain barrier, the higher the permeability (partition coefficient), the higher the amount of substance taken up by the CNS. As the (endothelial) capillary permeability increases, the amount of substance crossing the blood–brain barrier depends more and more on regional blood flow, to the extent that in a theoretical situation of maximum possible lipid solubility, the CNS uptake is linearly correlated with brain flow, as shown by Equation 6.74, which defines diffusion ability [237,239]:

$$P_S = Q \ln(1 - E), \tag{6.74}$$

in which P_S is ability to diffuse through the barrier, that is, the permeability surface product, Q is cerebral blood flow, and E is substance extraction.

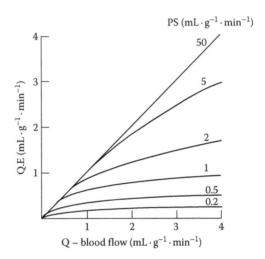

FIGURE 6.41

Theoretical curves based on Equation 6.74, also called the Crone–Renkin equation, in which product flow by extraction (Q.E) is compared with blood flow (Q). Q.E represents blood volume at any time (minute) at the barrier, and it is directly proportional to substance flow or the amount entering the brain per minute. Given that the equation assumes that back-diffusion is zero, the Q.E product is also proportional to unidirectional substance flow from blood to brain through the blood–brain barrier.

The graph in Figure 6.41, modified from Bradbury [240], shows the relation between cerebral blood flow and blood extraction into cerebral tissue for different P_S values.

Considering that the endothelial capillary surface is linear and that there is no back-diffusion from the brain back to the blood, the brain uptake of a given compound is not significantly affected by blood flow when this is higher than 0.5 mL g^{-1} min^{-1} and the P_S is lower than 0.2 mL g^{-1} min^{-1}. However, for P_S of 5 mL g^{-1} min^{-1} or higher (meaning high extraction and lipid solubility), brain uptake is chiefly determined by blood flow, and so it can be said that brain uptake is limited by regional cerebral blood flow. Consequently, this high liposolubility property has been used as the key factor in the development of radiopharmaceuticals for brain function studies, especially of regional blood flow, for example, ^{123}I-labeled iodoantipyrine and ^{85}Krypton. Another example of high liposolubility is ^{133}Xenon. In this case, there is overt back-diffusion through the barrier, because is neither metabolized in tissues nor does it bind to any cell structure unless its concentration in blood is the same as or higher than in the brain. After peak concentration (for the dose administered) is achieved in the brain, if there is no arterial input of ^{133}Xenon all molecules entering the bloodstream will be taken downstream (from arteries to veins) in proportion to regional capillary blood flow (washout rate).

6.5.3.1.3.2 Influence of Osmolarity or Osmotic Pressure For the examples of interest in this chapter, osmolarity explains the movement of water molecules through the blood–brain barrier. The P_S of the blood–brain barrier is hard to determine, even for radionuclide-labeled water. Water diffusion into the brain and from there to the bloodstream is partially dependent on blood flow and is totally dependent on it for relatively low flow levels [241]. Paulson et al. [242] calculated filtration coefficient or rate values (i.e., membrane osmotic permeability) higher than the diffusion permeability rate, with a 4:3 ratio. This led them to conclude that any osmotic pressure gradients across the blood–brain barrier are higher than potential hydrostatic pressure gradients. Any initial water filtration due to hydrostatic pressure will be immediately limited by an osmotic gradient of higher value and in the opposite direction. However, if for any reason there is a rupture in the blood–brain barrier, filtration of substances from the plasma to cerebral tissues (interstitial space) is possible, with or without proteins. This is most likely the mechanism underlying so-called vasogenic cerebral edema.

6.5.3.1.3.3 Influence of Binding to Plasma Proteins Binding to plasma proteins significantly influences pharmacokinetics and interaction between pharmaceuticals and substances in the bloodstream. Free and plasma protein-bound states are considered to have a significant influence on brain uptake. However, some authors still disregard this, even in relatively recent studies. In general, the drug's lipophilic portion readily crosses the blood–brain barrier, whereas nonlipophilic molecules stay in the bloodstream to be excreted or deposited in tissues or organs other than the brain. Besides this, the amount of the plasma protein-bound fraction also directly influences the passage through the blood–brain barrier. This is further complicated by the type of bond, that is, reversible or irreversible. A reversible bond introduces an important variable in the amount of drug available to cross the barrier (free fraction), which changes according to the free and protein-bound fraction concentration. In an irreversible bond, the concentration of the free fraction available to cross the barriers is constant.

However, there are some exceptions to this rule that are not fully understood. Despite having significant serum protein-bound fractions, the steroid hormones ^{3}H-progesterone and ^{3}H-testosterone show brain uptake percentages of between 80% and 100%, regardless of their serum protein-bound fraction concentration [243]. Another example is tryptophan, 90% of which is bound to plasma albumin. However, brain uptake after intracarotid bolus injection of ^{14}C-tryptophan is not influenced by the presence of albumin [244]. Seemingly, tryptophan was separated, as though sucked, from its binding protein (albumin) during the first passage through the cerebral capillary. Three mechanisms can be suggested to explain this:

1. Fast albumin dissociation due to strong brain uptake, with a consequent very fast reduction in the free fraction.

2. Differential changes in local physiological pH.

3. Metabolite displacement possibly due to antagonism, which can change the association constant of the tryptophan–albumin complex.

The exact mechanism is still unknown. Nevertheless, it seems that the dissociation constant for the tryptophan–albumin complex is significantly higher than the dissociation constant measured *in vitro* [245].

In fact, the explanation for both this event and what happens with thyroid hormones, which is similar, is still unclear. Thyroid hormones are polar compounds, but they enter the brain by a specific transport mechanism [246]. However, hormones are still introduced into cerebral capillaries with their serum protein-bound fraction. The hormones are disconnected from the proteins within capillaries by mechanisms that are also still unknown.

6.5.3.1.3.4 Other Influences Binding to erythrocytes (red blood cells) and endothelial sequestration can similarly reduce the brain uptake rate of lipophilic compounds.

Washout of molecules from the brain into the bloodstream and its rate of change are different from what occurs in extracerebral tissues. Here, high molecular weight molecules or suspended particles that cannot cross the capillary endothelial barrier from the interstitial fluid to blood are removed by lymphatic drainage. In the brain, washout of even low molecular weight polar solutes is difficult. These cannot cross the blood–brain barrier from the interstitial tissue to blood unless there is a specific transport mechanism through the endothelium. Nonetheless, there are nonspecific removal mechanisms for several polar solutes (nonlipophilic) that slowly penetrate the interstice, such as ^{35}S-sulphate and ^{14}C-inulin. There are three, nonmutually exclusive mechanisms [247] to explain the existence of convection in washout of polar solutes from the brain interstice to blood:

1. Secretion of interstitial fluid through the capillaries to the blood

2. Circulation through perivascular spaces

3. Wave movements, particularly in arterial spaces

6.5.3.1.3.5 Specific Transport within the Blood–Brain Barrier If lipophilicity were the only determining feature of the blood–brain barrier crossing rate to enter cerebral tissues, then the brain would be deprived of adequate amounts of several substrates it needs to survive, such as glucose and many amino acids. Due to their relatively low partition ratio, these substrates cannot cross the blood–brain barrier by simple (i.e., nonfacilitated) diffusion mechanisms. Monosaccharides, carboxylic acids, neutral amino acids, dicarboxylic acids, and some amines can cross the barrier with the help of active or facilitated, specific transport mechanisms. These are stereospecific, saturable transport mechanisms that can be inhibited by specific antagonists. They, therefore, depend on energy expenditure. So, for example, monosaccharides have a highly stereospecific transporter with high affinity for D-glucose. Peptides are another example, presenting as large, polar molecules [248,249]. Without an appropriate, specific transport mechanism across the blood–brain barrier, it would be impossible for peptides to play their important role in neurotransmitter synthesis or to be neurotransmitters themselves (Table 6.2).

6.5.3.2 Factors Influencing Cerebral Capillary Permeability

Rather than being defined by the permeability ratio, cerebral capillary permeability is usually defined by the above mentioned P_S product, because the exact endothelial area by unit of cerebral tissue weight is unknown. Permeability to liposoluble (lipophilic) substances is primarily influenced by regional cerebral blood flow distribution and, of course, by the P_S product. Consequently, permeability increases with flow, for instance, in hypercapnia and, thus, capillary vasodilatation and diminishes with regional cerebral blood flow reduction during hypocapnia, causing vasoconstriction [253]. Specific transport mechanisms, both facilitated and active, can be modulated by a very variable number of factors [254–257], which change according to brain tissue energy needs. Fasting [255] and porto-caval anastomosis [258] dramatically change selective transporters in the blood–brain barrier, most likely due to the marked metabolic changes they cause.

6.5.3.2.1 Importance to NM

The special selective permeability characteristic of the blood–brain barrier, in particular its lipophilicity, prompted the research for development of radioligands able to cross the barrier and show brain functions through uptake by CNS cells. However, some radiopharmaceuticals that do not cross an intact

TABLE 6.2

Examples of Transport Systems through the Blood–Brain Barrier

Substance Type (Example)	Transport (Maximum Capacity) (V_{max}) (μmols/g per min)	Michaelis Constant (Apparent) (K_m) (mM)
Monosaccharides [250] (D-glucose)	2–4	7–11
Monocarboxylic acids [251] (L-lactate)	90	2
Neutral amino acids [251,252] (L-leucine)	30–60	0.025–0.1
Basic amino acids [251] (L-arginine)	8	0.09
Amines [251] (choline)	11	0.34
Nucleosides [251] (adenosine)	0.75	0.025
Purines [251] (adenine)	0.05	0.01

blood–brain barrier are interesting for specific clinical applications, especially in oncology and neurological infectology.

After intravenous injection, sodium pertechnetate ($^{99m}TcO_4^-$) presents at the blood–brain barrier in its ionic form, hence it does not cross an intact blood–brain barrier. However, it is secreted into cerebrospinal fluid through the choroid plexi (blood–cerebrospinal fluid barrier), giving an idea of how this barrier works in the blood–cerebrospinal fluid direction. Sodium pertechnetate was the first radiopharmaceutical to mark an intact blood–brain barrier, but negatively that is, by showing the normal brain as an empty image and revealing uptake in areas of barrier disruption, as in cerebral infarction and tumors, in which sodium pertechnetate accumulates in extracellular fluid. Its main disadvantage is choroid plexi visualization, which in less-experienced hands can lead to false positive results. Therefore, whenever sodium pertechnetate is needed, its use should be preceded by the secretion blockade of choroid plexi, for instance, with sodium perchlorate. More recently, other radiopharmaceuticals have been used (^{99m}Tc-labeled glucoheptonate and diethylene triamine pentaacetic acid) that are not secreted by the choroid plexi and, thus, do not need blocking drugs. Thallium chloride (^{201}Tl) is another example of a radiopharmaceutical that does not cross an intact blood–brain barrier. In recent times, the clinical importance of these radiopharmaceuticals, especially thallium chloride, has gained a new life, because they can be used in the differential diagnosis of cerebral tumor recurrence and fibrosis after surgery and/or irradiation [259] and in the differential diagnosis between infection (cerebral toxoplasmosis) and intracerebral lymphoma in patients with viral immunodeficiency syndrome [260].

The lipophilicity of the blood–brain barrier that gives it selectivity prompted the development of neutral and lipophilic compounds able to cross an intact barrier. These compounds (radioligands) behave like chemical microspheres, that is, they cross the endothelial-basement membrane-astrocyte barrier from blood to brain to turn into polar compounds in the cerebral tissue, thus becoming unable to cross the barrier in the opposite direction.

This causes intracerebral retention directly proportional to tin (^{113}Sn)-radiolabeled microsphere distribution in the capillary network (Figure 6.42) and, thus, proportional to the regional cerebral blood flow distribution. This proportionality happens with highly varied levels of regional cerebral blood flow. In the white matter (lower blood flow levels), there is apparently an inflation of flow values by perfusion markers compared with microspheres. The correlation is better in the usual flow level range of gray matter (the regression line shows a significantly high correlation coefficient, $r = 0.924$).

These radiopharmaceuticals, generally called cerebral or regional cerebral blood flow perfusion markers (rCBF), have a clinical application with SPECT to examine patients with cerebral ischemia and in the differential diagnosis of dementia, especially frontotemporal dementia and Alzheimer's disease. ^{123}I-IMP (iodo-amphetamine), ^{99m}Tc-ECD (*l,l*-ethylcysteinate dimer),

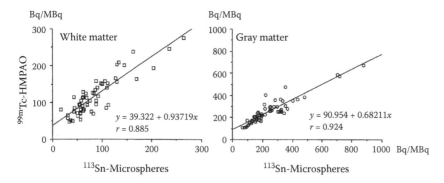

FIGURE 6.42

Comparison between distribution of a lipophilic cerebral blood flow marker (^{99m}Tc-hexa-metazime) and distribution of ^{113}Sn-microspheres in white and gray matter samples from dog brain (Adapted form D. C. Costa. *A Study of the First ^{99m}Tc-labelled Radiopharmaceutical for the Investigation of Cerebral Blood Flow in Man*. PhD Thesis, University of London, 1989.)

^{99m}Tc–hexametazime, and ^{99m}Tc–HMPAO (hexa-methyl-propylene-amine-oxime) are among the substances available on the market.

The physiological dynamics of cerebrospinal fluid circulation is especially well demonstrated by isotopic cisternography studies using ^{99m}Tc- or ^{111}In-labeled DTPA. After intrathecal injection (usually in the epidural space at the lumbar spine level), the radiopharmaceutical is distributed by the cisternae and eventually reabsorbed in the blood through the arachnoid villi or granulations. At the moment, there is no other imaging technique that can show the physiology of cerebrospinal fluid distribution and reabsorption as innocuously as isotopic cisternography. This technique serves not only to diagnose hydrocephalus in the differential diagnosis of intracranial hypertension [262] but also to predict the outcomes of potentially useful surgery.

The other clinical application of this isotopic cisternography technique, screening for cerebrospinal fluid loss [263], pertains to very fine anatomical resolution techniques, such as magnetic resonance. It is, however, still useful in cases where magnetic resonance gives rise to false negative results. However, more important is that the loss can be quantified with radiopharmaceuticals but not with magnetic resonance.

6.5.3.3 Neurotransmitters

In the CNS, brain, and beyond the blood–brain barrier, cell-to-cell, that is, interneuronal, interaction and communication are central to all information processing theories. The molecular neurobiology era, which has been growing for some years within neurosciences, is based on knowledge of neuronal transmission and the role of neurotransmitter molecules in the many brain functions. Neurotransmission requires the relevant neurotransmitters to be first synthesized at the presynaptic neuron and then transported to the neuron terminal to be ready to be released into the synaptic gap. There, neurotransmitters can activate receptors in the postsynaptic neuron membrane and trigger electrochemical processes that carry on the information chain. For this process to be effective, free neurotransmitter molecules in the synaptic gap must be removed, either by enzymes or by systems for reuptake into the presynaptic neuron, through its membrane and by specific mechanisms. At the presynaptic terminal, neurotransmitters are stored in vesicles ready to be freed and trigger a new transmission process. These general transmission rules are the same everywhere, regardless of the system in question: adrenergic, cholinergic, glutaminergic, dopaminergic, serotoninergic, etc.

Any interference with one or more of these steps from synthesis to enzymatic metabolism and presynaptic neuronal re-uptake will, of course, interfere with neurotransmission and its specific role.

6.5.3.3.1 *Synaptic Cleft*

The synaptic gap is the space between two neurons and it is approximately 10 to 50 nm wide (varies with the neurotransmission system). This gap could be considered to be almost virtual, because it is permanently filled by neurotransmitter molecules moving from pre- to postsynaptic neurons and vice versa. The synaptic gap communicates with the interstitial or extracellular fluid. Several explanations have been advanced for how transmission works properly without there being significant currents to allow the dissipation of the electrochemical information chain beyond the synapse. In 1967, De Robertis [264] tried to explain the existence of connection filaments between pre- and postsynaptic membranes, which would guide neurotransmitter molecules to the appropriate receptors. However, quite a number of other authors, using immunohistochemistry techniques, concluded that the synaptic gap is filled with a material of intermediate density composed of a wide range of proteins and sugars as mucopolysaccharides and glycoproteins [265]. This protein and sugar material would prevent neurotransmitters from leaving the synaptic gap by keeping the pre- and postsynaptic membranes connected. Glycoproteins have autoimmune activity and are subject to ongoing research. They seem to play an important recognition role during synaptogenesis. This is currently considered a process in permanent activity, which might explain the capacity for neuronal regeneration—now believed to be frequent in memorization processes and also as a reaction to severe, neuron-destroying tissue injury. Interneuronal synaptic plasticity is a fact that has been shown by several authors in multiple brain areas [266,267]. In addition to the interneuronal there is another type of synaptic cleft, at the neuroeffector junction, that is more peripheral and varies according to the tissues where it is located. This is present in gland tissue, skeletal muscle, arterial smooth muscle, intestine smooth muscle tissues, etc.

6.5.3.3.2 Receptors

The receptors are the specific sites in the postsynaptic membrane, and others, activated by neurotransmitters. Each system has its own specific receptors, divided into a number of classes and subclasses. As mentioned earlier, neurotransmitters reversibly bind to receptors. This bond may be of variable intensity, determined by the energy needed to undo receptor binding. In pharmacology, this energy is usually expressed as kilocalories per molecule. In biological systems, receptor attraction forces are weak (1–5 kcal/molecule), except for covalent forces. Therefore, these can be easily (and desirably) undone, that is, they are reversible. In general, there are four types of binding forces to receptors: covalent (50–150 kcal/molecule); ionic (~5 kcal/molecule); hydrogen bonds (2–5 kcal/molecule); and Van der Waals forces (~0.5 kcal/molecule). Neurotransmitter–receptor bonds may undergo interference from exogenous drugs and other endogenous substances with an affinity to the same receptor. Moreover, these may or may not be able to trigger a response, acting as agonists (eliciting a response, i.e., they are effective) or antagonists (not eliciting a response), respectively. Antagonists may still be competitive (affinity similar to transmitter), fighting similar to the neurotransmitter for receptor binding and showing comparable reversibility. If antagonist affinity for the receptor is higher than that of the neurotransmitter agonist, there is noncompetitive antagonism (irreversible). Radioligands have affinity for receptors without measurable pharmacological effects. Usually, they act as agonists or antagonists with more or less reversible, competitive binding. Therefore, and also according to their pharmacokinetic characteristics, SPECT and PET images should be obtained as far as possible in a balanced concentration of specific binding and nonspecific binding compartments and free radioligand as well as plasma.

In another section in this book, you will surely find explanations, definitions, and equations that will help you understand how to obtain quantitative values for radioligand uptake by receptors and other specific sites based on the multiple compartment theory (Figure 6.43).

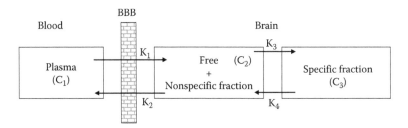

FIGURE 6.43
Diagram showing the three compartments, two of them within the brain, used for quantification of radioligand uptake by cerebral receptors. BBB = blood–brain barrier, C = concentration, K = crossing/diffusion/passage constants.

6.5.3.3.3   *Importance to NM and Neurosciences*

The best example of the importance of neurotransmission systems to clinical practice in NM is the *in vivo* characterization of neurochemical changes in Parkinson's disease. This CNS degenerative disease, which affects movement, is characterized by degeneration of nigrostriatal bundle neurons, whose soma lies in the mid-brain substantia nigra and which has dopaminergic terminals (which produce the dopamine neurotransmitter) in the striatal bundle (caudate nucleus and putamen). Ioflupane, a ^{123}I–labeled cocaine analog, specifically labels presynaptic terminals in the neuron striatum of the nigrostriatal bundle, which are significantly reduced in Parkinson's disease. This reduction is reflected in a marked reduction of ioflupane uptake by the striatum (putamen) contralateral to the hemibody showing signs and symptoms of parkinsonism. Further, due to the marked reduction in dopamine production by presynaptic neurons, postsynaptic dopamine receptors are more avid than normal (upregulation). SPECT with ^{123}I-labeled benzamide shows normal striatum uptake (mainly the putamen) with an image that dramatically contrasts with the negative putamen image obtained with ioflupane.

Many other radioligands can be developed to study other transmitter systems. The potential clinical importance of all of them has not yet led to the development of radiopharmaceuticals with a significant commercial interest.

Before concluding these thoughts on the problems and potential of methodologies that use radiopharmaceuticals, I would like to say that the future has more, and better, in store for us. Therefore readers, should neurosciences be your research interest, read the book [268] that we suggest as further reading. We believe that *Principles of Neuropsychopharmacology* by Robert S. Feldman, Jerrold S. Meyer, and Linda F. Quenzer is a really excellent work.

References

1. W. E. Adam. A general comparison of functional imaging in nuclear medicine with other modalities. *Sem. Nucl. Med.*, 17:3–17, 1987.
2. J. T. Bushberg, J. A. Seibert, E. M. Leidholdt, Jr., J. M. Boone. *The Essential Physics of Medical Imaging* (2nd Edition).Williams & Wilkins, 487–527, 2001.
3. D. J. Dowsett, P. A. Kenny, R. E. Johnston. *The Physics of Diagnostic Imaging*. Chapman & Hall Medical, London, pp. 66–79, 1998.
4. W. R. Hendee, R. E. Ritenour. *Medical Imaging Physics* (4th Edition). Wiley-Liss, 197–217, 2002.
5. G. L. Zeng, J. R. Galt, M. N. Wernick, R. A. Mintzer, J. N. Aarsvold. Single-photon emission computed tomography. In M. N. Wernick, J. N. Aarsvold, eds. *Emission Tomography: The Fundamentals of SPECT and PET*. Elsevier, San Diego, CA, 2004.
6. Y. Censor, D. Gustafson, A. Lent, H. Tuy. A new approach to the emission computerized tomography problem: Simultaneous calculation of attenuation and activity coefficients. *IEEE Trans. Nucl. Sci.*, NS-26:2775–2779, 1979.

7. A. Bronnikov. Approximate reconstruction of attenuation map in SPECT imaging. *IEEE Trans. Nucl. Sci.*, 42:1483–1488, 1995.
8. R. Ramlau, R. Clackdoyle, F. Noo, G. Bal. Accurate attenuation correction in SPECT imaging using optimization of bilinear functions and assuming an unknown spatially-varying attenuation distribution. *Z. angew. Math. Mech.*, 80:613–621, 2000.
9. L.-T. Chang. A method for attenuation correction in radionuclide computed tomography. *IEEE Trans. Nucl. Sci.*, NS-25:638–642, 1978.
10. A. Markoe. Fourier inversion of the attenuated X-ray transform. *SIAM J. Math. Anal.*, 15:718–722, 1984.
11. C. Metz, X. Pan. A unified analysis of exact methods of inverting the 2D exponential Radon transform, with implications for noise control in SPECT. *IEEE Trans. Med. Imaging*, MI-14:643–658, 1995.
12. F. Natterer. Inversion of the attenuated Radon transform. *Inverse Problems*, 17:113–119, 2000.
13. G. Gullberg, R. Huesman, J. Malko, N. Pelc, T. Budinger. An attenuated projector-backprojector for iterative SPECT reconstruction. *Phys. Med. Biol.*, 30:799–816, 1985.
14. B. M. W. Tsui, G. Gullberg, E. Edgerton, J. G. Ballard, J. Perry Jr., W. McCartney, G. Berg. Correction of nonuniform attenuation in cardiac SPECT imaging. *J. Nucl. Med.*, 30:497–507, 1989.
15. T. F. Lang, B. H. Hasegawa, S. C. Liew, J. K. Brown, S. C. Blankespoor, S. M. Reilly, E. L. Gingold, C. E. Cann. Description of a prototype of a emission–transmission computed tomography imaging system. *J. Nucl. Med.*, 33:1881–1887, 1992.
16. N. P. Rowell, N. J. Glaholm, M. A. Flower, B. Cronin, V. R. McCready. Anatomically derived attenuation coefficients for use in quantitative single photon emission tomography studies of the thorax. *Eur. J. Nucl. Med.*, 19:36–40, 1992.
17. S. Loncaric, W. Chang, G. Huang. Using simultaneous transmission and scatter SPECT imaging from external sources for the determination of the thoracic μ-map. *IEEE Trans. Nucl. Sci.*, 41:1601–1606, 1994.
18. J. W. Wallis, T. R. Miller, T. R. Koppel. Attenuation correction in cardiac SPECT without a transmission source. *J. Nucl. Med.*, 35:82P, 1994.
19. G. V. Heller, J. Links, T. M. Bateman, J. A. Ziffer, E. Ficaro, M. C. Cohen, R. C. Hendel. American Society of Nuclear Cardiology and Society of Nuclear Medicine joint position statement: Attenuation correction of myocardial perfusion SPECT scintigraphy. *J. Nucl. Cardiol.*, March/April, 229–230, 2004.
20. O. Tretiak, C. Metz. The exponential Radon transform. *SIAM J. Appl. Math.*, 39:341–354, 1980.
21. T. M. Bateman, S. J. Cullom. Attenuation correction single-photon emission computed tomography myocardial perfusion imaging. *Semin. Nucl. Med.*, 35:37–51, 2005.
22. M. T. Madsen. Recent advances in SPECT imaging. *J. Nucl. Med.*, 48:661–673, 2007.
23. R. J. Jaszczak, K. L. Greer, C. E. Floyd Jr., C. C. Harris, E. R. Colleman. Improved SPECT quantification using compensation for scattered photons. *J. Nucl. Med.*, 25:893–900, 1984.
24. K. Ogawa, Y. Harata, K. Ichiara, A. Kubo, S. Hashimoto. A practical method for position-dependent Compton-scatter correction in single-photon emission CT. *IEEE Trans. Med. Imaging*, 10:408–412, 1991.

25. M. A. King, G. J. Hademenos, S. J. Glick. A dual-photopeak window method for scatter correction. *J. Nucl. Med.*, 33:605–612, 1992.
26. S. A. Larsson. Gamma camera emission tomography. Development and properties of a multi-sectional emission computed tomography system. *Acta Radiol. Suppl.*, 363:30–32, 1980.
27. S. R. Meikle, B. F. Hutton, D. L. Bailey. A transmission-dependent method for scatter correction in SPECT. *J. Nucl. Med.*, 35:360–367, 1994.
28. B. F. Hutton, A. Osiecki, S. R. Meikle. Transmission-based scatter correction of 180° myocardial single-photon emission tomographic studies. *Eur. J. Nucl. Med.*, 23:1300–1308, 1996.
29. C.-M. Kao, P. La Rivière, X. Pan. Basics of imaging theory and statistics. In M. N. Wernick, J. N. Aarsvold, eds. *Emission Tomography: The Fundamentals of SPECT and PET*. Elsevier, San Diego, CA, 2004.
30. M. A. King, S. J. Glick, P. H. Pretorius, R. G. Wells, H. C. Gifford, M. V. Narayanan, T. Farncombe. Attenuation, scatter, and spatial resolution compensation in SPECT. In M. N. Wernick, J. N. Aarsvold, eds. *Emission Tomography: The Fundamentals of SPECT and PET*. Elsevier, San Diego, CA, 2004.
31. R. Lewitt, P. R. Edholm, W. Xia. Fourier method for correction of depth-dependent collimator blurring. *SPIE*, 1092:232–243, 1989.
32. W. Xia, R. M. Lewitt, P. R. Edholm. Fourier correction for spatially invariant collimator blurring in SPECT. *IEEE Trans. Med. Imaging*, 14:100–114, 1995.
33. B. M. W. Tsui, H.-B. Hu, D. R. Gilland, G. T. Gullberg. Implementation of simultaneous attenuation and detector response correction in SPECT. *IEEE Trans. Nucl. Sci.*, 35:778–783, 1988.
34. A. R. Formiconi, A. Pupi, A. Passeri. Compensation of spatial system response in SPECT with conjugate gradient reconstruction technique. *Phys. Med. Biol.*, 34:69–84, 1989.
35. G. L. Zeng, G. T. Gullberg. Frequency domain implantation of the three-dimensional geometric point-source correction in SPECT imaging. *IEEE Trans. Nucl. Sci.*, 39:1444–1453, 1993.
36. L. Itti, L. Chang, T. Ernst, F. Mishkn. Improved 3D correction for partial volume effects in brain SPECT. *Hum. Brain Map.*, 5:379–388, 1997.
37. D. W. Townsend, T. J. Spinks, T. Jones, A. Geissbuhler, M. Defrise, M. C. Gilardi, J. Heater. Three-dimensional reconstruction of PET data from a multi-ring camera. *IEEE Trans. Nucl. Sci.*, 36:1056–1065, 1989.
38. D. L. Bailey, M.-C. Gilardi, S. Grootoonk, P. E. Kinahan, C. Nahmias, J. Ollinger, D. W. Townsend, R. Trébossen, M. Zito. Quantitative procedures in 3D PET. In B. Bendriem, D. W. Townsend, eds. *The Theory and Practice of 3D PET*. Kluwer Academic, Dordrecht, 1998.
39. B. McKee, R. Clack, P. Harvey, L. Hiltz, M. Hogan, D. Howse. Accurate attenuation correction for a 3D PET system. *Phys. Med. Biol.*, MI-11:560–569, 1991.
40. De Lima, J. J. P. Radioisotopes in medicine. *Eur. J. Phys.*, 19:485–497, 1998.
41. P. J. Robinson, L. Kreel. Pulmonary tissue attenuation with computed tomography: Comparison of inspiration and expiration scans. *J. Comput. Assist. Tomogr.*, 3:740–748, 1979.

42. K. J. LaCroix, B. M. W. Tsui, B. H. Hasegawa, J. K. Brown. Investigation of the use of x-ray CT images for attenuation compensation in SPECT. *Nucl. Sci., IEEE Trans.*, 41(6):2793–2799, 1994.

43. S. C. Blankespoor, X. Wu, J. K. Kalki, et al. Attenuation correction of SPECT using X-ray CT on an emission–transmission CT system: Myocardial perfusion assessment. *IEEE Trans. Nucl. Sci.*, 43:2263–2274, 1996.

44. C. Burger, G. Goerres, S. Schoenes, et al. PET attenuation coefficients from CT images: Experimental evaluation of the transformation of CT into PET 511-keV attenuation coefficients. *Eur. J. Nucl. Med. Mol. Imag.*, 29:922–927, 2002.

45. C. Bai, L. Shao, A. J. Da Silva. CT-based attenuation correction for PET/CT scanners. *Proc. IEEE Nucl. Sci. Symp. and Med. Imag. Conf.*, Norfolk,VA, Nov 11–16, 2002.

46. P. E. Kinahan, D. W. Townsend, T. Beyer, et al. Attenuation correction for a combined 3D PET/CT scanner. *Med. Phys.*, 25:2046–2053, 1998.

47. T. Beyer. *Design, Construction and Validation of a Combined PET/CT Tomography for Clinical Oncology*. Doctoral thesis, University of Surrey, Surrey, UK, 1999.

48. M. Charron, T. Beyer, N. N. Bohnen, et al. Image analysis in oncology patients studied with a combined PET/CT scanner. *Clin. Nucl. Med.*, 25:905–910, 2000.

49. E. Kamel, T. F. Hany, C. Burger, et al. CT vs 68Ge attenuation correction in a combined PET/CT system: Evaluation of the effect of lowering the CT tube current. *Eur. J. Nucl. Med.*, 29:346–350, 2002.

50. Y. Nakamoto, M. Osman, C. Cohade, et al. PET/CT: comparison of quantitative tracer uptake between germanium and CT transmission attenuation-corrected images. *J. Nucl. Med.*, 43:1137–1143, 2002.

51. R. E. Alvarez, A. Macovski. Energy-selective reconstructions in X-ray computed tomography. *Phys. Med. Biol.*, 21:733–744, 1976.

52. B. H. Hagesawa, T. F. Lang, E. L. Brown et al. Object specific attenuation correction of SPECT with correlated dual-energy X-ray CT. *IEEE Trans. Nucl. Sci.*, NS-40:1242–1252, 1993.

53. B. E. Cooke, A. C. Evans, E. O. Fanthome, R. Alarie, A. M. Sendyk. Performance figure and images from the Therascan 3128 positron emission tomograph. *IEEE Trans. Nucl. Sci.*, 31:640–644, 1984.

54. G. Knoll. *Radiation Detection and Measurement* (3ª edição). John Wiley & Sons, New York, 1999.

55. R. D. Badawi, M. P. Miller, D. L. Bailey, P. K. Marsden. Randoms variance reduction in 3D PET. *Phys. Med. Biol.*, 44:941–954, 1999.

56. K. F Koral, N. H. Clinthorne, W. L. Rogers. Improving emission-computed-tomography quantification by compton-scatter rejection through offset windows. *Nucl. Instr. Meth. Phys. Res.*, A242:610–614, 1986.

57. C. C. Watson, D. Newport, M. E. Casey. A single scatter simulation technique for scatter correction in 3D PET. In P. Grangeat, J.-L. Amans, eds. *Three-dimensional Image Reconstruction in Radiology and Nuclear Medicine*. Kluwer Academic Publishers, Dordrecht, pp. 255–268, 1996.

58. J. M. Ollinger. Model-based scatter correction for fully 3D PET. *Phys. Med. Biol.*, 41:153–176, 1996.

59. A. S. Goggin, J. M. Ollinger. A model for multiple scatters in fully 3D PET. IEEE *Nuclear Science Symposium and Medical Imaging Conference Record*. Norfolk, VA, USA, 1994.

60. S. R. Cherry, H. Sung-Cheng. Effects of scatter on model parameter estimates in 3D PET studies of the human brain. *IEEE Trans. Nucl. Sci.*, 42:1174–1179, 1995.
61. C. W. Stearns. Scatter correction method for 3D PET using 2D fitted Gaussian functions. *J. Nucl. Med.*, 36:105, 1995.
62. D. L. Bailey, S. R. Meikle. A convolution-subtraction scatter correction method for 3D PET. *Phys. Med. Biol.* 39:411–424, 1994.
63. M. Bergström, L. Eriksson, C. Bohm, G. Blomqvist, J. Litton. Correction for scattered radiation in a ring detector positron camera by integral transformation of the projections. *J. Comput. Assist. Tomogr.*, 7(1):42–50, 1983.
64. S. Grootoonk, T. J. Spinks, T. Jones, C. Michel, A. Bol. Correction for scatter using a dual energy window technique with a tomograph operated without septa. *IEEE Nuclear Science Symposium and Medical Imaging Conference Record*, Santa Fe, 1991.
65. R. Trébossen, B. Bendriem, A. Fontaine, R. Rougetet, V. Frouin, P. Remy. Quantitation of clinical 3D PET studies with the ETM scatter correction, *IEEE Nuclear Science Symposium and Medical Imaging Conference Record*, San Francisco, CA, USA, 1995.
66. L. Shao, R. Freifelder, J. S. Karp. Triple energy window scatter correction technique in PET. *IEEE Trans. Med. Imaging*, 13(4):641–648, 1994.
67. M. Bentourkia, P. Msaki, J. Cadorette, R. Lecomte. Assessment of scatter components in multispectral PET imaging. *IEEE Conference Record. Nuclear Science Symposium and Medical Imaging Conference*, San Francisco, CA, USA, 1993.
68. C. S. Levin, M. Dahlbom, E. J. Hoffman. A Monte Carlo correction for the effect of Compton scattering in 3-D PET brain imaging. *IEEE Trans. Nucl. Sci.*, 42:1181–1185, 1995.
69. N. C. Ferreira, R. Trébossen, C. Lartizien, V. Brulon, P. Merceron, B. Bendriem. A hybrid scatter correction for 3D PET based on an estimation of the distribution of unscattered coincidences: implementation on the ECAT EXACT HR+. *Phys. Med. Biol.*, 47:1555–1571, 2002.
70. N. C. Ferreira. *Contribuição para a quantificação em Tomografia por Emissão de Positrões no modo 3D*. Tese de doutoramento, Universidade de Coimbra, 2001.
71. R.D. Badawi. *Aspects of Optimisation and Quantification in Three-Dimensional Positron Emission Tomography*. PhD, University of London, London, 1998.
72. M. E. Casey, E. J. Hoffman. Quantitation in positron emission tomography: 7. A technique to reduce noise in accidental coincidence measurement and coincidence efficiency calibration. *J. Comput. Assist. Tomogr.*, 10:845–850, 1986.
73. E. J. Hoffman, T. M. Guerrero, G. Germano, W. M. Digby, M. Dahlbom. PET system calibrations and corrections for quantitative and spatially accurate images. *IEEE Trans. Nucl. Sci.*, 36:1108–1112, 1989.
74. D.A. Chesler, C. W. Stearns. Calibration of detector sensitivity in positron cameras. *IEEE Trans. Nucl. Sci.*, 37:768–772, 1990.
75. M. Bergström, L. Eriksson, C. Bohm, G. Blomqvist, T. Greitz, J. Litton, L. Widén. A procedure for calibrating and correcting data to achieve accurate quantitative values in positron emission tomography. *IEEE Trans. Nucl. Sci.*, 29:555–557, 1982.
76. M. E. Casey, H. Gadagkar, D. Newport. A component based method for normalization in volume PET. In P. Grangeat, J. L. Amans, eds. *Proceedings of the 3rd International Conference on Three-Dimensional Image Reconstruction in Radiology and Nuclear Medicine*. Aix-les-Bains, pp. 67–71, 1995.

77. N. C. Ferreira, R. Trébossen, M.-C. Grégoire, B. Bendriem. Influence of malfunctioning block detectors on the calculation of single detector efficiencies in PET. *IEEE Trans. Nucl. Sci.*, 46:1062–1069, 1999.

78. H. Müller-Gärtner, J. Links, J. L. Prince. Measurement of radiotracer concentration in brain gray matter using positron emission tomography: MRI-based correction for partial volume effects. *J. Cereb. Blood Flow Metab.*, 12:571–583, 1992.

79. W. W. Moses, S. E. Derenzo, T. F. Budinger. PET detector modules based on novel detector technologies. *Nucl. Instr. Meth. Phys. Res. A*, 352:189–194, 1994.

80. M. Defrise, P. Kinahan. Data Acquisition and Image Reconstruction for 3D PET. In B. Bendriem, D. W. Townsend, eds., *The Theory and Practice of 3D PET*. Kluwer Academic, Dordrecht, 1998.

81. P. P. Bruyant. Analytic and iterative reconstruction algorithms in SPECT. *J. Nucl. Med.*, 43:1343–1358, 2002.

82. P. E. Kinahan, M. Defrise, R. Clackdoyle. Analytic image eeconstruction methods. In M. N. Wernick, J. N. Aarsvold, eds. *Emission Tomography: The Fundamentals of SPECT and PET*. San Diego, CA, Elsevier, 2004.

83. R. B. Blackman, J. Tukey. *Particular Pairs of Windows. In The Measurement of Power Spectra, From the Point of View of Communications Engineering*. Dover, New York, 98–99, 1959.

84. J. A. Fessler. Penalized weighted least squares image reconstruction for positron emission tomography. *IEEE Trans. Med. Imag.*, 13(2):290–300, 1994.

85. A. Macovski. Tomography. In T. Kailath, ed. *Medical Imaging Systems*. Prentice-Hall, Englewoods Cliffs, NJ, pp. 106–144, 1997.

86. L. A. Shepp, Y. Vardi. Maximum likelihood reconstruction for emission tomography. *IEEE Trans. Med. Imag.*, MI-1:113–122, 1982.

87. E. Veklerov, J. Llacer. Stopping rule for the MLE algorithm based on statistical hypothesis testing. *Med. Imag., IEEE Trans. Med. Imag.*, 6(4):313–319, 1987.

88. V. V. Selivanov, D. Lapointe, M. Bentourkia, R. Lecomte. Cross-validation stopping rule for ML-EM reconstruction of dynamicPET series: Effect on image quality and quantitative accuracy. *IEEE Trans. Nucl. Sci.*, 48(3-Part 2):883–889, 2001.

89. P. E. Kinahan, J. G. Rogers. Analytic three-dimensional image reconstruction using all detected events. *IEEE Trans. Nucl. Sci.*, NS-36:964–968, 1989.

90. H. M. Hudson, R. S. Larkin. Accelerated image reconstruction using ordered subsets of projection data. *IEEE Trans. Med. Imag.*, 13:601–609, 1994.

91. C. Michel, X. Liu, S. Sanabria et al. Weighted schemes applied to 3D-OSEM reconstruction in PET. *Nucl. Sci. Symp. Conf. Rec.*, Seattle, USA, 3:1152–1157, 1999.

92. P. E. Kinahan, J. G. Rogers. Analytic three-dimensional image reconstruction using all detected events. *IEEE Trans. Nucl. Sci.*, NS-36:964–968, 1989.

93. J. G. Colsher. Fully three-dimensional positron emission tomography. *Phys. Med. Biol.*, 20:103–115, 1980.

94. J.-S. Liow, S.C. Strother, K. Rehm, A. Rottenburg. Improved resolution for PET volume-imaging through three-dimensional iterative reconstruction. *J. Nucl. Med.*, 38:1623–1630, 1997.

95. J. Qi, R. M. Leahy, S. R. Cherry, A. Chatziioannou, T. H. Farquhar. High resolution 3D Bayesian image reconstruction using the small animal microPET scanner. *Phys. Med. Biol.*, 43:1001–1013, 1998.

96. P. E. Kinahan, C. Michel, M. Defrise, D. W. Townsend, M. Sibomana, M. Lonneux, D. F. Newport, J. D. Luketich. Fast iterative image reconstruction of 3D PET

data. In *Proceedings of the IEEE Nuclear Science Symposium and Medical Imaging Conference*. Anaheim, CA, 1918–1922, 1996.

97. C. A. Johnson, Y. Yan, R. E. Carson, R. L. Martino, M. E. Daube-Witherspoon. A system for the reconstruction of retracted.septa data using the EM algorithm. *IEEE Trans. Nucl. Sci.*, 42:1223–1227, 1995.

98. M. E. Daube-Witherspoon, G. Muehllehner. Treatment of axial data in three-dimensional PET. *J. Nucl. Med.*, 28:1717–1724, 1987.

99. R. M. Lewittt, G. Muehllehner, J. S. Karp. Three-dimensional image reconstruction for PET by multi-slice rebinning and axial image filtering. *Phys. Med. Biol.*, 39:321–339, 1994.

100. M. Defrise, P. E. Kinahan, D. W. Townsend, C. Michel, M. Sibomana, D. Newport. Exact and approximate rebinning algorithms for 3D PET data. *IEEE Trans. Med. Imag.*, 16:145–158, 1997.

101. L. Janeiro, C. Comtat, C. Lartizien, P. E. Kinahan, M. Defrise, C. Michel, R. Trebossen, P. Almeida. NEC-scaling applied to FORE+OSEM. *IEEE Nuclear Science Symposium Conference Record*, 2:717–721, 2002.

102. L. Janeiro. *Incorporating Accurate Statistical Modeling in PET Reconstruction for Whole-Body Imaging*. Tese de doutoramento, Universidade de Lisboa, 2007.

103. X. Liu, M. Defrise, C. Michel, M. Sibomana, C. Comtat, P. E. Kinahan, D. W. Townsend. Exact rebinning methods for three-dimensional PET. *IEEE Trans. Med. Imag.*, 18:657–664, 1999.

104. M. Defrise, X. Liu. A fast rebinning algorithm for 3D positron emission tomography using John's equation. *Inverse Problems*, 15:1047–1065, 1999.

105. M. Defrise. Fourier rebinning of time-of-flight PET data. *Phys. Med. Biol.*, 50:2749–2763, 2005.

106. A. K. Jain. *Fundamentals of Digital Image Processing*, Prentice-Hall, Englewood Cliffs, NJ, 1989.

107. R. C. Gonzalez, R. W. Woods, *Digital Image Processing*, 2nd Edition, Prentice-Hall, Englewood Cliffs, NJ.

108. D. Forsyth, J. Ponce, *Computer Vision: A Modern Approach*, Prentice-Hall, Englewood Cliffs, NJ.

109. L. G. Brown. A survey of image registration techniques. *ACM Comput. Surv*, 24(4):325–376, 1992.

110. M. R. Neuman, ed. *Medical Image Registration*. CRC Press, New York, 2001.

111. P. A. van den Elsen, E.-J. D. Pol, M. A. Viergever. Medical image matching—a review with classification. *IEEE Eng. Med. Biol. Mag.*, 12, 26–39, 1993.

112. J. B. A. Maintz. An overview of medical image registration methods. In: *Symposium of the Belgian hospital physicists association (SBPH/BVZF)*, Vol 12, pp. V:1-22, 1996/1997.

113. J. Maintz and M. Viergever. A survey of medical image registration. *Med. Image Anal.*, 2(1):1–36, 1998.

114. R. Wan, M. Li. An overview of medical image registration. In ICCIMA '03: *Proceedings of the 5th International Conference on Computational Intelligence and Multimedia Applications*, p. 385, Washington, DC, USA, IEEE Computer Society, 2003.

115. B. Zitova, J. Flusser. Image registration methods: A survey. *Image Vision Comput.*, 21(11):977–1000, 2003.

116. B. F. Hutton, M. Braun, L. Thurfjell, D. Y. H. Lau. Image registration: An essential tool for nuclear medicine. *Eur. J. Nucl. Med. Mol. Imag.*, 29(4):559–577, Apr 2002.

117. M. Holden, D. L. G. Hill, E. R. E. Denton, J. M. Jarosz, T. C. S. Cox, D. J. Hawkes. Voxel similarity measures for 3d serial MR brain image registration. *Information Processing in Medical Imaging*, 472–477, 1999.

118. R. B. Cattell, J. R. Hurley. The procrustes program: Producing direct rotation to test a hypothesized factor structure. *Behav. Sci.*, 7:258–262, 1962.

119. B. K. P. Horn. Closed-form solution of absolute orientation using unit quaternions. *J. Opt. Soc. Am. A*, 4(4):629–642, Apr 1987.

120. B. K. P. Horn, H. M. Hilden, S. Negahdaripour. Closed-form solution of absolute orientation using orthonormal matrices. *J. Opt. Soc. Am. A*, 5(7):1127, 1988.

121. P. D. Fiore. Efficient linear solution of exterior orientation. *IEEE Trans. Pattern Anal. Mach. Intell.*, 23(2):140–148, 2001.

122. Z. Wang, Z. Wang, A. Jepson. A new closed-form solution for absolute orientation. In A. Jepson, ed., *Proc. CVPR '94. IEEE Computer Society Conference on Computer Vision and Pattern Recognition*, pp. 129–134, 1994.

123. P. Schönemann. A generalized solution of the orthogonal procrustes problem. Psychometrika, 31(1):1–10, March 1966. available at http://ideas.repec.org/a/spr/psycho/v31y1966i1p1-10.html.

124. K. S. Arun, T. S. Huang, S. D. Blostein. Least-squares fitting of two 3-d point sets. *IEEE Trans. Pattern Anal. Mach. Intell.*, 9(5):698–700, 1987.

125. J. Ashburner K. J. Friston. Nonlinear spatial normalization using basis functions. *Hum. Brain Map.*, 7(4):254–266, 1999.

126. K. J. Friston, J. Ashburner, C. Frith, J. B. Poline, J. D. Heather, R. S. J. Frackowiak. Spatial registration and normalization of images. *Hum. Brain Map.*, 2:165–189, 1995.

127. P. A. Viola, W. M. Wells III. Alignment by maximization of mutual information. *Int. J. Comput. Vis.*, 24(2):137–154, 1997.

128. R. P. Woods, S. T. Grafton, C. J. Holmes, S. R. Cherry, J. C. Mazziotta. Automated image registration: I. General methods and intrasubject, intramodality validation. *J. Comput. Assist. Tomogr.*, 22(1):139–152, 1998.

129. J. V. Hajnal, N. Saeed, E. Soar, A. Oatridge, I. Young, G. Bydder. 1995. A registration and interpolation procedure for subvoxel matching of serially acquired MR images. *J. Comput. Assist. Tomogr.*, 19(2):289–296, 1995.

130. E. De Castro, C. Morandi. Registration of translated and rotated images using finite Fourier transforms. *IEEE Trans. Pattern Anal. Mach. Intell.*, 9(5):700–703, 1987.

131. M. McGuire. An image registration technique for recovering rotation, scale and translation parameters. Technical report, NEC Res. Inst. Tech. Rep., 1998.

132. B. S. Reddy, B. S. Reddy, B. N. Chatterji. An fft-based technique for translation, rotation, and scale-invariant image registration. *IEEE Trans. Image Process*, 5(8):1266–1271, 1996.

133. G. Wolberg, S. Zokai. Robust image registration using log-polar transform. In *Proc. IEEE Int. Conf. Image Processing*, 2000.

134. C. E. Shannon. A mathematical theory of communication. *The Bell System Tech J.*, 27, 1948.

135. A. Collignon, D. Vandermeulen, P. Suetens, G. Marchal. 3d multi-modality medical image registration using feature space clustering. In *Proceedings of the First International Conference on Computer Vision, Virtual Reality and Robotics in Medicine*, 905:195–204, 1995.

136. D. L. G. Hill, C. Studholme, D. J. Hawkes. Multi-resolution voxel similarity measures for MR-PET registration. *Proc. IPMI* 95, 187–198, 1995.
137. H. Luan, F. Qi, Z. Xue, L. Chen, D. Shen. Multimodality image registration by maximization of quantitative–qualitative measure of mutual information. *Pattern Recogn.*, 41(1):285–298, 2008.
138. W. Wells, P. Viola, H. Atsumi, S. Nakajima, R. Kikinis. Multi-modal volume registration by maximization of mutual information. *Med. Image Anal.*, 1(1):35–51, 1996.
139. J. B. A. Viergever, M. A. Pluim, J. P. W. Maintz. Mutual-information-based registration of medical images: A survey. *Med. Imaging, IEEE Trans.*, 22(8):986–1004, Aug. 2003.
140. L. Freire, A. Roche, J. Mangin. What is the best similarity measure for motion correction of fMRI time series. *IEEE Trans. Med. Imaging*, 21(5):470–484, 2002.
141. A.D. Fright, W.R., Linney. Registration of 3-d head surfaces using multiple landmarks. *IEEE Trans. Med. Imag.*, 12(3):515–520, Sep 1993.
142. R. P. Woods, J. C. Mazziotta, S. R. Cherry. MRI-PET registration with automated algorithm. *J. Comput. Assist. Tomogr.*, 17(4):536–546, 1993.
143. J. Stewart. *Calculus*. Brooks/Cole Publishing Co., Pacific Grove, CA, USA, 2007.
144. J. R. Shewchuk. An introduction to the conjugate gradient method without the agonizing pain. Technical report, Pittsburgh, PA, USA, 1994.
145. S. Henn. A Levenberg–Marquardt scheme for nonlinear image registration. *BIT Numer. Math.*, 43(4):743–759, 2003.
146. G. Taubin. An improved algorithm for algebraic curve and surface fitting. Computer Vision, 1993. *Proceedings Fourth International Conference on*, pp. 658–665, 11–14 May 1993.
147. T. van den Elsen, P. A. Napel, S. Adler, J. Hemler, P. F. Sumanaweera. A system for multimodality image fusion. Computer-Based Medical Systems, 1994. *Proceedings 1994 IEEE Seventh Symposium on*, pp. 335–340, 10–12 June 1994.
148. F. Maes, A. Collignon, D. Vandermeulen, G. Marchal, P. Suetens. Multimodality image registration by maximization of mutual information. *IEEE Trans. Med. Imag.*, 16(2):187–198, Apr 1997.
149. W.-H. Tsui, H. Rusinek, P. Van Gelder, S. Lebedev. Fast surface-fitting algorithm for 3d image registration. *Med. Imag. 1993: Image Process.*, 1898(1):14–23, 1993.
150. M. van Herk, H. M. Kooy. Automatic three-dimensional correlation of CT-CT, CT-MRI, and CT-SPECT using chamfer matching. *Med. Phys.*, 21(7):1163–1178, 1994.
151. J. L. Andersson, A. Sundin, S. Valind. A method for coregistration of pet and mr brain images. *J. Nucl. Med.*, 36(7):1307–1315, Jul 1995.
152. W. Press, S. Teukolsky, W. Vetterling, B. Flannery. *Numerical Recipes in C* (2nd Edition) Cambridge University Press, Cambridge, UK, 1992.
153. D. Whitley. A genetic algorithm tutorial. *Statist. Comput.*, 4:65–85, 1994.
154. D. Whitley. Genetic algorithms and evolutionary computing. *Van Nostrand's Scientific Encyclopedia*, 2002.
155. M. Mitchell. *An Introduction to Genetic Algorithms*. MIT Press, Cambridge, MA, USA, 1996.
156. S. Kirkpatrick, C. D. Gelatt, M. P. Vecchi. Optimization by simulated annealing. *Science*, Number 4598, 13 May 1983, 220, 4598:671–680, 1983.
157. V. Cerny. Thermodynamical approach to the traveling salesman problem: An efficient simulation algorithm. *J. Optim. Theory Appl.*, 45(1):41–51, January 1985.

158. N. Metropolis, A. W. Rosenbluth, M. N. Rosenbluth, A. H. Teller, E. Teller. Equation of state calculations by fast computing machines. *J.Chem. Phys.*, 21(6):1087–1092, 1953.

159. P. Thèvenaz, T. Blu, M. Unser. Image interpolation and resampling. In N. Bankman, ed., *Handbook of Medical Imaging, Processing and Analysis*, Academic Press, pp. 393–420, 2000.

160. C. Spitzer, K. Lehmann, T. M. Gonner. Survey: Interpolation methods in medical image processing. *Med. Imag., IEEE Trans.*, 18(11):1049–1075, 1999.

161. J. Duchon. Splines minimizing rotation-invariant semi-norms in sobolev spaces. Springer Lecture Notes in Math. 571, Springer-Verlag, Berlin, pp. 85–100.

162. J. Meinguet. Multivariate interpolation at arbitrary points made simple. *Zeit. Angew. Math. Phys. (ZAMP)*, 30(2):292–304, 1979.

163. A. Goshtasby. Registration of images with geometric distortions. *Geosci. Remote Sensing, IEEE Trans.*, 26(1):60–64, 1988.

164. K. Eriksson, A. P. Astrom. Bijective image registration using thin-plate splines. *Pattern Recognition, 2006. ICPR 2006. 18th International Conference on*, 3:798–801, 2006.

165. R. Bajcsy, S. Kovacic. Multiresolution elastic matching. *Comput. Vision Graph. Image Process*, 46(1):1–21, 1989.

166. G. E. Christensen, G. E. Christensen, R. D. Rabbitt, M. I. Miller. Deformable templates using large deformation kinematics. *IEEE Trans. Image Process*, 5(10):1435–1447, 1996.

167. M. Bro-Nielsen, C. Gramkow. Fast fluid registration of medical images. In *Proc. Visualization in Biomedical Computing (VBC'96)*, Hamburg, Germany, September, Springer Lecture Notes in Computer Science, 1131:267–276, 1996.

168. B. G. Siim, W. T. Laux, M. D. Rutland, B. N. Palmer, W. R. Wilson. Scintigraphic imaging of the hypoxia marker 99mTechnetium-labeled 2,2*-(1,4-Diaminobutane)bis(2-methyl-3-butanone) Dioxime (99mTc-labeled HL-91; Prognox): noninvasive detection of tumor response to the antivascular agent 5,6-Dimethylxanthenone-4-acetic Acid. *Cancer Res.*, 60:4582–4588, 2000.

169. J. C. W. Crawley. Quantitation of uptake of bromine-77 in the human brain. *Clin. Phys. Physiol. Meas.*, 5:121–124, 1984.

170. D. J. Yang, E. E. Kim, T. Inoue. Targeted molecular imaging in oncology. *Ann. Nucl. Med.*, 20:1–11, 2006.

171. N. Pandit-Taskar. Oncologic imaging in gynecologic malignancies. *J. Nucl. Med.*, 46:1842–1850, 2005.

172. A. Del Sole, A. Falini, L. Ravasi, L. Ottobrini, D. De Marchis, E. Bombardieri, G. Lucignani. Anatomical and biochemical investigation of primary brain tumours. *Eur. J. Nucl. Med.*, 28:1851–1872, 2001.

173. C. van de Wiele, C. Lahorte, W. Oyen, O. Boerman, I. Goethals, G. Slegers, ARA. Dierckx. Nuclear medicine imaging to predict response to radiotherapy. *Int. J. Radiat. Oncol. Biol. Phys.*, 55:5–15, 2003.

174. J. R. Buscombe, E. Bombardieri. Imaging cancer using single photon techniques. *Q. J. Nucl. Med.*, 49:121–31, 2005.

175. H. Kawai, J. Toyohar, H. Kado, T. Nakagawa, S. Takamatsu, T. Furukaw, Y. Yonekur, T. Kubota, Y. Fujibayashi. Acquisition of resistance to antitumor alkylating agent ACNU: A possible target of positron emission tomography monitoring. *Nucl. Med. Biol.*, 33:29–35, 2006.

176. B. Gagel, P. Reinartz, C. Demirel, H. J. Kaiser, M. Zimny, M. Piroth, M. Pinkawa, S. Stanzel, B. Asadpour, K. Hamacher, H. Coenen, U. Buell, M. J. Eble. 18F fluoromisonidazole and [18F] fluorodeoxyglucose positron emission tomography in response evaluation after chemo-/radiotherapy of non-small-cell lung cancer: a feasibility study. *BMC Cancer*, 6:51–59, 2006.

177. M. Yoshimoto, S. Kinuya, A. Kawashima, R. Nishii, K. Yokoyam, K. Kawai. Radioiodinated VEGF to image tumor angiogenesis in a LS180 tumor xenograft model. *Nucl. Med. Biol.*, 33:963–969, 2006.

178. J. Toretsky, A. Levenson, I. N. Weinberg, J. F. Tait, A. Uren, R. C. Mease. Preparation of F-18 labeled annexin V: A potential PET radiopharmaceutical for imaging cell death. *Nucl. Med. Biol.*, 31:747–752, 2004.

179. L. Aloj, A. Zannetti, C. Caracó, S. Del Vecchio, M. Salvatore. Bcl-2 overexpression prevents 99mTc-MIBI uptake in breast cancer cell lines. *Eur. J. Nucl. Med. Mol. Imag.*, 31:521–527, 2004.

180. S. Del Vecchio, M. Salvatore. 99mTc-MIBI in the evaluation of breast cancer biology. *Eur. J. Nucl. Med. Mol. Imag.*, 31(Suppl. 1):S88–S96, 2004.

181. S. DelVecchio, A. Ciarmiello, M. I. Potena, M. V. Carriero, C. Mainolfil, G. Botti, R. Thomas, M. Cerra, G. D'Aiuto, T. Tsuruo, M. Salvatorel. *In vivo* detection of multidrug-resistant (MDR1) phenotype by technetium-99m sestamibi scan in untreated breast cancer patients. *Eur. J. Nucl. Med.*, 24:150–159, 1997.

182. S. Kinuya, J. Bai, K. Shiba, K. Yokoyama, H. Mori, M. Fukuoka, N. Watanabe, N. Shuke, T. Michigishi, N. Tonami. 99mTc-sestamibi to monitor treatment with antisense oligodeoxynucleotide complementary to MRP mRNA in human breast cancer cells. *Ann. Nucl. Med.*, 20:29–34, 2006.

183. N. H. Hendrikse, E. J. F. Franssen, WTA. van der Graaf, W. Vaalburg, E. G. E. de Vries. Visualization of multidrug resistance in vivo. *Eur. J. Nucl. Med.*, 26:283–293, 1999.

184. J. R. Murren. Modulating multidrug resistance: Can we target this therapy? *Clin. Cancer Res.*, 8:633–635, 2002.

185. W. Guo, G. H. Hinkle, R. J. Lee. 99mTc-HYNIC-Folate: A novel receptor-based targeted radiopharmaceutical for tumor imaging. *J. Nucl. Med.*, 40:1563–1569, 1999.

186. I. Madar, B. Bencherif, J. Lever, R. F. Heitmiller, S. C. Yang, M. Brock, J. Brahmer, H. Ravert, R. Dannals, J. J. Frost. Imaging δ- and μ-opioid receptors by PET in lung carcinoma patients. *J. Nucl. Med.*, 48:207–213, 2007.

187. M. Mirowski, R. Wiercioch, A. Janecka, E. Balcerczak, E. Byszewska, D. Birnbaum, S. Byzia, P. Garnuszek, R. Wierzbickia. Uptake of radiolabeled morphiceptin and its analogs by experimental mammary adenocarcinoma: *In vitro* and *in vivo* studies. *Nucl. Med. Biol.*, 31:451–457, 2004.

188. C. Zeng, S. Vangveravong, J. Xu, K. C. Chang, R. S. Hotchkiss, K. T. Wheeler, D. Shen, Z. P. Zhuang, H. F. Kung, R. H. Mach. Subcellular localization of Sigma-2 receptors in breast cancer cells using two-photon and confocal microscopy. *Cancer Res.*, 67:6708–6716, 2007.

189. E. Bombardieri, F. Crippa, L. Maffioli, M. Greco. Nuclear medicine techniques for the study of breast cancer. *Eur. J. Nucl. Med.*, 24:809–824, 1997.

190. F. G. Blankenberg, H. W. Strauss. Nuclear medicine applications in molecular imaging. *J. Mag. Res. Imaging*, 16:352–361, 2002.

191. D. J. Hnatowich. Observations on the role of nuclear medicine in molecular imaging. *J. Cell. Biochem.*, Suppl. 39:18–24, 2002.

192. S. H. Britz-Cunningham, S. J. Adelstein. Molecular targeting with radionuclides: State of the science. *J. Nucl. Med.*, 44:1945–1961, 2003.

193. C. E. Deppe, P. J. Heering, S. Viengchareun, B. Grabensee, N. Farman, M. Lombés. Cyclosporine a and FK506 inhibit transcriptional activity of the human mineralocorticoid receptor: A cell-based model to investigate partial aldosterone resistance in kidney transplantation. *Endocrinology*, 143:1932–1941, 2002.

194. P. T. Fox, A. R. Laird, L. Lancaster. Coordinate-based voxel-wise meta-analysis: Dividends of spatial normalization. Report of a Virtual Workshop. *Human Brain Mapping*, 25:1–5, 2005.

195. P. Broca, *Bull. Soc. Anthropol. (Paris)* 1861, 2:235; *Bull. Soc. Anat. (Paris)*, 6:330,398, 1861.

196. S. S. Ketty. The nitrous oxide method for the quantitative determination of cerebral blood flow in man—Theory, procedure and normal values. *J. Clin. Invest.*, 27:476, 1948.

197. W. D. Obrist. Determination of regional cerebral blood flow by inhalation of 133-Xenon. *Circ. Res.*, 20:124, 1967.

198. N. A. Lassen. Regional cerebral blood-flow in stroke by Xe-133 inhalation and emission tomography. *Stroke*, 12:284, 1981.

199. G. Neves, S. F. Cooke, T. V. Bliss. Synaptic plasticity, memory and the hippocampus: A neural network approach to causality. *Nat. Rev. Neurosci.*, 9(1):65–75, 2008.

200. M. S. A. Graziano, C. S. R. Taylor, T. Moore. Complex movements evoked by microstimulation of precentral cortex. *Neuron*, 34:841–851, 2002.

201. B. M. Ramsden, C. P. Hung, A. W. Roe. Real and illusory contour processing in area V1 of the primate: A cortical balancing act. *Cerebral Cortex*, 11:648–665, 2001.

202. A. A. Vein, M. L. C. Maat-Schieman. Famous Russian brains: Historical attempts to understand intelligence. *Brain*, 131(Pt 2):583–590, 2008.

203. N. Bernasconi, S. Duchesne, A. Janke, J. Lerch, D. L. Collins, A. Bernasconi. Whole-brain voxel-based statistical analysis of gray matter and white matter in temporal lobe epilepsy. *Neuroimage*, 23(2):17–723, 2004.

204. E. Courchesne, H. J. Chisum, A. Cowles, J. Covington, B. Egaas, M. Harwood, S. Hinds, G. A. Press. Normal brain development and aging: Quantitative analysis at *in vivo* MR imaging in healthy volunteers. *Radiology*, 216:672–682, 2000.

205. J. G. Frontera. Evaluation of the immediate effects of some fixatives upon measurements of the brains of macaques. *J. Comp. Neurol.*, 109:417–438, 1958.

206. H. Stephan. Methodische studien uber den quantitativen vergleich architektonischer struktureinheite des gehirns. *Z. Wiss. Sool.*, 164:1–2, 1960.

207. S. M. Blinkov, I. I. Glezer. *The Human Brain in Figures and Tables: A Quantitative Hand-book*. Plenum Press and Basic Books, New York, NY, pp. 334–338, 1968.

208. K. Brodmann, *Vergleichende localizationlehr der Grosshirnrinde in ihren Prinzipien dargestellt auf Grund des Zellenbaues*. J.A. Barth, Leipzig, 1909.

209. J. Talairach, P. Tournoux. Co-planar stereotaxic atlas of the human brain. 3-Dimensional proportional system: An approach to cerebral imaging. Translated by Mark Rayport. Thieme Medical Publishers, Inc., Stuttgart - New York, NY, 1988.

210. J. L. Lancaster, D. Tordesillas-Gutiérrez, M. Martinez, F. Salinas, A. Evans, K. Zilles, J. C. Mazziotta, P. T. Fox. Bias between MNI and Talairach coordinates analysed using the ICBM-152 brain template. *Hum. Brain Map.*, 28:1194–1205, 2007.

211. D. C. Costa, C. Tannock, J. Brostoff. Brainstem perfusion is impaired in patients with myalgic encephalomyelitis/chronic fatigue syndrome. *Quart. J. Med.*, 88:767–773, 1995.

212. A. Greco, C. Tannock, J. Brostoff, D. C. Costa. Brain MR imaging in chronic fatigue syndrome. *Am. J. Neuroradiol. (AJNR)*, 18:1265–1269, 1997.

213. P. Ehrlich. *Das Sauerstoff-Bederfuis des Organismus: eine farbenalytische Studie.* Hirschward, Berlin, 1885.

214. P. Ehrlich. *Uber die Beziehungen von chemische Constitution, Vertheilung, und Pharmakologisher Wirkung. Collected Studies in Immunology.* (Reproduced and translated). Wiley, New York, 567–595, 1906.

215. L. Stern, R. Gautier. Rapports entre le liquide céphalo-rachidien et la circulation sanguine. *Arch. Int. Physiol.*, 17:138–192, 1921.

216. L. Stern, R. Gautier. 1992 Les rapports entre le liquide céphalo-raquidien et les éléments nerveux de l'axe cérébrospinal. *Arch. Int. Physiol.*, 17:391–448, 1921.

217. H. Davson. History of the blood–brain barrier concept. In E. A. Neuwelt, ed., *Implications of the Blood–Brain Barrier and its Manipulation.* Plenum, New York, 27–52, 1989.

218. H. Becker, G. Quadbeck. Tierexperimentelle tersuchungen uber die Funktsionswise der Blut-Hirnschranke. *Z. Naturforsch.*, 7B:493, 1952.

219. B. B. Brodie, H. Kurtz, L. S. Schanker. The importance of dissociation constant and lipid-solubility in influencing the passage of drugs into the cerebrospinal fluid. *J. Pharmacol.*, 130:20–25, 1960.

220. K. Welch. A model for the distribution of materials in the fluids of the central nervous system. *Brain Res.*, 16:453–468, 1969.

221. H. Spatz. Die Bedeutung der vitalen Fäbund für die Lehre vom Stoffaustausch zwischen dem Zentralnervensystem und dem übrigen Körper. *Arch. Psychiat. Nervenheilk.*, 101:267–358, 1934.

222. Riser. *Le liquide céphalo-rachidien.* Masson, Paris, 1920.

223. J. Wolff. Beiträge zur Ultrastruktur der Kapillaren im Zentralnervensystem. *Z. Zellforsch.*, 60:409–431, 1963.

224. C. D. Clemente, E. A. Holst. Pathological changes in neurons, neuroglia and blood–brain barrier induced by x-irradiation of the heads of monkeys. *Acta Neurol. Psychiatr. Scand.*, 71:66–79, 1954.

225. M. W. Brightman, T. S. Reese. Junctions between intimately apposed cell membranes in the vertebrate brain. *J. Cell Biol.*, 40:648–677, 1969.

226. D. C. Costa. The blood–brain barrier, in *Nuclear Medicine in Clinical Diagnosis and Treatment*, 3rd ed., Eds. P. J. Ell and S. S. Gambhir, Churchill Livingstone, Edinburgh, London, New York, chap. 89:1235–1246, 2004.

227. H. Davson. The blood–brain barrier. *J. Physiol.*, 255:1–28, 1976.

228. W. Oldendorf, H. Davson. Transport in the central nervous system. *Proc. Roy. Soc.*, 60:326–328, 1967.

229. P. A. Stewart, M. J. Wiley. Developing nervous tissue induces formation of blood–brain barrier characteristics in invading endothelial cells: a study using quail–chick transplantation chimeras. *Dev. Biol.*, 84:183–192, 1981.

230. R. C. Janzer, M. C. Raff. Astrocytes induce blood–brain barrier properties in endothelial cells. *Nature*, 325:253–257, 1987.

231. J.-H. Tao-Cheng, Z. Nagy, M. W. Brightman. Tight junctions of brain endothelium *in vitro* are enhanced by astroglia. *J. Neurosci.*, 7:3293–3299, 1987.

232. M. W. Brightman. The anatomic basis of the blood–brain barrier. In E. A. Neuwelt, ed., *Implications of the Blood–Brain Barrier and its Manipulation*. Plenum, New York, pp. 53–83, 1989.

233. C. Crone. Tight and leaky endothelia. In H. Ussing, N. Bindslev, N. A. Lassen, O. Sten-Knudsen, ed., *Water Transport Across Epithelia*. Munksgaard, Copenhagen, pp. 256–267, 1981.

234. C. Crone, S. P. Olesen. Electrical resistance of brain microvascular endothelium. *Brain Res.*, 241:49–55, 1982.

235. R. A. Hawkins, D. R. Peterson, J. R. Viña. The complementary membranes forming the blood–brain barrier. *IUBMB Life*, 54:101–107, 2002.

236. R. L. O'Kane, R. A. Hawkins. Na-dependent transport of large neutral amino acids occurs at the abluminal membrane of the blood–brain barrier. *Am. J. Physiol. Endocrinol. Metab.*, 285:1167–1173, 2003.

237. C. Crone. The permeability of capillaries in various organs as determined by use of the indicator diffusion method. *Acta Physiol. Scand.*, 58:292–305, 1963.

238. C. Crone. Lack of selectivity to small ions in paracellular pathways in cerebral and muscle capillaries of the frog. *J. Physiol.*, 353:317–337, 1984.

239. E. M. Renkin. Transport of potassium-42 from blood to tissue in isolated mammalian skeletal muscle. *Am. J. Physiol.*, 197:1205–1210, 1959.

240. M. W. B. Bradbury. *The Concept of a Blood–Brain Barrier*. Wiley, New York, 1979.

241. J. O. M. Eichling, M. E. Raichle, R. L. Grubb et al. Evidence of the limitations of water as a freely diffusible tracer in brain of the rhesus monkey. *Circ. Res.*, 35:358–364, 1974.

242. O. B. Paulson, M. M. Hertz, T. G. Bolwig et al. Filtration and diffusion of water across the blood–brain barrier in man. *Microvasc. Res.*, 13:113–124, 1977.

243. W. M. Partridge, L. J. Mietus. Transport of steroid hormones through the rat blood–brain barrier. Primary role of albumin bound hormone. *J. Clin. Invest.*, 64:145–154, 1979.

244. A. Yuwiler, W. H. Oldendorf, E. Geller et al. Effect of albumin binding and amino acid competition on tryptophan uptake into brain. *J. Neurochem.*, 28:1015–1023, 1977.

245. W. M. Pardridge, E. M. Landaw. Tracer kinetic model of blood–brain barrier transport of plasma protein bound ligands. Empiric testing of the free hormone hypothesis. *J. Clin. Invest.*, 74:745–752, 1984.

246. W. M. Partridge. Carrier-mediated transport of thyroid hormones through the rat blood–brain barrier. Primary role of albumin-bound hormone. *Endocrinology*, 105:605–612, 1979.

247. M. W. Bradbury. Transport across the blood–brain barrier. In E. A. Neuwelt, ed., *Implications of the Blood–Brain Barrier and its Manipulation*. Plenum, New York, pp. 119–136, 1989.

248. B. V. Zlokovic, D. J. Begley, D. G. Chain. Blood–brain barrier permeability to dipeptides and their constituent amino acids. *Brain Res.*, 271:65–71, 1983.

249. W. M. Pardridge, J. Eisenberg, J. Yang. Human blood–brain barrier insulin receptor. *J. Neurochem.*, 44:1771–1778, 1985.

250. A. Gjedde, M. Rasmussen. Blood–brain glucose transport in the conscious rat: Comparison of the intravenous and intracarotid injection methods. *J. Neurochem.*, 35:1375–1381, 1980.

251. W. M. Pardridge. Brain metabolism: A perspective from the blood–brain barrier. *Physiol. Rev.*, 63:1481–1535, 1983.
252. Q. R. Smith, Y. Takasato, D. J. Sweeney et al. Regional cerebrovascular transport of leucine as measured by the *in situ* brain perfusion technique. *J. Cereb. Blood Flow Metab.*, 5:300–311, 1985.
253. M. E. Phelps, S.-C. Huang, E. J. Hoffman et al. Cerebral extraction of N-13 ammonia: its dependence on cerebral blood flow and capillary permeability–surface area product. *Stroke*, 12:607–619, 1981.
254. J. E. Cremer, L. Braun, W. H. Oldendorf. Changes during development in transport processes of the blood–brain barrier. *Biochim. Biophys. Acta*, 448:633–637, 1975.
255. A. Gjedde, C. Crone. Induction processes in blood–brain transfer of ketone bodies during starvation. *Am. J. Physiol.*, 229:1165–1169, 1975.
256. J. H. James, J. Escourrou, J. E. Fischer. Blood–brain neutral amino acid transport activity is increased after portacaval anastomosis. *Science*, 200:1395–1397, 1978.
257. A. M. Mans, J. F. Biebuyck, R. A. Hawkins. Ammonia selectively stimulates neutral amino acid transport across blood–brain barrier. *Am. J. Physiol.*, 245:C74–C77, 1983.
258. G. S. Sarna, M. W. B. Bradbury, J. E. Cremer, et al. Brain metabolism and specific transport at the blood–brain barrier after porto-caval anastomosis in the rat. *Brain Res.*, 160:69–83, 1979.
259. R. B. Schwartz, P. A. Carvalho, E. Alexander III, J. S. Loeffler, R. Folkerth, B. L. Holman. Radiation necrosis versus high-grade recurrent glioma: Differentiation by using dual-isotope SPECT with ^{201}Tl and ^{99m}Tc-HMPAO. *AJNR*, 12:1187–1192, 1992.
260. D. C. Costa, S. Gacinovic, R. F. Miller. Radionuclide brain imaging in immunodeficiency syndrome (AIDS). *Q. J. Nucl. Med.*, 39:243–249, 1995.
261. D. C. Costa. *A Study of the First $^{99Tc^m}$-labelled Radiopharmaceutical for the Investigation of Cerebral Blood Flow in Man*. PhD Thesis, University of London, 1989.
262. G. Orefice, L. Celentano, M. Scaglione, M. Davoli, S. Striano. Radioisotopic cisternography in benign intracranial hypertension of young obese women. A seven-case study and pathogenetic suggestions. *Acta Neurol (Napoli)*, 14:39–50, 1992.
263. M. Primeau, L. Carrier, P. C. Milette, R. Chartrand, D. Picard, M. Picard. Spinal cerebrospinal fluid leak demonstrated by radioisotopic cisternography. *Clin. Nucl. Med.*, 13:701–703, 1988.
264. E. De Robertis. Ultrastructure and cytochemistry of the synaptic region. *Science*, 156:907–914, 1967.
265. G. D. Pappas. Ultrastructural basis of synaptic neurotransmission. In D. B. Tower, ed., *The Nervous System*, Vol. 1, Raven Press, New York, pp. 19–30, 1975.
266. C. W. Cotman, M. Nieto-Sampedro. Cell biology of synaptic plasticity. *Science*, 225:1287–1294, 1984.
267. R. L. Clem, T. Celikel, A. L. Barth. Ongoing *in Vivo* experience triggers synaptic metaplastcity in the neocortex. *Science*, 319:101–104, 2008.
268. R. S. Feldman, J. S. Meyer L. F. Quenzer. *Priciples of Neuropharmacology*. Sinauer Associates, Inc., Publishers, Sunderland, MA, 1997.

7

Systems in Nuclear Medicine

J. J. Pedroso de Lima and António Dourado

CONTENTS

7.1 Theory of Biological Systems: Introduction

7.1.1 Some Basic Systemic Concepts in Biology and Physiology

In this chapter, we will study the basic notions of systems theory that are most relevant to biological systems (including physiological ones). In the last few decades, the methods for modeling and analyzing biological systems have probably been one of the most relevant facets of the enormous development of mathematical biology.

We will study population models, compartmental models, and some principles of the kinetics involved in the system's dynamics and interactions, starting with some concepts of general systems theory that are very useful for understanding biological systems.

7.1.1.1 Systems, Elements, and Relations

System is a concept that pervades our daily life and even our imagination. However, it is not easy to define. First of all, a system is goal oriented. Second, a system is a collection of components. We may tentatively define a system as a set of related elements (components) interacting with one another in an orderly way to enable them to attain a goal.

Generally, we say that there is a relation between two elements, E1 and E2, if the behavior of E1 influences the behavior of E2 and if the behavior of E2 influences the behavior of E1. Reciprocity is the essence of a relation.

Each element has a set of characteristic properties (color, size, weight, temperature, etc.) that define its attributes. An attribute of an element is a characteristic quality such as intensity, speed of communication, etc. In the study of biological systems (and of systems in general), the evolution of the attributes over time is fundamental, because this is what expresses the system's dynamics and it is through this evolution that the system attains its goals.

If we think of an element as being some phenomenon of the natural or social world, or its abstract representation, having some attributes that evolve as a consequence of its own behavior in the system to which it belongs we have a definition of a system that is sufficiently general to embrace the aim of our study: biological and physiological systems. A system is said to be dynamic if its attributes change over time.

There are two key concepts in this definition of a system: relation and interaction. There is no system without them. They build it up.

A simple unique animal or vegetal cell is a very complex system with a large number of elements acting in a united and coordinated manner in order to carry out the actions needed for the life and reproduction of the cell.

A working automobile contains thousands of elements of different natures: fuel tank, engine, tires, electronic components, microprocessors, etc. Information (from multiple sensors and actuators to and from the microprocessors), mass (from fuel tank for the motor injectors), and energy (from the engine to the wheels, e.g., the mechanical energy) establish a multiplicity of connections among the automobile elements.

Mass, energy, and information are the three elements circulating in systems. They are the three pillars of modern science. In biological systems, the information circulates in the proteins and is in the genome. A relation between two elements is established by the flow of materials, information, and energy between them. So, communication is synonymous of relation, in this context.

A cell is a sophisticated and complex fabric, with many components subject to a well-defined set of relations. The same components subjected to a different set of relations may give rise to cells with different characteristics. Each component has a specific mission, and the interaction among them, combining their missions, makes life possible. The exchange of mass (e.g., nutrients), energy (e.g., heat), and information (through genes and proteins) are the basic conditions for the existence of the cellular system as a whole.

7.1.1.2 Feedback and Self-Regulation

Our nervous system senses the skin temperature and sends this information to the brain. The brain, knowing whether it is good or bad, sends the order to produce more heat (by consuming more calories) or to cool (by perspiration) to the appropriate organs. This is a regulating system similar to many others that are part of our daily environment (air conditioning, speed regulator in cars, temperature regulator in an oven, etc.), in which the essential element is the flow of information about the state of the system to some device whose mission is to keep the system in good condition. This is called feedback (from output to input). Feedback allows self-regulation, as the capacity of the whole system to keep itself in a viable state. The human body contains multiple intricate feedback paths that enable the self-regulation of the different organs and physiological systems essential to the preservation of life (sugar in blood, acid–base equilibrium, etc.).

Consider, for example, a two-population biological system, the predator–prey pair, where the prey is a herbivore and the predator is a carnivore. If the vegetation increases in the environment, the herbivore population increases exponentially; as a consequence, the carnivore population also increases exponentially, which leads to a fall in the herbivore population, because there are more predators. Therefore, the initial increase in the herbivore population retroacted through its relation with the carnivore population.

The sugar regulation in blood can be represented by Figure 7.1, where multiple feedback paths exist to allow the self-regulation of a healthy body.

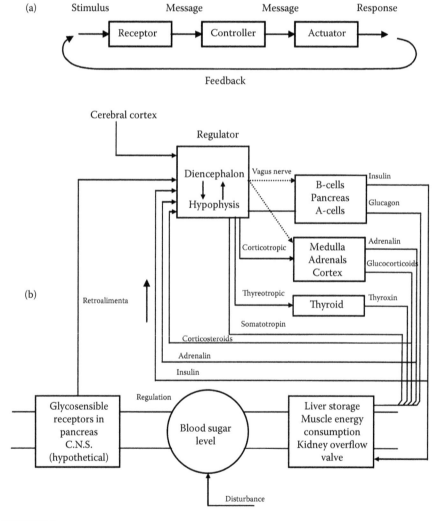

FIGURE 7.1
Feedback allows the control of blood sugar concentration. (Adapted from L. Bertalanffy. *General Systems Theory*, George Brazillier, NY, 1969.)

In self-regulation, the feedback happens with a negative sign and allows the comparison by the difference between two values: the desired value (the reference) and the observed (measured) value. This difference comprises the system error at that moment, and corrective actions are taken (by a controller) to reduce the error to zero (or nearly). It is thus, said to be a negative feedback.

7.1.1.3   Environment, Closed Systems, Open Systems, and Homeostasis

A system is always surrounded by its environment and communicates with it through its boundary. There is a dense relationship among the system elements, as observed in Figure 7.2.

The intensity of communication between the system and its environment varies from case to case. A chemical reaction in a closed vessel where reagents are mixed is described, from the outside perspective, as a closed system. The laws of thermodynamics apply to these systems.

However, if the communication between the system and its environment is intense, then we have an open system, exchanging materials, energy, and information with its environment through a boundary. A living cell is a typical case of an open system.

Sometimes, it is difficult to identity the boundary of an open system exactly, as it is permanently and continuously communicating with its environment.

Living beings are open systems: they constantly receive energy from and send energy to the environment. Their internal structure is permanently undergoing change and so they are never in a stationary state (in the sense that all the attributes of all its elements have a constant value over time); instead they are in a state of chemical and thermodynamic equilibrium, and their attributes display well-defined patterns over time, maybe with small transient changes. This equilibrium state is homeostasis and is illustrated in Figure 7.3. The concept was created by the French physiologist Claude Bernard (1813, 1878) [3], when he famously wrote, "La fixité du milieu intérieur est la condition de la vie libre," which today remains the concept that underpins homeostasis. The concept was afterward established by the

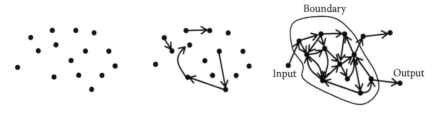

FIGURE 7.2
(a) A set of independent objects does not build a system. (b) A few scattered relations are not enough to make a system. It needs a dense collection of relations as seen in (c). (Adapted from R. Flood and E.R. Carson. *Dealing with Complexity, An Introduction to the Theory and Applications of Systems Science*, Plenum Press, 1993.).

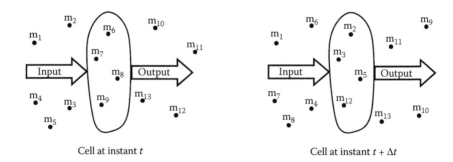

Cell at instant t Cell at instant $t + \Delta t$

FIGURE 7.3

Homeostasis. The inner components of the cell (m_i represents molecules) have been replaced between two successive instants of time, t and $t + \Delta t$, but the cell remains in a steady viable state. (Adapted from R. Flood and E.R. Carson. *Dealing with Complexity, An Introduction to the Theory and Applications of Systems Science*, Plenum Press, 1993.)

American physiologist Walter Cannon (1971–1945), in his book *The Wisdom of the Body* (1932), where he describes the general characteristics of homeostasis though his famous four propositions [4]:

1. Constancy in an open system similar to that represented by our bodies requires mechanisms which act to maintain this constancy. Cannon based this proposition on insights into the ways by which steady states such as glucose concentrations, body temperature, and acid–base balance were regulated.

2. Steady-state conditions require that any tendency toward change is automatically faced with factors that resist change. An increase in blood sugar results in thirst as the body attempts to dilute the concentration of sugar in the extracellular fluid.

3. The regulating system that determines the homeostatic state consists of a number of cooperating mechanisms simultaneously or successively acting. Blood sugar is regulated by insulin, glucagons, and other hormones that control its release from the liver or its uptake by the tissues.

4. Homeostasis does not occur by chance: it is the result of organized self-government.

7.1.1.4 *Entropy and Negentropy*

Things show a tendency toward disorder and disorganization; entropy measures this tendency. Entropy, as previewed by the second principle of thermodynamics, conveys the irreversible law of degradation of materials and energy: systems tend toward disorder.

This conclusion was stated by Clausius [5] with his famous principle of maximum entropy ("the entropy of the universe tends to a maximum"), or mathematically we have Equation 7.1, where S represents the entropy.

$$dS \geq 0 \tag{7.1}$$

However, there is an apparent contradiction in that homeostatic systems maintain order and improve organization. This was the problem addressed by Ilya Prigogine, who in 1945 proposed the theorem of the minimal production of entropy. This applies to stationary nonequilibrium states and explains the analogy relating the stability of thermodynamic equilibrium states (the classical notion) with the stability of biological systems similar to that expressed by homeostasis. Prigogine won the Nobel Prize for chemistry in 1977 [6]. His results are given by Equations 7.2 and 7.3.

$$dS = d_e S + d_i S, \tag{7.2}$$
$$dS \leq 0, \tag{7.3}$$

$d_i S$ represents the entropy internally produced by an irreversible degradation process (chemical reactions, diffusion, heat transfer, etc.) and $d_e S$ represents the entropy imported by the living open system, in the form of matter and energy. From Clausius' law, we have Equation 7.4:

$$d_i S > 0, \tag{7.4}$$

which, replaced by Equation 7.2, gives Equation 7.5.

$$d_e S \leq 0 \tag{7.5}$$

Due to Equation 7.5, we call entropy negative to $d_e S$ negentropy. Open systems import negentropy. Figure 7.4 illustrates the difference between a closed system and an open one.

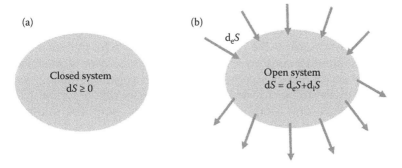

FIGURE 7.4
(a) Closed system (Clausius equation) and (b) open system (Prigogine equation).

As a consequence, the thermodynamic laws are only applicable to closed systems. Entropy thus, appears to be a force against homeostasis. With the aid of homeostasis, systems import materials, energy, and information from the environment; and with them, they offset the tendency toward disorder by creating and maintaining the order, resisting the second thermodynamic principle, and even being capable of evolving toward states of greater order and organization.

The structure of a system expresses the way its elements relate to each other and fixes the foundations of the processes evolving within it.

If we observe a system at successive instants of time, we notice that it evolves in some way. The behavior of the system is, in fact, this perception. This behavior is goal oriented if it can be understood and assessed with regard to some particular purpose.

In open systems, a certain final state may be reached from different initial states, following different trajectories. The final equilibrium state is said to have an equifinal value, and the system has the property of equifinality.

7.1.1.5 Autopoiesis, Adaptation, and Cybernetics

One living cell produces its own components, which, in turn, produce the actual cell. This is the autopoietic property of living systems: the ability to self-reproduce. Living systems are autopoietic, because they are organized so as to allow their own processes to generate the components needed for these same processes to continue.

A system is adaptive if it is able to adapt. Adaptation emerges as a need to respond to environmental changes in order to survive and keep the system performing well. It can be said that the Darwinian theory of evolution constitutes a theory of adaptation. The ability to adapt is fundamental when the environment frequently and profoundly changes. Adaptation requires the ability to regulate and control to cope with rapid changes and long-term changes (short-term and long-term regulation). Any living being requires short- and medium-term multiple regulation and control systems in order to survive, sometimes called the law of requisite variety.

In our lives, we often look at things only partially, seeing just some parts and forgetting to look at the holistic view. That is why the results of our actions are sometimes contrary to our expectations, because we dismissed or forgot the phenomenological complexity. We need systemic thinking; we need models and methodologies to deal with complexity. Without formal systemic views, only parts, only simplistic explications and simplistic solutions are attainable.

Cybernetics studies adaptation, regulation, and control. It has been defined as the science of control and communications in animals and machines (proposed by Wiener in his 1948 book *Cybernetics, or Control and Communication in the Animal and Machine*).

7.1.1.6   A Hierarchy of Systems

In a hierarchy of systems and control, we may define systems on several levels: we will have a hierarchy of systems. A metasystem is a system above other systems in a hierarchy. A subsystem is below another system. In humans, the brain or consciousness is the highest level metasystem, and the atom is at the lowest level. We may define a hierarchy: atom, molecule, cell, tissue, organ, functional system, ..., consciousness.

The hierarchical organization is a logical representation of the phenomena, systems, and subsystems. Going down hierarchy, we increase the resolution of the analysis and see more and more details. The choice of an appropriate resolution level for the analysis of a given phenomenon is crucial to finally deciding the questions we will work with in the study. The chosen level becomes the system in focus.

A system scientist must be simultaneously holistic, looking for the system as a whole, and a reductionist, that is, capable of understanding the system in its details, in each of its multiple parts.

Going up in a systems hierarchy, we find a property fundamental to all of systems theory: the whole is greater than the sum of the parts. This is the emergency property: the hypersystem has (sometimes unexpected) properties that do not exist in any of the subsystems, they just emerge when the hypersystem is built. This is the synergy effect.

Cells are grouped in different organs (kidney, liver, heart, lung, etc.), each with a specific irreplaceable function and with properties different from its cells. The organs together build up a body with emerging properties: they are organized through communication and (cybernetic) control into a hierarchy of body parts that allow the capacities of sight, hearing, smell, locomotion, and emotion, that is, of the human being. Neither is a human being an aggregate of body parts nor is a society an aggregate of human groups. Systems join together to make a hypersystem whose properties differ from those of the subsystems.

Emergence is a characteristic of phenomena that we cannot explain any other way. The emergence of a society is its culture.

Synergy is a concept used to describe the emergence of unexpected and interesting properties, for example, the theory of organizations (the synergy of group work, for instance).

According to Bertalanffy [1], we can define a system hierarchy as shown in Table 7.1.

During the last century, several metaphors appeared for systems, as the system theory was tentatively applied to human groups and societies.

First came the machine metaphor, looking to a closed system with a set of well-defined objectives and a rigid control structure to reach them (technological servomechanisms are a good example). This metaphor inspired the development of industrial systems and their management.

TABLE 7.1

A Hierarchy of Systems According to Bertalanffy [1]

Level	Description and Examples
Symbolic systems	Language, logic mathematics, sciences, arts, morals, etc.
Sociocultural systems	Populations of organisms (humans included); symbol-determined communities (cultures) in man only.
Humans	Symbolism; past and future, self and world; self-awareness as a consequence of communication by language.
Animals	Increasing importance of traffic of information (evolution of receptors, nervous systems, etc.); learning; beginnings of consciousness.
Lower organisms	"Plant-like" organisms: increasing differentiation of systems ("division of labour"); distinction of reproduction from individual functions.
Open systems	Fire, cells, and organisms in general.
Control mechanisms	Thermostat, servomechanism, and homeostatic mechanisms in living organisms.

After Prigogine, this vision was superseded by the organic metaphor, describing complex networks of elements and relations with a transformation affected by feedback.

However, the organic metaphor cannot be used to describe emotions. The brain metaphor naturally appeared.

The culture metaphor means that a network of values, beliefs, and norms can be included.

Finally, the political metaphor allows the interaction of networks of interests that individuals aim at reaching.

In our study of biological and physiological processes, we will concern ourselves with the development of computable mathematical models, that is, models which we can program in a computer and use to simulate, analyze, and predict the behavior of systems in the first two levels of the Bertalanffy hierarchy.

Our analysis will be based on the physical, chemical, thermodynamic, properties of phenomena. Using the appropriate laws, we obtain a set of differential equations that enable us to computationally describe and simulate the respective dynamic behavior. The differential Equations also serve to obtain very practical mathematical representation: the state equations both linear and nonlinear.

Some dynamic properties, such as stability and type of transient time response, are easily obtained from these representation tools.

Some nonlinear systems have special importance in the biological and physiological context: chaotic systems. As we will see, a system is chaotic if we cannot predict its behavior from observable initial conditions. We will look at some examples and a short introduction to its theory.

7.1.2 Mathematical Modeling and Analysis of Biological Systems: Population Models

Mathematical models of biological species find a wide spectrum of applications, ranging from environmental studies to disease propagation and viral epidemics, tumour growth, etc. These models, if realistic, have a wide practical application for the understanding of population dynamics and to forecast population evolution.

7.1.2.1 *The Growth of a Single Species' Population*

Let $N(t)$ be a population of a single species at time t, in a given environment.

At the next time instant, $t + \Delta t$, the population will be $N(t + \Delta t)$, given by the mass conservation equation: the population in the next period will be the population of the current period plus those born until that point, minus the dead ones, plus those entering from abroad, minus those leaving, as in Equation 7.6.

$$N(t + \Delta t) = N(t) + \text{the born during } \Delta t - \text{the dead during } \Delta t$$
$$+ \text{the imigrated during } \Delta t - \text{the emigrated during } \Delta t. \tag{7.6}$$

Manipulating Equation 7.6, and assuming that there are no migrations, one obtains Equation 7.7:

$$N(t + \Delta t) - N(t) = \Delta N(t) = \text{born during } \Delta t - \text{dead during } \Delta t. \tag{7.7}$$

Assuming now a birth rate a and a mortality rate b per time interval, the born and dead during Δt are $aN(t)\Delta t$ and $bN(t)\Delta t$, respectively. Substituting in Equation 7.1, we get Equation 7.8.

$$N(t + \Delta t) - N(t) = \Delta N(t) = aN(t)\Delta t - bN(t)\Delta t. \tag{7.8}$$

Dividing both sides of Equation 7.8 by Δt, we get Equation 7.9.

$$\frac{N(t + \Delta t) - N(t)}{\Delta t} = \frac{\Delta N(t)}{\Delta t} = aN(t) - bN(t). \tag{7.9}$$

Next, applying to Equation 7.9 the limit when Δt tends toward zero

$$\lim_{\Delta t \to 0} \frac{\Delta N(t)}{\Delta t} = aN(t) - bN(t). \tag{7.10}$$

we obtain the definition of the derivative of N with regard to t Equation 7.11.

$$\frac{dN(t)}{dt} = aN(t) - bN(t). \tag{7.11}$$

7.1.2.2 The Logistic Equation

Let us assume that there are no deaths according to Equation 7.11. Then the population will grow without limits, according to Equation 7.12

$$\frac{dN(t)}{dt} = aN(t). \tag{7.12}$$

The solution of Equation 7.12 is an exponentially growing function. In reality, a population in a given environment with finite resources cannot exceed a certain limit, when all the available resources are used to sustain the population. To take this into account, a limitative term is introduced in Equation 7.12, resulting in Equation 7.13,

$$\frac{dN(t)}{dt} = aN(t) - bN^2(t) = aN(t)\left[1 - \frac{b}{a}N(t)\right] = aN\left[1 - \frac{N}{k}\right], \tag{7.13}$$

known as the logistic equation or the Verhulst equation [7]. The constant a is the intrinsic growth rate, and $k = b/a$ is the maximum population value that the environment can support, the so-called carrying capacity. If $N = k$, the population remains constant.

We can also develop a discrete model, giving the population values only at well-defined times.

Let us suppose a biological species whose individuals live a whole number of time intervals (e.g., years) and are born and die at the end of the year. Initially, there are N_0 individuals. At the end of that year and the start of the next year, there are N_1 individuals; after two years, there are N_2 individuals; and at the end of k year, there are N_k individuals.

In principle, we can say that at the end of the first year, the population N_1 is proportional to N_0, through a proportionality constant A that depends on the environmental conditions (amount of food and water, climate, etc.), as in Equation 7.14

$$N_1 = AN_0. \tag{7.14}$$

If $A > 1$, the population increases and can be considered the reproduction rate; if $A < 1$, it diminishes until total extinction, and it is the mortality rate. If $A > 1$ is constant over the years, the population will grow without limit, leading to a population explosion.

However, if the population increases without limit, on the one hand, there will not be enough food and, on the other hand, any predators living in the same environment will find more individuals of the species in question, which will, thus, limit the population growth. This limiting effect can be introduced into the equation though a subtractive term, which is very low for small population values and very strong for high population values, for example,

$$N_1 = AN_0 - BN_0^2. \tag{7.15}$$

If B is so much less that A, A, $B \ll A$, the subtractive term will only be relevant for high values of N_0, corresponding to the desired effect. In the succeeding years, we will have, assuming that A and B are constant over the years,

$$N_2 = AN_1 - BN_{1^2}$$
$$N_3 = AN_2 - BN_{2^2}$$
$$\vdots \tag{7.16}$$
$$N_{k+1} = AN_k - BN_{k^2}.$$

The time will come when the population cannot grow any more, because it has reached its maximum values N_{max}. Let m be that year and $N_m = N_{max}$. The next year, the population will be Equation 7.17

$$N_{m+1} = AN_{max} - BN_{max^2}. \tag{7.17}$$

Forcing N_{m+1} to be positive, we will have Equation 7.18.

$$AN_{max} - BN_{max}^2 \geq 0 \Leftrightarrow N_{max}(A - BN_{max}) \geq 0 \Rightarrow N_{max} \leq \frac{A}{B}. \tag{7.18}$$

The relation A/B must be less than the maximum reachable population figure. Expressing the population as a fraction of its maximum values, as in Equation 7.19

$$x_k = \frac{N_k}{N_{max}} \Rightarrow N_k = x_k M_{max} \quad \text{for all values of } k, \tag{7.19}$$

and subtracting Equation 7.19 from Equation 7.17, we obtain Equation 7.20a.

$$N_{k+1} = AN_k - BN_k^2 \Rightarrow x_{k+1}N_{max} = Ax_kN_{max} - Bx_k^2N_{max}^2. \tag{7.20a}$$

Solving for Equation 7.20a for x_{k+1}, we get Equation 7.20.

$$x_{k+1} = Ax_k - Bx_k^2N_{max} \Rightarrow x_{k+1} = Ax_k - Bx_k^2\frac{A}{B} \Leftrightarrow x_{k+1} = Ax_k(1 - x_k). \tag{7.20b}$$

We have finally achieved the model for population growth, in the form of a nonlinear first-order difference Equation 7.21.

$$x_{k+1} = Ax_k(1 - x_k) = f(x_k) \quad x \in [0, 1], \tag{7.21}$$

where x_k is the population at year k, as a function of the maximum population. This is the famous logistic equation for the population model that we will further discuss later on.

Is it possible for the population to be constant over time in some equilibrium state? In these circumstances, we would have $x_{k+1} = x_k$, or,

$$x_{k+1} = Ax_k(1 - x_k) = x_k \Rightarrow Ax_k - Ax_k^2 = x_k \Rightarrow (A - 1 - Ax_k)x_k = 0$$

$$\Rightarrow \begin{cases} A - 1 - Ax_k = 0 \\ x_k = 0 \end{cases} \Rightarrow \begin{cases} x_k = \dfrac{A-1}{A} \\ x_k = 0 \end{cases} \Rightarrow \begin{cases} x_k = 1 - \dfrac{1}{A} \\ x_k = 0 \end{cases} \qquad (7.22)$$

If $A > 1$, we have two equilibrium states. If $A < 1$, only the origin can be an equilibrium state, because x_k cannot be negative. Note that $A > 0$.

7.1.2.3 The Lotka–Volterra Predator–Prey Model

In nature, there are many examples of predator–prey pairs: lion–gazelle, bird–insect, cat–rat, pandas–eucalyptus, etc. [13]. To describe the biological interaction between two species in the particular relation where one, the predator, eats the other, the prey, we can use the Lotka–Volterra model. For this purpose, let us make the following simplifying assumptions:

- The predator is totally independent of the prey, and this is its only food.
- The prey has unlimited food available.
- The prey has only one predator, the one that is being considered, and has no other environmental constraints.

In nature, the reality is more complicated; but with these assumptions, we can develop an easily understood model.

If there are zero predators, the prey has no limit on its growth and it grows exponentially. If $x(t)$ is the actual population at time t, its growth rate will be Equation 7.23

$$\frac{dx(t)}{dt} = ax(t). \qquad (7.23)$$

However, with the existence of predators, growth will be lower; the predators introduce a negative factor into the equation.

If $y(t)$ is the predator population at time t, each of its individuals has a certain probability of meeting prey. This probability is higher the more predators exist and also the more prey exists. So we will have the following:

- The rate of predator–prey contacts is jointly proportional to the size of both populations.

- A fixed portion of the contacts that occur results in the death of the prey.

So we will have Equation 7.24.

$$\frac{dx(t)}{dt} = ax(t) - bx(t)y(t). \tag{7.24}$$

Concerning the predator population, y, if there is no food (no prey), it will die at a rate proportional to its size, as in Equation 7.25,

$$\frac{dy(t)}{dt} = -cy(t), \tag{7.25}$$

because there is no energy for new births.

Fortunately for the predators, there is prey, giving them energy to reproduce themselves. For each predator–prey contact, there is a part resulting in food for the predator; and its population will evolve according to Equation 7.26,

$$\frac{dy(t)}{dt} = -cy(t) + px(t)y(t), \tag{7.26}$$

where p is a constant expressing the likelihood that the predator will eat the prey.

Joining Equations 7.24 and 7.26, we finally get the Lotka–Volterra model (Equation 7.27).

$$\frac{dx(t)}{dt} = ax(t) - bx(t)y(t)$$
$$\frac{dy(t)}{dt} = -cy(t) + px(t)y(t), \tag{7.27}$$

This model consists of two first-order nonlinear coupled differential equations that cannot be independently solved and for which it is not possible to find an analytical solution; only numerical solutions are possible.

It can be shown that there is a pair (x, y) for which $dt = dy/dt = 0$, that is, the populations are in equilibrium and do not vary with time. For that, we can write Equation 7.28

$$0 = ax(t) - bx(t)y(t)$$
$$0 = -cy(t) + px(t)y(t), \tag{7.28}$$

Let us solve this system of algebraic Equations 7.29

$$\begin{cases} [a - by(t)]x(t) = 0 \wedge x(t) \neq 0 \\ [-c + px(t)]y(t) = 0 \wedge y(t) \neq 0 \end{cases} \Rightarrow \begin{cases} [a - by(t)] = 0 \\ [-c + px(t)] = 0 \end{cases} \Rightarrow \begin{cases} y(t) = \dfrac{a}{b} \\ x(t) = \dfrac{c}{p} \end{cases}. \tag{7.29}$$

This state is called steady state or stationary state. Its value depends on the constants a, b, c, and p, the parameters of the model.

We will come back to this model later to analyze its phase trajectories.

There are more complete models of two populations. For example, in [8] we can find a model that includes competition (fighting for the same resources) among species, aggressive interaction (fighting each other, although not eating each other), symbiotic cooperation (mutual help), predator–prey relation, and weak–strong interaction (one species is better prepared to survive).

7.1.2.4 A General Model for the Interaction of Two Populations

Let us consider two biological populations interacting in a certain environment. They can be modeled by the two differential equations below [7]:

$$\dot{x} = ax + bxy$$
$$\dot{y} = cy + dxy, \tag{7.30}$$

with the four parameters a, b, c, and d (real numbers).

The linear terms ax and by describe the growth or decline of the populations $x(t)$ and $y(t)$ separately considered from one another. For $a > 0$, $x(t)$ grows exponentially; for $a < 0$, $x(t)$ decreases exponentially. For $b > 0$ or $b < 0$, the same happens with y.

The interaction between the two populations is modeled by the nonlinear terms bxy and dxy. The interaction depends on the number of contacts between the elements of x and the elements of y. The constants b and c express the likelihood of an interaction if there is a meeting. Several forms of interaction between the two populations can be described by this model, depending on the signs of the four constants a, b, c, and d, as seen in Table 7.2, assuming that a predator has only one kind of prey.

TABLE 7.2

Signs of the Constants in the General Model of Biological Interaction and Interaction That They Express

A	B	C	d	Interaction
−	+	+	−	Predator(x)-prey(y) models
+	−	−	+	Prey(x)-predator(y)
+	+	+	+	Mutualism or symbiotic models
+	−	+	−	Models of competion

If we now consider the logistic equation for each of the populations and, at the same time, their interaction, we will have Equation 7.31:

$$\dot{x} = a_1 x \left(1 - \frac{x}{K_1}\right) - \frac{b_{12}}{K_1} xy$$

$$\dot{y} = a_2 y \left(1 - \frac{y}{K_2}\right) - \frac{b_{21}}{K_2} xy,$$

(7.31)

where the constants a_1, a_2, b_{12}, and b_{21} have specified meaning.

7.1.2.5 Modeling Epidemic Phenomena

Let us now consider the case of an influenza epidemic. One infected individual is introduced into the population. The infection spreads by contact between an infected and a healthy person. The healthy person may or may not be infected by this contact. After some time (about two weeks in the case of flu), the infected person recovers and remains immune to the particular virus; so the person will not be infected again in our model. The following state diagram (Figure 7.5) illustrates these interactions.

The differential equations governing the dynamics of these populations are Equation 7.32

$$\dot{S} = -\beta IS + \gamma R$$

$$\dot{I} = \beta IS - \alpha I$$

(7.32)

$$\dot{R} = \alpha I - \gamma R.$$

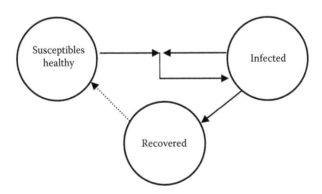

FIGURE 7.5
Interactions in the flu epidemic model.

If the immunity period is infinite, then $\gamma = 0$, and this model results in the Kermak–McKendrick model [7] given by Equation 7.33.

$$\dot{S} = -\beta IS$$
$$\dot{I} = \beta IS - \alpha I. \tag{7.33}$$

After a while, the population divides into three groups: the susceptible, the infected, and the recovered. The recovered can eventually become susceptible if their immunity ends (they are only immune for a certain time).

7.1.2.6 Physiological Systems

Physiological systems include the organs of animals, in general, and human ones, in particular. They are the most complex systems that nature has ever created. They are largely composed of various kinds of systems (mechanical, fluidic, chemical, etc.), which, by including living organs, acquire new properties through synergy.

Example 7.1   A Blood Vessel

Figure 7.6 represents a blood vessel.
Assume that the vessel is working in a steady state where all quantities are constant. In these conditions, it has to be that $Q_1 = Q_2$, as there is no blood accumulation because the volume is constant. Let $Q = Q_1 = Q_2$ be that blood flow.

To find a relation between Q, P_{11}, P_{22}, and V, two properties are considered: its resistance to the flow and its compliance with the pressure (which increases the volume of the vessel). To simplify the analysis, consider the following assumptions:

1. A vessel without compliance, rigid, of constant and well-known volume. For that we will have, as in a fluidic system, the flow equal to

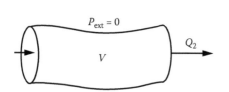

$V \triangleq$ Volume of the vessel
$Q_1 \triangleq$ Input flow
$P_1 \triangleq$ Pressure at the input
$Q_2 \triangleq$ Onput flow
$P_2 \triangleq$ Pressure at the output
P_{ext} (Outside pressure, the reference) $\triangleq 0$

FIGURE 7.6
A blood vessel. (Adapted from F.C. Hoppensteadt, C. S. Peskin. *Modeling and Simulation in Medicine and the Life Sciences*, 2nd ed., Springer-Verlag, 2004.)

the pressure difference at the two ends of the vessel divided by the fluidic resistance of the vessel, as in Equation 7.34.

$$Q = \frac{P_1 - P_2}{R}, \qquad (7.34)$$

where R is the vessel resistance; we have a resistive vessel here.

2. An elastic vessel without any fluidic resistance. In these conditions, the pressure at the input is equal to the pressure at the output, for any value of Q. The relation between the pressure P and the volume of the vessel can be approximated by Equation 7.35,

$$V = CP \Rightarrow C = \frac{V}{P}, \qquad (7.35)$$

where the constant C is the compliance of the vessel. We can also write Equation 7.36,

$$V = V_d + CP, \qquad (7.36)$$

where V_d is the volume of the vessel for zero pressure (dead volume). C is again the compliance of the vessel.

We can check the analogy of the following two situations:

Vessel: Due to the compliance C, the vessel enlarges with the pressure P until it reaches a fixed volume V, which depends on the compliance. Then we will have $V = CP$.

Capacitor (electric): Due to the capacity C, the capacitor charges when applied to a fixed voltage V, until its charge reaches a fixed value, which depends on the capacity. Then we will have $Q = CV$.

These two theoretical situations are reasonably close to reality. A real vessel has simultaneously resistance and compliance. The relation among the variables is not linear.

In the human body, the circulation separates the vessels of resistance from the vessels of compliance. The big arteries and veins are mainly compliance vessels, because small pressure differences are sufficient to produce the blood flow and their volumes change considerably. The small arterials (arterioles) and the small veins (venules) are mainly resistive vessels: their volumes change very little, and big pressure differences are generated there.

Considering both properties, we can conceive an electrical circuit equivalent to a vessel, with the analogies of Table 7.3, as observed in Figure 7.7. From the circuit laws, we obtain Equation 7.37.

$$\frac{dP_2}{dt} = -\frac{P_2}{RC} + \frac{P_1}{RC'}, \quad C\frac{dv_2}{dt} = \frac{v_1 - v_2}{R} \Rightarrow \frac{dv_2}{dt} = -\frac{v_2}{RC} + \frac{v_1}{RC'}, \qquad (7.37)$$

TABLE 7.3

Analogies among the Physiological and
Electric Systems and Variables

Vessel	Electrical Circuit
Pressure, P	V, Voltage
Volume, V	Q, Charge
Flow, Q	I, Current
Resistance, R	R, Resistance
Compliance, C	C, Capacity C

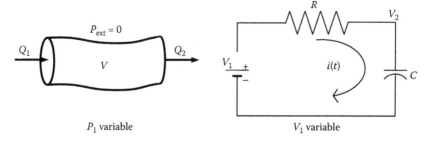

P_1 variable V_1 variable

FIGURE 7.7
Analogy between a vessel and an electrical circuit.

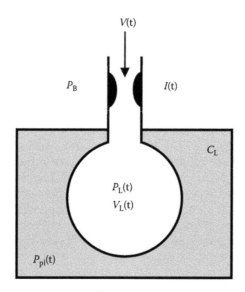

FIGURE 7.8
The biomechanics of inhalation. (Adapted from E.N. Bruce. *Biomedical Signal Processing and Signal Modeling*, John Wiley & Sons, Inc., 2001.)

Example 7.2   Biomechanical Model of Pulmonary Inhalation/Exhalation

Inhalation and exhalation can be described by a biomechanical system illustrated in Figure 7.8.

In a relaxed state, without any air flow, the pressure inside the lungs is equal to the atmospheric pressure, P_B. During inhalation, the respiratory muscles—the diaphragm and the thoracic muscles—contract and lower pleural pressure $P_{pl}(t)$ around the lungs inside the chest.

Due to their elasticity (described by the compliance C, the inverse of plasticity), the lungs expand as the transmural pressure decreases; and as a consequence, the pressure inside the lungs $P_L(t)$ momentarily decreases. The pressure in the fluidic resistance R_f of the respiratory tract falls, resulting in a flow of air to the inside of the lungs. As long as the muscles contract increasingly strongly, this process is repeated and inhalation continues. At the end of inhalation, the abdomen and chest muscles contract and force $P_{pl}(t)$ to exceed P_B (atmospheric pressure), thus expelling the air from the lungs. In resting respiration, the exhalation muscles do not need to contract, because the relaxation of the inhalation muscles alone is enough to make $P_{pl}(t)$ greater than P_B.

This biomechanical process can be mathematically described based on the physical properties of the physiological structures involved. It is a mixed fluidic-mechanical system.

The air flow depends on the pressure difference though the fluidic resistance R_f and on the resistance itself. If $I(t)$ represents the air flow rate of inhalation or exhalation, we will have Equation 7.38.

$$I(t) = \frac{P_B - P_L(t)}{R_f}, \quad \text{or,} \quad P_B - P_L(t) = I(t)R_f. \tag{7.38}$$

If $V(t)$ is the volume of the lungs during the time interval Δt, the volume variation will be Equation 7.39, assuming $I(t)$ to be constant during Δt, then Equation 7.39,

$$V_L(t + \Delta t) - V_L(t) = I(t) \quad \Delta t \Rightarrow \frac{V_L(t + \Delta t) - V_L(t)}{\Delta t} = I(t) \Rightarrow I(t) = \dot{V}_L(t), \tag{7.39}$$

Substituting Equation 7.39 in Equation 7.38, we get Equation 7.40.

$$P_B - P_L(t) = \dot{V}_L(t)R_f. \tag{7.40}$$

The compliance of the lungs C_L is defined by the relation between the volume of the lungs and the transmural pressure. Mathematically, it is Equation 7.41.

$$C_L = \frac{V_L(t)}{P_L - P_{pl}(t)}. \tag{7.41}$$

from which we get Equation 7.42.

$$P_L - P_{pl}(t) = \frac{V_L(t)}{C_L} \Rightarrow P_L = P_{pl}(t) + \frac{V_L(t)}{C_L}. \tag{7.42}$$

Now substituting P_L in Equation 7.42, we obtain Equation 7.43.

$$P_B - P_L(t) = \dot{V}(t)R_f \Rightarrow P_B - P_{pl}(t) + \frac{V_L(t)}{C_L} = \dot{V}(t)R_f$$

$$\Rightarrow P_B - P_{pl}(t) = \dot{V}(t)R_f + \frac{V_L(t)}{C_L}. \tag{7.43}$$

Since we consider the atmospheric pressure constant, we define the differential pressure (Equation 7.44),

$$P(t) = P_B - P_{pl}(t), \tag{7.44}$$

and

$$P(t) = \dot{V}_L(t)R_f + \frac{V_L(t)}{C_L}$$

$$\dot{V}_L(t) + \frac{1}{R_f C_L}V_L(t) = -\frac{1}{R_f}P(t) \tag{7.45}$$

or

$$\dot{V}_L(t) + aV_L(t) = bP(t)$$

Finally, Equation 7.45 is a first-order linear differential Equation relating the lungs' volume, V_L, with the relative intramural pressure, $P(t)$. From the systemic point of view, it can be represented by the block diagram in Figure 7.9.

Note, further, that Equation 7.46

$$\dot{V}(t) = I(t) \Rightarrow V(t) = V(0) + \int_0^t I(t)\, dt, \tag{7.46}$$

where $V_L(0)$ is the resting volume. Since it is a constant characteristic of each individual, it can be considered null; and $V_L(t)$, thus, represents the volume of variation in the lungs relative to $V_L(0)$.

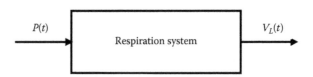

FIGURE 7.9
A systemic representation of the respiratory system.

7.1.2.7 Compartmental Analysis

Compartmental analysis in physiology makes it possible to monitor the distribution of biological substances inside the human body [11]. To define a correct dosage of medication, we need to know the rate of drug absorption by the different tissues and organs. Other natural substances (hormones, metabolic substrates, lipoproteins, peptides, etc.) have very complex distribution patterns. Radioactive tracers are used to discover these patterns. A tracer is mixed with the intended substance and the mixture is introduced into the body. Its concentration in the blood can be sampled at successive instants of time through radioactive analysis. This radioactivity can also be measured through radiation imaging. The physiologist can then compute the rate of absorption and release of a substance by a tissue, or the rate of degradation or elimination from the bloodstream (by the kidneys, for example) or from the tissues (by biochemical degradation).

In compartmental analysis, the body is divided into a set of homogeneous (well mixed) interconnected compartments that exchange substances between themselves and degrade those substances according to simple linear kinetics. Let us consider, for example, the case of two compartments, the first representing the circulatory system and the other representing all the tissues relevant to the analysis (not necessarily an organ or a single physiological entity). Figure 7.10 illustrates the situation.

The variables in Figure 7.10 have the following meanings:

- u_1, u_2 rate of injection of substance into compartments 1 and 2.
- K_{12} rate of flow of substance from compartment 1 to compartment 2.
- K_{21} rate of flow of substance from compartment 2 to compartment 1.
- K_{10} rate of degradation of substance in compartment 1.
- K_{20} rate of degradation of substance in compartment 2.

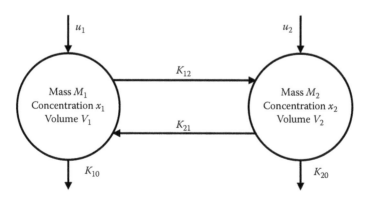

FIGURE 7.10
Two compartments exchanging substances between themselves and the environment.

- x_1 density of substance in compartment 1.
- x_2 density of substance in compartment 2.

Applying the principle of mass conservation, we will have

1. For compartment 1, Equation 7.47:

$$\frac{dM_1}{dt} = -K_{12}M_1 + K_{21}M_2 - K_{10}M_1 + u_1; \tag{7.47}$$

2. For compartment 2, Equation 7.48:

$$\frac{dM_2}{dt} = +K_{12}M_1 - K_{21}M_2 - K_{20}M_2 + u_2. \tag{7.48}$$

These equations can be written with regard to the concentrations x_1 and x_2. Since $x_1 = M_1/V_1$ and $x_2 = M_2/V_2$, we have Equations 7.49 and 7.50,

$$\frac{dx_1}{dt} = -k_1x_1 + k_{21}x_2 + w_1, \tag{7.49}$$

$$\frac{dx_2}{dt} = +k_{12}x_1 - k_2x_2 + w_2, \tag{7.50}$$

where $k_1 = K_{12} + K_{10}$, $k_2 = K_{21} + K_{20}$, $k_{21} = K_{21} V_2/V_1$, and $k_{12} = K_{12} V_1 V_2$.

7.1.2.8 State, State Space, and State Equations

Some elementary attributes of a system are very important and of themselves they provide a sufficiently precise image of the system behavior at a given instant. The temperature in a greenhouse, the speed of a car, the population of a species, the level of a reservoir, the concentration of a substance, etc., are examples of this. These attributes are called state variables, and they synthesize the past history of the system. For the mathematical representation, we define a vector x, called the state vector, whose n components $x_i, i = 1, \ldots, n$, are the state variables (Equation 7.51)

$$x = [x_1 \quad x_2 \quad \cdots \quad x_n]^\mathrm{T}. \tag{7.51}$$

As we will see, the state is a synthesis of the past history of the system: the present state results from all its past evolution. In a dynamic system, the state will change over time. If graphically represented, it traces a trajectory in the n-dimensional space, called the state space. If we can rigorously foresee the future behavior of the system when its exact initial state is known, we have a deterministic system; otherwise, the system is nondeterministic, probabilistic, or stochastic.

The representation in the state space is developed in the time domain. The state equations are derived from the differential equations describing the system; the state equations themselves are first-order differential equations. In the pulmonary inhalation or exhalation model, where the state variable x_1 is the value of the volume V_L, the system input u is $P(t)$, and the system output y (which we want to observe) is the pulmonary volume, we obtain Equation 7.52,

$$\dot{V}_L(t) + aV_L(t) = bP(t)$$

$$x_1 \triangleq V_L \quad u \triangleq P(t) \quad y \triangleq V_L$$

$$\dot{x}_1 = -ax_1 + bu$$

$$y = x_1$$

(7.52)

as the state equation, which is a first-order linear differential Equation with constant coefficients (and so time invariant), complete because it has an external input (also called exogenous input) $u(t)$ and an algebraic equation as the output equation. The state variable must express the memory of the system, and it is generally associated with the quantity of mass or energy stored in the system. In this example, this variable is the pulmonary volume (associated with the amount of air).

Example 7.3 Glucose metabolism

Glucose metabolism in blood can be approximated by the simplified Ackerman's model described below [12]:

$$\frac{dg}{dt} = -m_1 g - m_2 h + J(t)$$

$$\frac{dh}{dt} = -m_3 h + m_4 g + K(t),$$

(7.53)

where g is the deviation of the glucose level from the recommended one, h is the deviation of the insulin level from the recommended one, J represents the experimental rate of glucose infusion, and K is the experimental rate of insulin infusion. The parameters are m_1, m_2, m_3, and m_4 characteristic constants of each individual.

Choosing the state variables as $x_1 = g$, $x_2 = h$, which relate to the mass storage in the system, and defining the system inputs as $u_1 = J$, $u_2 = K$, we obtain Equations 7.54 and 7.55.

$$\frac{dx_1}{dt} = -m_1 x_1 - m_2 x_2 + u_1(t),$$

(7.54)

$$\frac{dx_2}{dt} = -m_3 x_1 + m_4 x_2 + u_2(t).$$

(7.55)

Rewriting Equations 7.54 and 7.55 in matrix form, we have Equation 7.56.

$$\begin{bmatrix} \dot{x}_1 \\ \dot{x}_2 \end{bmatrix} = \begin{bmatrix} -m_1 & -m_2 \\ -m_3 & +m_4 \end{bmatrix} \begin{bmatrix} x_1 \\ x_2 \end{bmatrix} + \begin{bmatrix} 1 & 0 \\ 0 & 1 \end{bmatrix} \begin{bmatrix} u_1 \\ u_2 \end{bmatrix}. \tag{7.56}$$

In the case of the two-compartmental models, we had obtained Equation 7.57.

$$\frac{dx_1}{dt} = -k_1 x_1 + k_{21} x_2 + w_1$$
$$\frac{dx_2}{dt} = +k_{12} x_1 - k_2 x_2 + w_2. \tag{7.57}$$

In matrix form, we have Equation 7.58,

$$\begin{bmatrix} \dot{x}_1 \\ \dot{x}_2 \end{bmatrix} = \begin{bmatrix} -k_1 & k_{21} \\ +k_{12} & -k_2 \end{bmatrix} \begin{bmatrix} x_1 \\ x_2 \end{bmatrix} + \begin{bmatrix} 1 & 0 \\ 0 & 1 \end{bmatrix} \begin{bmatrix} u_1 \\ u_2 \end{bmatrix}, \tag{7.58}$$

or, even more simply, Equation 7.59.

$$\dot{x} = Ax + B. \tag{7.59}$$

Example 7.4   Predator–Prey Model

In the Lotka–Volterra model analyzed earlier, we have Equation 7.60,

$$\frac{dx(t)}{dt} = ax(t) - bx(t)y(t)$$
$$\frac{dy(t)}{dt} = -cy(t) + px(t)y(t). \tag{7.60}$$

Choosing x_1 and x_2 as the state variables, the two populations (accumulation of mass), and substituting and simplifying the notation, we obtain Equation 7.61,

$$x_1 \triangleq x \text{ (prey)} \quad x_2 \triangleq y \text{ (predators)}$$
$$\frac{dx_1(t)}{dt} = ax_1(t) - bx_1(t)x_2(t) \quad \dot{x}_1 = ax_1 - bx_1 x_2 \tag{7.61}$$
$$\frac{dx_2(t)}{dt} = -cx_2(t) + px_1(t)x_2(t) \quad \dot{x}_2 = -cx_1 + px_1 x_2.$$

which are two state equations, time invariant (because their coefficients are constant over time). In this case, there is no external input; and as a consequence, the equations are homogeneous. They are also nonlinear, because they have products of state variables on the right-hand side.

It is common practice to designate the two populations as the output of the system, as that is what we want to observe. So we may have

$$\text{output 1}: \ y_1 = x_1$$
$$\text{output 2}: \ y_2 = x_2. \tag{7.62}$$

A similar development can be made for all the systems described by differential equations. If we have a differential equation of an order n greater than one, we reduce it to a set of n first-order differential equations. Then we can state that the generalized structure of a model for a continuous system, regardless of its nature, is composed of a set of n first-order differential equations, with m external inputs. Using a generic notation, we then have Equation 7.63,

$$\frac{dx_i}{dt} = f_i\,(x_1(t), \ldots, x_n(t), u_1(t), \ldots, u_m(t), \ t) \quad \text{with} \quad i = 1, \ldots, n, \tag{7.63}$$

where the f_i are continuous functions of their arguments. The initial conditions needed to define the initial state of the system (its memory) are in Equation 7.64.

$$x_i(t_0) = x_{i0}, \quad i = 1, \ldots, n. \tag{7.64}$$

We will have also a set of r output equations r outputs (Equation 7.65)

$$y_i(t) = g_i\,(x_1(t), \ldots, x_n(t), u_1(t), \ldots, u_m(t), t) \quad \text{with} \quad i = 1, \ldots, r. \tag{7.65}$$

where the measured output variables are expressed as functions of the n state variables and of the m input variables. There is no mandatory general relation between the dimensions of the state, input, and output vectors; but in most cases, $n \gg \max(m, r)$.

If f_i or g_i explicitly depends on t, for some i, as in Equation 7.66

$$\dot{x}_1 = -2x_1e^{-t} + 3u \tag{7.66}$$

then the system is time varying. If this is not the case, then the system is time invariant.

If there is no external input $u(t)$, the functions f_i have the single argument x_i, and the system is said to be autonomous [14] (depending only on itself).

In the functions g_i, the input $u(t)$ enters as an argument only exceptionally. The input influences the output through the state variables and not directly.

The outputs are, therefore, usually functions only of the state. The outputs are those system variables that we want to observe or compute.

In matrix notation, we can write Equation 7.67.

$$
\begin{aligned}
x &= [x_1, \ldots, x_n]^{\mathrm{T}} \quad \text{(state vector)} \\
y &= [y_1, \ldots, y_n]^{\mathrm{T}} \quad \text{(output vector)} \\
u &= [u_1, \ldots, u_m]^{\mathrm{T}} \quad \text{(input vector)} \\
x &= [x_{10}, \ldots, x_{n0}]^{\mathrm{T}} \quad \text{(initial state vector)},
\end{aligned}
\tag{7.67}
$$

and the vector of functions in Equation 7.68.

$$
f(x, u) = \begin{bmatrix} f_1 \\ \cdot \\ \cdot \\ \cdot \\ f_n \end{bmatrix} \qquad g(x, u) = \begin{bmatrix} g_1 \\ \cdot \\ \cdot \\ \cdot \\ g_r \end{bmatrix}.
\tag{7.68}
$$

The model in the state space is then defined by a state equation with initial conditions and an output Equation 7.69.

$$
\begin{aligned}
\dot{x} &= f(x, u), \quad x(t_0) = x_0 \\
y &= g(x, u).
\end{aligned}
\tag{7.69}
$$

If the functions f_i and g_i are linear and invariant, we have Equation 7.70,

$$
\begin{aligned}
f_i(x, u) &= a_{i1}x_1 + a_{i2}x_2 + \cdots + a_{in}x_n + b_{i1}u_1 + \cdots + b_{im}u_m, \quad i = 1, \ldots, n \\
g_i(x, u) &= c_{i1}x_1 + c_{i2}x_2 + \cdots + c_{in}x_n + d_{i1}u_1 + \cdots + d_{im}u_m, \quad i = 1, \ldots, r,
\end{aligned}
\tag{7.70}
$$

where a_{ij}, b_{ij}, c_{ij}, and d_{ij} are constant coefficients.

The state equation takes the matrix form Equation 7.71,

$$
\begin{aligned}
\dot{x} &= Ax + Bu \quad x(t_0) = x_0 \\
y &= Cx + Du,
\end{aligned}
\tag{7.71}
$$

which can be represented by the block diagram of Figure 7.11.

The matrixes in Equation 7.71 are

- $A = n \times n$ state matrix;
- $B = n \times m$ input matrix, or command matrix, or control matrix;
- $C = r \times n$ output matrix or observation matrix;

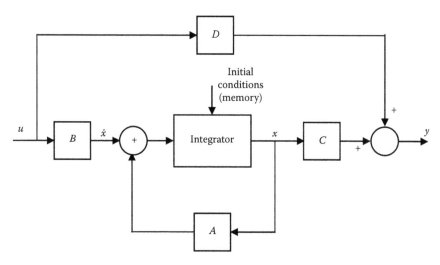

FIGURE 7.11
Block diagram of the state space representation. D is usually null.

- $D = r \times m$ feed forward matrix;
- $u =$ input vector, or command vector, or control vector, dimension m; and
- $y =$ output vector, dimension r.

In the SISO (Single Input Single Output) case, B is a column vector b, and C is a row vector c^T; and Equation 7.71 becomes Equation 7.72.

$$\dot{x} = Ax + bu \quad x(t_0) = x_0$$
$$y = c^T x + du,$$

(7.72)

However, Equation 7.71 can also be used for the SISO case.

If the matrixes A, B, C, or D contain elements that vary over time, the system is time varying (some coefficients a_{ij}, b_{ij}, c_{ij}, or d_{ij} vary over time); and we can then write $A(t)$, $B(t)$, $C(t)$, and e $D(t)$, where t is an explicit argument.

The matrix D expresses the instantaneous influence of the input in the outputs, before the influence of the state is noticed. That is why D is called feed forward matrix: its influence bypasses the differential systems and goes directly to the output. In normal physical systems, containing some source of inertia, this instantaneous influence does not exist, and D is null. Think about a car, for instance: changing the position of the accelerator (the input signal) does not cause an instant change in speed; the reaction time is always nonnull, unless the power of the engine is infinite ... There is no feed forward action when there is some inertia.

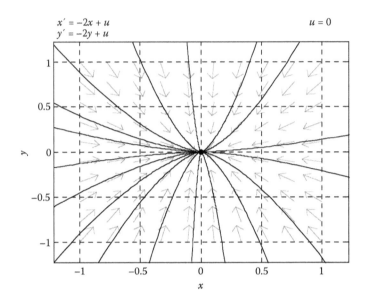

$$x' = -2x + u$$
$$y' = -2y + u$$

$$u = 0$$

FIGURE 7.12
Stable sink (eigenvalues: −2; −3).

The state space is the real space X^n with dimension n, a subspace of R^n. The state is a point in this space, and its coordinates are the state variables values.

When the state moves in the state space, it describes a trajectory, which is the state trajectory. Some authors call the bi-dimensional state space *phase space* and the trajectory in the phase space *phase trajectory* (we will come back to this later in this chapter).

The state equation is a set of first-order differential equations, where n is the number of state variables (the number of attributes needed to describe the memory of the system).

7.1.2.9 How to Choose the State Variables

The first step in the application of this theory to a concrete system is to choose the state variables. It should be stressed that there is neither a general rule nor a single way to do it. In practice, there are often several possible choices. When the state variables represent measured observable attributes, they are called physical variables. The choice of physical variables is generally based, as in previous examples, on the attributes of the elements storing energy or mass that synthesize the history of the system. In fact, since all physical phenomena are transformations of mass or energy, the present state of mass and energy stores in a given process is the result of everything that has happened to the system since its creation.

7.1.2.10   Solution of the Linear State Equation in the Time Domain

7.1.2.10.1   Solution of the Homogeneous State Equation

Consider the state equation of a linear invariant system Equation 7.73,

$$\dot{x} = Ax + Bu. \tag{7.73}$$

The homogeneous equation is obtained by making $u = 0$, resulting in an unforced (autonomous) system Equation 7.74.

$$\dot{x}(t) = Ax(t) \tag{7.74}$$

Integrating both sides of the equation, we get Equation 7.75

$$x(t) = \int_0^t Ax(t)dt + x_0. \tag{7.75}$$

However, the integral in Equation 7.75 is recurrent in $x(t)$, and we can write Equation 7.76:

$$
\begin{aligned}
x(t) &= \int_0^t Ax(t)dt + x_0 = \int_0^t A\left(\int_0^t Ax(t)dt + x_0\right)dt + x_0 \\
&= \int_0^t A\left(\int_0^t A\left(\int_0^t Ax(t)dt + x_0\right)dt + x_0\right)dt + x_0 = \cdots \\
&= x_0 + \int_0^t Ax_0\,dt + \int_0^t\int_0^t AAx_0\,dt + \int_0^t\int_0^t\int_0^t AAAx_0\,dt + \cdots \\
&= x_0 + Ax_0\int_0^t dt + A^2x_0\int_0^t\int_0^t dt + A^3x_0\int_0^t\int_0^t\int_0^t dt + \cdots \\
&= x_0 + Ax_0 t + A^2x_0\frac{t^2}{2!} + A^3x_0\frac{t^3}{3!} + \cdots + A^kx_0\frac{t^k}{k!} + \cdots \\
&= \left[1 + At + A^2\frac{t^2}{2!} + A^3\frac{t^3}{3!} + \cdots + A^k\frac{t^k}{k!} + \cdots\right]x_0.
\end{aligned}
\tag{7.76}
$$

Comparing Equations 7.76 and 7.77

$$e^{at} = 1 + \frac{at}{1!} + \frac{(at)^2}{2!} + \cdots + \frac{(at)^k}{k!} + \cdots . \tag{7.77}$$

we can write Equation 7.78:

$$x(t) = e^{At}x(t),$$ (7.78)

and, due to that, e^{At} is called the matrix exponential.

Note that the exponential of the matrix A is not made up of the exponentials of the elements of A.

The matrix exponential is called the state transition matrix, $\Phi(t)$, and we can write Equation 7.79.

$$x(t) = \Phi(t)x(0).$$ (7.79)

Knowing the initial condition, to compute the state in any future instant, it is sufficient to multiply the initial state by the state transition matrix. This gives it its name.

How can we compute e^{At}?

Computationally, we can calculate term by term, as in Equation 7.80,

$$e^{At} = I + At + A^2\frac{t^2}{2!} + A^3\frac{t^3}{3!} + \cdots + A^k\frac{t^k}{k!} + \cdots,$$ (7.80)

until the value of the next one is practically null.

However, it would be more useful to look for another solution that is easier to compute; and that would give us, moreover, some explicit indications about the properties of the system.

This solution exists and is based on the eigenstructure of the state matrix:

- The eigenvalues;
- The left eigenvectors; and
- The right eigenvectors.

Let us consider again the homogeneous equation 7.81

$$\dot{x}(t) = Ax(t).$$ (7.81)

One possible solution for this system is where $\dot{x}$ and x have the same direction in state space, differing only in their amplitudes by a proportionality factor. The solution will be of the form (Equation 7.82),

$$\dot{x} = \lambda x.$$ (7.82)

Substituting Equation 7.82 into Equation 7.81 leads to Equation 7.83.

$$\lambda x = Ax \Leftrightarrow [\lambda I - A]x = 0.$$ (7.83)

If some value $x \neq 0$ satisfies this equation, it is a nontrivial solution, and we will necessarily have Equation 7.84.

$$Q(\lambda) = \det[\lambda I - A] = 0 \Leftrightarrow |\lambda I - A| = 0. \tag{7.84}$$

If A has order n, $Q(\lambda)$ is the characteristic polynomial of A, and the resulting polynomial equation is the characteristic equation of A, with the form (Equation 7.85).

$$Q(\lambda) = \lambda^n + a_{n-1}\lambda^{n-1} + \cdots + a_1\lambda + a_0 = 0 \tag{7.85}$$

The roots of the characteristic equation are the eigenvalues of the matrix A. A root can be simple or repeated with a multiplicity p. A root can be real or complex. If roots are complex, they must appear in conjugate pairs, as the coefficients of the characteristic equation are real numbers.

When the roots are distinct, the polynomial $Q(\lambda)$ can be written as Equation 7.86.

$$Q(\lambda) = (\lambda - \lambda_1)(\lambda - \lambda_2) \cdots (\lambda - \lambda_n) \tag{7.86}$$

The product of the eigenvalues of the matrix A is equal to its determinant. Their sum equals the trace of the matrix (sum of the elements of the principal diagonal).

The nonnull vectors x that verify Equation 7.87

$$Ax = \lambda x, \tag{7.87}$$

are the right eigenvectors of A.

For each distinct eigenvalue there is a corresponding eigenvector, and the n distinct eigenvectors are linearly independent. If there is an eigenvalue of multiplicity p, the number of associated eigenvectors may vary from 1 to p, depending on the properties of the matrix.

Note that if x is an eigenvector, λx, for any $\lambda \in IR$, is too. It can be proved by the definition.

The left (or reciprocal) eigenvectors of the matrix A are the vectors w_i, (associated with the eigenvalues λ_i, $i = 1, \ldots, n$) verifying the Equation 7.88

$$w_i^T A = \lambda_i w_i^T \quad \text{ou} \quad w_i^T[\lambda_i I - A] = 0. \tag{7.88}$$

If w_i is an eigenvector, so is μw_i, for any $\mu \in R$.

It can be proved that the solution of the homogeneous state equation using the eigenstructure is given by Equation 7.89, if all the eigenvalues are simple and distinct.

$$x(t) = \left[\sum_{j=1}^{n} e^{\lambda_j t} . v_j w_j^T \right] x(0) = e^{\lambda_1 t} v_1 w_1^T x(0) + e^{\lambda_c t} v_2 w_2^T x(0) + \cdots$$

$$+ e^{\lambda_n t} v_n w_n^T x(0). \tag{7.89}$$

This result is full of information. It tells us that the state trajectory for a given initial condition is a weighted sum of exponentials of the eigenvalues of the A matrix (the terms $e^{\lambda t}$). The weightings depend on the eigenvalues. It can be said in this regard that the eigenstructure of the A matrix is the genome of the autonomous system.

What happens when one of the eigenvalues of the A matrix is positive? Its exponential tends to infinity with t, and the same happens for the term with which it contributes to the state evolution in Equation 7.89. As a consequence, if the initial condition is nonzero, even if very small, the state tends toward the infinite, even if all the other terms in Equation 7.89 tend toward zero. The system is unstable!

Here we come across the problem of the stability of the system with regard to the initial conditions. The eigenvalues of A are also called the modes of the system. The solution (Equation 7.89) of the state equation is called the modal solution, because it is expressed by the modes.

7.1.2.10.2 Solution of the Complete State Equation

The complete state Equation 7.90 has the input part

$$\dot{x} = Ax + Bu, \tag{7.90}$$

and it is not so easy to find its solution.

Consider the property of derivation of time-varying matrices (Equation 7.91).

$$\frac{d[M(t)N(t)]}{dt} = \dot{M}(t)N(t) + M(t)\dot{N}(t), \tag{7.91}$$

in which the derivative of a matrix is the matrix whose elements are the derivatives of the elements of the initial matrix, that is, $\dot{M} = \left[\dot{m}_{ij}\right]$

$$\frac{d}{dt}\left[e^{-At}x(t)\right] = e^{-At}\dot{x}(t) - Ae^{-At}x(t) = e^{-At}[\dot{x}(t) - Ax(t)]. \tag{7.92}$$

However, since,

$$\dot{x}(t) = Ax(t) + Bu(t). \tag{7.93}$$

then substituting Equation 7.93 into Equation 7.92, we have Equation 7.94.

$$\frac{d}{dt}\left[e^{-At}x(t)\right] = e^{-At}Bu(t). \tag{7.94}$$

Integrating both sides of Equation 7.94, we have Equation 7.95.

$$e^{-At}x(t) = \int_0^t e^{-A\tau}Bu(\tau)d\tau + K. \tag{7.95}$$

where K is a constant (matrix) of integration. Making $t = 0$ in Equation 7.95, we get Equation 7.96.

$$e^{-A0}x(0) = \int_0^0 e^{-A\tau}Bu(\tau)d\tau + K \Leftrightarrow x(0) = K. \qquad (7.96)$$

Multiplying both sides of Equation 7.96 by e^{At}, grouping the terms, we get Equation 7.97.

$$e^{At}e^{-At}x(t) = e^{At}\int_0^t e^{-A\tau}Bu(\tau)\,d\tau + e^{At}x(0). \qquad (7.97)$$

Multiplying the two matrix exponentials, adding the exponents $At - At = 0$, knowing that the exponential of the null matrix is the identity matrix, we obtain Equation 7.98.

$$Ix(t) = e^{At}x(0) + \int_0^t e^{A(t-\tau)}Bu(\tau)\,d\tau. \qquad (7.98)$$

As we can see in Equation 7.98, $x(t)$ is the sum of two contributions

$$x(t) = x_{zi}(t) + x_{zs}(t) \qquad (7.99)$$

one coming from the initial conditions (*zi*—zero input) and the other coming from the external input (*zs*—zero state).

7.1.2.10.3 Solution of the Two-Compartment State Equation

We have seen that the Equations of two compartments are Equations 7.100 and 7.101.

$$\frac{dx_1}{dt} = -k_1x_1 + k_{21}x_2 + w_1 \qquad (7.100)$$

$$\frac{dx_2}{dt} = +k_{12}x_1 - k_2x_2 + w_2 \qquad (7.101)$$

If we assume that a substance of mass M_0 is instantaneously introduced into the blood flow (bolus injection) at $t = 0$ and that it mixes quickly, we have $w_1 = 0$ e $w_2 = 0$. The homogeneous state equation will be Equation 7.102,

$$\begin{bmatrix} \dot{x}_1 \\ \dot{x}_2 \end{bmatrix} = \begin{bmatrix} -k_1 & k_{21} \\ k_{12} & -k_2 \end{bmatrix}\begin{bmatrix} x_1 \\ x_2 \end{bmatrix}. \qquad (7.102)$$

whose solution is Equation 7.103.

$$\mathbf{x}(t) = c_1\mathbf{v}_1 e^{-\lambda_1 t} + c_2\mathbf{v}_2 e^{-\lambda_2 t}$$

$$\mathbf{x}(t) = \begin{bmatrix} x_1(t) \\ x_2(t) \end{bmatrix} \quad e \quad c_i\mathbf{v}_i = \begin{bmatrix} a_i \\ b_i \end{bmatrix},$$

(7.103)

λ_1 and λ_2 are the eigenvalues of the A matrix, and the coefficients a_i and b_i depend on the right and left eigenvectors. The eigenvalues are Equation 7.104,

$$\lambda_{1,2} = -\frac{k_1 + k_2}{2} \pm \frac{\sqrt{(k_1 - k_2)^2 + 4k_{12}k_{21}}}{2}.$$

(7.104)

7.1.2.11 Nonlinear Systems, Singular Points, and Linearization

Many natural systems, particularly biological and physiological ones, are described by nonlinear differential equations. This poses the question whether the nonlinear systems can be approximated to linear ones, if we want to apply the linear systems theory. We call this mathematical operation linearization, and it is done at the points where the derivatives vanish, called singular points or singularities.

7.1.2.11.1 Singular Points or Singularities

In many practical situations, the ideal operating conditions are characterized by the constancy of inputs, states, and outputs, that is, a steady state or an equilibrium situation. Suppose that the system is described by the nonlinear model in Equation 7.105.

$$\dot{x} = f(x, u).$$

(7.105)

Linearization is mathematically represented around equilibrium points or singular points, also called steady states if the system is stable. We can say that the vector x_s ($n \times 1$) is a singular point (or an equilibrium point) corresponding to the constant input $u(t) = u_s$, if and only if Equation 7.106 is true.

$$f(x_s, u_s) = 0.$$

(7.106)

The corresponding equilibrium output is the vector y_s ($r \times 1$) such that

$$y_s = g(x_s, u_s).$$

(7.107)

Example 7.5 The Lotka–Volterra model

To compute the singularities of the Lotka–Volterra model in Equation 7.14, we eliminate the derivatives as in Equation 7.108.

$$0 = ax_1(t) - bx_1(t)x_2(t)$$
$$0 = -cx_2(t) + px_1(t)x_2(t). \tag{7.108}$$

Solving for x_1 and x_2 we obtain 7.109

$$\begin{cases} [a - bx_2(t)]x_1(t) = 0 \wedge x_1(t) \neq 0 \\ [-c + px_1(t)]x_2(t) = 0 \wedge x_2(t) \neq 0 \end{cases} \Rightarrow \begin{cases} [a - bx_2(t)] = 0 \\ [-c + px_1(t)] = 0 \end{cases}. \tag{7.109}$$

The two solutions of Equation 7.109 are Equations 7.110 and 7.111.

$$\Rightarrow \text{solution 1}: \begin{cases} x_2(t) = \dfrac{a}{b} \\ x_1(t) = \dfrac{c}{p} \end{cases}; \tag{7.110}$$

$$\Rightarrow \text{solution 2}: \begin{cases} x_2(t) = 0 \\ x_1(t) = 0 \end{cases}. \tag{7.111}$$

We can conclude that the Lotka–Volterra model has two singularities. We will see the consequences of this fact later.

7.1.2.11.2 *Linearization Methods: Approximation to the Taylor Series*

The linearization (of a model) of nonlinear equations in the neighborhood of a singular point defined by the triplet x_s, u_s, y_s consists of approximating the vectors of functions f, for example, to linear functions, in the neighborhood of that singular point. Analytical linearization is based on the approximation of the functions f_i and g_i to a Taylor series (in the neighborhood of the singular point), where Δx_i and Δu_i are the deviations with regard to the singular point, as in Equation 7.112.

$$f_i(x_{1s} + \Delta x_1, x_{2s} + \Delta x_2, \ldots + x_{ns} + \Delta x_n, u_{1s} + \Delta u_{1s}, u_{2s}$$
$$+ \Delta u_2, \ldots, u_{ms} + \Delta u_m)$$

$$= f_i(x_{1s}, x_{2s}, \ldots + x_{ns}, u_{1s}, u_{2s}, \ldots, u_{ms}) + \sum_{k=1}^{n} \frac{\partial f_i}{\partial x_k} \Delta x_k + \sum_{k=1}^{m} \frac{\partial f_i}{\partial u_k} \Delta u_k$$

$$+ \text{termos de ordem superior.} \tag{7.112}$$

For the output equations, in the same way we have Equation 7.113.

$$g_i (x_{1s} + \Delta x_1, x_{2s} + \Delta x_2, \ldots + x_{ns} + \Delta x_n, u_{1s} + \Delta u_{1s}, u_{2s}$$

$$+ \Delta u_2, \ldots, u_{ms} + \Delta u_m)$$

$$= g_i (x_{1s}, x_{2s}, \ldots + x_{ns}, u_{1s}, u_{2s}, \ldots, u_{ms}) + \sum_{k=1}^{n} \frac{\partial g_i}{\partial x_k} \Delta x_k + \sum_{k=1}^{m} \frac{\partial g_i}{\partial u_k} \Delta u_k$$

+ termos de ordem superior. (7.113)

Neglecting the terms of a higher order (meaning that the linearization will be valid only in a small neighborhood) and considering Equations 7.114 through 7.117, the linearization is almost ready.

$$A = \begin{bmatrix} \dfrac{\partial f_1}{\partial x_1} & \cdots & \dfrac{\partial f_1}{\partial x_n} \\ \vdots & \cdots & \\ \dfrac{\partial f_n}{\partial x_1} & \cdots & \dfrac{\partial f_n}{\partial x_n} \end{bmatrix} = F_x^T; \tag{7.114}$$

$$B = \begin{bmatrix} \dfrac{\partial f_1}{\partial u_1} & \cdots & \dfrac{\partial f_1}{\partial u_m} \\ \vdots & \cdots & \\ \dfrac{\partial f_n}{\partial u_1} & \cdots & \dfrac{\partial f_n}{\partial u_m} \end{bmatrix} = F_u^T; \tag{7.115}$$

$$C = \begin{bmatrix} \dfrac{\partial g_1}{\partial x_1} & \cdots & \dfrac{\partial g_1}{\partial x_n} \\ \vdots & \cdots & \\ \dfrac{\partial g_p}{\partial x_1} & \cdots & \dfrac{\partial g_p}{\partial x_n} \end{bmatrix} = G_x^T; \tag{7.116}$$

$$D = \begin{bmatrix} \dfrac{\partial g_1}{\partial u_1} & \cdots & \dfrac{\partial g_1}{\partial u_m} \\ \vdots & \cdots & \\ \dfrac{\partial g_p}{\partial u_1} & \cdots & \dfrac{\partial g_p}{\partial u_m} \end{bmatrix} = G_u^T. \tag{7.117}$$

The derivatives in Equations 7.114 through 7.117 are computed in the singular point, substituting the values of x and u with x_s and u_s, respectively.

The matrices of the first derivatives of the functions in order to each of their arguments are called Jacobians.

Substituting in the truncated Taylor series (in the first-order term), we obtain a linear model. The singularity itself is a point of the system and so respects the state Equation 7.118.

$$\dot{x}_s = f(x_s, u_s). \tag{7.118}$$

If we consider a point in the neighborhood of the singularity $x_s + \Delta x$, we will have Equation 7.119, because the derivative of the sum is equal to the sum of the derivatives.

$$\left(\overline{\dot{x_s + \Delta x}} \right) = \dot{x}_s + \Delta \dot{x}. \tag{7.119}$$

However, now approximating the Taylor series (only first-order terms), we have Equation 7.120.

$$(\dot{x}_s + \Delta \dot{x}) = f(x_s + \Delta x, u_s + \Delta u). \tag{7.120}$$

Taking into account Equations 7.114 through 7.117, we obtain Equation 7.121.

$$(\dot{x}_s + \Delta \dot{x}) = f(x_s, u_s) + A \, \Delta x + B \, \Delta u. \tag{7.121}$$

or the state and output equation in standard form (Equation 7.122).

$$\Delta \dot{x} = A \, \Delta x + B \, \Delta u$$
$$\Delta y = C \Delta x + D \, \Delta u. \tag{7.122}$$

The validity of Equation 7.122 can only be assumed when the terms of second and higher orders of the Taylor series are negligible, that is, when the deviations from the singular points are very small. To simplify the notation, we can represent the linearized system eliminating the Δ_s in Equation 7.122 but not forgetting that the input, state, and output variables represent deviations from the singular point and not absolute values. The matrices $A, B, C,$ and D are the Jacobians of f and g (in order to x and to u) given by Equations 7.114 through 7.117.

7.1.2.11.3 Types of Singularities

A linear system, except for the special case where $\det(A) = 0$, has only one singularity. If the input is null, the singularity will be the origin of the state space. Let us consider this case, without loss of generality (in the linear case, the input displaces the singularity in the state space but does not change its nature). The stability of the systems is determined, as we saw, by the eigenvalues of the state matrix A.

There are several types of singularities, some stable and some unstable [15]. Let us look at an example of each type.

1. Stable node (sink, attractor): when the eigenvalues of A are negative and distinct, the state trajectory approaches the singularity without oscillations, as seen in Figure 7.12, for a second order system. The state equations in the example are Equation 7.123. These trajectories, called phase curves, were computed using pplane7 [16] in

test

Matlab®. The arrows indicate the direction of the evolution of the trajectory.

$$\dot{x} = \begin{bmatrix} -2 & 0 \\ 0 & -3 \end{bmatrix} x + \begin{bmatrix} 1 \\ 1 \end{bmatrix} u \tag{7.123}$$

$u = 0.$

2. Stable focus (focus attractor): when the eigenvalues of A are complex conjugates with negative real parts, the trajectory tends toward the focus with oscillations. Figure 7.13 shows the trajectories for the system of Equation 7.124.

$$\dot{x} = \begin{bmatrix} 0 & 1 \\ -3 & -1 \end{bmatrix} x + \begin{bmatrix} 0 \\ 1 \end{bmatrix} u \tag{7.124}$$

$u = 0.$

3. Unstable source (repelling source, unstable node): both the eigenvalues are real positive. The trajectory is repelled without oscillations.

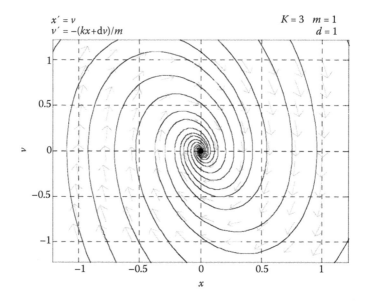

FIGURE 7.13
Stable focus (eigenvalues: $-0.5000 + 1.6583\,i$; $-0.5000 - 1.6583\,i$).

The system is unstable, given by Equation 7.125 and Figure 7.14.

$$\dot{x} = \begin{bmatrix} 2 & 0 \\ 0 & 3 \end{bmatrix} x + \begin{bmatrix} 1 \\ 1 \end{bmatrix} u$$

$$u = 0.$$

(7.125)

4. Unstable focus (repelling focus): when the eigenvalues of A are complex conjugates with a positive real part. The trajectory is repelled with oscillations, as shown in Figure 7.15, for the system (Equation 7.126).

$$\dot{x} = \begin{bmatrix} 0 & 1 \\ -3 & 1 \end{bmatrix} x + \begin{bmatrix} 0 \\ 1 \end{bmatrix} u$$

$$u = 0.$$

(7.126)

5. Center: when the eigenvalues are both imaginary. The trajectories are closed curves, periodic, around the center, as seen in Figure 7.16, for the system (Equation 7.127).

$$\dot{x} = \begin{bmatrix} 0 & 1 \\ -3 & 0 \end{bmatrix} x + \begin{bmatrix} 0 \\ 1 \end{bmatrix} u$$

$$u = 0.$$

(7.127)

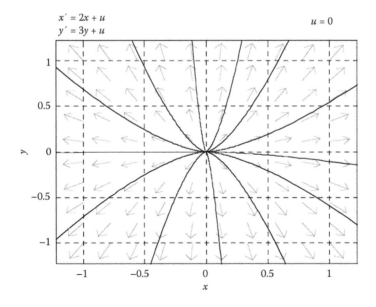

FIGURE 7.14
Unstable node (eigenvalues: 2;3).

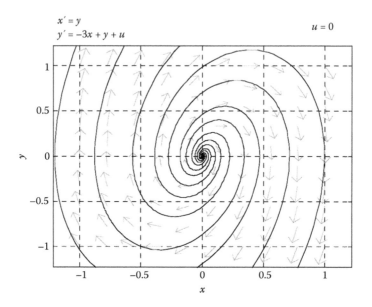

FIGURE 7.15
Unstable focus (eigenvalues: 0.5000 + 1.6583 i; 0.5000 − 1.6583 i).

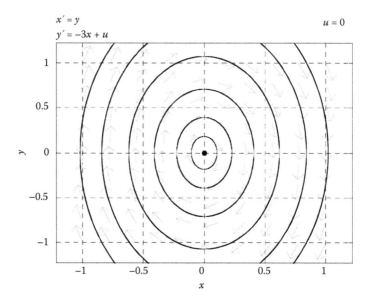

FIGURE 7.16
Center (eigenvalues: +1.73 i; −1.73 i).

A linear oscillator has the differential equation 7.128

$$\frac{d^2y}{dt} + ay = 0, \tag{7.128}$$

where a is a positive constant.

Choosing $x_1 = y$ and $x_2 = dy/dt$ as state variables, Equation 7.129 is the state equation for Equation 7.128.

$$\dot{x}_1 = x_2$$
$$\dot{x}_2 = -ax_1. \tag{7.129}$$

The phase plane of Equation 7.129 is drawn in Figure 7.17. It is similar to Figure 7.16, which is an oscillator in which $a = 3$, as can be seen in Equation 7.129.

6. Saddle point: when one eigenvalue is positive real and the other is negative real, as in the system, which is seen in Equation 7.130 and Figure 7.18.

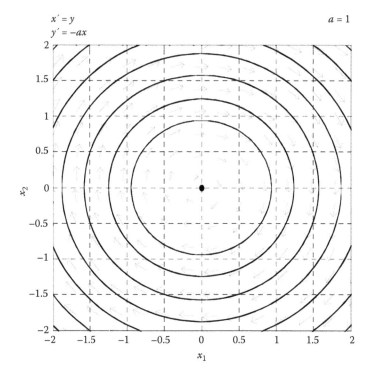

FIGURE 7.17
Phase plane of the oscillator (Equation 7.128).

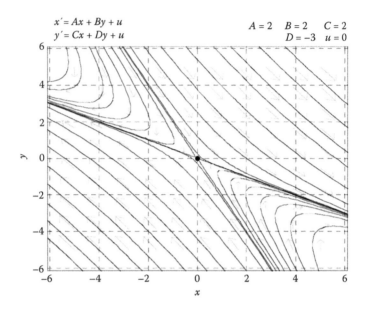

FIGURE 7.18
Saddle point (eigenvalues: 1; -2).

$$\dot{x} = \begin{bmatrix} 2 & 2 \\ -2 & -3 \end{bmatrix} x + \begin{bmatrix} 0 \\ 1 \end{bmatrix} u$$

$$u = 0.$$

(7.130)

7.1.2.11.4   *Nonlinear Systems with Several Singularities and Local Stability*

Example 7.6   The Lotka–Volterra Model

In Example 7.5, we compute the singular points of the Lotka–Volterra model, to give Equations 7.110 and 7.111. Let us consider the nonnull singularity. Linearizing the nonlinear differential equations around it, the Jacobian Equation 7.131 is obtained.

$$A = \frac{\partial f}{\partial x} = \begin{bmatrix} a - bx_2 & -bx_1 \\ px_2 & -c + px_1 \end{bmatrix}_{\left[\frac{c}{p}, \frac{a}{b}\right]}$$

$$= \begin{bmatrix} 0 & -b\frac{c}{p} \\ p\frac{a}{b} & 0 \end{bmatrix}.$$

(7.131)

Calculating the eigenvalues of A, its characteristic polynomial is given by Equation 7.131,

$$|\lambda I - A| = \begin{vmatrix} \lambda & +b\frac{c}{p} \\ -p\frac{a}{b} & \lambda \end{vmatrix} = \lambda^2 + pb\frac{a}{b}\frac{c}{p} = \lambda^2 + ac, \qquad (7.132)$$

whose roots are (Equation 7.133):

$$|\lambda I - A| = 0 \Leftrightarrow \lambda^2 + ac = 0 \Rightarrow \lambda = 0 \pm j\sqrt{ac}. \qquad (7.133)$$

For given initial populations of each species, how will they evolve? The singularity is a center, because the eigenvalues are imaginary conjugates. Using pplane7, the phase curves of Figure 7.19 are obtained, for $a = 0.4$; $b = 0.01$; $c = 0.3$; and $p = 0.005$.

The two singular points are clearly visible in Figure 7.19: one saddle point (the origin, eigenvalues 0.4 and -0.3) and a center (eigenvalues $+0.3464i$ and $-0.3464i$). The dark surface represents a trajectory that is gradually opening in the basin of attraction of the origin.

7.1.2.11.5 Uncertainty in Initial Conditions

Let us now see what happens at the boundary of the regions of convergence for a certain singularity. Take the example of the nonlinear system (Equation 7.134).

$$\dot{x} = (x - 1)^2 - u$$
$$\dot{y} = (y - 2)^2 - u. \qquad (7.134)$$

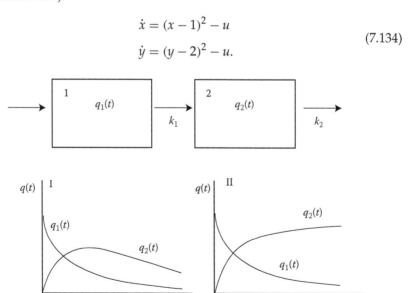

FIGURE 7.19
Phase curves for Lotka–Volterra model.

For the input $u = 1$, this system has 4 singularities, given by Equation 7.135,

$$x_S^1 = \begin{bmatrix} 2 \\ 3 \end{bmatrix} \quad x_S^2 = \begin{bmatrix} 2 \\ 1 \end{bmatrix} \quad x_S^3 = \begin{bmatrix} 0 \\ 3 \end{bmatrix} \quad x_S^3 = \begin{bmatrix} 0 \\ 1 \end{bmatrix}; \tag{7.135}$$

to which the following Jacobians correspond Equation 7.136.

$$A = \begin{bmatrix} 2 & 0 \\ 0 & 2 \end{bmatrix} \quad A = \begin{bmatrix} 2 & 0 \\ 0 & -2 \end{bmatrix} \quad A = \begin{bmatrix} -2 & 0 \\ 0 & 2 \end{bmatrix} \quad A = \begin{bmatrix} -2 & 0 \\ 0 & -2 \end{bmatrix}. \tag{7.136}$$
$$\;\; \text{repeller(source)} \qquad \text{seal repeller} \qquad \text{seal repeller} \qquad \text{attractor (sink)}$$

Drawing the phase plane with pplane7, the interesting Figure 7.20 is obtained, exhibiting the different types of singularities.

The line $y = 3$ is a decision boundary for the singularity [0.3]. Above 3, the trajectories diverge to infinity; whereas under 3, they converge to the stable node [0.1]. We can see this boundary in more detail, as in Figures 7.22 and 7.23.

For example, in Figure 7.21, the line $y = 3$, between $x = -2$ and $x = 0$, is a boundary.

Reducing the scale of the vertical axis, we obtain Figure 7.22.

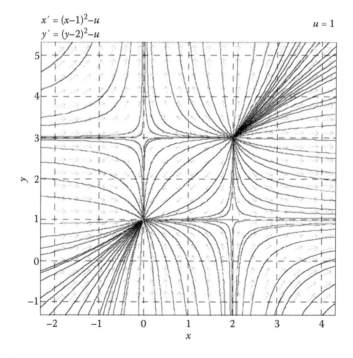

FIGURE 7.20
Phase plane trajectories for system (Equation 7.134). It has two saddle points, one stable node and one unstable node (Equation 7.136).

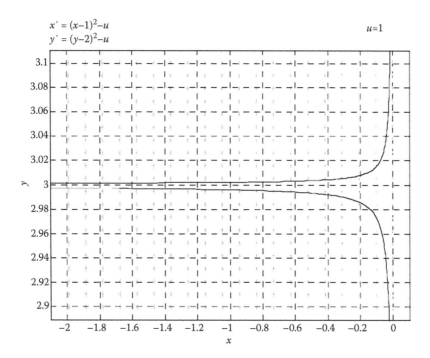

FIGURE 7.21
Phase trajectories for the system (Equation 7.134) around a boundary.

Reducing the vertical scale still further, we can again see a well-defined boundary. In Figure 7.23, the vertical interval shown is [2.9999 3.0001].

Here we may raise the following question: if the initial conditions are measured by an instrument and if the measurement error is greater than 0.0001, is it possible to predict the behavior of this nonlinear system?

No, it is not. This illustrates a characteristic of the chaotic behavior of nonlinear systems: the impossibility of predicting their behavior. An infinitesimal difference in the initial conditions may make the difference. This illustrates the famous butterfly effect first used by E. N. Lorenz in 1992 [15]: "Predictability: Does the Flap of a Butterfly's Wings in Brazil set off a Tornado in Texas?" which afterward has been repeated in multiple contexts and geographic versions. This does not mean that the system is stochastic. The system is, in fact, deterministic, but the uncertainty in the initial conditions prevents us from predicting its evolution.

The future of a chaotic system is indeterminate, although the system is deterministic.

The trajectories that define a boundary are called separatrices, and they divide the state space into regions of different dynamic modes.

The set of points giving rise to trajectories leading to a certain singular point is called the basin of attraction of that point, by analogy with the concept of a hydrographic basin.

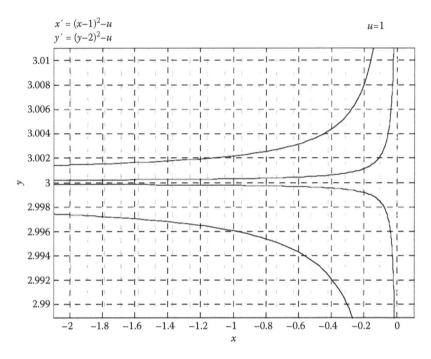

$x' = (x-1)^2 - u$
$y' = (y-2)^2 - u$ $u=1$

FIGURE 7.22
More detail on the boundary. Note the different scale of the vertical axes with regard to Figure 7.21.

7.1.2.11.6 Bifurcations and Chaos

Let us look again at the logistic equation of the population growth of a species, repeated in Equation 7.137, where A is the fertility rate of the population.

$$x_{k+1} = Ax_k(1 - x_k) = A(x_k - x_k^2), \quad x \in [0, 1], \qquad (7.137)$$

This model has its two singularities defined by Equation 7.138

$$\begin{cases} x_k = 1 - \frac{1}{A} \\ x_k = 0. \end{cases} \qquad (7.138)$$

The convergence toward one or the other depends on the initial state. When $A < 1$, the single possible singularity is the origin, $x_k = 0$ (the species vanishes).

We can implement this equation in Simulink [17], as in Figure 7.24, using the memory block as being equivalent to a pure time delay.

Simulating with several values of A, we obtain the results shown in Figure 7.25.

The system exhibits a strange behavior: its period depends on A, the fertility rate. As A increases, the population is periodic with a period depending on A. For A in the interval:

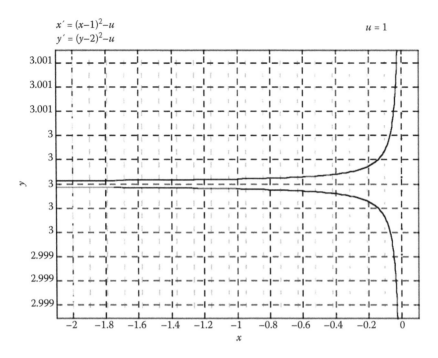

$$x' = (x-1)^2 - u$$
$$y' = (y-2)^2 - u$$

$u = 1$

FIGURE 7.23
An infinitesimal difference in the initial conditions produces very different system behaviors: in one case, it goes to infinity; whereas in another case, it goes to a stable node.

[3; 3.34495] oscillates with period of 2 generations;

[3.4495; 3.54408] oscillates with period of 4 generations; and

[3.54408; 3.56440] oscillates with period of 8 generations.

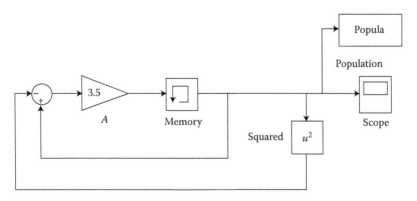

FIGURE 7.24
The logistic discrete equation implemented in Simulink.

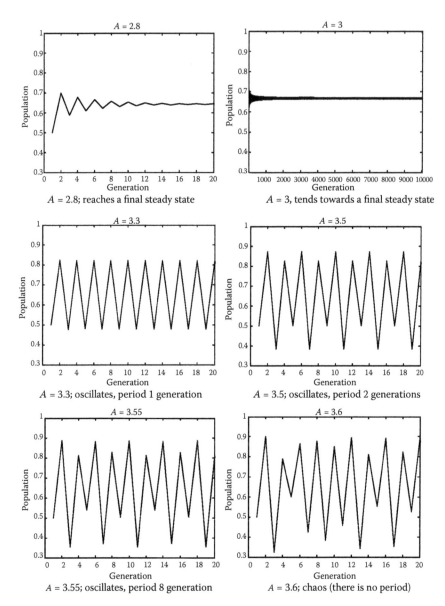

FIGURE 7.25
Evolution of the populations for different values of *A*.

Let us now perform the following graphical construction:

For each value of *A*, we calculate the values of the population *x* that belong to the respective cycle. These values can be obtained by simulating in Simulink, for example, and then reading the values of the population vector that compose one period afterward in the workspace. For example,

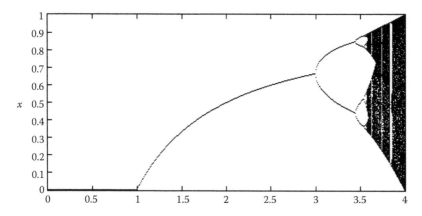

FIGURE 7.26
Bifurcation map of the logistic function of population growth.

for $A = 3.3$, we will have $x = 0.8236$ or $x = 0.4794$; for $A = 3.5$, it will be $x \in \{0.3828;\ 0.5009;\ 0.8269;\ 0.8750\}$; and for $A = 0.55$, $x \in \{0.3548;\ 0.81265;\ 0.54049;\ 0.88168;\ 0.37034;\ 0.82781;\ 0.50601;\ 0.88737\}$. For values of A less than 3, we will have only the steady state value; for A less than 1, the steady state is the origin. Graphically, we obtain the famous bifurcation diagram of the logistic function, as seen in Figure 7.26.

If we amplify the final part of the figure, between 3 and 4, interestingly we obtain a figure in which the same patterns in Figure 7.26 are repeated but on a different scale. If we amplify in this new figure again, we would obtain another figure with the same patterns on another scale. Therefore, by increasing the detail of the observation, we always get the same pattern as in Figure 7.27. This is a characteristic of fractals.

We find that as A increases, the oscillation period is duplicated until periodic behavior is no longer seen. Behavior in which there is no repetition of a pattern is chaotic. A point on the A axis in Figure 7.27 where there is a duplication of the period is called a bifurcation point.

7.1.2.11.7 Feigenbaum Numbers

Let A_n be the value of A at which the period $2n$ originates the period $2n + 1$. The number Equation 7.139

$$\delta_n = \frac{A_n - A_{n-1}}{A_{n+1} - A_n},\qquad (7.139)$$

is called the Feigenbaum delta. We can compute it for different values of n and obtain Table 7.4.

The point after which there is no more periodic behavior and chaotic behavior appears is called the accumulation point.

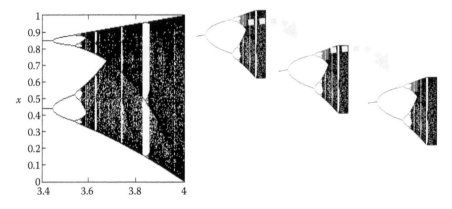

FIGURE 7.27
The fractal characteristics of the bifurcation map in Figure 7.26. (http://www.vanderbilt
.edu/AnS/psychology/cogsci/chaos/workshop/BD.html).

Feigenbaum studied the evolution of his delta for several systems at the
Los Alamos National Laboratory, USA; and in 1975, he found the following
result for all of them [15]:

$$\lim_{n \to \infty} \delta_n = 4.66920161 \ldots \tag{7.140}$$

According to Hilborn [15], this number is destined to belong to the Pantheon
numbers of physics (as does π, the golden number $(\sqrt{5}-1)/2$, etc.), and it
has been considered a universal constant.

TABLE 7.4

The Sequence of Feigenbaum Numbers

n	A_n	δ_v
1	3	
2	3.44949	
3	3.54409	4.751479915
4	3.564407	4.656199242
5	3.568.750	4.668428309
6	3.569692	4.664523044
7	3.569891	4.688442211
8	3.569934	...
9	3.569943	
10	3.5699451	
11	3.56994556	
Infinity	3.56994567	4.669201...

7.2 Tracer Kinetics Modeling

7.2.1 Introduction

NM methodologies have progressively achieved a central role in the research of physiological and pathological processes, in the discovery or assessment of drugs, and in the diagnosis and treatment of a variety of medical conditions. Recent developments in human physiology have been firmly based on the molecular approach of metabolic processes and on the use of radiotracer kinetic methods to study the associated functions, inspired by paradigms of biological and pharmaceutical sciences.

Radiotracers are distributed and located in the body as a result of physical–chemical interactions with living tissues in the course of specific physiological processes, and they can be used to convey information. However, the complex chain of events of various kinds that involves the dynamics of radiotracers is a sequence of processes governed by laws, which are generally well known. These processes may consist of mechanisms such as ionic exchange, diffusion, metabolic incorporation, binding to receptors, immunological reactions, etc.

The PET and SPECT are imaging techniques that use minimal amounts of radiolabeled molecules as tracers to study molecular pathways and molecular interactions, *in vivo*, without interfering with the processes being studied.

The detection of changes in the tracer distribution and concentration in biological tissues is the direct aim of the NM experimental method. By processing series of maps of tracer concentrations in a particular region of interest, with reasonable spatiotemporal resolution (millimeters and seconds), the characteristic biochemical and physiological dynamic parameters can be ascertained with good approximation.

Important characteristics of a good tracer are affinity, selectivity, adequate distribution in physiological compartments, and low participation in nonspecific bonds.

Quantitative approaches to the function of human organs and tissues based on modeling or data analysis are the goal in most of the studies on molecular functional imaging. The role of modeling is to apply appropriate analysis paths to data in order to evaluate the physiological and kinetic behavior of tracers.

Although the models used in NM involve different analysis of the systems, they generally lead to differential or integral equations or to differential equation systems. The solutions of these equations yield expressions that are supposed to recreate the experimental data and the behavior of the system in relation to the tracer used.

Broadly speaking, every methodology in NM is a systems theory exercise, that is, an input function is used to study a system through an induced response.

System here means an entity with broad latitude in its complexity, theoretically linear, shift time invariant, and able to supply a specific response to stimuli.

In NM, the input function is basically the radiopharmaceutical administered and the detected response, that is, the output function, the result of both the transport and the biological processing of the tracer. This methodology has been extensively used to study the biokinetics, biodistribution, and occupancy of thousands of radioactive molecules.

NM has progressed as a result of advances at each of the stages of this process, that is, advances in the input functions due to the introduction of new radiotracers, advances in the technology and reliability of the detectors to improve the output, and, finally, the increase in the capacity to extract and process the resulting data.

Recent PET imaging science has synthesized new molecules to be used in genetically engineered or transplanted tissues, in animal models. They have been developed to provide molecular information in human diseases, that is, to assess the molecular and genetic basis of normal cellular function and the transformations associated with diseased cells.

New and powerful statistical algorithms have been developed to extract spatiotemporal information from PET data sets in medical specialities such as oncology, neurology, and cardiology.

The development of novel radioactive tracers, multimodality, 3D acquisition, the creation of new data and image processing tools, etc., are all part of this promising effort to characterize the molecular basis of health and disease through imaging.

The extraordinarily high sensitivity of PET techniques ($\sim 10^{-9} - 10^{-12}$ M) and a specificity that reaches molecular level make them unique in the *in vivo* studies of internal organs, particularly with specific molecules present in the body at very low concentrations ($<10^{-8}$ M). For example, extrastriatal dopamine D1 receptors present in the brain at concentrations of approximately 10^{-9} M can be measured with PET using ^{11}C-NNC756 [21].

7.2.2   Concepts in Tracer Kinetics Modeling

The study of the movements of molecules within organisms, as well as the immediate consequences of such movements, is often called biokinetics.

The methods used to study molecular kinetics, particularly in pharmacological applications, have benefited from developments in a number of areas, particularly in nuclear methods and data analysis.

The purpose of NM imaging techniques (PET, SPECT, and planar scintigraphy) is to map physiological processes *in vivo*. With these techniques, information on local and temporal variations in functional activity of a given labeled molecule are detected as image contrast and analyzed either qualitatively or using quantitative approaches of varying degrees of complexity, based on the intrinsic or statistical properties of the system under observation.

Functional parameters can be estimated from dynamically acquired NM data by a number of different techniques such as reconstructing a temporal sequence of images, generating regions of interest (ROIs) overlaid on the image sequence, drawing time–activity curves (TACs), and subsequently fitting the parameters using appropriate relationships.

Theoretically, PET allows the dynamic study of any organic molecule labeled with a positron emitter, either an organic element (^{11}C, ^{15}O, ^{13}N) generating molecules chemically identical to the native ones (isotopic labeling) or an element not present in native molecules (e.g., ^{18}F to substitute H), leading to tracer molecules that are different from native ones (nonisotopic labeling); and these will, therefore, be discriminated at a later stage of the metabolic chain.

In SPECT, almost all of the molecules used as tracers are different from the native molecules, because they have to be labeled with nonbiological γ emitter elements (^{99m}Tc, ^{201}Tl, ^{123}I, etc). The radioisotopes of low Z elements such as the biological C, O, H, and N do not have any usable artificial γ ray emitter radioisotopes, because they are pure β emitters (β^+ or β^-).

Apart from the advantage of isotopic labeling, PET detection efficiency and spatial resolution are generally higher than in SPECT. Further, accurate attenuation correction is easy in PET, and it is difficult or almost impossible in SPECT.

Temporal constraints in SPECT measurements are also a common source of difficulties in studying dynamic processes.

An additional problem in SPECT is related to the gantry movement during data acquisition. Since the projections at different angles occur at different times during acquisition, they correspond to different tracer distributions, an effect that is maximized in fast dynamic studies. Images reconstructed from these incompatible projections can contain artifacts that lead to errors in the estimation of kinetic parameters.

Most of this chapter is dedicated to PET techniques, because they are the most representative in tracer applications.

Models are currently used in data analysis and in research projects on the biokinetics of radiopharmaceuticals.

A model uses the magnitudes and physical dimensions of the biological phenomena being studied and may make it possible to mathematically predict the behavior of an endogenous molecule or drug and to evaluate its concentration and its active metabolites at different stages and locations, including nonaccessible pools of the biological system.

A model is a tool for studying and understanding phenomena and, simultaneously, it is a shift from the experimental field to a phase of theoretical analysis and interpretation. Models are used to gain a better knowledge of biological systems; test new ideas, hypotheses, or other logical processes; evaluate future consequences of the observed phenomena; secure control of processes in specific situations, etc.

The aim of modeling in the particular case of NM is to apply analysis tools that are appropriate for SPECT, PET data, or biological sample counting in order to evaluate the physiological kinetics of tracers in the body and to study the function of its organs and tissues.

Modeling allows the prediction of the metabolic steps of an endogenous molecule or drug and the evaluation of its concentration and its active metabolites in nonaccessible compartments of the biological system.

First principles of chemistry and physics can be used to quantitatively model and simulate static and dynamic physiological processes that are usually described only qualitatively. A model may help us to both understand the phenomena qualitatively and analyze the process quantitatively, using mathematical methods.

Modeling techniques may try to represent a phenomenon exhaustively, that is, by being its replica, or just translate its functional performance in a simplified or partial way by showing the molecular paths by which to derive parameters of interest, such as binding potentials.

Most systems in which we use models to try to understand the involvement of tracer molecules are in a steady or quasi-equilibrium state. This state exists due to homeostatic control that keeps the concentration of products constant, but it does not correspond to true chemical equilibrium. Examples of this quasi-equilibrium are the tight control of body electrolyte concentrations, blood glucose level, even blood cell counts, and many other populations of molecules or cells. Metabolism and other dynamic processes have to occur for a steady state to be maintained.

Some models are essentially deterministic. These models assume that the system is governed by known physical–chemical laws, and the values of the parameters to be introduced are obtained from the measured samples.

Other systems are stochastic in nature. The statistical laws of the system as well as the functional parameters have to be assessed from collected data.

Compartmental analysis, functional mathematical models, and systems theory have been widely used in studies with radionuclides since the 1960s [22].

In modeling, a compartment may be an anatomical, physiological, chemical, or physical subdivision of the system under investigation. A steady state is generally assumed for the system, and it is also held that the tracer is rapidly mixed in compartments so that the specific activity is uniform throughout the compartment at any time. The deterministic approach may be an approximation in one or more respects. In fact, the steady state condition is frequently not achieved; sometimes, instant dilution cannot be considered to prevail; the reaction in some processes is not determined just by a constant but by a statistical distribution of constants; the order of the reactions is not known; and the pathological model may not coincide with the normal one. Compartmental models are the most common tracer kinetic models used in both SPECT and PET studies. These models are formulated by using differential equations that describe exchanges between the system compartments, and they can provide accurate measurements of rates of exchange and binding potentials, which are proportional to the concentration of available receptor sites and other quantities.

In most cases, the complexity of physiological systems makes it difficult to directly associate a predefined biological kinetic model with a tracer. An analysis of the dynamic processes that are plausibly observable within the constraints of heterogeneity of the object is necessary. This analysis has to take into account the detector resolution, the detector overall spatiotemporal response, and the statistical quality of the data. For a well-defined kinetic model to be accepted as a simplifying tool from which quantitative biological information on exchange and targeting can be accessed, we need a sound knowledge of the biological fate of the tracer. This is related to the exchange mechanisms that take place in the tissues and the contribution of the different exchange paths.

As an example, brain studies with PET and SPECT require radiotracers with high affinity for neurotransmitter receptors and transporters. However, the peripheral metabolism of radiotracers often leads to lower radioactivity in the brain. Some metabolites nonspecifically bind to receptors, whereas others even bind preferentially to other target receptors.

The *a priori* analysis of kinetic data to derive the identifiable kinetic processes before implementing a model may involve powerful preprocessing analytical methods, some of which are still at a developmental stage. Examples of such methods are cluster analysis, factor analysis, spectral analysis, and fractal analysis [23–25].

Mainly, as a result of the possibilities offered by PET in brain studies and the large number of radiotracers and radioligands that can be used to probe molecular binding sites, kinetic modeling is nowadays a vast and highly specialized field. Estimation of quantitative biological images from complex 4D spatiotemporal data sets requires the application of appropriate image processing and tracer kinetic modeling techniques.

Extending the analysis of ROIs and time or activity curves to voxel level, parametric images can be generated where the voxel index represents a parameter (e.g., glucose metabolism or blood perfusion) instead of tracer concentration as in the original PET images. Mathematical modeling plays an important role in the conversion of the original images into local reaction rates, that is, parametric images.

Interest has recently been increasing in the formation of parametric images that individually model the kinetic behavior of each volume element (voxel) in the image domain under analysis. This approach is more appropriate when the volume cannot be effectively segmented into homogeneous regions that would be modeled with a single kinetic parameter set.

The merging of information from different modalities or from different tracers can also be very useful in identifying structure or function relationships or assessing the normal and disease states of a particular organ.

7.2.3 Transport of Tracers and Localization Mechanisms

Tracer studies are based on the principle that the mass (or activity) of tracer delivered to an organ or tissue is proportional to perfusion (blood flow/mass) to the organ or tissue, that is,

$$\frac{\text{Mass of tracer in A}}{\text{Blood flow in A}} = \frac{\text{Mass of tracer in B}}{\text{Blood flow in B}} = C^{te}. \qquad (7.141)$$

If tissue B is taken as a reference with known blood flow for a given activity of tracer in B, then blood flow to tissue A can be determined by measuring the activity of tracer in A.

The use of ^{13}N-ammonia for imaging the relative myocardial blood flow is based on the above principle.

The uptake of a tracer by a tissue perfused by blood also depends on tissue extraction and clearance flow. Extraction can be defined for steady state or for first-pass conditions.

The former (net extraction, E_n) is a relative difference between the arterial and venous tracer concentrations (C_A and C_V) of an organ or tissue in steady state, that is,

$$E_n = (C_A - C_V)/C_A \qquad (7.142)$$

In the latter (unidirectional extraction, E_u), only the amount of tracer extracted from blood to tissue is taken into account. The amount of tracer delivered back from tissue to blood is not included. Equation 7.141 still applies but now the concentration C_V refers to first pass.

If there is no utilization of the tracer, that is, if the tracer transferred to the tissue is returned to blood, the E_n is zero. This situation applies to inert gases.

E_n is generally larger than the net extraction fraction. An exception to this general rule occurs with O_2. Virtually, all oxygen extracted by tissue is metabolized; thus, the net and unidirectional extraction fractions are the same. For all other substances, a major portion of what is extracted by the tissue is transported back to the blood.

After intravenous injection, radiopharmaceuticals are transported to the capillaries by blood flow (perfusion) and move out of the vascular compartment, mostly across the capillary walls by several passive or active transport mechanisms.

From the interstitial space, the radiopharmaceuticals can move across the cellular membrane; and, in the cell, they participate in biochemical reactions or bind to receptors.

In passive transport, the movement of the tracer is determined by diffusion forces, following the direction of the concentration gradient across the membrane, either in the membrane phase or in tiny passages between the capillary endothelial cells.

For a constant tracer concentration gradient, the net flux in passive diffusion depends only on the diffusion coefficient and on the concentration gradient.

Another type of passive transport is facilitated transport, which is performed by special carriers that exist in the membrane and permits a selective and controlled uptake of substrates, especially amino acids and glucose.

Facilitated transport is generally considered a three-stage process: the substrate bonding with carrier molecules on one side of the membrane; the assisted movement of the substrate–carrier complex; and the release of the substrate on the other side. For low concentrations of labeled tracers, this process can be regarded as linear and similar to catalyzed enzyme kinetics with competitive inhibition.

In active transport, tracers are carried against concentration gradients, thanks to energy supplied by the adenosine triphosphate (ATP) present in the membrane. The known pump mechanism that regulates the Na^+ and K^+ concentrations on either side of the cellular membrane and the tubular reabsorption of glucose in the kidneys are examples of active transport mechanisms.

The literature on the transport and uptake of elements and compounds after their administration into the body is vast [25].

It is known that the localization of elements in tissues after injection is closely related to their specific electronic configuration.

On the left-hand side of the periodic table, vertical relations predominate, and the tissue distribution of elements of a particular periodic family is similar. However, on the right-hand side of the periodic table, horizontal relations tend to predominate, especially for the transition metals. The tissue distribution of an element is similar to neighboring elements with almost the same atomic weight.

As a rule, anions, including halogens, oxygenated or halogenated ions of groups IV, V, and VI and the transition metals, are rapidly eliminated by the kidneys. In contrast, the retention time of metallic cations with the inner orbitals completely occupied by electrons tends to be prolonged in the skeleton and soft tissues.

The main properties of an element that determine tissue localization and excretion are oxidation or valence state, for the pH of the blood (pH=7.4); relative solubility in aqueous media or, if insoluble, the size of the colloidal particle; and tendency to be incorporated into organic compounds or specific proteins.

The biological fate of compounds is determined by many factors, and it is much more complex than for elements. After intravenous injection, most radiopharmaceuticals leave the vascular compartment, mainly through capillaries. An average man has about 4×10^{10} capillaries with an exchange area of around $1000\,m^2$ and a volume that contains, at rest, less than 5% of the total blood circulation.

The uptake of a drug by an organ or tissue depends on the fraction of cardiac output received by the organ and on there being receptor sites with a specific biochemical affinity for the extraction and concentration of the labeled compound. In addition, factors related to capillaries such as their number per cubic millimeter of tissue and the type of capillary endothelium within the organ are also important. There are three main groups of capillaries in the body: continuous, fenestrated, and discontinuous. Capillaries with continuous, nonfenestrated endothelium occur in the muscles, fat, connective tissue, and lungs.

The endothelium of fenestrated capillaries may be closed as in the brain, gastrointestinal tract, and endocrine glands or open as in the heart or renal glomerula where the effective pore radius is 45–50 A°. Capillaries with discontinuous endothelium allow large molecules to leak into the perivascular space; they are found in the sinusoids of the reticuloendothelial organs—liver, spleen, and bone marrow.

Other factors that are important in the localization of drugs within the body are molecular size and shape in plasma, degree and strength of protein binding in blood and tissues, liposolubility, and specific cellular mechanisms.

After administration of a radioactive tracer, slow passive diffusion processes due to concentration gradients through permeability barriers may prevail. When this occurs, the tracer can be used to measure parameters within its area of localization, such as volume, or those related to the barrier, such as permeability and dynamics.

After administration, a radiopharmaceutical passes through a series of membranes before reaching the target tissue, and it is often important to know the permeability of these membranes for the different tracers.

In general terms, the potential energy gradients responsible for the passive transport of noncharged radioactive molecules in biological media result from concentration and pressure. The latter is usually the most important factor in dilution processes.

The distribution and localization of radiopharmaceuticals is not a simple process, as a rule; it tends to be a complex chain of processes with multiple interactions and probably involving several of the phenomena described next.

To reach a particular cell, after administration, a radiopharmaceutical must penetrate a series of membranes. These are predominantly lipidic and may include the capillary endothelial membrane, the gastrointestinal membrane, and even membranes surrounding intracellular organelles.

The more common processes involved in the movement and distribution of tracer molecules after the injection are briefly considered next.

- Ion exchange, which consists of removing ions from the native site in the organ of interest by exchange with identical radiotracer ions that are concentrated in the organ. Fluoride, phosphate, and thallium ions are initially extracted from circulating plasma, mainly by a process of ion exchange.

- Metabolic incorporation, in which the radioactive tracer follows the same biochemical pathways as those of the natural, nonradioactive substrate and is, thus, actively incorporated during the metabolic processes. Labeled glucose and radioactive iodide are typical examples; the first is used to detect local glycolysis and the second is used to study the production and use of thyroid hormone.

- Binding to receptors, in which the radiotracer (ligand) interacts with receptors that are macromolecules (glycoproteins), to produce characteristic complexes. Ligands are often regarded as consisting of two parts, one that contains the essential binding site and a second part that does not intervene in the reaction and can be modified to a certain extent. When a radionuclide is introduced into a ligand molecule to produce a radiopharmaceutical, incorporation should take place in the modifiable part of the molecule. After a radioligand leaves the vascular pool, it reversibly binds to nonspecific binding sites in the interstitial and cellular spaces and subsequently diffuses to the specific sites. Many receptors have been investigated, including polypeptide hormones, acetylcholine, etc.

- Enzyme-catalyzed reactions. Most chemical reactions in living tissues are catalyzed by enzymes and can be described by Michaelis–Menten kinetics. For the tracer, the equation is linear with regard to its concentration; linearity holds even when the reactions do not strictly follow Michaelis–Menten kinetics, as long as the concentration of tracer is low compared with the natural substance.

- Labeled monoclonal antibodies, derived from cultured hybridoma cells, can recognize and interact with single antigens and can potentially identify and target specific groups of cells such as blood and tumor cells.

A major problem in applying the technique to imaging is tissue specificity, which may differ only slightly between unaffected tissues and tumor cells and so give rise to poorly specific or nonspecific binding ratios.

- When gaseous radiopharmaceuticals are administered by inhalation, they pass into the pulmonary venous circulation from the lungs and can freely exchange between the lungs and tissues. In conventional NM studies, the gases used are principally isotopes of xenon (^{133}Xe, ^{127}Xe) and ^{81m}Kr, which are chemically inert, and in PET ^{13}N and ^{11}CO$_2$ or ^{10}CO$_2$ are gases with great potential.

Other processes such as phagocytosis, pinocytosis, and capillary trapping may be involved when radioactive particles or cells are used.

7.2.4 Compartmental Models

It is simple and often useful to assume that an organism consists of a number of linked and interacting units called compartments, which are studied as kinetic compartmental models.

A continuous exchange of matter and energy occurs between such compartments and between some of these and the environment. These exchanges are the physical–chemical processes related to absorption, distribution, synthesis, degradation, and excretion. The interrelationships between these compartments are so close that generally the perturbation of one particular exchange can induce modifications in several compartments of the system. Illness is an alteration of one or more of these exchanges or of their regulatory mechanisms [26,27].

Compartmental models originated in the field of pharmacokinetics, and they are a widely used mathematical tool for analyzing data from PET and SPECT. A compartment is not necessarily a physical volume or space but, in a generic sense, it is an amount of a chemical compound that can be kinetically studied in the same way as a well-defined homogeneous mass of the same product. In living organisms, there are many compartments, either physically well separated or theoretically defined.

Radiopharmaceuticals may achieve their initial bio-distribution by simple dilution, after a fast injection, in an accessible compartment taken as being strategic for the experiment. The length of time a tracer stays in a compartment depends on the specific biochemical and biophysical processes in which it will be involved. The evolution over time of the amount of tracer present in a compartment after a fast injection is often the time–activity function obtained by external detection (residue detection).

A compartment model that perfectly fits the NM experimental data is the gold standard for kinetic studies [28,29].

In compartmental models, when the metabolite-corrected arterial blood curve and dynamic data are available from the time of injection to the time when all important changes in tracer kinetics have occurred and the data have good statistics, it may be possible to estimate perfusion, blood volume in tissue vasculature, metabolism rate constants, and specific binding potentials. In practice, this situation mostly applies to simple compartmental models, because the mathematical complexity drastically increases as the number of compartments grows. Moreover, in most cases, there is no need to measure all the parameters, but only those characteristic parameters that determine the property under study.

Common problems in compartmental analysis are the development of plausible models for particular biological systems and of analytical theories for each class of systems.

The metabolite-corrected arterial plasma curve is the input function to the compartment model. Although plasma is not a compartment of the model itself, it is usually counted as such.

A common important category of compartment models is those in which the rates of all transfers between compartments are first-order rate constants. This is a deterministic constraint based on experience and it has a wide application.

7.2.4.1 Single-Compartment Models and Multicompartment Models

7.2.4.1.1 Single-Compartment Models

Consider a simple model for the injection of a radiopharmaceutical into the bloodstream and its disappearance. The blood pool is taken to be a single compartment of volume V; the blood concentration of tracer is $c(t) = [q(t)/V]$, (in activity per volume unity); and $i(t)$ is the tracer input rate into blood (input activity per time unit), for $t \geq 0$.

The disappearance of the radiopharmaceutical is assumed to occur according to a first-order reaction with effective constant λ_{ef}.

Between times t and $t + dt$, effective elimination will reduce the tracer activity in the compartment from $q(t)$ to $q(t)(1 - \lambda_{ef}\, dt)$, and input will increase it by the amount $i(t)dt$ and so, the activity at time $t + dt$ is

$$q(t + dt) = q(t) - q(t)\lambda_{ef}\, dt + i(t)\, dt \tag{7.143}$$

Taking the function $q(t)$ as continuous,

$$q(t + dt) = q(t) + q'(t)\, dt + \ldots\ldots \tag{7.144}$$

and, neglecting terms of second order and above,

$$q'(t) = -\lambda_{ef} + i(t) \text{ or} \tag{7.145}$$

$$q(t) = q_0 e^{-\lambda_{ef}t} + \int_0^t e^{-\lambda_{ef}(t-s)}i(s)ds \tag{7.146}$$

and, dividing by V,

$$c(t) = c_0 e^{-\lambda_{ef}t} + \int_0^t e^{-\lambda_{ef}(t-s)}i(s)ds/V \tag{7.147}$$

If there is no input for $t > 0$, that is, $i(t) = 0$ and, at $t = 0$, $c(0) = c_0 = q_0/V$, then,

$$q(t) = q_0 e^{-\lambda_{ef}t} \quad \text{and} \quad c(t) = c_0 e^{-\lambda_{ef}t}. \tag{7.148}$$

An example of this situation (Figure 7.28) is the instantaneous injection of activity $q_0 = c_0 V$, at $t = 0$, of a diffusible tracer, with instantaneous mixing in the compartment.

If the concentration in the compartment for $t = 0$ is zero, that is, $c_0 = 0$, but $i(t)$ is not null, then

$$c(t) = \int_0^t e^{-\lambda_{ef}(t-s)}i(s)\, ds/V \tag{7.149}$$

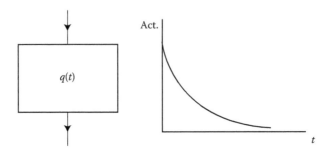

FIGURE 7.28
Activity/time curve after injection of activity q_0 at $t = 0$ in an open single-compartment system.
Instantaneous mixing is assumed.

that is, the convolution of the input function with the response of the system
to an instantaneous injection of unitary activity.

If $i(t)$ has a constant value, $i(t) = k$, then the integration of Equation 7.149
gives

$$c(t) = \frac{k}{\lambda_{ef} V}(1 - e^{-\lambda_{ef} t}). \tag{7.150}$$

The plot of this function is given in Figure 7.29.

An example of this situation is the continuous perfusion of ^{99m}Tc-DTPA for
the determination of renal clearance.

In a compartment where instantaneous, homogeneous mixing occurs and
sampling is allowed, if at $t = 0$ an instantaneous tracer injection of activity q_0 is
performed and $\lambda_{ef} = 0$, the volume of distribution can be easily determined
if the sample activity and volume injected are known. If q_0 is the activity
administered in volume v and if, after homogenization, a sample of volume
v' and activity a' is collected, then the concentration in the compartment is

$$c = \frac{a'}{v'} = \frac{q_0}{V+v} \approx \frac{q_0}{V}$$

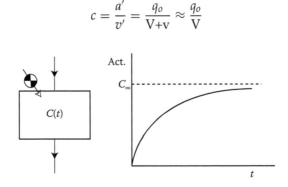

FIGURE 7.29
Activity/time curve during continuous perfusion in a single-compartment open system.

and

$$V = \frac{q_0 - cv}{c} \approx \frac{q_0}{c} \qquad (7.151)$$

q_0 and c are known. The approximation holds for the case of $V \gg v$.

This situation is known as the dilution principle.

It assumes that no significant activity decline occurs, and it seems to be a theoretical conservation principle for the activity in biological systems. In practice, even the samples collected very shortly after the injection (essential to ensure mixing in the blood) are affected by error due to radioactive decay and biological elimination. After collecting blood samples at different times and performing linear curve fitting on the plot of activity concentration versus time, the activity at zero time can be determined by extrapolation and the corrected value can be introduced into Equation 7.150. For an exponential decrease, as in Figure 7.28, extrapolation is easy, because fitting in a semilog scale is linear.

We can easily relate clearance (Cl), volume (V), and half-time ($T_{1/2}$) in a single-compartment model:

Elimination rate (g/mL) $= \lambda_{ef} Q = \lambda_{ef} Vc = Clc$ with $Cl = \lambda_{ef} V$ and

$$T_{1/2} = 0.693/\lambda_{ef} = 0.693\ V/Cl \qquad (7.152)$$

Typical examples of compartmental volume determination are plasma and red cells: the first using labeled human serum albumin and the second using labeled autologous red cells (Figure 7.28).

7.2.4.1.2 Two-Compartment Models

7.2.4.1.2.1 Open Two-Compartment Series System
An instantaneous injection of tracer is performed in compartment 1 (Figure 7.30) at $t = 0$, and instantaneous mixing will give an even concentration q_0. The rate constants k_1 and k_2 are for the irreversible exchanges from compartment 1 to 2 and from 2 to the outside, respectively. The equations for the variations of the tracer concentrations in 1 and 2 are

$$\frac{dq_1}{dt} = -k_1 q_1 \qquad (7.153)$$

and

$$\frac{dq_2}{dt} = k_1 q_1 - k_2 q_2. \qquad (7.154)$$

Integration of Equation 7.153 is immediate:

$$q_1(t) = q_0 e^{-k_1 t} \qquad (7.155)$$

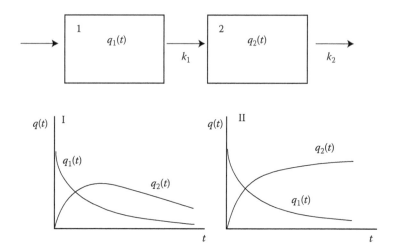

FIGURE 7.30
Activity variations in compartments 1 and 2 after instantaneous injection of radiotracer in the
first. Case $I \rightarrow k_2 \neq 0$; case $II \rightarrow k_2 = 0$.

Integration of Equation 7.154 gives

$$q_2(t) = q_0 \frac{k_1}{k_2 - k_1} \left(e^{-k_1 t} - e^{-k_2 t} \right) \tag{7.156}$$

Graphs of Equations 7.155 and 7.156 can be seen in Figure 7.30, for the
general case of $k_2 \neq 0$ (plot I) and for $k_2 = 0$ (plot II).

An example of case I is, in a first approximation, ^{99m}Tc-Deshida is a radiopharmaceutical for liver studies. After injection, it is taken up from the plasma
by liver parenchymal cells and eliminated with bile to the intestine. For case
II, and liver studies again, but now with radiocolloids such as ^{99m}Tc-S, and
assuming no fixation by the spleen and bone marrow.

7.2.4.1.2.2 Closed Two-Compartment System (One-Tissue Compartment Model)
This simple model describes the bidirectional exchange of tracer between
two compartments, assuming no loss of tracer. A radioactive bolus of activity
Q is instantaneously injected into compartment 1 at time $t = 0$. Activity in
compartments 1 and 2 is $q_1(t)$ and $q_2(t)$, respectively.

The Equations that characterize transfer rates in the system (Figure 7.31)
are

$$\frac{dq_1}{dt} = -k_1 q_1 + k_2 q_2 \tag{7.157}$$

$$\frac{dq_2}{dt} = -k_1 q_1 + k_2 q_2 \tag{7.158}$$

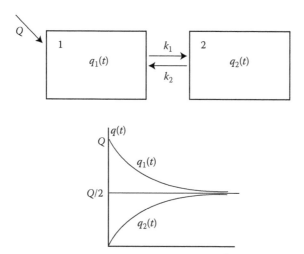

FIGURE 7.31
Closed two-compartment system in equilibrium. The plots relate to $k_1 = k_2$, and volumes are equal in both compartments.

There is no output of tracer from the two-compartment system. Then $q_1 + q_2 = Q$ and deriving, we get,

$$\frac{dq_1}{dt} = -\frac{dq_2}{dt}$$

Showing that with $q_1(t)$ being a decreasing function, then, as expected, function $q_2(t)$ must be growing, which means an increasing amount of tracer in 2. Initial conditions are

$$q_1 + q_2 = Q \quad \text{and} \quad q_2 = 0$$

and also

$$\frac{dq_1}{dt} = -k_1 Q \quad \text{and} \quad \frac{dq_2}{dt} = k_1 Q.$$

The solutions for these Equations for the stated conditions are

$$q_1(t) = q_0 \left\{ 1 - \frac{k_1}{k_2 + k_1} \left[1 - e^{-(k_1 + k_2)t} \right] \right\} \tag{7.159}$$

and

$$q_2(t) = q_0 \left\{ \frac{k_1}{k_2 + k_1} \left[1 - e^{-(k_1 + k_2)t} \right] \right\} \tag{7.160}$$

Activity will exponentially decrease in 1 and exponentially increase in 2 until steady state is attained. In both cases, the rate of variation is determined by the sum of the two constants k_1 and k_2.

Where $k_1 = k_2$ and the volumes of the two compartments are equal, functions $q_1(t)$ and $q_2(t)$ tend toward the value of $Q/2$, as is seen in Figure 7.31.

7.2.4.1.3 Three-Compartment Models

7.2.4.1.3.1 Three-Compartment Models in Equilibrium
The activity variations in the three compartments of the system in Figure 7.32 are

$$\frac{dq_1}{dt} = k_{21}q_1 - k_{13}q_1 \tag{7.161}$$

$$\frac{dq_2}{dt} = k_{12}q_1 - k_{24}q_2 \tag{7.162}$$

$$\frac{dq_3}{dt} = k_{13}q_1 - k_{34}q_3 \tag{7.163}$$

Time or activity functions for the compartments under study are obtained by integration,

$$q_1 = q_o \exp[-(k_{12} + k_{13})t]$$
$$q_2 = \{\exp[-(k_{24})t] - \exp[-(k_{12} + k_{13})t]\} \tag{7.164}$$
$$q_3 = \{\exp[-(k_{34})t] - \exp[-(k_{12} + k_{13})t]\} \tag{7.165}$$

The measured values of count rates on kidneys are, in fact,

$$z_1 = q_1 + h_1q \tag{7.166}$$
$$z_2 = q_2 + h_2q \tag{7.167}$$

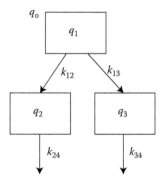

FIGURE 7.32
Three-compartment system, with no equilibria and two parallel compartments.

where h_1 and h_2 indicate the amount of blood measured in the ROIs of each kidney

$$q_2 = \frac{q_0}{k_{12} + k_{13} - k_{24}} \{k_{12} \exp[-(k_{24})t] + [h_1(k_{12} + k_{13} - k_{24}) - k_{12}]$$
$$\times \exp[-(k_{12} + k_{13})t])\} \tag{7.168}$$

$$q_3 = \frac{q_0}{k_{12} + k_{13} - k_{34}} \{k_{13} \exp[-(k_{34})t] + [h_2(k_{12} + k_{13} - k_{34}) - k_{13}]$$
$$\times \exp[-(k_{12} + k_{13})t]\} \tag{7.169}$$

This is a simplified nonexact model to describe ^{123}I-Hippuran clearance from plasma by the kidneys. It does not give results that accurately fit the real situation. It is generally accepted that after intravenous injection of ^{123}I-Hippuran its plasma concentration is described by a summation of two exponentials as a result of diffusion to a peripheral compartment. The possibility of a third, very fast exponential term immediately after injection was also considered.

7.2.4.1.3.2 Open Mammillary Systems In mammillary models, of which the system shown in Figure 7.33 is the simplest, compartments are linked to a central open compartment. Concentrations in compartments 1 and 2 are $q_1(t)$ and $q_2(t)$, respectively, and a radioactive bolus is added to compartment 1. The constant k_{10} is the fractional excretion rate from compartment 1 to a third compartment.

The rates of change of activity are

$$\frac{dq_1}{dt} = -k_{10}q_1 - k_{12}q_1 + k_{21}q_2 \tag{7.170}$$

$$\frac{dq_2}{dt} = k_{12}q_1 - k_{21}q_2 \tag{7.171}$$

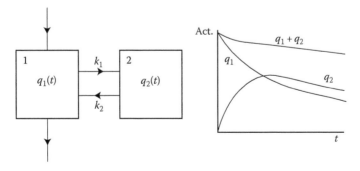

FIGURE 7.33
Open mammillary system with two compartments.

At $t = 0$:

$$q_1 = q_0; \quad q_2 = 0; \quad \frac{dq_1}{dt} = -k_{10}q_1 - k_{12}q_1 \quad \text{and} \quad \frac{dq_2}{dt} = k_{12}q_1$$

Then by integration we get

$$q_1(t) = q_0 \left(\frac{k_{21} - s_1}{s_2 - s_1} e^{-s_1 t} + \frac{k_{21} - s_2}{s_1 - s_2} e^{-s_2 t} \right) \tag{7.172}$$

$$q_2(t) = q_0 \frac{k_{21}}{s_2 - s_1} (e^{-s_1 t} + e^{-s_2 t}) \tag{7.173}$$

with

$$s_1 s_2 = K_{10}K_{12} \quad \text{and} \quad s_1 + s_2 = K_{10} + K_{12} + K_{21} \tag{7.174}$$

The metabolism of plasma proteins and the uptake of $^{99m}\text{T}_c\text{O}_4^-$ by the thyroid have been studied using this model.

The open mammillary model can be used to evaluate renal clearance; it consists of an intravascular compartment that reversibly exchanges with an extravascular compartment and irreversibly exchanges with the urinary compartment (Figure 7.34).

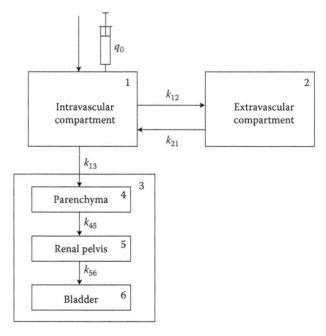

FIGURE 7.34
Renal clearance model. An intravascular compartment reversibly exchanges with an extravascular compartment and nonreversibly exchanges with the urinary compartment.

Intravascular compartment (1) represents the amount of tracer that is exchangeable with renal parenchyma and extravascular space. Urinary compartment (3) represents the tracer that is cleared by the kidneys, and it is related to the renal pelvis and bladder. Extravascular compartment (2) contains the tracer that has not been cleared, that is, the tracer in the extra-renal tissue and the tracer that is bound to other molecules.

Equations for compartments (1) and (2) have been already obtained.

The urinary compartment (3) includes all the tracer activity in urine, without distinguishing whether it is in the renal pelvis, ureters, or bladder (3). The consideration of these three spaces in the mammillary model considerably complicates the equations.

When the tracer is injected into the intravascular compartment through a peripheral vein, the initial distribution of the tracer is not uniform in the body; but this situation is rapidly attenuated over time, (3) as the blood circulates. In a densely vascularized region, a time or activity curve shows a fast initial rise that quickly decreases. The peak amplitude of this curve varies with the anatomic region, injection site, and injection speed.

Strictly compartmental analysis cannot be applied to the initial phase of a renogram, because the basic assumption of uniform distribution of the tracer, implicit in compartmental analysis, is not confirmed.

After this phase, the amount of tracer in the intravascular compartment begins to fall (Figure 7.33) as a consequence of uptake by the kidneys (represented by k_{13}) and diffusion to the extravascular space (represented by k_{12}).

With increasing amounts of tracer in the extravascular compartment and the continued elimination of tracer from the blood, after a time the direction of exchange is reversed (represented by k_{21}), with a maximum being attained in extravascular concentration before it begins to fall (Figure 7.33).

Constant k_{56} is related to the urine production rate.

In a real situation, the crossing of the tracer through the renal parenchyma is characterized by a transit time, t_0, whose introduction complicates the system's equations.

The solutions of the differential equations for the amounts of tracer in the renal parenchyma, in the renal pelvis, and in the bladder need to take into account a time delay t_0 such that $t < t_0$

$$q_4 = 1 - A_3 \exp(-s_1 t) - A_4 \exp(-s_2 t)] \tag{7.175}$$

$$q_5 = 0 \tag{7.176}$$

and

$$q_6 = 0 \tag{7.177}$$

and when $t > t_0$:

$$q_4 = A_3[1 - \exp(-s_1 t_0)] \exp[-s_1(t - t_0)] - A_4[1 \exp(-s_2 t_0)] \exp[-s_2(t - t_0)] \tag{7.178}$$

$$q_5 = A_7 \exp[-s_1(t - t_0)] + A_8 \exp[-s_2(t - t_0)] - A_9 \exp[-s_3(t - t_0)] \tag{7.179}$$

and

$$q_6 = 1 - A_{10} \exp[-s_1(t - t_0)] - A_{11} \exp[-s_2(t - t_0)] + A_{12} \exp[-s_3(t - t_0)]$$
$$(7.180)$$

where s_3 is related to k_{56}.

The curves showing the amount of tracer versus time in the compartments considered above are shown in Figure 7.35.

7.2.4.1.3.3 Mammillary System with Multiple Equilibria The model in Figure 7.36a was proposed to explain ^{99m}Tc-MAG3 kinetics in human kidneys, targeting the acquisition and processing capacities of modern gamma cameras.

The four compartments of the model are plasma with fast distribution; extrarenal extravascular space; and the left and right kidneys, which can reversibly exchange with plasma. The only irreversible movement of the tracer is from kidneys to bladder.

This model combines the advantages of plasma clearance methods with the processing potential of the new gamma cameras, making the quantification of differential renal function of the two kidneys possible. Figure 7.36b represents a gamma camera image, with the areas marked for kidneys and background.

Figure 7.36c shows a model with tracer exchange to the interstitial fluid that can also reversibly exchange with a third compartment. This compartment can represent bone tissue in a study with [18]F.

These models can be clinically useful but need a mathematical analysis that is complex and extensive, which is of little interest in the present context.

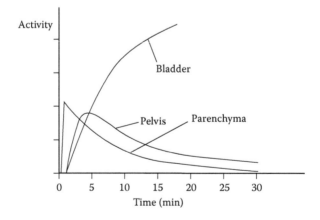

FIGURE 7.35
Parenchymal curves ($q4$) of the renal pelvis ($q5$) and bladder ($q6$), obtained for $t_0 = 2$ min and $k_{56} = 1$ min^{-1}.

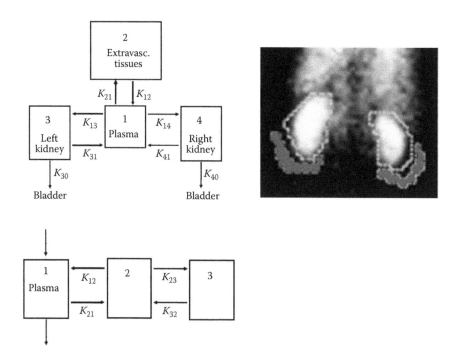

FIGURE 7.36
(a) Four-compartment model for kinetic studies after ^{99m}Tc-MAG3 injection. (b) Gamma camera image showing areas marked for kidneys and background. (c) Model with exchange of tracer to interstitial fluid, with a reversible exchange with a third space.

7.2.5 Tracers, Volumes, and Flows in Dilution Systems

The variables generally used in kinetic studies in biological systems are the volume V, (probably the tracer distribution volume), time t (frequently the mean transit time of a tracer in a system), the mass of tracer M in mg or the activity in MBq, the tracer concentration, C in mg/mL or mBq/mL, and the flow of tracer in mg/min. These variables are related to each other by the definitions themselves or through simple equations resulting from known principles such as mass conservation.

A system in which a tracer is distributed and is intended to give information about volumes, flows, transit times, and the general dynamics of the liquid in the system is called a dilution or distribution system.

The biological processes involving the behavior of a particular tracer in a particular space depend on the mean time the tracer takes to go through it (mean transit time) or the mean time the tracer stays in that space (mean residence time). It is useful to analyze these quantities.

Let us suppose we have a tube with a liquid passing along it and two very well-collimated radiation detectors D_1 and D_2 seeing small volumes dV_1 and dV_2 at the input and output of the tube (Figure 7.37a).

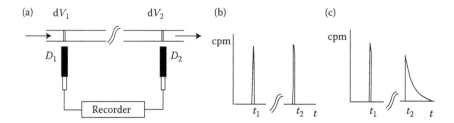

FIGURE 7.37
(a) Tube with a liquid flow and two radiation detectors applied to small volumes dV_1 and dV_2.
(b) Responses of the two detectors D_1 and D_2 as a function of time are obtained in the same
record. (c) Responses of the detectors if a bolus of a radioactive tracer is injected at the entrance
of the tube and assuming the flow in the tube is laminar.

Supposing that a single, small radioactive particle is injected at the entrance
of the tube, and the particle follows its path without incident, the responses of
the two detectors as a function of time in the same plot are as in Figure 7.37b.

The small particle has passed through the tiny volumes seen by the detectors at times t_1 and t_2, respectively. This experiment with just one particle is
probably the only situation in tracer dynamics in which the output curve is
not only predictable but also equal to the input curve.

If we inject not a particle but a bolus of radioactive tracer miscible in the liquid stream and assume the flow regime in tube is laminar, the output response
is different from the input one, as shown in Figure 7.37c. It was assumed in this
example that the input is a very fast injection which we call impulse function
(or impulse injection), or delta function.

The input and output curves are different but the output is predictable,*
that is, if the input is an impulse function and the flow is laminar we can
predict what the output curve will be like.

In most of the situations of interest to biology and medicine, the systems are
much more complex than single tubes. The injected particles can not only have
a range of different paths but they can also be drawn by a range of different
velocities of flow within the system. So these will have to be considered to be
chaotic systems in which the processes are random in nature. We are interested
in the overall statistical properties of the population of particles, and we are
dealing with density functions and statistical distributions. Contrary to the
preceding cases, one cannot deterministically predict the output curve.

Mean transit time is the average time necessary for tracer molecules to move
between two points, which may or may not include a compartment.

Mean residence time is the average time tracer molecules stay in a space
where they have been deposited or dragged to.

Both these times depend on the flow that drags the tracer and on the volume
of the dispersing system.

---

* The response function for a catheter of length L and minimum transit T is $h(t) = \left(\frac{T}{2t^2}\right)$.

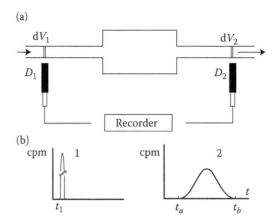

FIGURE 7.38
(a) Residue detection. (b) Responses of detectors D_1 and D_2, in a real system. The output curve is a widened curve showing the transit times of the particles distributed between times t_a and t_b.

Transit times are generally obtained by measuring at input and output, by detection or direct sampling (Figure 7.38b). Since the tracer molecules can have very different paths in a dilution system, their output concentration varies with time to give a transit time curve (or a histogram in the case of sampling). In other words, a time or activity or time or concentration curve is obtained, with a profile that depends on the input function profile and on the characteristics of the dilution system and that can be used to calculate the mean transit time and volume.

Residence times are obtained by residue detection, with which the total tracer activity that remains in the system is measured, generally by external detection (Figure 7.38a). This type of detection is convenient and often the only possibility available, but it has the disadvantage in that an accurate definition and exclusive detection of the regions of interest are sometimes hard to achieve, frequently leading to the need to make further corrections.

Suppose that for a spike of radioactive tracer in a real system curve 2 is obtained at the output (Figure 7.38b).

The particles injected at time t_1 in an impulse injection came out as a widened curve showing the transit times of the particles distributed between times t_a and t_b. There is a statistical distribution of transit times.

We can see from the shape of the curve that the transit times which occurred most often were those located about half way between t_a and t_b. The output curve is an indication of how probable (frequent) a given value of transit time has been.

If we call dV_2 the element of volume seen by the output detector 2, F the flow through the system $C_2(t)$ the tracer concentration at the output then the amount of tracer present in dV_2 between t and $t + dt$ is

$$dq_2(t) = dV_2 C_2(t) \tag{7.181}$$

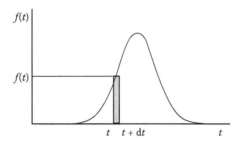

FIGURE 7.39
The area of the shaded rectangle is $f(t)\,dt = dq_2$ and it represents the elementary amount of the label that has crossed the system between instants t and $t + dt$.

Defining a function $f(t) = f(t) = (dq_2/dt)$ which, for every value of t represents the rate of change of the amount of tracer in volume dV_2, is also

$$dq_2(t) = f(t)\,dt \qquad (7.182)$$

In Figure 7.39, the area of the shaded rectangle $f(t)\,dt = dq_2$ represents the elementary amount of tracer that has passed the system between times t and $t + dt$.

The area under $f(t)$ gives the total amount Q of injected label

$$\int_0^\infty f(t)\,dt = \int dq_2 = Q \qquad (7.183)$$

Instead of $f(t)$, it is more convenient to use another function (normalized) given by

$$h(t) = \frac{f(t)}{Q} = \frac{dq_2}{Q}\frac{1}{dt} \qquad (7.184)$$

The product $h(t)\,dt$ is not the absolute amount of tracer that has crossed the system between times t and $t + dt$ but the fraction of the total amount of injected tracer that has crossed the system between t and $t + dt$. Of course, if we carry out an injection such that $Q = 1$, the area under the curve equals unity, the curve is then normalized and $f(t) = h(t)$. The function $h(t)$ is called the frequency function of transit times or unit response function of the system. The dimensions of $h(t)$ are $[T]^{-1}$.

After an instantaneous injection, the fraction of the total amount of label that has left the system through 2 until time t, as a function of time, can be obtained by

$$H(t) = \int_0^t h(t)\,dt \qquad (7.185)$$

We call $H(t)$ the cumulative frequency function (Figure 7.40b). The fraction of the total amount of labeled tracer that has left the system by time t is the area under $h(t)$, between time zero and t, because the total area under $h(t)$ is the unit

$$H(\infty) = \int_0^\infty h(t)\, dt = 1 \qquad (7.185a)$$

An obvious relationship is

$$\frac{d}{dt}(H(t)) = h(t) \qquad (7.186)$$

Transit times and residence times are closely related to each other. For an impulse input, the activity or time curve obtained by external detection at the output that represents the transit times distribution can be converted into the corresponding retention function $r(t)$, that is, the activity or time curve distribution of residence times (Figure 7.40c). In every time instant, the fraction of the total tracer injected that remains inside the system is 1 minus the fraction already eliminated $H(t)$, leading to

$$r(t) = 1 - \int_0^t h(t)\, dt = \int_0^\infty h(t)\, dt - \int_0^t h(t)\, dt = \int_0^\infty h(t)\, dt \qquad (7.187)$$

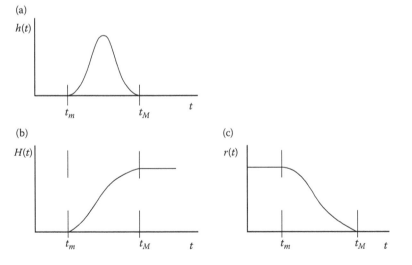

FIGURE 7.40
(a) Frequency function of transit times. (b) $H(t)$ is the cumulative frequency function. It represents the fraction of the total amount of tracer that leaves the system up to time t. (c) $r(t)$ is the retention function, or the fraction of the total amount of tracer within the system as a function of time, that is $1-H(t)$.

Then for every time instant t, the integration of curve $h(t)$ between t and infinity is a value of $r(t)$, the retention function. Also true is

$$\frac{d}{dt}(r(t)) = -h(t) \quad \text{and} \quad h(t)\,dt = -d\,r(t). \tag{7.188}$$

The retention function is the residue curve for a unitary impulse input, and it can be converted to the transit time distribution through differentiation and multiplication by -1.

Let us return to the definition of mean transit time for the case of an impulse injection of an activity Q in a dilution system and enter with Equation 7.187 for residue detection. Then

$$\bar{t} = \frac{\int\limits_{0}^{\infty} tQh(t)\,dt}{\int\limits_{0}^{\infty} Qh(t)\,dt} = \frac{\int\limits_{r(0)}^{0} -t\,dr}{\int\limits_{r(0)}^{0} -dr} = \frac{\int\limits_{0}^{\infty} r(t)dt}{r(0)} \tag{7.189}$$

The mean transit time of a tracer, or the corresponding mean residence time, respectively through or in an organ, space, or compartment is the quotient of the total area under the residue curve and its initial height ($t = 0$ abscissa) (Figure 7.41).

There is another way of obtaining the cumulative frequency function of transit times, apart from the integration of $h(t)$.

If the tracer is introduced into the system at a constant rate, after a sharp rise (step function $i(t)$), the output will be the cumulative frequency function (Figure 7.42).

If the input is a δ function, by definition the output is $h(t)$. Then, since

$$\delta(t) = \frac{d}{dt}[i(t)] \quad \text{and} \quad h(t) = \frac{d}{dt}[H(t)], \tag{7.190}$$

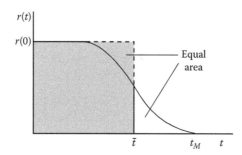

FIGURE 7.41
The mean transit time of a tracer equals its mean residence time, and it is the quotient of the total area under the residue curve and its initial high ($t = 0$ abscissa).

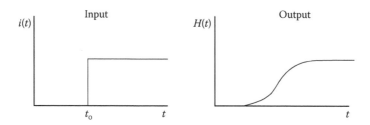

FIGURE 7.42
If the tracer is introduced into the system at a constant rate, after a sharp rise (step function $i(t)$), the output will be the cumulative frequency function.

we conclude that, by differentiation of the functions in Figure 7.42, at the input and output, the same functions are obtained as in the case of an instantaneous injection (Figure 7.43).

For a dilution system in which there is an impulse function of a tracer that is exponentially cleared, the detected residue function of the system is of the type $Q \exp(-\alpha t)$. The mean residence time is $\bar{t} = 1/\alpha$.

If the residue curve is a double exponential with equation

$$Q(t) = Q_1 \exp(-\alpha_1 t) + Q_2 \exp(-\alpha_2 t),$$

the mean residence time will be

$$\bar{t} = \frac{Q_1/\alpha_1^2 + Q_2/\alpha_2^2}{Q_1/\alpha_1 + Q_2/\alpha_2} \tag{7.191}$$

7.2.5.1 The Stewart–Hamilton Principle

Let us consider a dilution system perfused by a constant flow F and that at $t = 0$ an instantaneous injection of an amount Q of tracer is administered.

The output function is $Qh(t)$, and the amount of label leaving the system between times t and $t + dt$ is $dq = Qh(t)dt$. However, dq can also be given by

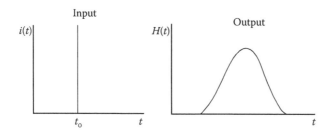

FIGURE 7.43
If input is a δ function, $\delta(t) = \frac{d}{dt}[i(t)]$ then the output is $h(t) = \frac{d}{dt}[H(t)]$.

the product of the elementary volume of liquid that leaves the system in the same time interval, $dV = F\,dt$, multiplied by the tracer output concentration $C_0(t)$:

$$dq = F\,dt\,C_0(t)\,dt$$

Then

$$FC_0(t)\,dt = Q\,h(t)\,dt. \qquad (7.192)$$

Integrating this equation between 0 and ∞, and considering the initial conditions and that $\int_0^\infty h(t)dt = 1$

$$F = \frac{Q}{\int_0^\infty C_0(t)dt} \qquad (7.193)$$

which expresses the Stewart–Hamilton principle that used to be a familiar equation for blood flow determination.

7.2.5.2 Volume Calculation

Let us suppose that in a labyrinth system, at instant $t = 0$, a liquid flow was initiated and simultaneously there was a step function $U(t)$ of a tracer, accompanying the liquid displacement. It is assumed that there is no transition regime in the flow, that is, a stationary situation prevails from the beginning.

With these constrains, a function $H(t)$ will be detected at the output.

For each value of t $d[H(t)]$ is the fraction of the injected tracer with transit times between t and $t + dt$. The value of $d[H(t)]$ is also the fraction of the total number of possible paths in the labyrinth with lengths whose crossing times have to be between t and $t + dt$. Then the flow at the output that corresponds to the paths with these particular lengths is $F\,d[H(t)]$, and the volume of this set of channels is

$$dV = F\,t\,d[H(t)]. \qquad (7.194)$$

The total volume of the system corresponds to all groups of paths with length corresponding to transit times between 0 and ∞ being, using Equations 7.189 and 7.193,

$$V = F\int_0^\infty t\,h(t)dt \qquad (7.195)$$

Considering the integral on the right-hand side of Equation 7.184 and the definition of the mean value of a function is

$$\frac{\int_0^\infty t\,h(t)dt}{\int_0^\infty h(t)dt} = \bar{t} \qquad (7.196)$$

that is, the mean transit time. Then Equation 7.197

$$V = F\bar{t}. \tag{7.197}$$

The volume of the system is equal to the product of the flow and the mean transit time of the tracer through the system. Equality has been proved for a stationary situation.

7.2.6 Distribution Systems and the Convolution Integral

Assume a series of two distribution systems in steady state (Figure 7.44) and that there is an impulse injection of a tracer, $\delta(t)$, at the entrance of the first member, at $t = 0$. Let the frequency functions of transit times for the members of the series be $h_1(t)$ and $h_2(t)$, respectively.

Each liquid particle that leaves the series between time t and $t + dt$ needs a time between τ and $\delta\tau + d\tau$ to cross the first element of the series and a time between $(t - \tau)$ and $(t - \tau) + d(t - \tau)$ to cross the second one.

The fraction of the total number of particles that crosses the first element between times τ and $\tau + d\tau$ is $h_1(\tau)d\tau$.

If at time τ, a δ function is introduced at the input of the second element a shifted frequency function of transit times would be obtained at the output $h_2(t - \tau)$.

All the element volume $h_1(\tau)d\tau$ leaving the first element can be considered a δ function of small amplitude that is introduced at time t in the second element, that is, $h_1(\tau)d\tau d(t - \tau)$, meaning a δ function with amplitude $h_1(\tau)d\tau$ applied at instant $t = \tau$.

The output for this function, which is the output of the series, is

$$h_1(\tau)h_2(t - \tau)\, d\tau \tag{7.198}$$

If all plausible τ values are considered for all t values, that is, if we consider all the transit time values through element 1 which, added to possible transit times through 2, make t, and if we calculate all the products $h_1(t)\, d\tau$ assumed as δ functions for the second element, with outputs $h_1(\tau)h_2(t - \tau)d\tau$, the sum of these output functions is the frequency function of the transit times through

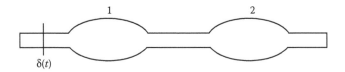

FIGURE 7.44
Series of two distribution systems.

the series, then

$$h(t) = \int_0^t h_1(\tau)h_2(t-\tau)\,d\tau \tag{7.199}$$

If, for the same conditions, instead of considering the particles that leave the first element at time τ, we consider those that need the same time τ to cross the second element, then these particles need time $(t-\tau)$ to cross the first element:

$$h(t) = \int_0^t h_2(\tau)h_1(t-\tau)d\tau \tag{7.199a}$$

The right-hand sides of both Equations 7.198 and 7.199 describe the same $h(t)$.

The physical meaning of this is that if we rearrange the system in such a way that the second element is now the first one, and if we inject a unit impulse at the entrance of the new system the output will have the same frequency function as in the former situation. Equations 7.199 and 7.199a are called convolution integrals, and we have just proved that the sequence of the functions under convolution is immaterial. The simplified notation for convolution of two functions was introduced on the last member of Equation 7.199b.

$$h(t) = \int_0^t h_1(\tau)h_2(t-\tau)d\tau\delta = h_1(\tau) * h_2(t) \tag{7.199b}$$

We can then write,

$$h(t) = h_1(\tau) * h_2(t) = h_2(\tau) * h_1(t). \tag{7.200}$$

One property which could be anticipated from the equations just described is that the convolution of one frequency function on another is itself a frequency function.

Three time characteristics of frequency functions are additive:

1. The shortest transit time through the series must be the sum of the shortest transit times through every member of the series.
2. The longest transit time through the series must be the sum of the longest transit times through every member of the series.
3. The mean transit time through the series must be the sum of the mean transit times through every member of the series.

This last property is not obvious, but it can be proved using some mathematics.

Another important property derives from characteristic number 3 that, if in a series of two elements, we know the flow and the mean transit time through either member then we can find the volumes of the members.

The flow through the series can be written:

$$F = \frac{V1}{\bar{t}1} = \frac{V2}{\bar{t}2} = \frac{V}{\bar{t}} \tag{7.201}$$

where V_1, V_2, and $V = V_1 + V_2$ are the volumes of the elements and the total volume, and $\bar{t}_1$, $\bar{t}_2$ and $\bar{t} = \bar{t}_1 + \bar{t}_2$ are the mean transit times through the elements and the total mean transit time.

Since

$$\bar{t}_1 = \bar{t} - \bar{t}_2$$

and F is known, then

$$V_1 = F(\bar{t} - \bar{t}_2)$$

and similarly

$$V_2 = F(\bar{t} - \bar{t}_1). \tag{7.201a}$$

Among the distribution systems in series found in the body are the pulmonary and systemic circulations.

When samples are collected by means of a catheter, a new distribution system is introduced into the series, the catheter itself.

7.2.6.1 Determination of Flow

We do not need to know the shape of the input function to find the flow and probably the transit time, if the injection starts and ends during the course of the experiment. It is

$$FC_0(t) = i(t)*h(t). \tag{7.202}$$

We can write $i(t) = q\, f(t)$, where q is the total quantity of tracer injected during the experiment and $f(t)$ is the frequency function of the transit times of injection, then

$$F = q\frac{f(t)*h(t)}{C0(t)} \tag{7.203}$$

If $f(t)*h(t)$ is the convolution of two frequency functions, then it is a frequency function and its integral from zero to infinity is unity.

Integrating the above equation for all times

$$F\int_0^\infty C0(t)\, dt = q\int_0^\infty f(t)*h(t)dt \tag{7.204}$$

or

$$F = \frac{q}{\int_0^\infty C_0(t)dt} \qquad (7.205)$$

Thus, one can calculate the blood flow from the total quantity injected and from the area under the tracer time concentration curve, no matter over what period of time the tracer was injected. However, we cannot calculate the volume of the system without knowing more about the input function. Unless the injection is truly instantaneous, the mean transit time of the concentration function at the output is not the same as the frequency function of transit times through the system.

The true mean transit time through the system is the difference between the mean transit time of the concentration function at output and the mean time of the tracer input

$$\bar{t} = \bar{t}_o - \bar{t}_i \qquad (7.206)$$

7.2.7 Regeneration of the Frequency Function of Transit Times by Deconvolution

We are now going to consider some applications of convolution and deconvolution methods.

These are important operations concerning the applicability of the input–output problems of systems to any biological system (not only to dilution systems), provided that certain conditions are satisfied by the system (linearity as well as time invariance, that is, constant behavior over time).

Figure 7.45 is a schematic representation of the functions involved in a process in which a system responds to stimuli.

Function $h(t)$ is the response of the system to a unit impulse, that is, a delta function of unit amplitude at zero time, or in the case of NM, for example, a very fast injection of unit activity at the beginning of the experiment.

The convolution of functions $h(t)$ and $I(t)$ is represented by the integral

$$R(t) = \int_0^t I(t - \tau)h(\tau)d\tau = h(t)*I(t)$$

where $h(t)$ is an operator that converts an object or input function $I(t)$ into an output function $R(t)$. In the above equation, t is the sole variable, and τ is used as an auxiliary variable.

There can be any variables and they may be applied in convolution in 2D or multidimensional spaces.

Through convolution, a response function can be calculated if the input function and the unit response function are known (Figure 7.46).

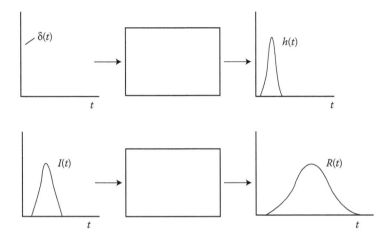

FIGURE 7.45
$h(t)$ is the response of the system to a unit impulse, transit time distribution (or retention function or residence times function). $I(t)$ is the input function and $R(t)$ is the response function (or a residue function).

In most cases, however, the important problem is the inverse of convolution, that is, $I(t)$ is wanted when $R(t)$ and $h(t)$ are known, or $h(t)$ is wanted when $R(t)$ and $I(t)$ are known. In such cases, the unknown integrand function is said to be deconvoluted from the known functions, and the necessary mathematical procedures are called deconvolution. Deconvolution is, then, an operation defined as the inverse of convolution.

The shape of the activity dilution curve of an organ depends on the organ, the blood flow, and also the shape of the tracer bolus applied. In principle, the deconvolution of these curves will remove the effects of the bolus shape, but in many practical situations the results have been shown to be too unstable for reliable use. Continuous deconvolution after fitting with known curves often leads to integral equations with general solutions only in constrained

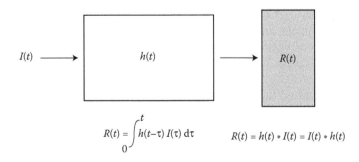

$$R(t) = \int_0^t h(t-\tau)\, I(\tau)\, d\tau \qquad R(t) = h(t) * I(t) = I(t) * h(t)$$

FIGURE 7.46
Input $I(t)$ and response $R(t)$ functions of a system portrayed by a unit response function $h(t)$.

conditions or somewhat special cases. In metabolic processes, the organ is often regarded as the system, the input as the time-activity function, and the response (or retention function) as the time-activity function in the organ itself. Using deconvolution, the residue function of the organ can be recovered.

Biomedical uses of 1D and 2D deconvolution in dynamic and metabolic studies have been reported since the 1950s, particularly in cardiovascular, renographic, brain, kidney, and gastroenterology applications; in image processing; and in reconstruction and restoration algorithms.

Discrete deconvolution is carried out through polynomial division, matrix division, fast Fourier transform algorithms, and numerical function minimization methods. When noise affects the data, the deconvolution procedures are strongly perturbed. Deconvolution is an ill-conditioned procedure; this means that small errors in data obtained for the two measured functions can give rise to large deviations in the solution. The perturbation due to noise in these techniques can be studied by adding known noise content to the functions under deconvolution and using these data to develop efficient methods of noise filtering.

Assuming an output concentration $C_0(t)$ in response to an arterial concentration at input $C_a(t)$ and that these functions are known, then

$$C_0(t) = \int_0^t C_a(t - \tau)h(\tau)d\tau \tag{7.207}$$

Now we want to calculate $h(t)$.

The Laplace transform of Equation 7.207 is

$$C_0(s) = C_a(s)H(s) \tag{7.208}$$

and the inverse transform of $H(s)$ is

$$L^{-1}\{H(s)\} = L^{-1}\left\{\frac{C_0(s)}{C_a(s)}\right\} = h(t). \tag{7.209}$$

In general, $C_0(s)$ and $C_a(s)$ are only empirically determined and have no recognizable, simple, analytical form for formal inverse transformation.

There are two ways of overcoming this difficulty. In the first one, the convolution integral is considered a simple summation and $h(t)$ is calculated either numerically or on a digital computer. In the second, the curves $C_a(t)$ and $C_0(t)$ are empirically fitted to arbitrary formal expressions that can be more easily deconvoluted.

The convolution integral in summation form becomes (Figure 7.47)

$$C_0(t) = \int_{k=1}^t h(t + 1 - k)C_a(k) \tag{7.210}$$

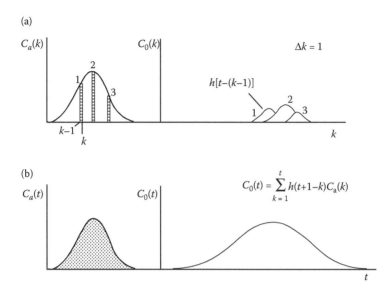

FIGURE 7.47
(a) Input function is decomposed in a sequence of time shifted delta functions represented by small rectangles whose area is equal to the mean value of the ordinate. To each of these functions corresponds an output numbered as 1, 2, and 3. (b) The summation of all input delta functions is $C_a(t)$ and of all the output curves is $C_0(t)$.

where k takes only integer values for integer values of t.

The convolution process can be visualized if the input function $I(t)$ is decomposed in an infinite number of time shifted delta functions with amplitude equal to the value of the function in each time and considering as final result the summation of the responses to all these delta functions.

Most of the situations in practical NM need a discrete approximation of convolution. When the functions involved in convolution are sampled functions, the convolution integral becomes

$$R(k) = \sum_{n=0}^{k} h(k-n)I(n) \tag{7.211}$$

where $R(k)$, $h(k-n)$, and $I(n)$ are the sampled values of the functions. Equation 7.211 can be written in matrix form as

$$R = H \times I \tag{7.212}$$

where I is a triangular matrix with elements $(k+1)$ $(k+1)$ with $n+1$ line consisting of terms $I(n)$, $I(n-1)$, $I(n-2)$,..., $I(0)$ followed by $(k-n)$ zeros. H and R are column matrixes with $k+1$ elements.

In Figure 7.47a, the input function is decomposed into a sequence of delta functions represented by small rectangles whose area is equal to the mean value of the ordinate. To each of these functions corresponds an output function numbered as 1, 2, and 3. The summation of all input delta functions is $C_a(t)$ and the summation of all the output curves is $C_0(t)$.

For example,

$$C_0(t) = h(1)C_a(1) \tag{7.213}$$

$$C_0(2) = h(2)C_a(1) + h(1)C_a(2) \tag{7.214}$$

$$\vdots$$

$$C_0(t) = h(t)C_a(1) + h(t-1)C_a(2) + \ldots + h(1)C_a(t) \tag{7.215}$$

This gives a system of t simultaneous linear equations, the solution of which yields t values for h.

The fitting procedures have included several functions (sum of exponentials, Fourier series, etc.) to fit the frequency functions to formal expressions.

7.2.8 Data Analysis and Models in PET Studies

Immediately after reconstruction, SPECT and PET images are static maps of values proportional to activity in units of counts per second per pixel.

These values are generally converted to units of counts per second per milliliter of tissue by calibration with a known volume of tracer activity.

NM images can be used as static, dynamic, or parametric entities [30–32].

When the biological processes under analysis evolve slowly over time, static images may be good enough to supply the wanted information (e.g., tumor detection). A color look-up table is often used to assign a color to each pixel value in static images.

When functional information is wanted, dynamic imaging acquisition is required. For example, it may be necessary to estimate the parameters of a compartmental model that supposedly fits data from PET or SPECT. These kinetic parameters are important, because they quantify physiological processes.

In dynamic imaging, series of scans are acquired by continuous scanning, often for long times; and after the reconstruction of sequences of PET images, some kind of spatiotemporal quantification is associated with the activity data.

One common approach, for a sequence of PET or SPECT images, is to get the variation over time of the mean value of pixels in ROIs that have been previously rescaled and aligned. To help with the definition of anatomical ROIs in the brain, the higher-resolution MRI or CT is commonly used. The TACs are generated with data points that correspond to the mean pixel value in a specific ROI at a given time.

These TACs can be further normalized by plotting, from point-to-point, the ratio of the values of one ROI TAC to the values of another ROI TAC that is taken as a reference.

For quantification some reference is needed, and this can be the tracer concentration in arterial plasma (blood sampling required) or some other tissue (a different area of the same image).

In PET studies, the tracers are introduced into the body by intravenous injection or inhalation. The tracers are mixed with blood in the heart chambers to give a nearly constant concentration in arterial blood (model input). The concentration of the tracer delivered to the tissue capillaries can be obtained from any peripheral artery, and the amount of tracer delivered to the tissue is proportional to the blood flow (perfusion).

Some of the possibilities for finding the input function to an organ are by external measurement in the left ventricle, aorta, or large artery as a function of time (i.e., the respective TAC). This method, which requires fast response equipment to cope with the temporal sampling requirements of the particular input function, is also limited to those cases in which the left ventricle or a great vessel is located within the limits of the image [33,34].

Statistical methods that will be discussed later can also be used to effectively obtain these input functions.

Depending on their biochemical properties, the radiotracers used in PET studies are grouped in two broad categories.

The first category comprises the nonspecific radiotracers that trace a biochemical path and lead to the measurement of the extraction parameters of a tissue which describe its uptake and metabolism. Examples of these radiotracers are $^{15}OH_2$, which is an inert, freely diffusible tracer used to measure cerebral blood flow; ^{18}FDG, which traces the initial phases of glucose metabolism but does not follow Kreb's cycle after phosphorylation and is, therefore, effectively retained in cells, permitting the evaluation of glucose metabolism in tissue; and ^{18}F-fluoromisonidazole, which is a bioreductor drug that follows an intracellular path of reduction and can be used to localize viable hypoxic tissue. The kinetics of the radiopharmaceuticals of this kind can be evaluated through simple systems consisting of one or two compartments apart from plasma.

The second category includes specific radiotracers involved in interactions with a receptor, a carrier, or a specific interaction site. Examples of these radiotracers are ^{11}C-flumazenil, which is an antagonist with high affinity and selectivity for central benzodiazepine receptors, and ^{11}C-SCH23390, an antagonist with high affinity and selectivity for dopamine D1 receptors. Both radiotracers are used to study alterations in the density and affinity of central receptors. The behavior of the radiopharmaceuticals in this category is generally assessed by means of three-compartment models: free, nonmetabolized ligand in plasma; free ligand in tissues; and ligand specifically bound to tissues.

In tracer studies, the data acquisition methodology and the recommended data analysis method depend on the aim and type of study being performed.

Different approaches have been used for the kinetic modeling of images from SPECT and PET tracer studies, with the purpose of estimating quantitative biological parameters.

Despite their many differences, these methods—mostly only used for brain PET studies—can be roughly placed into general categories; for example, in those based on data (data-driven), where no models or compartments are assumed, and those that use models (model-driven) which tend to assume a compartmental system.

The data-driven methods require no previous decision about a model associated with the system (Figure 7.48).

After reconstruction, the images are used to draw ROIs and obtain TACs that will be subjected to one of several graphical procedure sequences. These correspond to the path 3-4-5-6 in Figure 7.48.

Tracer ROI curves can also be used for modeling. This corresponds to the path 3-4a in Figure 7.48 and 4a-5a-6a in Figure 7.49.

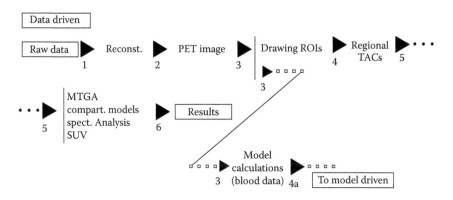

FIGURE 7.48
Data-driven methods for data analysis in NM.

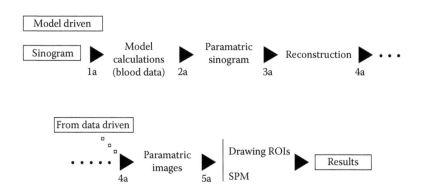

FIGURE 7.49
Model-driven methods of analyzing NM data.

The ROI TACs are fitted to mathematical models to estimate parameters that are biochemically or physiologically meaningful (path 4-5). These, when the input function is known, can be rate constants, receptor densities, or blood flows. The input function is usually the arterial plasma activity that is sampled during the PET study and counted in the well counter. This approach allows comparison of the model parameters in PET studies.

Multiple time graphical analysis methods and spectral analysis are generally used to quantify the dynamics of the processes.

The MTGA is a series of data-driven graphical procedures used to quantify PET data that involve the study of the distribution and accumulation of radioactive tracers in organs and tissues over time. The technique is an alternative to compartmental models, as it does not assume an imposed chain of predetermined events and it is suitable for both homogeneous and nonhomogeneous tissues.

The data are usually acquired as a series of dynamic frames, beginning with the tracer injection and continuing until either the statistics of the images is compromised by prolonged decay or there is a limitation in memory storage capacity.

The MTGA methods (Patlak and Logan plots) employ a transformation of the data such that a linear regression of the transformed data yields macro system parameters which are computed with better reproducibility from fewer data than the separate rate constants [35,36]. By plotting these macroparameters, the functional parameters are directly estimated from the slope of the linear phase of the curves obtained.

Some negative points about these methods are the uncertainty about the choice of the point in the plot when linearity begins, the possibility of bias introduced by statistical noise [29], and the failure to return any information about the underlying compartmental structure.

Spectral analysis [35] characterizes the system's impulse response function (IRF) as a positive sum of exponentials and uses nonnegative least squares to fit a set of these basic functions to the data. The macro system parameters of interest are then calculated as functions of the IRF [34,35]. Spectral analysis also returns information on the number of tissue compartments evident in the data and is defined as a transparent technique.

Schmidt [30] showed that for the majority of plasma input models the observation of all compartments led to only positive coefficients; and, as such, the spectral analysis [35] solution using nonnegative least squares is valid.

The ROI-based methods may be further classified into linear and nonlinear methods. The former transform the data so that the parameters of interest can be estimated by linear regression methods, whereas nonlinear techniques generally estimate the kinetic parameters by iterative minimization.

The two different types of analysis (Patlak and Logan) are used according to the type of situation.

The Patlak-plot applies to irreversibly binding radiotracers and involves creating an ordinate and an abscissa using combinations of the input function and image data at each of the observed time points for every PET frame. A TAC is obtained for a particular ROI.

For each time point, the ROI values represent the overall concentration of radioactivity in the tissue, which can differ from the tracer concentration.

A TAC is created for the input function $Cp(t)$, the concentration of tracer in the plasma, by interpolating the measured plasma curve to the middle of each frame.

For every image frame, the ratio of the ROI value in that frame divided by the plasma concentration is plotted as a function of the integral of the plasma curve (from $t = 0$ to the current time) divided by the plasma concentration at that time point, that is,

$$\text{ROI}(t)/C_p(t) \quad \text{versus} \quad \int_0^t C_p(t)dt/C_p(t) \qquad (7.216)$$

where $\text{ROI}(t)$ is the concentration of a tracer in a tissue ROI, t is the time elapsed from injection to a particular frame, and $Cp(t)$ is the concentration of tracer in plasma. In these instances, the net influx rate K_i (min^{-1}) equals the slope of the Patlak plot, that is, $k_1 \times k_3/(k_2 + k_3)$.

This method can be thought of as a way of scaling the tracer concentration in a region of tissue by the capacity of the tissue to absorb the tracer over time.

However, it was noticed that this plot becomes linear at those time instants when the transport of tracer trapped into the compartment is essentially unidirectional.

Patlak showed that the slope of a straight line fitted to this linear region is equivalent to the influx constant, K_i, for the system. K_i can be defined as the ratio of the total amount of tracer accumulated in a tissue region after an infinite amount of time, divided by the integral of the plasma TAC from $t = 0$ to infinity.

Graphical methods are easy to perform and are generally considered more robust than kinetic analysis with (full) compartmental models, especially for noisy data sets.

The traditional linear regression model takes into account only the errors in the Y variable.

The bias in the (reference input) Gjedde–Patlak slope may be diminished by using a linear regression model that accounts for errors in both variables [32]. This method can also take into account separate weights for both variables.

In another work (1985), Patlak has shown that the technique still applies when the tracer is simply "nearly" irreversibly bound and that a tissue reference region, with no specific tracer uptake, can be used as input function instead of the measured blood TAC.

This analysis is used in neurotransmitters ([^{18}F]-F-DOPA, [^{18}F]-FMT, [^{11}C]-raclopride, etc.), metabolism markers ([^{18}F]-FDG), and even for cardiac applications.

The Logan plot is a graphical linearization method that applies to reversibly bound tracers which have a constant net efflux that is nonnegligible when compared with the constant influx. The approach involves creating a plot with the following coordinate axis:

$$\int_0^T ROI(t)dt/ROI(t) \quad versus \int_0^T C_p(t)dt/ROI(t) \tag{7.217}$$

where $ROI(t)$ is the concentration of a tracer in a tissue ROI, t is the time elapsed from injection to a particular frame, and $Cp(t)$ is the concentration of tracer in plasma [33].

Similar to the Patlak plot, the slope and intercept of a Logan plot have distinct interpretations depending on the actual model one chooses to ascribe to the underlying system.

The straight line in a Logan plot should be fitted through those frames that are linear; these frames usually occur somewhere in the middle of the plot. This is another way of saying that data should be collected over a long enough period to see a net efflux of tracer from the ROI.

A formal analysis of the Logan plot shows that it is valid for an arbitrary number of compartments for both plasma and reference input models when the data are free from noise.

7.2.9 Parametric Analysis

Parametric analysis is required when the studies involve complex analysis and advanced statistical methods to localize the regional distribution of functions, particularly in situations in which input function is affected by noise. Parametric images incorporate the highest degree of quantification and require highly developed statistical methods that are mostly used in brain studies.

In parametric analysis, information is obtained up to the image resolution level, that is, the parameters are calculated for each image voxel and the results are represented as maps of the parameter.

The values bound to the image voxels are the physiological parameters studied (perfusion, glucose consumption, receptor density, etc.).

Existing methods for computing kinetic parametric images work by first reconstructing a sequence of PET images and then estimating the kinetic parameters for each voxel in the imaged volume. Special computer programs convert the dynamic information to parametric functional information.

Parameter estimates are obtained from *a priori* specified compartmental structures using one of a variety of least-squares fitting procedures: linear least squares; nonlinear least squares; generalized linear least squares; weighted integration; or base function techniques [37,38].

Image realignment, nonlinear automatic spatial normalization, segmentation, coregistration, smoothing, development of statistical parametric maps, development of posterior probability maps, and Bayesian estimates of parameters from generalized linear models (GLM) are steps that are included in these methods' data processing steps.

The major advantages of parametric images are the better low contrast detection, a better ROI definition, and their allowing the use of powerful SPM* (statistical parametric mapping) algorithms.

Disadvantages that have been pointed out for this type of analysis are the fact that it is similar to processing a black box, the distribution of regional values is frequently distorted, and it is computationally more intensive.

7.2.10 Monte Carlo Simulation

In NM, the development of new imaging devices, reconstruction algorithms, correction and optimization techniques, acquisition protocols, and description of time variable phenomena, such as detector or source movements, are frequently based on appropriate simulations, particularly using Monte Carlo techniques similar to those used in particle physics.

Quite a lot of Monte Carlo simulation packages have been used for either SPECT or PET, with different advantages and disadvantages and varying degrees of success.

Accurate and versatile general-purpose simulation packages such as Geant3 from CERN, EGS4 from SLAC, and MCNP from the Los Alamos National Laboratory contain well-validated physics models, geometric modeling tools, and efficient visualization utilities, but they require a major effort to be tailored to PET and SPECT.

On the other hand, dedicated Monte Carlo codes developed for PET and SPECT suffer from a variety of shortcomings. For example, SimSET, developed at the University of Washington, is one of the most powerful dedicated codes for PET and SPECT simulations; it precisely and efficiently models physical phenomena and basic detector designs [36]. However, the range of detector geometries that can be simulated is limited: for instance, a detector ring cannot be subdivided into individual crystals. In addition, neither SimSET nor other dedicated codes explicitly account for time, which limits their use for modeling time-dependent processes such as the movement of tracers.

Clearly, there is a need for a Monte Carlo tool that readily accommodates complex scanner geometries while retaining the comprehensive physical-modeling abilities of the general-purpose codes. To meet this demand, an

---

* SPM (statistical parametric mapping) is a statistical technique designed to examine the differences in brain activity registered during functional studies in neuroimaging with PET and MRI.

international collaboration of physicists at centers in different countries has developed the simulation toolkit Geant4, which has recognized advantages over its predecessors.

References

1. L. Bertalanffy. *General Systems Theory*, George Brazillier, NY, 1969.
2. R. Flood and E.R. Carson. *Dealing with Complexity, An Introduction to the Theory and Applications of Systems Science*, Plenum Press, New York, 1993.
3. http://en.wikipedia.org/wiki/Claude_Bernard.
4. http://en.wikipedia.org/wiki/Walter_Bradford_Cannon.
5. http://www-groups.dcs.st-and.ac.uk/~history/Mathematicians/Clausius.html.
6. http://nobelprize.org/nobel_prizes/chemistry/laureates/1977/prigogine-autobio.html.
7. G. Vries, T. Hillen, M. Lewis, J. Müller, B. Schönfisch. *A Course on Mathematical Biology. Quantitative Modeling with Mathematical and Computational Methods*, SIAM, Philadelphia, 2006.
8. http://www.math.montana.edu/frankw//ccp/modeling/continuous/twovars/body.htm.
9. F.C. Hoppensteadt, C.S. Peskin. *Modeling and Simulation in Medicine and the Life Sciences*, 2nd ed., Springer-Verlag, London, 2004.
10. E.N. Bruce. *Biomedical Signal Processing and Signal Modeling*, John Wiley & Sons, Inc., Hoboken, New Jersey, 2001.
11. C. Cobelli, D. Foster, G. Toffolo. *Tracer Kinetics in Biomedical Research*, Kluwer Academic, NY, 2000.
12. D.A. Linkens. *Biological Systems, Modelling and Control*, Inspec/IEE, London, 1979.
13. L. Edelstein-Keshet. *Mathematical Models in Biology*, SIAM Classics in Applied Mathematics, Philadelphia, 2005.
14. C.T. Chen, *Systems and Signals Analysis*, 2nd Ed, Saunders College Publ, Philadelphia, 1994.
15. R.C. Hilborn. *Chaos and Nonlinear Dynamics*, Oxford University Press, Oxford, UK, 1994.
16. http://math.rice.edu/~dfield/matlab7/pplane7.m.
17. *The Mathworks*, Inc., Natick, MA 01760, www.mathworks.com.
18. C. Halldin, C. Foged, L. Farde, P. Karlsson, K. Hansen, F. Grønvald, C.G. Swahn, H. Hall, G. Sedvall. [11C]NNC 687 and [11C]NNC 756, dopamine D-1 receptor ligands. Preparation, autoradiography and PET investigation in monkey, *Nucl. Med. Biol.*, 20(8):945–53, 1993.
19. D. Garfinkel. Computer modeling complex biological systems, and their simplifications, *Am. J. Physiol.*, 239: R1–R6, 1980.
20. R.E. Carson. The development and application of mathematical models in nuclear medicine, *J. Nucl. Med.*; 32:2206–2208, 1991.
21. J.T Kuikka, J.B. Bassingthwaighte, M.M. Henrich, L.E. Feinendegen. Mathematical modelling in nuclear medicine, *Eur. J. Nucl. Med.*; 18: 351–362,1991.
22. C.B. Sampson Ed. *Textbook of Radiopharmacy. Nuclear Medicine—A Series of Monographs and Texts*, Ordon & Breach Science Publ, Colchester, 1990.

23. E.M. Landaw, J.J. DiStefano III., Multiexponential, multicompartmental, and non-compartmental modeling, II. Data analysis and statistical consideration, *Am. J. Physiol.*; 246:R665–R677, 1984.
24. K. Zierler. A critique of compartmental analysis, *Annu. Rev. Biophys. Bioeng.*, 10:531–562, 1981.
25. M. Laruelle. Modelling: when and why?, *Eur. J. Nucl. Med.*, 26:571–572, 1999.
26. B.W. Reutter, G.T. Gullberg, R.H. Huesman. Kinetic parameter estimation from attenuated SPECT projection measurements. In *Conference Record of the 1997 IEEE Nucl. Sci. Sympos. and Med. Ima.g Conf.*, Albuquerque, NM, pp. 1340–1344, 1997.
27. T.F. Budinger, R.H. Huesman, B. Knittel, R.P. Friedland, S.E. Derenzo, Physiological modeling of dynamic measurements of metabolism using positron emission tomography. In: T. Greitz et al. *The Metabolism of the Human Brain Studied with Positron Emission Tomography*. Raven Press, New York, NY, 1985.
28. A.A. Lammertsma. Radioligand studies: imaging and quantitative analysis. *Eur. Neuropsychopharmacol.*, 12:513–516, 2002.
29. M. Slifstein, M. Laruelle. Effects of statistical noise on graphic analysis of PET neuroreceptor studies. *J. Nucl. Med.*, 41:2083–2088, 2000.
30. K.C. Schmidt, F.E. Turkheimer. Kinetic modeling in positron emission tomography, *Q. J. Nucl. Med.*, 46:70–85, 2002
31. M.N. Wernick, J.N. Aarsvold. *Emission Tomography*, Elsevier, Amsterdam, 2004.
32. C.S. Patlak, R.G. Blasberg. Graphical evaluation of blood-to-brain transfer constants from multiple-time uptake data. Generalizations, *J. Cereb. Blood. Flow. Metab.*, 5:584–590, 1985.
33. J. Logan. Graphical analysis of PET data applied to reversible and irreversible tracers, *Nucl. Med. Biol.*, 27:661–670, 2000.
34. R.N. Gunn, S.R. Gunn, V.J. Cunningham. Positron emission tomography compartmental models, *J. Cereb. Blood Flow Metab.*, 21:635–652, 2001.
35. V.J. Cunningham, T. Jones. Spectral analysis of dynamic PET studies. *J. Cereb. Blood Flow Metab.*, 13:15–23, 1993.
36. Z.J. Wang, Q. Peng, K.J.R. Liu, Z. Szabo. Model-Based Receptor Quantization Analysis for PET Parametric Imaging, *Eng. in Med. and Biol. Soc. IEEE-EMBS 2005.* 27th Ann. Int. Conf. Vol., Shanghai, Sept. 2005, pp. 5908–5911, 2005.
37. Statistical parametric mapping, 2009. http://www.fil.ion.ucl.ac.uk/spm/.
38. Allison, J., Amako, J., Apostolakis, K., et al., Geant4—Developments and applications, *IEEE Trans. Nucl. Sc.* 53(1), 270–278, 2006.

8

Dosimetry and Biological Effects of Radiation

Augusto D. Oliveira and J. J. Pedroso de Lima

CONTENTS

8.1 Introduction

In living matter, atoms and molecules maintain their individual and collective configurations as a result of very specific energy interactions, which are subject to complex physical–chemical and biological rules. Living matter at all levels can be understood as a set of active processes that organize matter and energy within ordered complex systems.

Radiation interactions with living matter release energy in biological tissues through processes that increase the entropy of the biological system, often beyond the limits that the organisms can tolerate, leading to the harmful effects of radiation.

This perturbation in the organization of energy and matter can lead to a great variety of biological consequences, ranging from local recoverable situations to complex irreversible modifications. Such biological effects are related to the energy of molecular bonds and of the reactions in living processes.

In living matter, the orders of magnitude for molecular and atomic binding energies are from a fraction to tens of eV; those for molecular vibrational states are tenths of eV; and, for molecular rotational states, they are hundredths of eV.

The lowest ionization potentials of elements, molecules, and radicals with biological interest lie between 11 and 14 eV. Examples include 11.24 eV for C, 13.54 eV for H, 13.57 eV for O, 14.24 eV for N, 13.12 eV for CH_4, 11.42 eV for C_2H_2, 10.56 eV for C_2H_4, 11.62 eV for C_2H_6, and 9.96 eV for $-CH_3$.

Radiation acts on matter by transferring its energy to the medium (kinetic energy in the case of particles and electromagnetic energy in the case of photons). In the case of photons, the ionization results mainly from secondary electrons generated by photoelectric, Compton, and pair production effects.

Atomic and molecular energy levels in living tissues can be perturbed by energy delivered by radiation; and as a consequence, this can initiate a chain of various possible events, depending on the type and the energy transferred, which can include energies below the minimum ionization potential (11 eV), down to the thermal energies.

Some characteristic examples of biological interest are the energies necessary to break covalent (ca. 5 eV) and hydrogen bonds (0.1–0.2 eV); the energies necessary to separate molecular dipoles (0.01–0.03 eV); the dielectric relaxation energies of proteins and water (0.4 and 0.2 eV, respectively, at 25°C); and the energy values for inducing photo-conduction (1–4 eV).

Below the minimum molecular rotation levels, the quantized energy levels in living tissues are so numerous and close to each other that energy absorption from radiation can be considered a continuum process, unlike the discrete events which occur at high energies (such as the photoelectric effect, Compton effect, etc.).

Energy from the absorption in tissue of photons having thermal energy is shared by a huge number of electrons of the medium and directly results in a bulk increase in temperature.

Particularly important types of radiation, relevant to biological action, are the so-called ionizing radiations that are those which are capable of inducing ionization in biological structures. Within this group, ionizing radiations include short wavelength electromagnetic radiations (remote UV rays, x- and γ-rays, and cosmic photons), particles, and heavy ions.

When considering ionization within biological tissues, it is important to consider the fact that ionizing radiation has to deliver an average of ~34 eV in tissue to produce an ion pair. This means that ionizing radiation together with ionization induces other atomic and molecular processes that do not involve an electrical charge, and these events are the most probable. In fact, for every ionization event that occurs, about 20 eV, that is, nearly twice the biological ionization potential, are dispensed in excitations of energy levels, either atomic or molecular, and other processes, all requiring energy below the ionization potential. This corresponds to the delivered 34 eV minus the 11–14 eV necessary to produce ionization.

Ionization creates an electrical imbalance, which, in the first stage, causes the newly created ion to attempt to regain electrical stability by combining with the nearest available opposite charge. This generally causes alterations incompatible with the surrounding cell structure and leads to cell damage or death.

The most sensitive target for radiation effects in tissues is the deoxyribonucleic acid (DNA) molecule, and it is accepted that the effects on this molecule are the end process of a chain of physical and chemical events initiated by the ionizing interaction and followed by biological changes of various types. However, recently, several nontargeted effects have been identified and are under intense research, such as the bystander effect, in which cells that are not directly exposed to radiation show the same effects as exposed cells.

The timing of the most important events that take place when ionizing radiation impinges on biological tissue (at time $t = 0$) is accepted to be about 10^{-15} s for ejection of secondary electrons (ionization) or atomic excitation, about 10^{-10} s for formation of radical ions, about 10^{-5} s for formation of free radicals, and 10^{-15}–10^{-3} s for breaking molecular bonds. The species produced free radicals, molecules, and ions; diffuse in the medium; and can disturb the normal chemical intracellular reactions. Over time, varying from seconds to hours, cellular lesions may appear, whereas stochastic biological effects may only become evident years later [1].

In NM, the main interests are in the ionizing and excitation effects of high-energy photons (γ- and x-rays) or of the resulting secondary electrons.

8.2 Interaction of Radiation with Matter: Photons

8.2.1 Introduction

There are five types of interactions that must be considered with regard to applications of radiation physics in medicine and radiological protection problems:

1. Compton effect
2. Photoelectric effect
3. Pair production
4. Coherent or Rayleigh scattering
5. Photonuclear interaction

The first three of these are the most important, because they lead to energy transfer to electrons, which, by Coulomb interactions along their track, impart the energy to matter. Rayleigh scattering is elastic; the photon only changes the direction of their track by a small angle, without loss of energy. The photonuclear interaction is only significant for energies above a few MeV, where it can cause problems in radiation protection resulting from the production of neutrons and subsequent induction of radioactivity.

The photoelectric effect is dominant at low energies, whereas the Compton effect is more important at intermediate energy values, and, finally, pair production is the main interaction at higher energies. For materials of low atomic number, Z, for example, carbon, air, water, human tissue, etc., the domain zone of the Compton effect is very wide, extending from approximately 20 keV to 30 MeV.

In Figure 8.1, the relative cross sections are shown, as an example, for water for energies up to 200 keV. Water is often used as a material whose properties are equivalent to those of biological tissue.

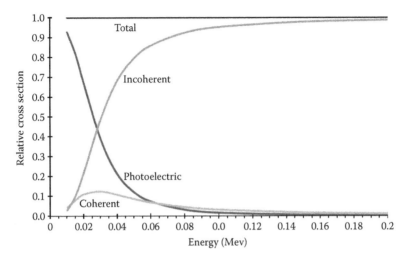

FIGURE 8.1
Graphical representation of cross-sections, normalized to unity, which includes the photoelectric effect, the coherent scatter, the incoherent or Compton scattering, and the total sum. The pair production exists only for energies above 1022 keV.

8.2.2 Cross Section

Within the context of ionizing radiation, the word *interaction* is applied to processes in which the energy and/or the direction of the radiation are changed. These processes are stochastic; therefore, we can only speak in terms of probabilities of the occurrence of interaction. These probabilities can be expressed in terms of cross-sections or interaction coefficients.

Let us consider a layer of a given material inside of which N scatter centers exist per unit of volume. Let us consider an infinitesimal layer thickness dx and consider a front surface with area A. The total number of scattering centers within the layer is $NA\,dx$.

Since the atoms and molecules are not points, we can consider that a collision or scattering occurs when the incident particles collide inside a spherical surface, of radius R, centered on the atoms or molecules. Therefore, there will be an effective collision surface, $\sigma = \pi R^2$, for each of the scatter centers, and the total collision surface for the layer dx will be $NA\,dx$.

The probability of collision for an incident particle with a scatter center in the layer dx, of frontal surface A, is defined as

$$dw = \frac{\text{total collision surface}}{\text{frontal surface}}, \tag{8.1}$$

that is,

$$dw = \frac{NA\sigma\,dx}{A} = N\sigma\,dx. \tag{8.2}$$

The interaction probability, per unit length of the particle track, is

$$\frac{dw}{dx} = N\sigma. \tag{8.3}$$

The quantity σ is termed the *cross-section*, and from the above discussion, we can express the probability of the occurrence of a given process in terms of the cross-section. If the process is one of absorption, scattering, etc., we obtain a cross section for absorption, scattering, etc.

The quantity $\mu = N\sigma$ is termed the *linear attenuation coefficient* and has units of (m^{-2}), because N is the number of atoms per unit volume (m^{-3}); whereas σ has the units (m^2) per atom.

The quantity, λ, defined by

$$\frac{1}{\lambda} = N\sigma, \tag{8.4}$$

is called the mean free path (the mean value of the space traveled by the particle between two consecutive collisions).

Let us consider a layer of a material with thickness x, and let n be the number of particles with a given energy incident in the layer. For a given layer thickness, dx, the variation of the number of the particles will be dn and leads to

$$dn = -\mu n \, dx. \tag{8.5}$$

By integration, we obtain*

$$n = n_0 \exp(-\mu x). \tag{8.6}$$

Sometimes, it is convenient to use the expression

$$n = n_0 \exp\left(-\left(\frac{\mu}{\rho}\right)\rho \times x\right). \tag{8.7}$$

The coefficient (μ/ρ) is termed the *linear mass attenuation coefficient* and has the units $(m^2 \, g^{-1})$.

---

* Resolution of Equation 8.5

$$dn = -\mu n \, dx$$

$$\frac{dn}{n} = -\mu \, dx$$

$$\ln n = -\mu x + \ln C$$

$$n = C \exp(-\mu x) \ (x = 0),$$

$$n_0 = C \exp(0) = C$$

$$n = n_0 \exp(-\mu x)$$

In general, the cross section depends on the scattering angle, and we can also define the differential cross-section,

$$d\sigma = \sigma(\theta)d\Omega, \tag{8.8}$$

which considers the scatter within a solid angle $d\Omega$, which forms an angle θ with the incident direction. The quantity $\sigma(\theta) = d\sigma/d\Omega$ gives us the angular distribution of the cross section and, therefore, a measure of the anisotropy of the process.

One can calculate σ, the total cross section, from $\sigma(\theta)$, if the phenomenon is azimuth independent, with regard to the incident direction, hence

$$d\Omega = 2\pi \sin\theta\, d\theta. \tag{8.9}$$

We have

$$\sigma = 2\pi \int_0^\pi \sigma(\theta) \sin\theta\, d\theta, \tag{8.10}$$

expressing σ in (m^2) per atom or nucleus.*

8.2.3 Compton Effect

The description of the Compton effect can be divided into two parts: kinematics and cross section. The first relates to energies and angles of the particles involved in a Compton interaction. In both cases, it is usually assumed that the electron in a collision is a free and stationary electron. This assumption is not rigorous, because the electrons, which can occupy several energy levels, are moving around and are bonded to the atomic nucleus. However, the resulting errors are of little consequence in radiation physics as applied to medicine under conditions of high atomic number and low energy due to the competing preponderance of the photoelectric effect, where the electron bonds are more important for the Compton interaction.

A photon of energy $\hbar w$ and linear momentum $\vec{k}$ can collide with a free and stationary electron (Figure 8.2).

The electron is ejected with energy T and angle θ in relation to the direction of the incident photon. The scattered photon has energy $\hbar w'$, linear momentum $\hbar w'/c$, and angle ϕ in the plane defined by the direction of the incident photon and the direction of the ejected electron.

The assumption of a free electron means that the kinematic relationship is independent of the atomic number of the medium.

---

* The unit often used in nuclear physics for cross-section is 1 barn $= 10^{-24}$ cm^2.

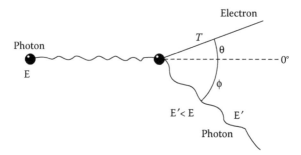

FIGURE 8.2
Kinematics of the Compton effect.

One can obtain the following kinematic equations for the Compton effect,

$$\hbar w' = \frac{\hbar w}{1 + (\hbar w/m_0 c^2)(1 - \cos \phi)},$$
(8.11)

$$T = \hbar w - \hbar w',$$
(8.12)

$$\cot \theta = \left(1 + \frac{\hbar w}{m_0 c^2}\right) \tan \left(\frac{\phi}{2}\right),$$
(8.13)

where $m_0 c^2$ (rest mass energy of the electron) is 0.511 MeV; and the energies $\hbar w$, $\hbar w'$, and T are expressed in MeV.

8.2.4 Thompson Scattering

In the Thompson description, both the incident photon and the scattered photon have the same energy. The electron does not acquire kinetic energy as a result of the elastic collision.

Thompson pointed out that considering the differential cross-section per electron for a photon scattered at an angle ϕ, per unit of solid angle, one can write

$$\frac{d_e \sigma_0}{d\Omega_\phi} = \frac{r_0^2}{2}(1 + \cos^2 \phi),$$
(8.14)

in units of ($\text{cm}^2 \text{ sr}^{-1}$ per electron), where

$$r_0 = \frac{e^2}{m_0 c^2} = 2.818 \times 10^{-13} \text{ cm},$$
(8.15)

which is named the *classic radius* of the electron; and the subscript e in $_e\sigma$ refers to a cross-section per electron.

We can conclude that there is a forward–reverse symmetry, meaning that Equation 8.14 has the same value for $\phi = 0°$ and $\phi = 180°$ and has half of this value for $\phi = 90°$.

The total cross-section per electron, $_e\sigma$, is obtained by integration of Equation 8.14 over all possible scattering angles. A further simplification can be introduced by assuming cylindrical symmetry and integrating over $0 \le \phi \le \pi$ and $d\Omega_\phi = 2\pi \sin \phi \, d\phi$.

$$_e\sigma = \int_{\phi=0}^{\pi} d_e\sigma_0 = \pi r_0^2 \int_{\phi=0}^{\pi} (1 + \cos^2 \phi) \sin \phi \, d\phi, \tag{8.16}$$

$$_e\sigma = \frac{8\pi r_0^2}{3} = 6.65 \times 10^{-25} \text{ cm}^2/\text{electron}. \tag{8.17}$$

This leads to a value independent of the energy.

It is well known that this cross section can be viewed as an effective target area and is numerically equal to the probability of occurrence of a Thompson scattering when a photon crosses a layer that contains one electron per cm^2.

8.2.5 Klein–Nishina Cross Section for the Compton Effect

In 1928, Klein and Nishina applied the Dirac relativistic theory of the electron to the Compton effect in order to obtain better values for the cross section.

The value obtained by Thompson (Equation 8.17), which is independent of the energy, has an error that can be up to a factor of 2 for energies of the order of 0.4 MeV. Klein and Nishina have had success in reproducing the values experimentally obtained, even assuming free electrons, initially at rest.

The differential cross-section for the scattering of a photon over an angle ϕ, per unit of solid angle and per electron, is given by

$$\frac{d_e\sigma_0}{d\Omega_\phi} = \frac{r_0^2}{2} \left(\frac{\hbar w'}{\hbar w}\right)^2 \left(\frac{\hbar w}{\hbar w'} + \frac{\hbar w'}{\hbar w} - \sin^2 \phi\right), \tag{8.18}$$

where the energy $\hbar w'$ is given by Equation 8.11 and the subscript e in $_e\sigma$ refers to the cross-section per electron.

Equation 8.18 is graphically shown in Figure 8.3.

It can be seen that the highest probabilities occur for angles near the incident axis. For backscattering, that is, angles between 90° and 270°, the highest value for the probability is at exactly 180°.

For lower energies, $\hbar w' = \hbar w$, and Equation 8.18 is reduced to the Thompson expression (Equation 8.14).

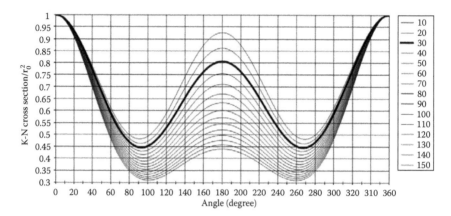

FIGURE 8.3
Klein–Nishina cross-section for several values of the energy in the range 10–150 keV.

The Klein–Nishina total differential cross section per electron is obtained by integration of Equation 8.18 over all scattering angles,

$$_e\sigma = 2\pi \int_{\phi=0}^{\pi} \frac{d_e\sigma}{d\Omega_\phi} \sin\phi \, d\phi \tag{8.19}$$

8.2.6 Corrections of the Bonding Energy of the Electrons

In the theory of gamma radiation transport, the bond energy of the electrons is, in general, approximately ignored or treated. The justification for this is that for elements of low Z, the bonding energies in the K shell are small when compared with the photon energies; whereas for materials with high Z (with rather high bonding energy in shell K), these electrons are a small fraction of the total.

In the Compton effect, a photon collides with one of the atomic electrons, transferring some of the momentum. When the electron acquires enough energy, it escapes from the atom or is excited to an unoccupied atomic level. If the energy is very small for any of these possibilities, then the electron remains bonded in its orbital, and the atom reacts as a single entity. In general, as the bonding energies are small compared with the photon energies, the electron is released, leaving the excited ion with an excitation energy equal to the bonding energy of the shell from which the electron was removed. After the previous process, the atom returns to the ground state through the emission of fluorescent photons and Auger or Coster-Kroning electrons. The Compton effect is similar to the photoelectric effect, as in both interactions there is energy transfer, leading to the appearance of an energetic electron and an ionized atom.

In general, as has already been mentioned, ionization or excitation of the atom is not considered in the energy transfer process in the Compton interaction. It is assumed that the bond energy of the electron is less than the energy acquired in the collision; and, therefore, the ionization energy of the atom is ignored. This approach was considered by Klein and Nishina when they assumed that there is a free electron.

When the bond energy in the atom is considered, the treatment of the collision becomes considerably more complex.

There are two conditions where the ionization energy of the atom is not small compared with the kinetic energy of the electron. These are (a) for all photon energies where a scratching collision occurs and (b) for photons with low energy where, for all collision angles, the bond energy cannot be neglected [2]. These conditions depend on the atomic number and the shell where the collision occurs.

The bonding corrections are often treated in the Waller–Hartree or impulse approximations [3], which take into account not only the K shell but also the atomic electrons.

These calculations involve the application of a multiplicative factor, termed the *incoherent scatter function*, $S(q, Z)$, to the expression of the Klein–Nishina differential cross section [4]

$$\frac{d\sigma(\theta)}{d\Omega} = S(q, Z)\frac{d\sigma^{KN}(\theta)}{d\Omega} \tag{8.20}$$

where q is the momentum transferred. In the impulse approximation, $S(q, Z)$ represents the probability of an atom leading to any excited or ionized state as a result of a sudden impulsive action that supplies a (recuo) momentum q to an atomic electron.

The expression *impulse approximation* results from the assumption that the interaction between both the radiation field and the atomic electron happens in such a short time that the electron sees a constant potential.

The calculation of $S(q, Z)$, which depends on knowledge of the atomic wave function, can be made for hydrogen and for other atoms using several methods of approximations based on the models of Thomas–Fermi, Hartree, and others.

Ribberfords and Carlsson [3] calculated σ_{en} using the impulse approximation. They pointed out that the tables of Hubbel [5] and Storm and Israel [6] have considerable errors for photon energies less than 200 keV, particularly for materials of high atomic number.

Experimental results shows that both approximations, impulse and Waller–Hartree, can give considerable errors for atomic numbers $Z > 50$. Though for higher Z values, these approximations need more refinements, they can be used for low Z materials.

8.2.7 The Photoelectric Effect

Theoretical analysis of the photoelectric effect is difficult, because the wave functions of the final states, which are solutions of the Dirac relativistic equation of the scattered electron, can only be obtained in the exact form as an infinite sum of partial waves. In addition, for high energies, a large number of terms are needed.

This is more complex than the Compton effect due to the existence of the bonding energy of the electron, and there is no simple equation, such as that of Klein–Nishina discussed earlier.

The wave function of the initial state for the bond electron is, in general, assumed to be an electron state in a central potential of the nucleus, and the effect of all the remainder electrons in the atom are taken in account as if the field of the nucleus has an electric charge $Z - S_i$. The constant S_i is named the *shielding constant* and represents the decrease of the nuclear electric charge arising from the presence of all the electrons of the atom. For further analysis, see, for example, Davisson [7].

In the photoelectric effect, a photon disappears and an electron is ejected from the atom. Instead of looking at this interaction as an effect between a photon and an electron, it must be seen as a process between a photon and an atom. In fact, complete absorption between only a photon and a free electron cannot occur, because the linear momentum will not be conserved [8].

The nucleus absorbs the momentum but acquires relatively little kinetic energy due to its large mass. It is clear that the photoelectric effect can only occur if the incident photon has energy greater than the bonding energy of the electron that will be removed. The hole created by the ejected electron is filled by electrons from any outer orbital, which can simultaneously produce fluorescence radiation, emission of Auger electrons, or both.

The competition between the emissions of a fluorescent K photon and the emission of Auger electrons is described by Y_K, termed the *fluorescence yield* of the K shell, which is defined by the number of K photons emitted per hole in the shell K. The probability that a photon K will be emitted is nearly 1 for elements with high Z and almost zero for materials with low Z.

In the photoelectric effect, the photoelectron acquires energy $\hbar w - E_s$, where E_s is the bonding energy of the shell from where the photon was ejected. In the filling of the holes created, on average, part of the energy, E_s, is emitted as characteristic radiation, whereas part is deposited through Auger electrons. For this reason, the mean energy transferred (to electrons), $\bar{E}_{tr}$, in the photoelectric process is given by

$$\hbar w - E_s < \bar{E}_{tr} < \hbar w. \tag{8.21}$$

For materials with high Z, this becomes more complicated [9]. For tissue equivalent materials of interest in radiological applications, it is much simpler, because in these materials the bonding energies of the K shell are very small (approximately 500 eV); therefore, the photoelectron acquires almost all the photon energy, meaning that $\bar{E}_{tr} = \hbar w$ and the coefficients of transfer and attenuation can be considered to be equal. Since the photoelectric process is important only for low energies, the "bremsstrahlung" produced by the ejected electrons can be neglected, and then the absorption and transfer coefficients are numerically identical.

The published tables are based on experimental results, complemented by theoretical interpolations for other energies and absorption media, and give satisfactory results for several regions of energy values.

8.2.8 Pair Production

Pair production is a process whereby a photon disappears, producing an electron and a positron. Often, this is mentioned as a process of materialization.

The pair production occurs only in a field of a Coulombic force, generally near an atomic nucleus; however, it can occur in a field of an atomic electron. This latter process is named as triplet production, because the host electron, which furnishes the Coulombic field, also acquires significant kinetic energy due to the conservation of momentum. This results in emission of two electrons and a positron from the site of interaction.

Obviously, a minimum of $2mc^2 = 1022\,\text{MeV}$ will be needed for pair production to occur in the nuclear field. Since we are more interested in energies below 1 MeV, this process will not be further developed.

8.2.9 Rayleigh Scattering

Rayleigh scattering, also called resonant scattering by an electron, is an atomic process where an incident photon is absorbed by a weakly bound electron. The electron is excited to a state of higher energy, and a second photon of the same energy is emitted; whereas the electron returns to its initial state, such that there is no excitation. The recoil of the scattered photon is actually taken by the whole atom, with a very small transference of energy, so that the energy lost by the photon is negligible.

This process is elastic. The atom is neither excited nor ionized. Some authors [9] have described Rayleigh scattering as a cooperative phenomenon that involves all the electrons of the atom. The photons are scattered by the bound electrons in a process where the atom is neither excited nor ionized. The process occurs mainly at low energies, and for materials of high Z, in the same region where the effects of the bonding electrons influence the Compton cross section.

Since the scattering of the photon results from the reaction of the whole atom, it is often termed *coherent scattering*.

Rayleigh scattering does not contribute to the kerma or the absorbed dose, because there is no energy furnished to any charged particle, and no ionization or excitation is produced.

8.2.10 Photonuclear Interaction

In a photonuclear interaction, an energetic photon (>MeV) enters and excites a nucleus that will emit a proton or a neutron. The protons contribute to the kerma. However, the relative amount remains below 5% of that due to the pair production; and, therefore, in general, these are not considered in dosimetry. Neutrons do already have practical importance, because they produce problems in radiation protection. This happens in some clinical electron accelerators, where energies involved can be greater than 10 MeV.

Since we are, in general, interested in energies below 1 MeV, we will not make any further considerations about photonuclear interaction.

8.2.11 Interaction Coefficients

For x- and γ-rays, the mass attenuation linear coefficient is often used. This is given by

$$\frac{\mu}{\rho} = \frac{\tau}{\rho} + \frac{\sigma_C}{\rho} + \frac{\sigma_{coh}}{\rho} + \frac{\kappa}{\rho} \tag{8.22}$$

whose units are (m^2 kg^{-1}) and the components refer to photoelectric effect, Compton effect, coherent scatter, and pair production, respectively.

The mass energy transfer coefficient is given by

$$\frac{\mu_{tr}}{\rho} = f_\tau \frac{\tau}{\rho} + f_C \frac{\sigma_C}{\rho} + f_\kappa \frac{\kappa}{\rho} \tag{8.23}$$

also in units of (m^2 kg^{-1}), and where the components refer to photoelectric effect, Compton effect, and pair production, respectively. The weights $f_\tau, f_C,$ and f_κ are conversion factors which indicate for each interaction the fraction of the photon energy that, eventually, is converted into kinetic energy of electrons and is dissipated in the medium through losses in collisions, ionizations, and excitations. The coherent scattering is excluded, because in this process there is no transference of energy.

The mass energy absorption coefficient for x- and γ-rays is given by

$$\frac{\mu_{en}}{\rho} = \frac{\mu_{tr}}{\rho}(1-g) \tag{8.24}$$

in (m^2 kg^{-1}), where g represents the fraction of the energy of secondary charged particles (for photons, these charged particles are electrons) that is emitted as bremsstrahlung.

For compounds and mixtures, we have,

$$\left(\frac{\mu}{\rho}\right)_{\text{mixture}} = \sum_i f_i \left(\frac{\mu}{\rho}\right)_i, \tag{8.25}$$

$$\left(\frac{\mu_{\text{tr}}}{\rho}\right)_{\text{mixture}} = \sum_i f_i \left(\frac{\mu_{\text{tr}}}{\rho}\right)_i, \tag{8.26}$$

where f_i is the relative fraction of the element i of the compound or mixture. We can also write,

$$\left(\frac{\mu_{\text{en}}}{\rho}\right)_{\text{mixture}} = \sum_i f_i \left(\frac{\mu_{\text{en}}}{\rho}\right)_i, \tag{8.27}$$

or in the exact form

$$\left(\frac{\mu_{\text{en}}}{\rho}\right)_{\text{mixture}} = \left(\frac{\mu_{\text{tr}}}{\rho}\right)_{\text{mixture}} (1 - g_{\text{mixture}}). \tag{8.28}$$

8.3 Interaction of Radiation with Matter: The Electrons

8.3.1 Introduction

As mentioned earlier, interactions of the photon with matter will eject electrons from atoms.

At a first step, the photons transmit their energy to electrons, which will then transfer their energy to matter; this second step is named deposition of energy within matter. In the following part, we will describe the processes of energy deposition.

The main difference between the interactions of photons and electrons is that, in general, photons undergo a relatively small number of interactions (something between 2 and 3 and 20 and 30, depending on their energy) involving relatively large losses of energy; whereas the electron suffers a large number of interactions (thousands), each involving only a very small amount of energy. As a first approximation, we can consider that the electrons continuously lose energy along their track.

8.3.2 Radiative Energy Loss

According to classic electromagnetic theory, it is well known that a charged particle, such as the electron, submitted to acceleration (curvilinear tracks in Coulombic fields, for example) emits radiation. Traditionally, the radiation produced by electrons submitted to acceleration is named *bremsstrahlung*, or braking radiation. This type of radiation is called x-rays when it is produced in an x-ray tube.

The energy lost by an electron is approximately proportional to its kinetic energy. The fraction of energy lost by an electron through this process is only 1% of the total loss due to interactions within the medium, for electrons with energy of the order of 1 MeV. Only for electrons with energy greater than 100 MeV does the radiative process dominate. In a material of high atomic number, for example, lead, the situation is different and the radiative process exceeds any other energy loss process even for 10 MeV.

8.3.3 Loss of Energy by Collision

Collisions with the atomic electrons are the most important mechanisms of loss of energy in matter, resulting in excitation or ionization of the material traversed.

The loss of energy, dE, along the track dl (dE/dl) is proportional to the electron density of the material traversed.

Applying quantum mechanical theory, one obtains,

$$\left(\frac{dE}{dl}\right)_{col} \propto \frac{q^2 \rho (Z/A)}{v^2} \ln(1/I), \tag{8.29}$$

where q is the electric charge, ρ is the density of the material, Z/A is the quotient between the atomic number and the atomic mass, v is the velocity, and I is the mean ionization potential.

8.3.4 Stopping Power

The quotient (dE/dl) is known as the *linear stopping power* of a material for a charged particle with energy E.

As already mentioned, the loss of the energy of the electrons has two main components, one due to losses by collisions and the other due to radiative losses (we will not consider here losses of energy by nuclear reactions).

We can write,

$$S = \left(\frac{dE}{dl}\right)_{col} + \left(\frac{dE}{dl}\right)_{rad}, \tag{8.30}$$

that is, the total stopping power is equal to the sum of the collision stopping power and the radiative stopping power.

Mass expressions are often used, dividing the stopping power by the density of the material, ρ.

Since the stopping power is proportional to the density, the mass stopping power is independent of the density.

In dosimetric issues, the restricted mass stopping power, $(S/\rho)_\Delta$, is often used, where losses of energy greater than Δ are not considered.

8.3.5 Linear Energy Transfer, L_Δ, LET

Considering the effects of electrons in matter and specifically with relation to the biological effects of radiation, in general, we are more interested in how the energy is deposited in the material irradiated rather than in how the particle loses its energy. Some collisions of the electrons produce new electrons that have sufficient energy to escape from the track of the initial electron leading to small tracks. These electrons are called δ-rays, and their energy is not deposited in the neighborhood of the initial electron track.

The energy locally deposited can be determined by ignoring all the losses of the initial electron energy that produce δ-rays with energy above a given value Δ. Therefore, L_Δ, the restricted linear energy transfer is defined that equals the stopping power if we restrict consideration to energy losses less than Δ, that is,

$$L_\Delta = \left(\frac{dE}{dl}\right)_\Delta . \tag{8.31}$$

Generally speaking, Δ is expressed in keV. If $\Delta = \infty$, then there are no restrictions to the loss of energy, and one can write $L_\Delta = S_{col}$, which is termed *unrestricted linear energy transfer*.

8.3.6 Range and CSDA

We can define the range R of a given charged particle in a given medium as the mean value, p, of the length of the track traveled until rest (neglecting thermal movement).

Additionally, we can also define the projected range, $\langle t \rangle$, of the charged particle of a given type and initial energy, in a given medium, as the expected value of the greatest penetration depth, t_f, of the particle along its initial direction. In Figure 8.4, the concepts of p and t_f are outlined.

Another important concept, available in tables published by several authors, is the continuous slowing down approximation (CSDA) range.

For electrons, the CSDA range is calculated using the equation

$$R_{CSDA} = \int_0^{T_0} \left(\frac{dT}{dx}\right)^{-1} dT. \tag{8.32}$$

For materials with low atomic number, Z, the value of t_{max} is comparable to the CSDA range, which has useful consequences for the application of tables of the range of electrons. It may be possible that these quantities will not be useful to describe the electron penetration due to the large number of interactions.

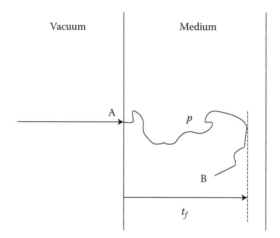

FIGURE 8.4
Outline of the concepts of track length, p, and greatest penetration depth, t_f. Notice that t_f is not necessarily the depth of the final point B.

8.3.7   Mean Energy Spent per Ion Pair Produced, *W*

In radiation dosimetry, the methods that use the ionization of matter are often employed due to the very large number of ion pairs formed in a gas by only one charged particle track. It will be useful to introduce a quantity named mean energy spent in a gas per ion pair produced. This quantity is represented by the symbol, W, and is given by

$$W = \frac{E}{N} \tag{8.33}$$

where E is the kinetic energy of the charged particle and N is the mean number of ion pair produced when the particle energy is totally dissipated in a gas. As previously mentioned, the electrons can lose energy through breamsstrahlung. The ions produced by bremsstrahlung radiation should be included in N. The unit of this quantity is energy, J or keV.

Fortunately, in electron or photon dosimetry, the W value for electrons is constant for the majority of gases for low-energy values. A parameter of major importance in dosimetry is the mean energy spent in a gas per unit electric charge produced, whose value is

$$\left(\frac{W}{e}\right)_{ar} = 33.97 \text{ JC}^{-1}. \tag{8.34}$$

8.4   Radiation Dosimetry Quantities

The fundamental concepts of classical dosimetry are nowadays well established by ICRU [10,11]. However, new developments have been accomplished in a field termed *microdosimetry* or *nanodosimetry*. Radiation dosimetry, sometimes called *classical dosimetry*, is based in the fact that absorption of radiation energy in a medium irradiated with indirect ionization radiation, such as photons (γ- or x-ray), can be described as a two-step process. In a first step, the energy of uncharged particles is transferred to electrons as kinetic energy, through interactions with atoms, resulting in the ejection of those electrons from atoms. In the second step, the released electrons impart their energy to matter along their path.

In classical dosimetry, the quantity kerma, K (kinetic energy released to matter per unit mass), is used to describe the first step. The second step is described by the quantity absorbed dose.

Kerma, K, is the quotient of the energy transferred to charged particles per unit of mass at a point of interest, including radiative-loss energy, so that units are J kg^{-1}, with the special name gray (Gy). Kerma rate is the time derivative quotient of kerma per unit time interval, so that the units are gray per second (Gy s^{-1}).

A more fundamental quantity can be defined, which is the fluence. *Fluence* is defined as the quotient of dN by da, where dN is the number of particles incident on a sphere of cross-sectional area da, thus

$$\Phi = \frac{dN}{da}. \tag{8.35}$$

The units are (m^{-2}). It can be shown that this definition can be generalized, by expressing as the lengths of the particle trajectories in a given volume [12]:

$$\Phi = \frac{dl}{dV}. \tag{8.36}$$

where dl is the sum of the particle trajectory lengths in the volume dV.

To develop classical dosimetry from fluence to absorbed dose, let us consider a narrow beam of photons, with energy E_0, impinging in a half-extended medium. Let us consider a point of interest at depth L, within the medium.

The number of primary or un-collided photons reaching depth L is

$$N_{E_0}(L) = N_{E_0}(0)e^{-\mu(E_0,Z)L} \tag{8.37}$$

where $N_{E_0}(L)$ is the number of particles at depth L, and $N_{E_0}(0)$ is the number of incident particles. The coefficient of linear attenuation, $\mu(E_0, Z)$, is defined for a given energy E_0 and material with atomic number Z.

Using the definition of Equation 8.35, the fluence equation for the beam is, in units (m^{-2}),

$$\Phi_{E_0}(L) = \Phi_{E_0}(0)e^{-\mu(E_0)L}. \tag{8.38}$$

The energy fluence is

$$\Psi = \Phi E \tag{8.39}$$

in units of ($J\,m^{-2}$). Therefore,

$$\Psi_{E_0}(L) = \Psi_{E_0}(0)e^{-\mu(E_0)L}. \tag{8.40}$$

Studies of the deposition of radiation energy in matter [13] led to the definition of two components of kerma: collision kerma, K_c, and radiative kerma, K_r.

$$K = K_e + K_r \tag{8.41}$$

As mentioned above, kerma describes the energy transferred from uncharged particles (photons) to charged particles (electrons).

An electron released in matter imparts its kinetic energy in a sinuous path with many more collisions than photons with the same energy. These interactions can be of two types:

1. Collision interactions. Coulomb-force interactions with atomic electrons resulting in the local dissipation of energy as ionization or excitation in or near the electron track.

2. Radiative interactions. Bremsstrahlung or braking radiation is emitted, and the photons produced will carry their energy far away from the charged-particle track.

Since we are only interested in the energy deposited at the point of interest near the charged-particle track, the radiative component of kerma can be neglected.

The value of kerma and collision kerma can be obtained using appropriate coefficients*:

$$K = \Psi \times \left(\frac{\mu_{tr}(E,Z)}{\rho} \right) \tag{8.42}$$

and

$$K_c = \Psi \times \left(\frac{\mu_{en}(E,Z)}{\rho} \right) \tag{8.43}$$

The coefficients are, for K, the mass energy (transfer coefficient) and, for K_c, the mass energy (absorption coefficient), and ρ is the density of the medium. Both are functions of the energy of the photons and the atomic number of the material.

Collision kerma is defined as the quotient of the energy transferred to charged particles per unit mass at the point of interest, excluding radiative loss energy.

We should note that the difference between kerma and collision kerma is the bremsstrahlung. In fact, the relationship between mass energy (transfer coefficient) and mass energy (absorption coefficient) is

$$\frac{\mu_{en}}{\rho} = \frac{\mu_{tr}}{\rho}(1 - g) \tag{8.44}$$

---

* Tables of this coefficients can be found in the site: http://physics.nist.gov.

where g represents the average fraction of secondary-electron energy which is lost in radiative interactions, that is, bremsstrahlung production and (for positrons) in-flight annihilation.

For the absorbed dose, it is not possible to write a mathematical expression directly related to the energy fluence of photons, such as Equations 8.42 and 8.43. Since the absorbed dose is the energy deposited by electrons liberated in matter, this energy is not directly related to the fluence of the incident radiation, such as photons. The so-called kerma approximation is often used, which means that it is assumed that electrons are in equilibrium or can be considered as if the energy of the photons transferred to electrons is deposited at the point of origin of the electrons: in other words, electrons do not carry away any energy; they deposit their energy at their site of origin. It can be shown for the kerma approximation that the collision kerma equals the absorbed dose [13]. In fact, absorbed dose is the quotient of the energy imparted to matter per unit mass. The units of absorbed dose are the same as those for kerma (Gy). It must be recognized that absorbed dose represents the energy per unit mass that remains in the matter at the point of interest, which produces any effects attributable to the radiation. Often, it is assumed that biological effects are proportional to the absorbed dose; however, sometimes this assumption is not applicable. Consequently, the absorbed dose is viewed as the most important quantity in radiological physics. Later in this section, the concept of absorbed dose will be further discussed.

Returning to the photon beam, if the kerma approximation is valid, then the collision kerma equals the absorbed dose

$$D = K_c, \qquad (8.45)$$
$$\text{\small cpe}$$

in which *cpe* means charged particle equilibrium.

From Equation 8.40 and considering Equations 8.43 and 8.45, we can write,

$$D_{E_0}(L) = D_{E_0}(0)e^{-\mu(E_0)L}. \qquad (8.46)$$

The term $D_{E_0}(L)$ means the absorbed dose at a point of depth L for a narrow beam of photons with energy E_0. Actually, Equation 8.46 only refers to the attenuation of un-collided photons, often called *primary radiation* and does not consider the scattered component, often termed *secondary radiation*, which can be very important from both dosimetric and image quality points of view.

In short, as the photons of the primary beam (un-collided photons) move forward in depth, they suffer interactions and will be renamed *secondary photons*. Likewise, electrons eventually ejected from atoms are also considered as secondary radiation. To quantify the transition from primary to secondary photons, it is necessary to use an exponential, whose behavior depends on the linear attenuation coefficient.

For the purpose of dosimetry and, more specifically, for measurements in radiation fields, it may be often convenient to use the kerma rate in free space, called the *free-in-air kerma rate*, with units of (Gys^{-1}).

In the 1920s, the first studies led to the description of a unit, called the *roentgen*, which is the "electrostatic unit of charge in 1 cm^3 of air." It must be noted that at the time of the definition of the unit roentgen, there was no special name yet for the related physical quantity. Over the years, this became the exposure, which is now defined as the total electrical charge of ions of one sign, produced per unit mass. It is important to notice that exposure is not a dose quantity but a descriptive parameter for the characterization of radiation fields consisting of x-rays and γ- rays only. The definition of the concept of exposure in air imposes a limitation on its application. It only refers to x-ray and low-energy gamma-charged particles. It cannot be used for the description of neutron fields.

The quantity exposure is the ionization equivalent to the collision kerma, in air, for γ- and x-ray radiation. The exposure, X, can be calculated from fluence by the expression:

$$X = \Psi \times \left(\frac{\mu_{en}(E, Z)}{\rho} \right) \left(\frac{\bar{w}}{e} \right)^{-1}_{air} \tag{8.47}$$

where $\bar{w}$ is the mean energy expended, in a gas, per ion pair formed and e is the charge of the electron, so that $(\bar{w}/e)_{air}$ is the mean energy expended, in a gas, per unit electric charge formed.

From Equations 8.43 and 8.47, we can determine the relationship between collision kerma and exposure:

$$(K_c)_{air} = X \times \left(\frac{\bar{w}}{e} \right)_{air} \tag{8.48}$$

Equation 8.48 means that the energy imparted to matter is obtained from the charge produced by radiation times the energy expended per unit charge. The constant $(\bar{w}/e)_{air}$ can be used as a constant, independent of the photon energy above some keV. For air, it is usual to use the value

$$(\bar{w}/e)_{air} = 33.97 \, JC^{-1} \tag{8.49}$$

so that

$$(K_c)_{air} = X \times 33.97 \tag{8.50}$$

The kerma can be measured with high precision by the so-called free-in-air chambers, which are specifically designed ionization chambers filled with atmospheric air. However, in practical applications, such as radiation protection, radiobiology, or radiation therapy, the absorbed dose in other nongaseous materials such as tissue or water is often needed, although these cannot be easily directly measured. Therefore, a method is required to convert air kerma into absorbed dose in other media. This can be done by the so-called kerma approximation in which the absorbed dose is assumed to be equal to the collision kerma. If the collision kerma in a medium A equals the absorbed dose in the same medium, then, the kerma approximation can be used; hence, for a given constant energy fluence of photons, the absorbed dose in a medium B is related to that of a medium A by

$$D_B = D_A \frac{(\mu_{en}/\rho)_B}{(\mu_{en}/\rho)_A} \tag{8.51}$$

Equation 8.51 reflects the special importance of the mass energy-absorption coefficients μ_{en}/ρ in applied photon dosimetry [14].

An alternative way to define absorbed dose is [15]:

$$D = \lim_{m \to 0} \frac{\varepsilon}{m} \tag{8.52}$$

The absorbed dose is a very well-defined quantity. However, some restrictions apply to its use. Absorbed dose is, from Equation 8.52, a measure of energy imparted, ε, to a medium divided by the mass, m, of the medium, ε/m [8]. Let us consider Figure 8.5. "When m is large enough to cause significant attenuation of the primary radiation (e.g., photons), the fluence of charged particles in the mass element under consideration is not uniform. This causes the ratio ε/m to increase as the size of the mass m is decreased. As m is further reduced we will find a region in which the charged particle fluence is sufficiently uniform that the ratio ε/m will be constant. It is in this region that the ratio ε/m represents absorbed dose. Thus expectation value, $\bar{\varepsilon}$, of the energy imparted over an appropriate size mass element must be used to determine absorbed dose. If m is further decreased from the region of constant ε/m, we will find that the ratio will diverge... Hence, the determination of

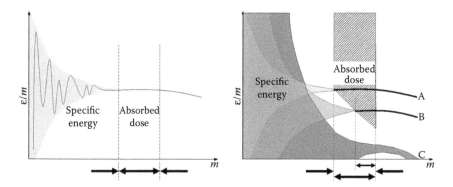

FIGURE 8.5
Left: Energy density as a function of the mass for which energy density is determined. The curve showed is one of the possible curves obtained if we choose a mass element at a random site inside the exposed medium. Right: Conjecture about the effect of decreasing dose. The zone with a pattern is where absorbed dose has physical meaning.

absorbed dose also requires that the mass element m be large enough so that the energy deposition is caused by many particles and many interactions... It must be realised that the macroscopic quantities defined using the differential notation imply that a limiting process, as described above, has occurred" [8].

The question that arises is: What is the appropriate sized mass that can be used to calculate absorbed dose? Even without a quantitative answer to this question, we can expect that a limited zone in the space $\{m, \varepsilon/m\}$ exists, where we can speak about absorbed dose (Figure 8.5 on the right, zone with a pattern). Another problem is: What happens when the dose decreases? Our conclusion is that it is not possible to define absorbed dose below a given value, because we fall into a stochastic zone as is shown in Figure 8.5 on the right-hand side, for example, with the shadowed zone labeled as C. An alternative way to enunciate this problem is: What dose value is high enough to fulfil the condition of continuity required by the dose concept?

These conclusions are relevant to the discussion of the linear no-threshold (LNT) hypothesis. Our conclusion is that dose–effect relationship at low doses is meaningless due to the failure of the absorbed dose concept [16]. The stochastic quantity that must be used is the specific energy (imparted), z, which is the quotient of ε by m, thus

$$z = \frac{\varepsilon}{m} \tag{8.53}$$

The units are also $(J\ kg^{-1} \equiv Gy)$ previously mentioned and named gray [11]. This quantity is fundamental in microdosimetry.

Radiation scattering has already been mentioned. In a radiation field, this can be defined: the primary radiation, which is the un-collided component and the secondary radiation, which means those particles resulting from collisions. Therefore, absorbed dose at a given point is the sum of primary, $D^{(p)}$, and secondary, $D^{(s)}$, components:

$$D^{(t)} = D^{(p)} + D^{(s)} \tag{8.54}$$

To describe secondary radiation, the buildup factor is often used, which can be defined for fluence, exposure, absorbed dose, etc.

For absorbed dose, the definition of the buildup factor is

$$B(\mu L) = \frac{D^{(t)}}{D^{(p)}} = \frac{D^{(p)} + D^{(s)}}{D^{(p)}} = 1 + \frac{D^{(s)}}{D^{(p)}} \tag{8.55}$$

where μ is the attenuation coefficient and L is the depth in the medium.

Equation 8.46 refers only to primary photons; therefore, to include the buildup factor from Equation 8.55, it can be shown that

$$D^t(L) = D^{(p)}_{E_0}(0)B(\mu L)e^{-\mu(E_0)L} \tag{8.56}$$

Let us note that if $L = 0$, then this equation becomes

$$D^t(0) = D_{E_0}^{(p)}(0)B(0) \tag{8.57}$$

meaning that $B(0)$ is the backscatter factor, $BSF = B(0)$.

8.5 Radiation Protection Quantities

In this section, we will review the radiation protection quantities.

The dose equivalent is defined as follows:

$$H = QD. \tag{8.58}$$

This quantity is the product of the absorbed dose, D, with the quality factor, Q.

The equivalent dose is defined as

$$H_T = \sum_R w_R D_{T,R} \tag{8.59}$$

and $D_{T,R}$ is the average of absorbed dose in a tissue or organ, T, for a radiation type, R, and w_R is a radiation weighting factor.

The effective dose is

$$E = \sum_T w_T H_T. \tag{8.60}$$

in which w_T is the tissue weighting factor. The values of the weighting factors can be found in ICRP-103 [19].

There are also other quantities, termed *operational quantities*, and they are used in individual monitoring [10,11,17].

8.6 Quality Factor, Weighting Factors, and Relative Biological Effectiveness

The ejected electrons may have sufficient energy to cover distances in tissue that can reach a few millimeters and leave a trail of ions. By definition, specific ionization is the number of ion pairs formed per unit distance traveled by the particle.

In radiobiology and radiation protection, when radiation traverses an absorbing tissue, an important quantity is the linear rate of energy loss of charged particles (or the density of energy absorption by the medium) due to ionization and excitation processes.

LET is a measure of the spatial density of energy absorption. It is the mean energy lost by the radiation particle, due to collisions with electrons, per unit distance traversed in material.

In radiation dosimetry, LET is used as the basic physical quantity in terms of radiation biological effectiveness. This may not be absolutely true, as there is evidence that other factors, such as pH variation in the medium or dose rate, may also be important.

In radiobiology, radiation is often referred to as low LET (e.g., photons or electrons) and high LET (e.g., α-particles). Recognizing the importance of the LET (often given only as L), the parameter quality factor was introduced, and this is a function of the radiation LET.

$$Q = Q(L) \tag{8.61}$$

$$Q(L) = \begin{cases} 1 & (L \leq 10) \\ 0.32L - 2.2 & 10 < L < 100 \\ 300/\sqrt{L} & L \geq 100 \end{cases} \tag{8.62}$$

The units of L are (keV μm^{-1}), whereas the quality factor is dimensionless.

In order to simplify the radiation protection quantities, the ICRP introduced weighting factors for radiation and tissues [32], and these have been recently revised [19]. The determination of the values of the weighting factors was guided by the concept of Relative Biological Effectiveness (RBE).

In 1931, a determination was reported for the first time of the RBE of x- and γ-rays. In radiobiology, RBE equals the ratio of the absorbed doses of two types of radiation that produce the same specific effect. The RBE was first used, in radiation protection, as a weighting factor; but complications arose due to their dependence on dose, dose rate, fractionation, and the cells or tissues in which the effect is being assessed. Consequently, in 1959, the ICRU recommended that the term *RBE* be used only in radiobiology and the quality factor was proposed [18].

8.7 Energy Degradation in Matter

Physical interactions occur in the first 10^{-15} s. Between 10^{-11}s and 10^{-6}s, free radicals, molecules, and ions will be produced and distributed, which will change the normal chemical reactions within cells exposed to radiation. After several years, biological effects can appear [1].

Radiation dosimetry aims at quantifying the energy imparted or the energy deposited by radiation within biological tissues through the main radiological protection quantity: the absorbed dose. Biological effects can be stochastic or deterministic, depending on its value. In their new fundamental recommendations [19], ICRP is proposing that deterministic effects will be named *tissue reactions*.

Biological effects are often correlated with absorbed dose in tissues or cells at risk. However, other factors are also important, such as type of radiation, energy spectra, spatiotemporal distribution of energy deposited, the total number of cells exposed, or radiosensitivity of cells. As an attempt to foresee biological effects, such as carcinogenesis or cell killing, the assumption is often accepted that risk is proportional to absorbed dose. However, some radiobiological experiences very often reveal rather complex relationships [20].

Radiation dosimetry and protection are directly related to the hierarchy of life: from molecules, passing through cells up to organisms and continuing to population, community, ecosystems, and biosphere. As far as the ecosystem level is concerned, studies on environmental radioactivity have to be carried out; whereas for smaller-sized systems, for example, absorbed dose at human organ levels, we are mainly concerned with medical applications of radiation, such as radio-diagnostics or radiotherapy. Radiobiology deals with effects in tissues, cells, or subcellular structures and the main target of radiation, DNA. Molecular biology is also very important in helping to seek the biophysical mechanism behind radiation effects. From the dosimetry point of view, we will be concerned with microdosimetry or nanodosimetry. The cell is the fundamental unit of life and can be seen as a fundamental unit for radiation dosimetry if we are interested in cellular dosimetry.

Considering the several orders of magnitude of size on going from molecules to organisms, several questions arise about the definition of dose, if we consider that dose is the expectation value of energy imparted over an appropriate size mass element. In fact, if we increase the dimensions of a cell to the dimensions of the moon, then a portion of the DNA is equivalent to a person walking on the moon. What is the relationship between a tissue dose, a cell dose, and the DNA dose? The answer to this question is not trivial.

Biological cells, tissues, organs, or living beings are complex systems. Let us look at life as a process that organizes matter and energy in an orderly system. From that point of view, we can see life as a complex set of processes that decreases the entropy of the system. The interaction of radiation with matter supplies energy through interaction processes that the biological system cannot handle, leading to the lethal effects of radiation. The effects due to the exposure of radiation are changes in the structure and behavior of these biologically complex systems. The exposure of organic or inorganic matter to radiation leads to changes in the target, due to the enormous statistical complexity that arises from the point of view of radiation physics. Any radiation beam can also be regarded as a complex system due to its interaction with matter, which leads to statistical complexity at the level of energy deposition. We are dealing with two complex entities: a biological entity, the target (which can be also inorganic), and a physical entity, the radiation beam or field from sources. Since Shannon's work [21], we know that entropy is the physical quantity suitable to describe such complex systems. Shannon defined the entropy of a probability distribution, which emerges from information theory. This can be applied in radiation physics, dosimetry, and protection in a straightforward fashion [16,22].

Let us consider a Gaussian distribution, with null average and standard deviation σ:

$$p(x) = \frac{1}{\sqrt{2\pi e}} \exp\left(-\frac{x^2}{2\sigma^2}\right) \tag{8.63}$$

In Figure 8.6, 4 distributions are shown with different values of the standard deviation.

Defining the entropy of a probability distribution, $p(x)$, as

$$S = -\int_{-\infty}^{+\infty} p(x) \ln p(x)dx \tag{8.64}$$

and using Equation 8.63, it follows that

$$S(x) = \ln(k\sigma) \quad \text{with} \quad k = \sqrt{2\pi e} \tag{8.65}$$

which means that the entropy of a Gaussian distribution is a function of the standard deviation. In Figure 8.6, the entropy values for p1, p2, p3, and p4 are $S1(\sigma) = 2.11$, $S2(\sigma) = 2.52$, $S3(\sigma) = 3.03$, and $S4(\sigma) = 3.72$, respectively. In

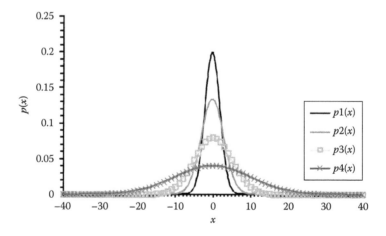

FIGURE 8.6
Graphical representation of 4 Gaussians, p1(x), p2(x), p3(x), and p4(x), with standard deviations, σ1 = 2, σ2 = 3, σ3 = 5, and σ4 = 10, respectively, and μ = 0.

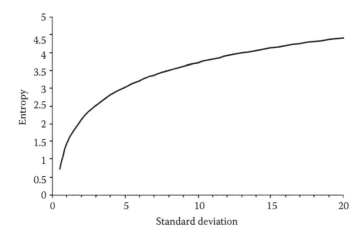

FIGURE 8.7
Entropy as a function of the standard deviation for a Gaussian with zero average.

Figure 8.7, the entropy, $S(\sigma, \mu = 0)$, is shown as a function of σ, for a probability distribution Gaussian with.

In the case of discrete probabilities, we have

$$S = -\sum_i^n p_i \ln p_i \quad \text{for} \quad p_1, \ldots, p_i, \ldots, p_n \tag{8.66}$$

The concept of entropy allows us to obtain new insights in radiation physics
and to seek links between physics and biology.

Let us assume that a monoenergetic narrow photon beam impinges on a
half-extended water medium. If all photons have the same energy value, then
the probability of finding a photon with that energy is, obviously, one. Application of Equation 8.66 with $p_i = 1, \forall i$, gives the entropy value $S = \ln 1 = 0$.
This is expected for such a highly ordered set of photons, which presents no
spread in energy distribution of the primary or un-collided photons.

Using Monte Carlo simulations, the set of energy deposition points due
to the interaction of radiation with matter was determined at the kerma
approximation, which means that only the photon interactions are considered; this assumes that the energy transferred to electrons is locally deposited
at their points of origin. The narrow beam simulated has 100,000 photons of
energy 50 keV, with three interactions being allowed: coherent and incoherent scattering and photoelectric absorption. At this energy, pair production is
negligible.

The points of the first interaction for each photon of the beam are shown in
Figure 8.8.

In each of the points shown in Figure 8.8, one of the three interactions
can occur: photoelectric, Compton, or Rayleigh. For energies large enough,

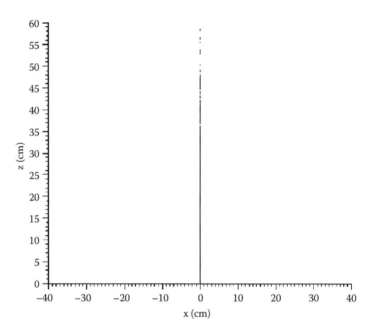

FIGURE 8.8
First points of interaction in water for a 50 keV photon beam in water.

pair production can also occur. The Rayleigh interaction, also called *coherent scattering*, is elastic and does not deposit energy. For the photoelectric interaction, all of the photon energy is deposited at the point of interaction. In the Compton interaction, also termed *incoherent scattering*, the photon deposits a small fraction of its energy. This leads to the energy spectrum shown in Figure 8.9. This type of spectrum shows the degradation of the energy of primary photons.

In the spectrum of Figure 8.9, two groups of lines can be identified: a single line at 50 keV, due to the photoelectric effect (total absorption of photon energy) and a group of lines due to the Compton interaction (small amounts of energy absorbed). The Compton interaction mainly results in small values of energy deposited, and it is one of the principal causes of image degradation.

Application of Equation 8.66 gives the entropy value $S(j = 0) = 2.25$.

From the point of view of image formation, the component of the primary energy, 50 keV, obtained by the photoelectric effect, is the desirable process leading to a signal suitable for the formation of the image. The low energy group, due to Compton interactions, corresponds to noise in the image, which must be reduced as far as possible.

The data of Figure 8.9 can be normalized to obtain a probability distribution of the energy deposited in the first interaction. It is often named the primary interaction of the beam in the target medium. We can apply the entropy to the normalized distribution of energy deposited as a statistical parameter to

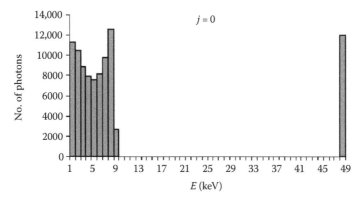

FIGURE 8.9
Energy spectrum of the energy deposited in water due to the first interactions of a photon beam with energy 50 keV. The number of photons is related to the total number of photons simulated, which was 100,000. Entropy of the spectrum, $S(j = 0) = 2.25$.

describe the spread of the energy values. Obviously, a Gaussian is not a good probability distribution to fit the data in Figure 8.9. How do we statistically describe a distribution like this? Gaussian parameters, such as mean and standard deviation, are not enough. The entropy is a suitable statistical descriptor of the complexity of the energy deposited in matter.

In spite of the wide use of absorbed dose, new approaches have been introduced. One such new approach results from the photon track and studies the probability, $P(E, j)$, that j secondary interactions will occur during the degradation of a photon of energy E [23]. The probability, $P(E, j)$, can be estimated by the relative frequency of the number of secondary interactions. The probability $P(E, j)$ introduces the step j in the energy degradation of the primary photon. In Figure 8.10, Monte Carlo results are shown for the estimation of this probability for a 50 keV photon beam in water. Note that the probability of a photon to suffer more than 16 or 18 interactions vanishes.

The pattern of the points of energy deposited for the second interaction is rather different from that for the first interactions shown in Figure 8.8, as can be seen in Figure 8.11.

The energy spectrum for the points of Figure 8.11 is shown in Figure 8.12.

At step $j = 0$ (Figure 8.8), there is a significant component in the energy distribution at 50 keV, meaning that some of the incident photons lose all of their energy in the first interaction, by the photoelectric effect. For other values of j, such as in Figure 8.11, whose energy distribution is shown in Figure 8.12, there also exists a small component at 50 keV due to the photoelectric absorption of photons that previously suffered only coherent scattering, which does not modify the energy of the photons in the interaction. Obviously for $j = 0$ (primary interactions), all the interactions have $x = y = 0$; whereas for $j \neq 0$, some complexity appears in the pattern of energy deposition. After the first interaction, a third group appears in the energy spectra just to the left of

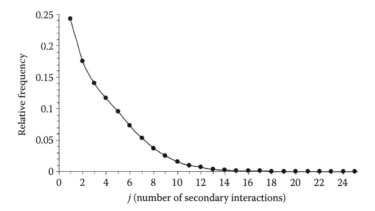

FIGURE 8.10
Relative frequency of the number of secondary interactions, which is an estimative of the probability $P(E, j)$, obtained for a Monte Carlo simulation of a narrow beam of 100,000 photons with energy of 50 keV, in water.

the photoelectric peak. It can be shown that the origin of this group is the backscattering from deeper layers in the medium irradiated [24]. Applying Equation 8.66 to the data of Figure 8.12 gives $S(j = 1) = 2.53$, which is higher than $S(j = 0)$.

Amazingly, after nine interactions, the line of the primary beam is still visible (Figure 8.13), meaning that some photons suffered 8 elastic collisions before total absorption by photoelectric effect.

The energy spectrum corresponding to Figure 8.13 is shown in Figure 8.14.

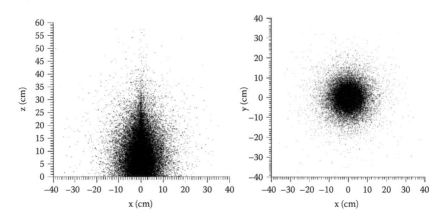

FIGURE 8.11
Patterns of the points of the second energy interactions in water for a photon beam in water with primary energy of 50 keV.

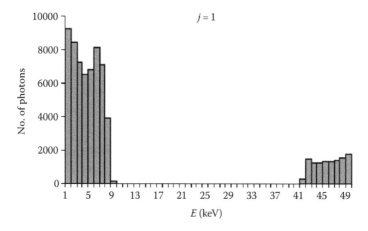

FIGURE 8.12
Energy spectrum of the energy deposited in water due to the second interactions of a photon beam with energy 50 keV. The number of photons is related to the total number of photons simulated, which was 100,000. The entropy value is $S(j = 1) = 2.53$.

Obviously, for dose calculation, all the steps of the photon track have to be simultaneously considered, which means that the set of points obtained for all values of j has to be considered. From Figure 8.10, it can be concluded that the probability of having photons with more than 18 secondary interactions, $j = 18$, is vanishingly small. Equation 8.66 leads to $S(j = 8) = 2.88$.

The results of the calculation of the entropy of all spectra for each step of the photon track evolution lead to the results shown in Figure 8.15.

The energy spectrum obtained as a superimposition of all the values of j is shown in Figure 8.16. In fact, the spectrum of this figure is the total

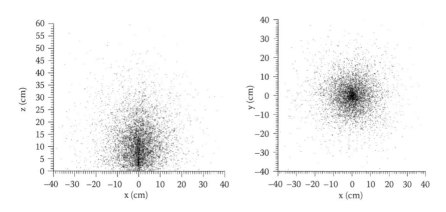

FIGURE 8.13
Patterns of the points of the ninth interactions (the first primary interaction followed by 8 secondary interactions, $j = 8$) in water for a photon beam in water with primary energy of 50 keV.

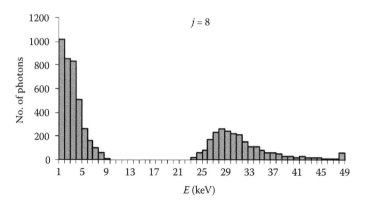

FIGURE 8.14
Energy spectrum of the energy deposited in water due to the ninth interactions of a photon beam with energy 50 keV. Note: The number of photons is related to the total number of photons simulated, which was 100,000. The entropy value is $S(j = 8) = 2.88$.

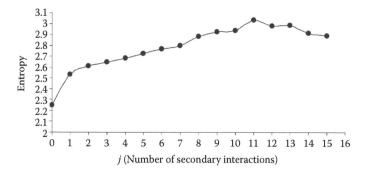

FIGURE 8.15
The entropy of the energy deposited in water as a function of the photon track step.

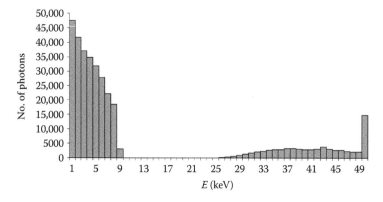

FIGURE 8.16
Energy deposition spectrum for 50 keV primary photons deposited in water in a half-extended geometry with a narrow beam, due to the first 8 steps of the photon track.

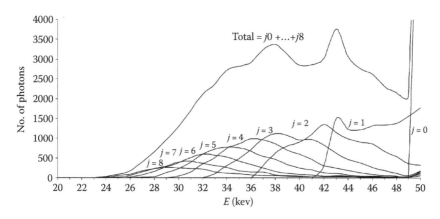

FIGURE 8.17

Details of the high energy region of the spectrum in Figure 8.16. The component $j = 0$ has nonzero value exclusively at 50 keV, meaning total deposition of primary energy in the first interaction of the photon by photoelectric effect. The remainder of the spectrum is a sum of the several steps j. Each component j suffers a displacement toward the low energy as j increases.

energy deposited due to a primary beam of photons with 50 keV. However, the concept of photon track with steps j allows us to conclude that there exists a specific energy and spatial distribution associated with each j value.

Figure 8.17 shows only the high energy group; the lower energy group from Compton scattering is omitted. It can be seen how the total energy spectrum is decomposed into several components associated with each step of the photon track, each one having a well-defined displacement toward the lower energy region.

As pointed out earlier [24], the components shown in Figure 8.17 mainly include the photons backscattered in deeper layers.

8.8 Monte Carlo Simulation

The interaction of radiation with matter can be simulated using the well-known method of Monte Carlo. Monte Carlo simulation can be seen as a black box to which we have to provide some input data, such as details of geometry of radiation source, target and medium, type of radiation, energy, and direction of radiation flight. The black box diagram of a Monte Carlo code is shown in Figure 8.18.

Monte Carlo methods of radiation transport for photons will be summarized next following the ideas of Chan and Doi [25].

As mentioned earlier, we must provide a set of input data, which in some codes have the form of a file and in others are introduced by dialog windows.

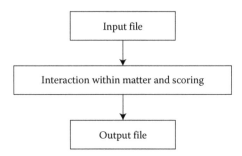

FIGURE 8.18
Black box of a Monte Carlo code.

Once we have defined the geometry of the problem with a source and a detector, the simulation can start.

1. A primary photon is emitted from the source. Its initial direction of flight is randomly chosen from the spatial distribution of the beam or source. The photon energy is chosen according to the spectrum emitted by the photon source.

2. The free path length of the photon is sampled from the exponential distribution function; and a potential collision site is determined from the free path length, the previous direction of flight, and coordinates.

3. If the point is outside the scattering medium and the radiation detector, the photon history ends and the computer returns to step 1 for other photon history.

4. If the collision site lies within the medium, the type of interaction is chosen from the possible processes already mentioned earlier.

5. For the photoelectric effect, all of the photon energy is absorbed, the photon history ends, and the computer returns to step 1.

6. If coherent scattering takes place, the photon is scattered with its original energy at an angle sampled from Thomson equation modified by the form factor. No energy is released. The next free path length is determined.

7. If incoherent scattering occurs, the new direction of flight is sampled from the Klein–Nishina formula modified with the incoherent-scattering function, using a uniform distribution of the azimuthal angle. The energy of the recoil electron is absorbed at the collision site (meaning the kerma approximation). The free path length of the photon with the remainder energy is determined. If the energy of the scattered photon is less than a preset cut-off energy, then the scattered photon is assumed to be absorbed at the collision site in a photoelectric effect. A new photon history starts from step 1.

8. Eventually, characteristic x-rays of all elements in the medium have to be considered. In that case, those characteristic photons start a new

history. The energy of the previous photon subtracted by the energy of the characteristics photon is absorbed in the collision point.

9. If the kerma approximation is not valid, the electron tracks are also simulated.

The development of Monte Carlo methods is outside of the scope of this book.

8.9 Conventional and Cellular Dosimetry for Radionuclides

The number of photons emitted by a radioactive source is termed the *source strength*, which may be different from the number of atoms that disintegrate. The source strength, S, and the activity, A, are related by the expression $S = A \times F$, where the factor F is the number of photons emitted per decay, which can be less than one. For Cesium-137, 94.1% of the β-decay results in Barium-137 m and 5.9% of the decays results in the fundamental state of Barium-137. Competition between internal conversion electrons and γ-ray emission for Barium-137m transition to fundamental state leads to 89.9% of γ-ray emission. Thus, for each decay of Cesium-137, the photon emission of 0.662 MeV has a frequency of $F = 0.899 \times 0.941 = 0.846$.

Considering dose calculation, we have previously mentioned the concept of radiation attenuation and the linear attenuation coefficient (see Equation 8.37). However, time and distance must also to be considered. For the time, there is a direct proportionality. In fact, if we consider the dose rate at a given point,

$$\dot{D} = \frac{dD}{dt},$$
(8.67)

then the dose at the same point, for an exposure of time t is

$$D = \dot{D} \times t$$
(8.68)

In relation to the distance, consider a sphere of radius r and a radioactive source at the center of the sphere emitting S photons.

The fluence (photons per unit area) at the surface of the sphere is

$$\Phi = \frac{S}{4\pi r^2}$$
(8.69)

Considering two spheres with radius and r_1 and r_2, and using Equation 8.69, the fluence ratio for the two spheres is given by

$$\frac{\Phi_1}{\Phi_2} = \frac{r_2^2}{r_1^2} \qquad (8.70)$$

As previously mentioned, the absorbed dose can be calculated from fluence, leading to the replacement of Φ by D in Equation 8.70. Making $r_1 = 1\,\text{m}$, $r_2 = r$, and $D_2 = D$ and if the dose rate at one meter, $\dot{D}_1$, is known, the relationship becomes

$$D = \frac{\dot{D}_1}{r^2} \times t. \qquad (8.71)$$

The dose is proportional to the inverse of the square distance from the point source to the point of interest.

Given the dose rate at one meter per unit activity, Equation 8.71 allows simple calculations, for example, on radiation protection around radioactive sources.

If we take into account scattered radiation by the buildup factor, this becomes

$$D(r) = \frac{\dot{D}_1}{r^2} \times t \times B(\mu r) \qquad (8.72)$$

where r is the distance from source to the point of interest.

If there is a homogeneous medium between the radiation source and the point of interest, the exponential attenuation must be included, leading to

$$D(r) = \frac{\dot{D}_1}{r^2} \times t \times B(\mu r) \times e^{-\mu(E,Z)r}. \qquad (8.73)$$

For air, we often use $B(\mu(air)r) \times e^{-\mu(air)r} \approx 1$, which simplifies some radiation shielding calculations. If the air attenuation is neglected and there exists a slab shield of thickness x, then Equation 8.73 becomes

$$D(r) = \frac{\dot{D}_1}{r^2} \times t \times B(\mu x) \times e^{-\mu(E,Z)x}. \qquad (8.74)$$

To determine the value of $\dot{D}_1$ for radionuclides, we can use the air kerma rate constant, Γ_δ. This factor is also called the k factor, specific γ-ray constant, or exposure rate constant in the last case used, obviously, to calculate the exposure instead of dose. It is defined for each radionuclide.

Assuming the kerma approximation, then the dose rate $\dot{D}_1$ can be calculated by the expression

$$\dot{D} = \frac{\Gamma_\delta A}{l^2}, \qquad (8.75)$$

where A is the activity of the source and l is the distance from a point source. The index δ means that only the photons with energy greater than δ are considered. If $A = 1\,\text{Bq}$ and $l = 1\,\text{m}$, then

$$\dot{D}_1 = \Gamma_\delta. \tag{8.76}$$

Therefore, Γ_δ is the kerma rate at 1 m from a gamma source of activity 1 Bq. The constant Γ_δ has to have units $(\text{m}^2\,\text{Gy}\,\text{Bq}^{-1}\,\text{s}^{-1})$ in order to give $\dot{D}_1$ in units $(\text{Gy}\,\text{s}^{-1})$.

Let us consider the geometry outlined in Figure 8.19.

To calculate the dose at the target point, t, due to the source at point s, we can define the photon dose point kernel. The dose point kernel is a function that represents the dose at a given distance from an isotropic point source, in a homogeneous and infinite medium, which can be written as

$$P(r, E) = \frac{\dot{S}}{4\pi r^2}\left(\frac{\mu_{en}}{\rho}\right) tB(\mu r)e^{-\mu r} \tag{8.77}$$

where $r = |\vec{r}_s - \vec{r}_1|$ and $\dot{s}$ is the source strength in photons per second. The coefficients in Equation 8.77, already earlier mentioned, are functions of the energy and the atomic number of the medium. In general, a gamma source emits photons of several energies so that it must be considered as an integral in terms of energy:

$$P(r) = \int_E P(r, E)\,dE. \tag{8.78}$$

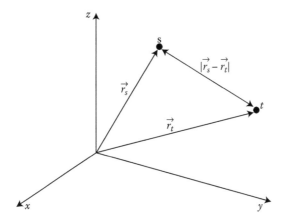

FIGURE 8.19
Isotropic point source at point s and the point of interest t, within a homogeneous and infinite medium.

After this integration, the point kernel can be defined as a function of the distance from the point source,

$$P(r) = P\left(|\vec{r}_s - \vec{r}_1|\right). \tag{8.79}$$

The point kernel is used in more general and complex geometry, such as volume source and volume target. This is the case for NM, where dosimetric calculations are made at the organ level or, in some cases, at the cellular level. For the case of the organ level, the organ source and the organ target must be considered. This geometry is outlined in Figure 8.20.

The concept of point kernel was introduced for use in the volume integration:

$$D(T \leftarrow S) = \frac{1}{V_1} \int_{V_s} dV_s \int_{V_t} dV_t P\left(|\vec{r}_s - \vec{r}_1|\right) \tag{8.80}$$

where $D(T \leftarrow S)$ means the dose in the target volume T, due to a source in volume S.

In suitable models such as those involving spherical geometry, for example, spherical source and spherical target, theoretically the integration (Equation 8.80) can be analytically solved.

The point kernel can be evaluated for any type of emission (photons, electrons, α-particles, etc.) and any homogeneous medium.

The calculation of Equation 8.80 at an organ or tissue level, where the geometry is rather more complex than the spherical case and the medium is non homogeneous, is only possible by Monte Carlo simulation, which was briefly reviewed in this chapter.

Due to the importance of absorbed dose calculations during the administration of radionuclides in the human body, several years ago, a scheme

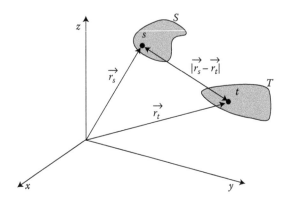

FIGURE 8.20
Volume source and volume target illustrating the general geometry for dose calculations, for example, in nuclear medicine.

known as MIRD (medical internal radiation dose) scheme was developed by the MIRD committee.

The advantage of this scheme is that the integration (Equation 8.80) is solved for several pairs of human organs, and the results are presented in tabular form or implemented in software packages. The integral equation 8.80 is replaced by the equation

$$D(O_t \leftarrow O_s) = \tilde{A}_s S(O_t \leftarrow O_s). \tag{8.81}$$

In this equation, the absorbed dose in the target organ, O_t, due to cumulative activity, $\tilde{A}_s$, in the source region O_s is simply obtained by multiplying the cumulated activity by an S-value, which is the absorbed dose at the target organ per unit cumulative activity in the source region.

All the complexity of the integration (Equation 8.80) is hidden in the simplicity of the S-value. All we have to do is calculate the cumulated activity, $\tilde{A}_s$:

$$\tilde{A}_s = \int A(t)dt \tag{8.82}$$

The S-value is defined as

$$S(O_t \leftarrow O_s) = \sum_i \frac{\Delta_i \phi_i(O_t \leftarrow O_s)}{m_t} \tag{8.83}$$

where m_t is the mass of the target region, Δ_i is the mean energy emitted per nuclear transition, and $\phi_i(O_t \leftarrow O_s)$ is the fraction of energy emitted from the organ source that is absorbed in the target region for the ith radiation component.

Next there are defined radiation protection quantities for application in internal dosimetry.

The committed equivalent dose for an organ or tissue, T, is represent by $H_T(\tau)$.

Since the radionuclide irradiates the body organs after its incorporation, this quantity is defined as the integral in time τ of the equivalent dose rate in the tissue of interest. It is defined as the total committed equivalent dose in the organ or tissue for the time τ after the incorporation:

$$H_T(\tau) = \int_{t_0}^{t_0+\tau} \dot{H}_T(t)dt \tag{8.84}$$

When the time of integration is not mentioned, a time of 50 years for adults or 70 years for children is assumed.

The committed effective dose, $E(\tau)$, is obtained, by multiplying the committed equivalent dose for each organ or tissue by the respective tissue weighting factor. This quantity is defined as the sum:

$$E(\tau) = \sum_t H_T(\tau) w_T \tag{8.85}$$

To end this short review of the radiation protection quantities, it is often useful to have a measure of the exposure of a population, which is made by the collective effective dose.

The effective collective S, is defined by the next integral:

$$S = \int_0^\infty E \frac{dN}{dE} dE \tag{8.86}$$

or by the sum

$$S = \sum_i \bar{E}_i N_i, \tag{8.87}$$

where $\bar{E}_i$ is the average effective dose for the subgroup i of a population. The time interval is not explicitly specified during which the population were exposed, resulting in the corresponding effective dose. Therefore, that time must be expressly specified in the presentation of effective dose values.

The requirement of homogeneous medium and point isotropic kernel has already been mentioned. However, the MIRD scheme is not limited to organ or suborgan dosimetry. In fact, a lot of useful applications have more recently appeared in cellular dosimetry, for example, in targeted radiotherapy, where dose at the cellular level is a major concern.

Adaptation of the formalism shortly summarized in Equations 8.81 through 8.83 for organs can be straightforwardly applied to cells or subcellular regions, by simply replacing the reference of organs to a reference of cells and adapting the formalism for absorbed dose calculation to multiple source regions, so that Equation 8.81 becomes

$$D(C_t \leftarrow C_s) = \sum_s \tilde{A}_s S(C_t \leftarrow C_s). \tag{8.88}$$

At the cellular level, the concepts of self-dose and cross-dose are often used, so that the mean dose of a cell cluster is a sum of the self-dose and the cross-dose. Similarly, we can define S-values for self-dose and cross-dose, S_{self} and S_{cross}, respectively.

Differences between organ dosimetry, often called conventional dosimetry, and cellular dosimetry are that in conventional dosimetry, no distinction is made between the cells and the medium. The radioactivity is assumed to be homogeneously distributed throughout the volume considered. This implies that the dose to the target cell nucleus is the same as that to the overall organ or to any given subvolume. In cellular dosimetry, the approach of Equation 8.88 can be further extended to subcellular regions, such as cell membrane, cytoplasm, and cell nucleus. Conventional dosimetry predicts poorly absorbed dose at the cell nucleus, which can be of great importance for small range particles, such as Auger electrons.

A drawback of this cellular approach lies in the validation of cross-dose due to difficulties in available data for comparison of results. In fact, calculations are made defining a given cell cluster. Cell clusters can differ in the number of cells, the cell and nucleus size, and the packaging of cells within the cluster. Sometimes, cells are assumed to be spheres. A typical approach at the cellular level assumes the last geometric details for a given cell cluster [26], allowing the evaluation of S_{cross}, $S_{selfNuc}$, $S_{selfCyt}$, and S_{selfMb}, where the index refers to cross, nucleus, cytoplasm, and cell membrane. Other authors [27] calculate intracellular S-values for cell–cell, cell–cell surface, nucleus–nucleus, nucleus–cytoplasm, and nucleus–cell surface, where the first region is the target and the second region is the source.

Different number of cells, cell size, or cell arrangement (package) make any comparison difficult. It seems apparent that there is a need to define some fundamental quantities at the cellular level in order to allow comparison between different calculation methods and clusters. With this scope, and considering that the cell is the basic unit of life, the basic unit of cellular dosimetry can be defined as a pair of cells, one of them with radionuclide incorporated and the other clean (Figure 8.21).

For the basic unit of cellular dosimetry, we can define several S-values:

In the earlier definitions of all the S-values, only the cell and the nucleus are considered as targets. The source can be the cell (C1), the cytoplasm (Cy1), the cell surface (CS1), and the cell nucleus (N1). Additional factors affecting the S-values are RC1, RN1, RC2, RN2, and R12, which are the radius of cell 1, radius of cell nucleus 1, radius of cell 2, radius of cell nucleus 2, and

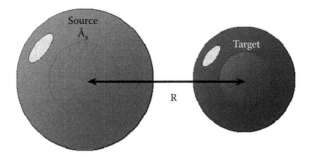

FIGURE 8.21
The basic unit of cellular dosimetry can be defined as a pair of cells. A contaminated one as source and a clean one as target.

TABLE 8.1

S-Values between Two Cells Including
Subcellular Regions

Self-dose	Cross-dose
$S(C1 \leftarrow C1)$	$S(C2 \leftarrow C1)$
$S(C1 \leftarrow Cy1)$	$S(C2 \leftarrow Cy1)$
$S(C1 \leftarrow CS1)$	$S(C2 \leftarrow CS1)$
$S(C1 \leftarrow N1)$	$S(C2 \leftarrow N1)$
$S(N1 \leftarrow C1)$	$S(N2 \leftarrow C1)$
$S(N1 \leftarrow Cy1)$	$S(N2 \leftarrow Cy1)$
$S(N1 \leftarrow CS1)$	$S(N2 \leftarrow CS1)$
$S(N1 \leftarrow N1)$	$S(N2 \leftarrow N1)$

intercellular distance, respectively. The intercellular distance can be defined as the distance between the cell centers. As can be seen, cellular dosimetry involves a lot of parameters that contribute to the difficulties already mentioned in comparisons between different methods and cell clusters. In Table 8.1, the need appears for a rigorous definition of all the parameters involved.

8.10 Physics of the Biological Effects of Radiation

Radiation dosimetry quantifies the energy deposited by radiation in biological tissue through the main quantity in radiation protection: the absorbed dose.

Deterministic or stochastic effects appear, depending on its value. From the physical point of view, the absorbed dose can be determined independently of the biological situation. From the biological point of view, the studies of the effects in the living matter, for a given localization and absorbed dose value, are made independently of the physical action.

Very often, the biological effects of ionizing radiation are correlated with absorbed dose values in tissues or cells at risk. In spite of that, other physical factors are also related, such as the radiation type, the energy spectrum, the spatial and time distribution of the energy deposited, the total number of cells irradiated, or different biological responses to radiation qualities.

In the attempt to estimate biological effects, for example, carcinogenesis, it is often assumed that the risk is simply proportional to the dose received. However, it is well known that biological experiments reveal a much more complex relationship [20].

In a simplified approach, we can consider that radiation protection studies the results of the interaction between a physical agent (radiation) and man (biological medium) or the environment. If we seek a field of knowledge that is multidisciplinary, embracing a vast set of areas, we can arrive at, for example, the characterization of the biological effects of radiation.

It is considered that two distinct types of biological effects exist: deterministic and stochastic. Edwards and Lloyd [28] and Doll [29] publish two review papers about deterministic and stochastic effects, respectively. From a pragmatic point of view, the effects that can be diagnosed in a medical examination within the few weeks after radiation exposure are certainly deterministic effects, such as skin burns or reduction in the levels of the blood cells, which, in extreme situations, can be fatal within weeks or months.

Perhaps, the best distinction between deterministic and stochastic effect is to mention the mechanisms of their production. In general, the stochastic effects (mainly cancer) are caused by nonlethal mutations in cells, whereas deterministic effects are caused by cell death.

The scheme shown in Figure 8.22 describes the processes related to the effects of ionizing radiation in living tissues.

The occurrence and proliferation of modified cells can have influences from modifications through other agents that act before the irradiation. These influences are common and can include exposure to other carcinogenic agents. As already indicated, in general, there are two types of radiation effects: stochastic and deterministic (Figure 8.23).

The replacement of dead cells is a natural process of maintenance of tissues. Through sophisticated feedback mechanisms, the rates of production of new cells are changed to compensate the rate of loss of mature cells. For example,

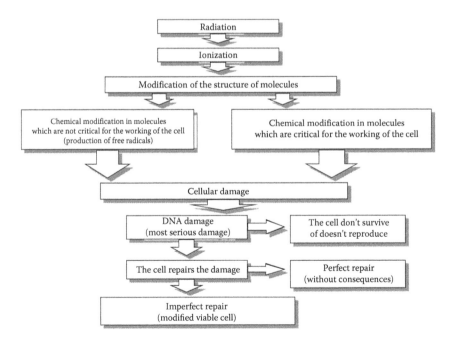

FIGURE 8.22
Radiation effect in living tissues.

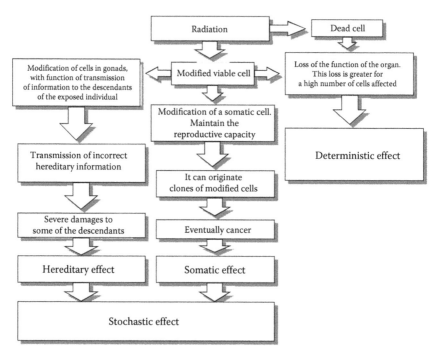

FIGURE 8.23
Stochastic and deterministic effects of radiation.

the external layer of the skin is replaced approximately every 6 weeks; and blood cells, depending on the type, are replaced with half-lives varying from some days to some years. The body has the capacity to tolerate the death of some additional cells due to external agents, such as radiation or chemicals, because it is able to replace cells at a higher velocity than under normal conditions.

The deterministic effects occur when the number of dead cells exceeds the capacity for replacement, that is, when the equilibrium state between production and death is perturbed by excess of the latter. The target organs and tissues then stop working properly, eventually leading to biological modifications with serious consequences.

One important parameter is the dose rate. For low LET radiation, the dose received at low dose rate or in fractions of time has very different consequences when the same dose is received at high dose rate, in a short period of time.

How do we relate the biological effects with the radiation that impinges on the tissues? There is a need to quantify, somehow, the energy deposited in the irradiated medium. Radiation dosimetry is the discipline of radiation physics which allows that quantification. The absorbed dose is the fundamental quantity, in such a way that it is used as the name of the discipline.

The absorbed dose is the expected value of the deposited energy in matter, per unit of mass, at a given point of interest,

$$D = \frac{d\varepsilon}{dm}. \tag{8.89}$$

Even today, it is common practice to measure the radiation effects as a function of absorbed dose, in which the dose rate is used as a parameter. However, there is an assumption in this approach that is sometimes ignored or neglected. The energy deposited has a constant value in all the biological tissue. Nowadays, it is well known that this assumption is, in general, incorrect, particularly for low doses, which are important in radiation protection [30].

Any model concerning the biological response should consider the physical, chemical, and biological factors. The physical factors are, among others, associated with the spatial pattern of the points of energy deposition, at a scale of the order of the nanometer (dimensions of the DNA molecule). Beyond the direct action, the chemical factors are related with the capacity for the production of free radicals, mainly in water, in the vicinity of the DNA molecule. Biologically, we should consider the repair and the probability of lethality of DNA lesions [31].

Nevertheless, research at biological and molecular levels still maintains the practice of expressing biological effects as a function of absorbed dose. Many of the models of biological response are defined through a dose–effect relationship. Some studies with cultures of cells of mammals indicate that the survival of the cells is a function of the absorbed dose, described by survival curves.

As an example, for the deterministic effects, it is accepted that the risk of a radiation exposure has a sigmoid relationship (see ICRP-60 [32], ICRP-103 [19] or [28]). Edwards and Lloyd [28] proposed a relationship between the risk, R, and absorbed dose, D, as follows:

$$R = 1 - \exp(-H), \tag{8.90}$$

$$H = \ln 2 \left(\frac{D}{D_{50}}\right)^V, \tag{8.91}$$

where H is named Hazard function and is related with the dose through the parameters D_{50} and V. The parameter D_{50} is the average lethal dose and signifies that it is expected that 50% of the exposed individuals shows the effect in question; V is related with the speed of rise.

An artificial threshold of dose is defined from 10% of the value of the sigmoid function. Note that this threshold value is defined for deterministic effects and is not considered for stochastic effects.

Radiation dosimetry is of fundamental importance for the study of biological effects and can then be developed in accordance with biological needs. The following figures were obtained from the data of Edwards and Lloyd [28].

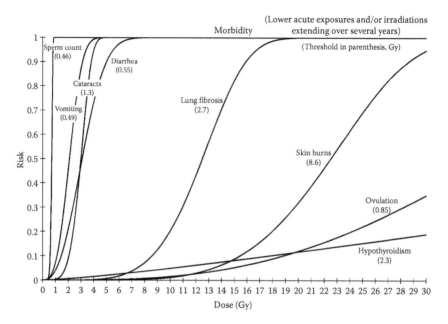

FIGURE 8.24
Biological effect of radiation (morbidity). In parenthesis: dose threshold.

It can be seen, for example, that vomiting and diarrhea are the first symptoms with dose thresholds of 490 and 550 mGy, respectively (Figure 8.24).

However, we must also consider the mortality due to radiation exposure, where the dose thresholds have higher values (Figure 8.25).

Knowing typical values for some of the serious effects of the radiations, it is interesting to compare these values with the dose limits. This comparison is made in the following figures:

In the graph of Figure 8.26, a vertical line is included, which corresponds to the all-body annual dose limit for radiation workers, 20 mSv, which means that for photons we have 20 mGy, or 0.02 Gy. As can be seen, the scale was reduced to a maximum of 1 Gy, instead of the 30 Gy of Figure 8.24. In Figure 8.27, the scale is reduced still further to a maximum of 0.55 Gy = 550 mGy. As can be seen, the dose threshold for the decrease in the sperm count has a value 23 times higher than the current dose limit for workers; whereas for vomiting, the dose threshold is 24.5 times higher than the dose limit; and diarrhea appears with a threshold 27.5 times higher than the dose limit. This difference of values still increases more when these threshold values are compared with the dose limits for the members of the public, which is 1 mSv, equal to 1 mGy for photons, significantly smaller than the thresholds of 460, 490, and 550 mGy for sperm count, vomiting, and diarrhea, respectively.

With data of this type, it is possible to trace scenarios of the consequences of a radiological accident, in particular when there is release of radioactivity

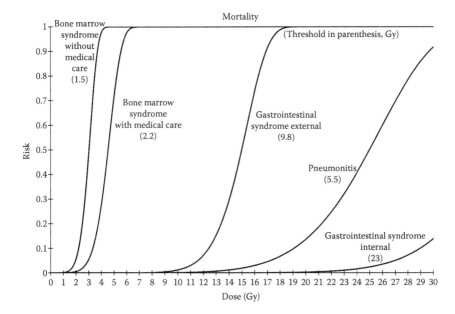

FIGURE 8.25
Biological effects of radiation (mortality).

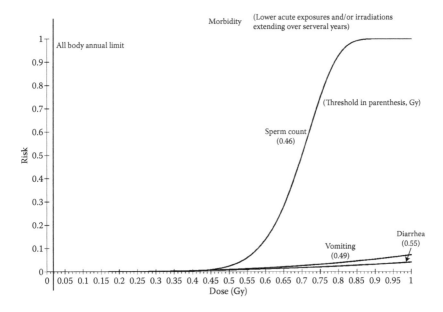

FIGURE 8.26
Comparison of some radiation effects with the dose limits for stochastic effects.

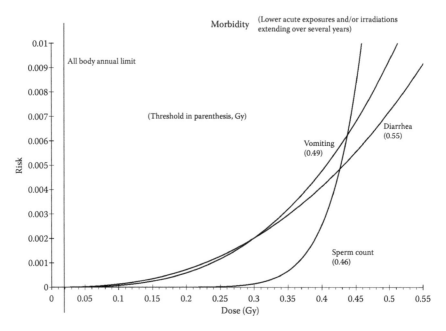

FIGURE 8.27
Comparison of deterministic effects with the dose limit for stochastic effects.

in the environment or when an incident results in high exposures of a group of individuals.

As can be seen in the figures just cited, a fundamental element in the forecast of the consequences from the irradiation of a population is the determination of the absorbed dose, which allows us to determine the associated risks.

As can be appreciated, these graphs have a multidisciplinary characteristic; in the abscissa axis, we found a physical quantity, and the ordinate axis represents the biological effect. Therefore, this type of graph has biophysical characteristics.

One important component of radiation protection is the determination of the risks related to radiation exposure. This subject is polemic due to the definition of risk with supporters and detractors. The defence of this type of analysis is found in Pochin [33].

This presents, for example, the importance that these studies have had at the level of radiological surveys on populations, allowing the conclusion, for example, that the risk is larger than the benefit for ages less than 30 years in stomach cancer surveys. In another example, Pochin [33] pointed out that the production of hydroelectric energy has a death risk almost 2.5 times higher than the nuclear production for a GW per year of produced electricity.

Notice that the problem of the risk associated with radiation exposures and radiation dosimetry always appears to be connected, in a global perspective of radiological protection. For a consultation on the risk philosophy in radiation protection, see Lindell [34].

8.10.1 The Cell as a Target of Ionizing Radiation

The cell is one of the important targets of ionizing radiation; its structure and function have, naturally, influenced the evolution of radiation dosimetry. Due to this, several physical considerations will be made, related with this complex target, which is the biological tissue. The cell is the fundamental unit of life. It is the structural unit constituent of plants and animals and, therefore, is an important target that cannot be neglected in the study of the mechanisms of radiation effects and radiation protection. Due to the complexity of biological matter, radiation dosimetry also tends to be complex, for example, through microdosimetry. Consider the following question: Is the cell the fundamental target in radiation dosimetry or should we consider only the cell nucleus, or, on the other hand, is the DNA molecule the most important target? Will other cell constituents also be important such as enzymes, ribosomes, cellular membranes, centrosomes, etc.? Are these questions for radiobiology or are they also for radiation dosimetry?

8.10.2 Composition, Shape, and Size

Although, often only implied, several approaches toward radiation protection are directly related with the hierarchy of life, which can be presented in the following way: molecules, organelles, cells, tissues, organs, organisms, populations, communities, ecosystems, and biosphere. Related to this hierarchy, we find, for example, studies in environmental radioactivity whose main focus are ecosystems, populations, and communities, where there are also included epidemiological studies of the radiation effects or the determination of the radioactivity in several types of soils or rivers and their transference to living organisms, or the spread of radioactive clouds in atmosphere, etc. Studies of the absorbed dose in living matter, in particular in the human body, in their organs, or tissues, are the goal, for example, of several surveys to determine the dose in medical practices, such as radioadiagnostics or the rigorous determination of absorbed dose in radiotherapy. The radiobiological study of the radiation effects in tissues, cells, and their constituents and molecular biology is applicable, in studying the DNA molecule, an important target of ionizing radiation. In this latter field, studies have appeared of the biophysical mechanisms of the action of ionizing radiation, from which has emerged the subdiscipline of radiation dosimetry, microdosimetry, which extends to nanodosimetry.

As such, the fundamental unit of life, the cell, and its constituents cannot be ignored in radiation protection.

Approximately 70–80% of the cell is water. For this reason, water is often used as a human tissue equivalent. However, the diversity of structures found in cells and the metabolic capacities, growth, and reproduction are so complex that the above-mentioned approximation can only be accepted as an intermediate step toward the full understanding of the biological effects of ionizing radiation, which is a fundamental goal of radiation protection.

In radiation dosimetry, an immediate question emerges about the size of cells. The diversity of cell sizes is rather large, such that finding typical sizes is only a first approximation.

Table 8.2 shows the dimensions of typical biological molecules and cells to give an idea of the size scales involved.

Life at all levels can be seen as a process that organizes matter and energy into ordered systems. It can be considered, at another level, as a process of decreasing the entropy of the system. The interaction of radiation with matter supplies energy through processes which increase the entropy of the biological system, sometimes at levels that the organism cannot reorganize, leading to lethal radiation effects. The concept of entropy allows new insights of radiation physics, as have been shown above. It can eventually be a link connecting physics and biology. Systems theory in the future will also play an important role in radiation protection.

How far can we ignore the question of the spatial scale? How can we not question the concept of uniform irradiation, in which we assume that

TABLE 8.2

Typical Sizes of Biological Molecules and Cells

0.1 nm diameter of hydrogen atom
0.8 nm amino acid
2 nm diameter of DNA helix
10 nm thickness of cell membrane
11 nm ribosome
25 nm microtube
50 nm nuclear pore
100 nm virus
200 nm centriole
200 to 500 nm lysosome
1–10 nm generic size of prokaryotes
1 μm diameter of a human nerve cell
3 μm mitochondria
6 (3–10) μm cell nucleus
4 μm leukocyte (white blood cell)
9 μm erythrocyte (red blood cell)
10–30 μm generic size of eukaryotic animal cells
20 μm liver cell
90 μm amoeba
100 μm (0.1 mm) human ovum

in a uniformly irradiated region the value of the absorbed dose is the same everywhere within the region? It seems difficult, following the above considerations, to accept the concept of uniform irradiation, where the absorbed dose value is the same for any point inside the volume exposed.

For a man with average height and weight, there are about 10^{14} cells. In addition to their normal metabolism, cells undergo four more processes:

1. Merge
2. Differentiate
3. Die
4. Divide

The process of division produces new cells in number and for a period of time, which is determined by the needs inherent to the organism and by the internal and external environmental circumstances. In a mammal, there are active centers for cellular division, for example, the skin, the regions of blood formation in bones, or some intestinal cells. Radiosensitivity is associated with the reproducibility of cells.

8.11 Action Modes. Target Theory Models: Survival Curves

The biological effects of radiation can be studied at different levels of biological organization, such as molecular, subcellular, cellular, organ or tissue, whole body, and population. Examples of the effects of ionizing radiation at a molecular level are damage to DNA or proteins; at a subcellular level, they are injury on cell membranes or chromosomes; at a cellular level, they are inhibition of cell division or loss of function; at a tissue or organ level, they are lesions in GI or hematopoietic systems; at the whole body level, they include cancer or death; and, finally, at the population level, they are the genetic or chromosomal mutations.

The inactivation of individual biological structures by the action of ionizing radiation, resulting from a discrete and localized transfer of energy, is one of the postulates in model development for the study of the effects of radiation on cells. In these models, two possibilities of interaction of ionizing radiation with molecular structures relevant in cellular function must be considered (Figure 8.28): direct action and indirect action, the latter resulting from the participation of intermediary products.

Direct action concerns the effects that occur when radiation directly hits molecules or microstructures of vital importance in the cells. It is normally described on the bases of a so-called target theory.

Target theory is the mathematical development of direct action with experimental grounds, but without taking into account the biological mechanisms of cellular attack. Target theory sets on two postulates:

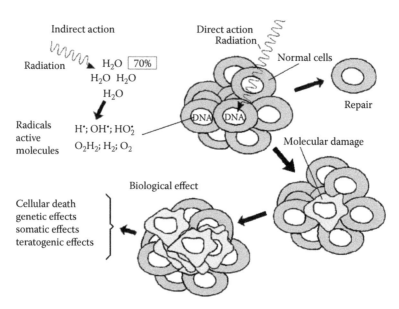

FIGURE 8.28
The two possibilities of interaction of the ionizing radiation with molecular structures relevant in cellular function: direct action and indirect action, resulting from the participation of intermediary products (radicals and chemically active molecules).

- The statistical nature of energy deposition; and
- The existence of a direct relationship between the number of hits and the final biological effects.

A problem inherent to this method is the difficulty to relating biological effects with specific physical or chemical causes [35].

According to target theory, the probability of a radiobiological event occurring depends on the probability that the primary physical action has taken place in specific zones of the cell. This is the concept of sensitive zones of the cell that correspond to the vital structures already mentioned.

The number of target molecules is generally small when compared with the total number of molecules in the cell; but the effects can be important if the molecules involved are enzymes or molecules from genes or chromosomes.

It is known, for example, that for some cellular enzymes, the number of molecules in a cell is minimal, limited to a few units, but their action is decisive. Inactivation of molecules of such enzymes can irreparably compromise the function in which they participate.

It was already known, using sampling and observation of cellular alterations after irradiation, that for a particular preparation in constant conditions, if the intensity of radiation is decreased but its quality is maintained, the lesions observed were no less severe but they occurred in a smaller number.

Target theory of cellular death was proposed by Lea in 1946 and experimentally developed using data from inactivation of microorganisms and active biomolecules [36].

In the implementation of the theory, a set of variables capable of altering the inactivation properties of radiation, such as temperature, type of radiation, chemical composition of the medium, etc., were taken into account.

The statistical treatment of experimental data and its systematic analysis led to the knowledge of the minimal energy necessary to damage a particular sensitive zone as well as its volume.

There are more complex situations than those resulting from a simple impact. Other effects can occur if a second photon collides with the molecule within a time of 10^{-8} to 10^{-4} s after the first hit and before the molecule has lost the excitation energy received.

Target theory has been shown to be applicable in a large number of clonogenic* survival situations in mammals, but its application as a general theory has always had some opposition [37].

Proliferation by cellular division is an essential property of cells in any biological system. The alterations in this capacity after irradiation with ionizing radiation results from chemical alterations in the genome and is of fundamental interest in radiotherapy and radioprotection.

In general terms, in the absence of radiation or any other external active entity, the factors that limit the cellular division capacity in healthy cells are the shortage of nutrients, the increase in cellular concentration, and the presence of regulatory systemic factors.

Apart from direct action, an indirect action that takes place between the initial physical events and the biological phenomena, which is equally capable of producing molecular alterations, also has to be taken into account.

In fact, radiation can act on the intracellular fluids or the extracellular environment, producing active radicals that, in turn, act upon the cells.

Water contributes to about 70% of total mass in living beings and reaches more than 80% in embryonic and young tissues. It is the main intervenient in indirect action.

Biological damage resulting from low LET radiation is mainly due to indirect action. About two thirds of biological damage, produced by x-rays, results from indirect action.† It is, therefore, of relevant interest to consider the products that arise from the action of ionizing radiation on water. The formation process of these intermediary structures is called *water radiolysis*.

---

* Cells starting from a single cell are able to indefinitely proliferate and form large colonies called *clonogenic*. Tumor cells indefinitely proliferate in cellular cultures.
† Assuming low levels of chronic irradiation from external sources, strong probabilities exist of repairing such biological damage without errors.

8.11.1 Water Radiolysis

The reaction

$$H_2O \rightarrow H_2O^+ + e^- \tag{8.92}$$

represents the ionization of the water molecule; the recombination of the ions produced is not probable.

These ions will lead to further reactions

$$H_2O^+ \rightarrow H^+ + OH^\bullet \tag{8.93}$$

whereas the free electron will combine with another water molecule.

$$e^- + H_2O \rightarrow H_2O^- \rightarrow H^\bullet + OH^- \tag{8.94}$$

$H^\bullet$ and $OH^\bullet$ radicals can diffuse and react with each other or initiate oxidation-reduction reactions with other molecules of the medium.*

In the case where the radicals react with each other, the probable reactions are

$$H^\bullet + H^\bullet \rightarrow H_2 \tag{8.95}$$

$$OH^\bullet + OH^\bullet \rightarrow H_2O_2 \tag{8.96}$$

$$OH^\bullet + H^\bullet \rightarrow H_2O \tag{8.97}$$

with the two radicals in Equation 8.97 having chemical activity.

In summary, about 10^{-7} s after the passage of an ionizing particle in water and around its track, $H^\bullet$, $OH^\bullet$, H_2O_2, and H_2 are found. The relative proportions of these intermediary reaction products depend on radiation TLE.

What happens next (from 10^{-7} to 10^{-3} s) depends on a number of factors, such as water purity.

Of special interest is the presence in water of dissolved oxygen. When this is present, free electrons can react with the O_2 molecule

$$e^- + O_2 \rightarrow O_2^- \tag{8.98}$$

which reacts with a water molecule, leading to

$$O_2^- ; + H_2O \rightarrow OH^- + HO_2^\bullet. \tag{8.99}$$

A reaction between the radical $H^\bullet$ and O_2 can also occur:

$$H^\bullet + O_2 \rightarrow HO_2^\bullet \tag{8.100}$$

---

* Free radicals are chemical species with a free valence and are very reactive. These radicals can react with molecules such as DNA and lipid membranes.

This radical can react with another $HO_2^\bullet$ or with $H^\bullet$ giving

$$HO_2^\bullet + HO_2^\bullet \rightarrow H_2O_2 + O_2 \qquad (8.101)$$

$$HO_2^\bullet + H^\bullet \rightarrow H_2O_2 \qquad (8.102)$$

The $HO_2^\bullet$ radical is less reactive as an oxidant than $OH^\bullet$, but it can diffuse to longer distances. The presence of this radical and a higher H_2O_2 concentration are the main consequences of the presence of O_2 in the solution.

Other reactions are possible by combination of the radicals available when oxygen is present.

The set of phenomena that make up the strengthening of radiation effects resulting from O_2 presence is designated *oxygen effect*.

This effect can be quantified, for a given biological action, through the oxygen enhancement ratio (OER), which is defined as the ratio between absorbed dose in anoxic conditions (D_1) and dose in good oxygenation conditions (D_2) necessary to have the same biological effect in a given tissue, or

$$OER = \frac{D_1}{D_2}\delta \qquad (8.103)$$

8.11.2 Effects in Aqueous Solutions

Let us consider an aqueous solution of RH molecules and that radicals such as $H^\bullet$, $OH^{\bullet-}$, and $HO_2^\bullet$ appear in solution by radiolysis.

The following reactions can occur:

$$RH + H^\bullet \rightarrow R^\bullet + H_2 \qquad (8.104)$$

$$RH + OH^\bullet \rightarrow R^\bullet + H_2O \qquad (8.105)$$

$$RH + O_2H^\bullet \rightarrow R^\bullet + H_2O_2 \qquad (8.106)$$

or

$$RO^\bullet + H_2O \qquad (8.107)$$

If the O_2 concentration in the medium is high, a reaction with radical $R^\bullet$ can occur, with peroxide radical as the product.

$$R^\bullet + O_2 \rightarrow ROO^\bullet \qquad (8.108)$$

This can react with an RH molecule

$$ROO^\bullet + RH \rightarrow ROOH + R^\bullet \qquad (8.109)$$

The radical $R^\bullet$ can, in turn, react with O_2 to produce another peroxide radical, in a process that repeats itself in a chain reaction.

Other reactions are possible, particularly in the presence of oxygen.

8.11.3 Ionizing Radiation, Ionization Mechanisms, and Biological Action

For a better understanding of damage resulting from ionizing radiation action on DNA, which is organized into chromosomes in the cell, we will first briefly recall the cellular division cycle.

The cycle of cell division in somatic cells is divided into two parts: interphase and mitosis.

During the interphase part, which occupies most of the duration of the cellular division cycle, the amount of cellular DNA is duplicated as a result of synthesis of replicas of existing chromosomes.

Interphase is divided into three parts: G1 (first intermediary stage), where synthesis of numerous proteins occurs; S (synthesis), during which a new copy of the entire molecule of DNA is produced, before cell division; and G2 (second intermediary stage), during which the integrity of the chromosome is tested. The process of synthesis S is separated from mitosis by intermediary stages G1 and G2, during which normal activity of the cell is maintained.

Mitosis, which lasts from 1 to 2 h, is the process of cell division and includes not only the division of the nucleus but also that of the cytoplasm (cytokinesis).

The duration of G1 determines the frequency of cell division, ranging from hours to more than a hundred days, depending on the type of cells.

The synthesis S lasts about 8 h, and G2 lasts about 4 h.

Mitosis, in turn, is divided into four parts: the prophase (P), during which the nuclear membrane disappears; the metaphase (M), during which the mitotic spindle is formed and where chromosomes fit in; the anaphase (A), during which the chromosomes migrate over the spindle to opposite parts in the cell; and telophase (T), in which nuclear and cell membranes reappear, contributing to the emergence of two separate cells.

Cell sensitivity to radiation varies during the different phases of the cell cycle. The cells of mammals are more resistant to radiation at the end of the S phase and more sensitive toward the end of G2 and mitosis.

The probability that a given molecule will be affected by direct action increases with its dimensions.

At a molecular level, the biological effects of radiation, direct and indirect, are rare events, random and independent of each other, and are, thus, covered by Poisson statistics.

It should be noted that, while accepting that DNA is a critical target for the effects that involve changes in the integrity of the processes on cell division, there is no evidence of an exclusive involvement of this molecule.

The cells, after irradiation, may die trying to start the process of cell division. They can divide, causing aberrant forms. They may be unable to divide, though physiologically functioning, behaving similar to normal cells and keeping alive for long periods. The cells can divide as apparently normal ones till a generation of sterile daughter cells occurs. They may, ultimately, divide normally or with minor changes in division, such as cycles with different times from normal [38].

Although it is known today that target theory is not entirely appropriate to explain what happens in a system as complex as a cell, its contribution is important for the study of the situation in which the mechanisms of recovery do not alter the initial damage in a significant way.

Directly ionizing radiations, with high LET (α particles, heavy ions, and neutrons), produce most of their biological damage by direct action. However, for any type of radiation and, in particular, with ionizing energy photons, a strong probability of interaction with water molecules occurs, leading to the production of ions, radicals, and other chemically reactive structures.

The processes leading to the initial cellular damage take place in times from 10^{-17} to 10^{-5} s. Possible further steps in the chain of events after irradiation, which occur on a time scale from minutes to decades, are damage repair; cellular death; and genetic, somatic, or teratogenic effects.

As a first approach in the development of models to simulate the effects of ionizing radiation, biological tissues can be considered as dilute solutions (suspensions) resulting from the large amount of water in their composition. Indirect effects are, consequently, to be expected.

For a radiation field with a given fluence, the probability that a particular target (molecule or structure) will be on the path of radiation and be affected by direct action increases with the increase of the target dimensions. DNA with molecular mass of $6-8\times10^6$ u and a main role in cellular life is, obviously, a critical target.

There is strong circumstantial evidence to indicate that DNA is the principal target for the biological effects of radiation, including cell death, mutation, and carcinogenesis. This has been observed and scored as a function of dose in experiments where the DNA was denatured and the supporting structure stripped away.

DNA is a large molecule that presents the well-known double helix structure, as seen in Figure 8.29.

Nucleotides, the two long chains on DNA molecule, consist of deoxyribose (S) and phosphates (P) that wind around each other in a spiral. The two long intertwined strands are held together by the bases, which form cross links between the long strands in the same manner as the treads in a stepladder. There are four different bases, two purines: adenine (A) and guanine (G), and two pyrimidines: cytosine (C) and thymine (T).

The base cross-links are formed when two of these bases, one of which is attached to each long strand, join together. The cross-linked bonds are specific, with adenine exclusively coupling with thymine (A–T) and cytosine coupling only with guanine (C–G). The bits of information are coded in triplets of various combinations of A–T and C–G cross links, and the sequence of these triplets determines the genetic information contained in a DNA molecule (Figure 8.29b).

Cells in the human body can be divided into two very general groups: somatic and germinative cells. The latter are the direct agents for species reproduction, and the former consist of all the remaining ones. Each normal

(a)   (b)

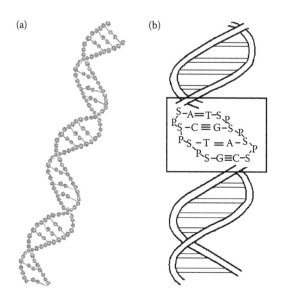

FIGURE 8.29
DNA double helix structure. The nucleotides, the strands of the double helix, consist of alternating deoxyribose (S) and phosphate (P) units. Two pairs of complementary bases, adenine (A) and thymine (T), and guanine (G) and cytosine (C), link the two nucleotide chains.

cell contains a complete set of chromosomes that consists of DNA and has the templates for human beings. In humans, cells contain 23 pairs of chromosomes. Germinative cells (spermatic cells and ovules) contain 23 single chromosomes, one half that is in somatic cells.

Genetic information can be altered by many different chemical and physical agents, called mutagens, which disrupt the sequence of bases in a DNA molecule Figure 8.30a. The breaks in an intact DNA molecule, produced by the action of ionizing radiation, can be of several different types and have different biological consequences with regard to an eventual cellular death or another type of damage. We can classify the damage:

1. One single-strand break (Figure 8.30b).
2. Double breaks with the ruptures well apart (Figure 8.30c).
3. Breaks close to each other, in different strands (Figure 8.30d).

If the information content of a somatic cell is altered, as is schematically represented in Figure 8.31, then its descendents may show some sort of an abnormality. If this altered information occurs in a germ cell that is subsequently fertilized, then the new individual may carry a genetic defect or a mutation.

A mutation resulting from alterations, such as those schematically shown in Figure 8.31a–e (base elimination, base substitution, hydrogen bond

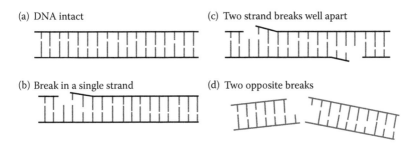

FIGURE 8.30
Single- and double-strand breaks in DNA molecule. (a) Intact DNA molecule. (b) One single-strand break. (c) Double break with the ruptures well apart; the repair efficiently occurs. (d) Breaks close to each other in different strands.

break, single-strand break, and double-strand break, respectively), is called a point mutation, because it results from damage to one point on one gene.

In intact DNA, however, single-strand breaks are of little biological consequence as far as a cell killing is concerned, because they are readily repaired,

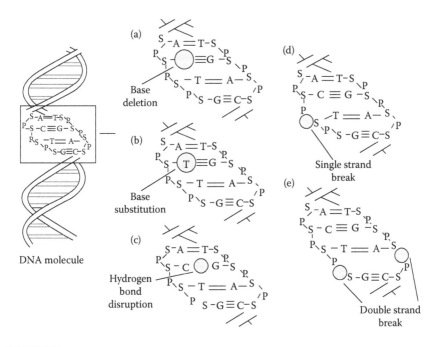

FIGURE 8.31
Mechanisms of DNA damage by radiation: (a) Base deletion. (b) Base substitution. (c) Hydrogen bond disruption. (d) Single-strand break. (e) Double-strand break.

over the course of a few hours, with the opposite strand being used as a template (Figure 8.30b).

However, if the repair is incorrect, it may result in a mutation.

If both strands of the DNA are broken and the breaks are well separated (Figure 8.30c), again repair readily occurs, as the two breaks are separately handled. In contrast, if the breaks on the two strands are opposite each other, or separated by only a few base pairs (Figures 8.30d and 8.31e), this may lead to a double-strand break. Most of the important biological effects of ionizing radiations are a direct consequence of the rejoining of two double-strand breaks.

When a double-strand break occurs, permanent genetic damage can arise in the next cellular division in the form of chromosomal aberrations. In most of these double-strand breaks, the fragments reunite with the original configuration, and the only result may be a point mutation at the site of the original break. In a small fraction of breaks, however, the broken pieces do not reunite, thereby giving rise to an aberration.

If this happens and one of the broken fragments is lost, the daughter cell does not receive the genetic information contained in the lost fragment when the cell divides. Another serious consequence after chromosomal breakage, especially if two or more chromosomes are broken, is the interchange of the fragments among the broken chromosomes and the production of aberrant chromosomes. Cells with such aberrant chromosomes usually have impaired reproductive capacity as well as other functions, with serious consequences in the next mitosis. Several forms of abnormal chromosomes are known, depending on where the damage occurred along the strand and how the damaged pieces connected or failed to connect to other chromosome fragments. Many of these chromosomal abnormalities are lethal: the cell either fails to complete mitosis the next time it tries to divide or fails within the next few divisions. Other abnormalities allow the cell to continue to divide, but they may contribute to the multistep process that sometimes leads to cancer many cell generations later.

Figure 8.32 illustrates two ways in which breaks in two separate chromosomes may rejoin. In the upper half of the figure, translocation occurs, generating abnormal chromosomes. Translocation, in general, is not lethal but may generate cells that will probably be active in cancer promotion.

In the lower half of the figure, the two broken chromosomes rejoin in such a way that a dicentric and two acentric fragments are formed. This exchange type aberration is lethal to the cell; the fragment with no centromere will be lost during cell division, whereas the aberrant chromosome with two centromeres will make a normal mitosis impossible.

In mammalian cells, the DNA double-strand breaks are mainly repaired by a junction of nonhomologous terminal parts.

Nonrepaired or badly repaired damage in phases before replication (G0-G1) can lead to chromosome aberrations.

Damage of nonrepaired or not well repaired in phases posterior to replication (end of S and G2) can lead to aberrations in chromatides.

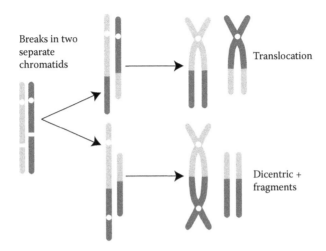

FIGURE 8.32
Chromosome aberrations. Translocation is, in general, not lethal but may release an oncogene. Dicentric fragments are generally not viable.

Aberrations resulting from asymmetric transpositions (dicentric and rings) are generally lethal.

Aberrations resulting from symmetric transpositions (translocations and eliminations) resulting from badly repaired DNA damage can lead to carcinogenesis.

The available techniques to study DNA double-strand breaks are not sensitive enough to be used as biological dosimeters in case of radiation accidents.

Cell survival is evaluated through cell clustering capacity. The inability of cells to make colonies in cultures in vitro is considered cellular death (reproductive death), and cellular survival is evaluated by the existence of this capacity.

Suspensions of isolated normal nonirradiated cells, sown in a gelified nutrient culture medium, at a large fraction, give rise after 1–2 weeks of incubation to macroscopically visible colonies that can be fixed, dyed, and counted.* A clustering efficiency is defined by the ratio between the number of clusters observed and the initial number of cells in plaque.†

This same operation carried out with cells after irradiation with ionizing radiation shows a decrease in the clustering efficiency. The evaluation of effects is performed using pedigree analysis.

---

* Isolated, normal cells after tripsinization are clonogenic.
† Clustering efficiency varies with cellular lines and cell density.

8.12 Mathematical Models of Cellular Survival in Ionizing Radiation Fields

Several mathematical equations have been proposed to describe the loss of reproductive capacity observed in experiments with cells in culture irradiated with ionizing radiation.

Cell survival curves are graphical representations of the fraction of survival cells (S) as a function of the absorbed dose. Survival curves show different shapes for different cells and for different types of radiation.

Certain types of radiation-sensitive cells, such as hematopoietic stem cells, show linear survival curves, on a log-linear scale. This profile is also a general rule for cells irradiated with high LET radiation (Figure 8.33, curve 1)

Cells irradiated with low LET radiation show survival curves that either present a variable dimension shoulder followed by a steep falloff (Figure 8.33, curve 2) or have no shoulder (nonlinear with slope increasing with dose) (Figure 8.33, curve 3). It is generally possible to obtain equations that fit these curves.

The inactivation of individual biological structures by ionization radiation is one of the accepted postulates in the development of models for the action of radiation on living matter.

At the level of the events that occur in cell sensitive zones by direct or indirect action of radiation, the eventual biological effects of radiation result

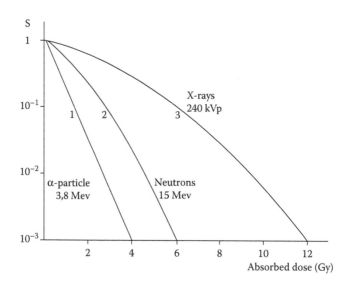

FIGURE 8.33
Survival curves are linear for radiation of high TLE (1) or present the shapes (2) and (3) for low TLE.

from discrete energy transfer, corresponding to rare events (hits),* which are random and independent of each other. To be damaged, a cell needs to be hit at least once, but a great deal of possibilities can be considered.

These processes are subjected to Poisson statistics, as they result from their own characteristics.

Two distinct approaches have been used to develop cell survival equations: one based on target theory, with several possible models, and a second one assuming a linear-quadratic model.

8.12.1 Multiple Target Models (Target Theory Models)

8.12.1.1 One Sensitive Region n Hits

Assuming that the conditions of Poisson statistics are valid, let us suppose that the average number of hits in a specific sensitive region of a cell in a population, in a field of radiation, is m. The probability that a particular cell receives n hits in its sensitive region is

$$P(n) = \frac{m^n e^{-m}}{n!}.$$ (8.110)

Obviously, m is proportional to the dose D and $m = aD$ where a is a constant.

$$P(n) = \frac{(aD)^n e^{-aD}}{n!}.$$ (8.111)

If n is also the number of hits necessary to inactivate the cell, all the cells receiving less than n hits will survive. Then the survival probability is

$$P(n) = P(0) + P(1) + P(2) + \cdots + P(n-1).$$ (8.112)

If N is the number of surviving cells after dose D and N_0 is the initial number of cells in the irradiated sample, then the probability of surviving can be written:

$$\frac{N}{N_0} = P(0) + P(1) + P(2) + \cdots + P(n-1).$$ (8.113)

---

* The number of atoms/cc tissue is about 10^{23}, a value that is much bigger than the number of atoms hit by radiation.

Applying Equation 8.111 to the terms on the right-hand side of Equation 8.113, it comes

$$N = N_0 e^{-aD}(1 + aD + \frac{(aD)^2}{2!} + \cdots + \frac{(aD)^{n-1}}{(n-1)!}), \qquad (8.114)$$

$$N = N_0 e^{-aD} \sum_{k=0}^{n-1} \frac{(aD)^k}{k!}. \qquad (8.115)$$

If $n = 1$, then

$$N = N_0 e^{-aD}. \qquad (8.116)$$

Equation 8.116 is a decreasing exponential function, which is the simplest case of cellular survival. It is seen in cases of irradiation of isolated cells in vivo or in culture (bacteria, animal cells, etc.) for high LET radiation. In this case, the data can be fitted by a straight line in a semilogarithmic plot. The constant a has the dimensions of the reciprocal of absorbed dose. Substituting $a = 1/D_o$, Equation 8.116 can be written as

$$S = \frac{N}{N_0} = e^{-D/D_0}, \qquad (8.117)$$

where S is the fraction of surviving cells for dose D.

$D_0 = 1/a$ is the mean lethal dose, that is, the average dose received by the cell population, assuming a potentially infinite life and that death is exclusively due to radiation. D_0 is proportional to the resistance of cells to radiation, and $1/D_0$ increases when the radio-sensitivity increases. D_0 is also the dose that reduces the initial population by the factor $1/e = 0.37$.

The semilog representation Equation 8.117 is a straight line with slope $-a$, the negative reverse of the mean lethal dose D_0, which fully characterizes the process.

In the present case, where inactivation occurs with just one hit, D_0 is also the dose that corresponds to one hit per sensitive region (average value). The total number of hits equals the number of sensitive regions, that is, of irradiated cells. D_0 varies from a few Gy in mammalian systems to values of the order of thousands Gy in virus.

The mean lethal dose D_{50} is also used, and t is the dose for which $S = 0.5$. It is easily shown that

$$D_{50} = 0.693/D_0 \qquad (8.118)$$

Assuming that the hits correspond to ionization processes and that w is the ionization mean energy, for an absorbed dose D, the mean energy delivered in the sensitive region of volume V and specific mass ρ is $DV\rho$.

For dose, D_0 is

$$D_0 V\rho = w.$$

Then, the volume of the sensitive region is

$$V = w/D_0\rho \qquad (8.119)$$

Values of V for tissues and cells are known. For mammalian tissues, V is of the order of 10^{-23} m^3.

An accurate value of w can be difficult to know, as it depends on the LET of the radiation and on the different interaction cross sections with the atoms of the medium.

When the ions produced are well separated in the medium and we have the situation of cellular death with a single hit, then a possible model for studying the process is the ionization in a gas with an atomic number identical to that of the tissue or suspension irradiated. This is the case of air and some biological tissues that have $w \approx 34$ eV.

When the situation of cells with a single sensitive region and a single hit, as in the case of Equation 8.116, is analyzed, the same result can be obtained starting from the equation

$$dN = -a\,N\,dD \qquad (8.120)$$

that is, the simple observation that the number of hits dN, that is supposed to occur for dose dD, increases when N and dD increases.

8.12.1.2 Multiple Sublethal Sensitive Zones/One-Hit Model

Most mammalian cells have several sensitive zones. We can postulate that for cellular death to occur, all the sensitive zones have to be hit (sublethal sensitive zones). Additionally, we assume that sensitive zones are inactivated with a single hit.

The probability that a sensitive zone is not struck is given by the survival equation for one zone one hit (Equation 8.117). The inverse probability to this one, that is, the probability that the zone is hit, is then

$$p = 1 - e^{-aD}. \qquad (8.121)$$

If the cell has h sensitive zones, the probability of these being struck is

$$p_h = \left(1 - e^{-aD}\right)^h. \qquad (8.122)$$

The survival probability of an irradiated cell with h sensitive zones is the inverse probability to p_h, that is,

$$S_h = 1 - \left(1 - e^{-aD}\right)^h. \qquad (8.123)$$

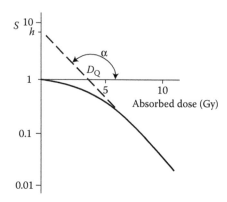

FIGURE 8.34
Plot of Equation 8.123.

This is the survival equation of a cell with h sensitive zones (multitarget/single-hit model) that, in Cartesian coordinates, is a sigmoid.

Expanding Equation 8.123 by applying the binomial theorem* and assuming that D has large values, we obtain

$$S_h = he^{-aD} = he^{-D/D_h}, \qquad (8.124)$$

with $a = 1/D_h$. As can be seen in Figure 8.34 in a semilogarithmic plot, the survival equation is linear for high doses. The fraction of cells surviving is plotted on a logarithmic scale against dose on a linear scale.

The extrapolation of this linear portion back to the ordinate (S axis) gives h, the number of sensitive zones per cell Figure 8.4).

The value D_Q is the absorbed dose that corresponds to $S = 1$ in the extrapolated straight line. D_Q is called the quasi-threshold dose and is a measure of the threshold length that precedes the linear portion of the survival curve for the situation multitarget or single-hit model.

This coefficient is related with the intrinsic capacity of cells to be repaired from sublethal lesions, and it gives an indication of the damage that has to be accumulated before cell death occurs.

Applying logarithms to Equation 8.124, we obtain

$$\ln S_h = -D/D_h + \ln h, \qquad (8.125)$$

or in decimal logarithms

$$\log_{10} S_h = -2.304\, D/D_h + \log_{10} h. \qquad (8.126)$$

---

* A binomial of power n can be expanded using the equation $(x + a)^n = C^n; pa^p x^{n-p}$ with n and p integer and $0 < p < n$.

Then, when $S_h = 1$

$$\log_{10} 1 = 0 = -2.304\, D_0/D_h + \log_{10} h \tag{8.127}$$

and

$$D_0 = D_h \log_{10} h/2.304, \tag{8.128}$$

or

$$D_0 = \ln h. \tag{8.129}$$

Using Equations 8.126 and 8.129, the linear portion of Equation 8.124 is

$$\log_{10} S_h = -2.304(D - D_0)/D_h. \tag{8.130}$$

The slope of tangent to the curve of Equation 8.123 (8.123), for dose D, is

$$\frac{dS_h}{dD} = -ahe^{-aD}\left(1 - e^{-aD}\right)^{h-1}. \tag{8.131}$$

For $D = 0$, the slope $\frac{dS_h}{dD}$ of function S_h is zero. At this point, curve $S(D)$ is parallel to the Ds axis, meaning the absence of effects for very small doses.

Some experimental sigmoidal curves do not fit the model of equation 8.123, simply because for $t = 0$, the slope of tangent is not zero. Most experimental survival curves have an initial slope, whereas the multitarget or single-hit model predicts no initial slope.

8.12.1.3 Mixed Model

A single-target single hit or multitarget single hit model, also termed a *mix model*, gets closer to the experimental situation discussed earlier. This model includes one lethal sensitive region and n sublethal sensitive regions.

Cellular death will occur either by one hit in the lethal region or by hits in all the n sublethal regions. For these conditions, the survival probability of one cell is

$$S_n = \frac{N}{N_0} = e^{-D/D_0}\left[1 - \left(1 - e^{-D/D_n}\right)\right]. \tag{8.132}$$

D_0 is the mean lethal dose of lethal zone, and D_n is the mean lethal dose for sublethal regions.

The slope of this function for every value of D is

$$\frac{dS_n}{dD} = -1/D_0 e^{-D/D_0}\left[1 - \left(1 - e^{-D/D_n}\right)^n\right] - n/D_n e^{-D/D_0}\left(1 - e^{-D/D_n}\right)^{n-1}. \tag{8.133}$$

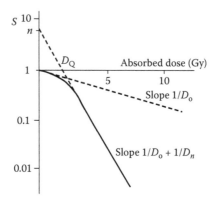

FIGURE 8.35
Survival curve for the mixed model, Equation 8.132.

For $D = 0$,

$$\frac{dS_n}{dD} = -1/D_0. \tag{8.134}$$

The plot of $\log_{10} S_n$ versus D is given in Figure 8.5.

For high values of D, the slope of the curve in Figure 8.35 becomes constant and equal to $-(1/D_0 + 1/D_n)$. The initial slope $-1/D_0$ indicates the action of sudden death due to hits on lethal target. Intersection with the ordinate of the extrapolation of the linear portion of Equation 8.132 gives the value n, the number of sublethal regions.

Considering two samples with distinct cell populations, satisfying the model of Equation 8.132, but with different numbers of sublethal zones, which we can call a and b, the corresponding survival curves 1 and 2 would have intersections on the ordinates at points a and b, respectively (Figure 8.36).

The mixed model frequently applies to cell populations of mammals irradiated with low LET radiation. Typical values of D_0 (lethal target) are 4.5–5 Gy and for D_n (sublethal death), they are 1–2 Gy.

For high TLE radiation, such as particles, neutrons, deuterons, etc., the curves are frequently simple exponential survival curves.

In a sample with two distinct sublethal populations and assuming that the mixed mammalian model prevails, the survival curve looks similar to Figure 8.37.

Finally, for situations in which the number of hits necessary to produce cell death is added to several cell populations with sensitive lethal and sublethal zones, the survival curve could present the form of Figure 8.38 with no linear portion for the doses with interest.

The major problem with the single-target, single hit or multitarget, single-hit model is that there are three parameters, too many to handle in practical situations.

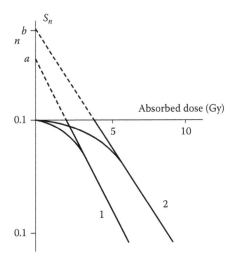

FIGURE 8.36
Survival curves for two distinct cellular populations satisfying model Equation 8.132.

A mathematically simpler model with fewer unknown parameters is needed.

8.12.1.4 The Linear–Quadratic Model

Since there is strong evidence that most of the important biological effects of ionizing radiation are direct or indirect consequences of DNA double-strand breaks, it makes sense to analyze the action of radiation on cells to develop a model based on the direct measurement of chromosome damage. The development of models to simulate chromosome damage and its consequences is an alternative to studying survival curves using target theory models [39,40].

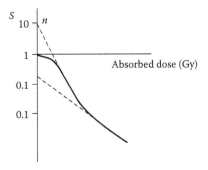

FIGURE 8.37
Survival curve for the mixed model with two distinct sublethal populations.

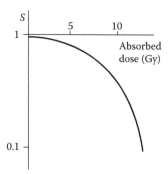

FIGURE 8.38
Survival curve for a complex situation with several cell populations with different sensitive lethal and sublethal zones the number of hits necessary to produce cell death is variable.

The probability of DNA double-strand breaks strongly depends on the ionizing power of the radiation used, as can be seen in Figure 8.39. As an example, the probability of a double break occurring with electrons is much smaller than with α particles of the same energy.

The simplified models for DNA damage from ionizing radiation, shown above in Figures 8.30 and 8.31, recognizes two types of damage. In type B damage, a single particle breaks only one strand (Figures 8.30b and 8.31d). The chromosome only suffers a double break if another particle breaks the other strand close to the first break before repair has taken place (which is not the case in Figure 8.30c). In type A damage, the ionizing particle, in a single

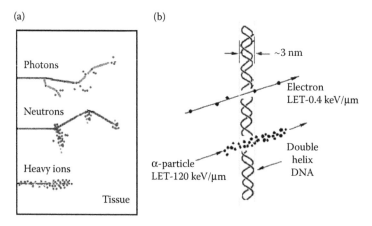

FIGURE 8.39
(a) Ionization produced by photons, neutrons and heavy ions. (b) Schematic representation of relative dimensions of ADN and ionizing range of an α particle (120 keV/μm) and an electron (0.4 keV/μm).

hit, breaks both strands of the DNA, and the chromosome is split into two fragments Figure 8.30d and 8.31e).

The probability of type A damage is proportional to the dose. The average number of cells with type A damage, after dose D, is $m = \alpha D = D/D_0$, and the probability of survival (with no damage) is the Poisson probability $P(0) = e^{-m} = e^{-\alpha D}$.

For double-strand break with type B damage, one strand is damaged by one ionizing particle, and the other is damaged by a second ionizing article. The probability of fragmenting the DNA molecule is, therefore, proportional to the square of dose. The average number of molecules with type B damage is βD^2, and the survival curve for type B damage alone is $e^{-\alpha D^2}$.

At low doses of x- or γ-rays, double-strand breaks are the consequence of the action of a single secondary electron. This behavior is represented by the dashed straight line in Figure 8.40 and corresponds to type A damage. At higher doses, the two chromosome breaks result from two separate electrons. The survival curve bends when the quadratic, type B damage component dominates, as is shown in Figure 8.40.

For x- or γ-rays, the linear-quadratic model for cell survival includes A and B type damage and is

$$S = e^{-(\alpha D + \beta D^2)}, \tag{8.135}$$

where S is the fraction of cells surviving dose D, and α and β are constants. This model assumes that there are two components to cell killing by radiation, one that is proportional to the dose and a second that is proportional to the square of the dose.

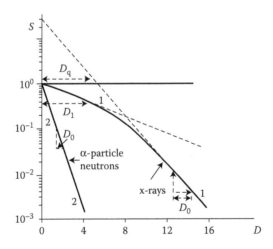

FIGURE 8.40
The important parameters in L–Q curve. The curve is described by the initial slope (D_1), the final slope (D_0), and a parameter that represents the width of the shoulder D_q. Curve 1: Cell survival under x- or γ-rays (A and B type damage). Curve 2: Type A damage for high LET radiation.

Since for equal components of A and B, the damage is

$$e^{-\alpha D} = e^{-\beta D^2}, \tag{8.136}$$

this becomes

$$D = \alpha/\beta, \tag{8.137}$$

The ratio α/β is the dose at which linear and quadratic components of cell killing are equal.*

For low LET radiation, α/β varies between 1.0 and 2 Gy. For densely ionizing (high LET) radiation, such as α-particles or low-energy neutrons, the cell-survival curve is a straight line from the origin; that is, survival approximates to an exponential function of dose. The greater the LET the higher the constant α, that is, the greater the slope of the straight line. In this case, the survival curve can be described by just one parameter, the slope.

The linear-quadratic curve continuously bends but is a good fit to experimental data for the first part of survival curve. The curve is described by the initial slope (D_1), the final slope (D_0), and a parameter that represents the width of the shoulder (D_q).

Constant α represents the probability of α-type damage, which is irreparable. Constant β represents the probability that independent, repairable β-type events have combined to produce lethal events, for example, double-chromosome breaks.

The linear-quadratic model is currently the model of choice for cell survival curves in a substantial number of situations (Figure 8.40).

8.12.1.5 Comparison of the L–Q and Target Theory Models

Survival curves of irradiated cell cultures from different tissues and of tumors show a wide variation in their shapes. Frequently, data are not precise enough to allow a comparison between the fits of L–Q and target theory models.

Survival curves for high LET radiations, such as α-particles, neutrons, and very slow electrons, are linear in both cases (irreparable damage dominates). For low LET radiation, the L–Q curves keep the curvature even at very high doses, whereas target theory curves become linear at high doses.

For x- and γ-rays and other poorly ionizing radiations, survival curves have a shoulder at low doses with a characteristic initial slope. At higher doses, these curves tend to either straighten or continue curved, and survival data can be fitted by either the target theory or the L–Q models.

The advantage of the L–Q model is that it uses only two parameters, α and β, comparing with three in the case of target theory models. The initial slope of the survival curve is represented by either $1/D_0$ or α. Repair is represented by either D_q (target theory) or β (L–Q model). The ratio α/β is the dose at which α-damage equals β-damage.

---

* α has dimensions of reciprocal of dose; β has dimensions of reciprocal of square of dose.

8.13 Nontargeted Effects Complementary to Direct Action of Ionizing Radiation for Low Doses

Biological systems generate defence and biological response mechanisms when subjected to ionizing radiation.

For high doses of ionizing radiation, the associated biological effects are relatively well known and include cellular death (apoptosis and necrosis), somatic mutations (especially those that result in cancer), and the mutations at germinative cells (transmissible genetic defects).

It is found that the biological response for low doses of ionizing radiation cannot be obtained from extrapolation of high dose data.

Evidence and accumulated experimental results exist, indicating that for small doses the action of ionizing radiation on tissues occurs through two simultaneous processes. The first corresponds to direct aggression on DNA with identical laws to those applicable to high doses, and a second consists in several forms of complementary action with characteristics that, in some cases, are still far from completely understood. The most important of these processes, called nontargeted, engulf the DNA repair mechanisms, genomic instability, bystander effects, and adaptive response.

It is generally accepted that damage to DNA can be the most devastating event in cells irradiated with ionizing radiation and that the nonrepaired or badly repaired double-strand breaks in DNA molecule are the damage responsible for the most severe biological processes. The consequences of biological damage in DNA led to the long-accepted paradigm of cytotoxic action, as inducer of mutagenesis and malignancy occurring in cells after ionizing radiation irradiation. Target theory postulates that a direct collision (excitation or ionization) of radiation with a critical cellular constituent, generally chromosome DNA, can produce a biological effect. In particular, the incidence of a microbeam* of ionizing radiation with a nucleus of a cell can rapidly induce cellular death.

Recent studies, however, question the paradigm of genetic alterations being exclusively associated with direct DNA damage after exposure to ionizing radiation.

Experiments have been carried out showing that cells not directly exposed to ionizing radiation may exhibit responses identical to those subjected to direct irradiation.

The collision of a microbeam of ionizing radiation with the nucleus of a cell may induce analogous effects in neighboring nonirradiated cells [41]. This process is known as the bystander effect and indicates the phenomena in which irradiated cells transmit damage signals to nonirradiated cells, leading to genetic effects in neighboring, nondirectly hit cells.

---

* These studies are now possible, thanks to new techniques of microbeaming that target thin beams ($\sim$2 μm^2) at cells with an accuracy of the order of 1–2 μm.

Two possible mechanisms have been suggested to explain transmission in the bystander effect: cellular communication by junctions through intercellular spaces and processes depending on soluble factors produced in the medium [42,43].

Bystander effects depend on several factors such as radiation quality, dose and dose rate, cell culture conditions, and cellular lines; and they can be studied according to several perspectives (clonogenic activation, apoptosis and micronucleus, mutations, genetic expression, chromosome aberrations, etc.)

Genomic instability is a process induced by radiation that consists on the transmission of signals to the descendants of irradiated cells, which, eventually, many generations after, leads to the occurrence of genetic effects, such as mutations and chromosome aberrations in descendants distant in time from the irradiated cells [44].

Adaptive response, another phenomenon, consists on the cellular response to an external aggression through an intrinsic mechanism that generates resistance to a new aggression.

The signals transmitted by the hit cells disturb the normal function of neighboring cells (bystander effects) and stimulate them to send new signals that hit those initially hit and their neighbors. The signals sent by the bystander cells can help in repairing hit cells or start the process that leads them to commit suicide.

A conclusion that comes from the characteristics of the effects of low ionizing radiation doses in tissues under normal conditions is that individual cells cannot be considered to be individual isolated entities in most of the tissues in multicellular organisms.

Apart from the effects referred to earlier, the interpretation of the final expression of cellular damage, for low doses, also has to consider the redistribution, restocking, and mobilization cellular processes.

On the basis of the present knowledge, one cannot yet state whether these effects increase the health risk or otherwise. A better understanding of the mechanisms involved is necessary to evaluate possible action with regard to radiological protection or applications in radiotherapy.

References

1. Wright, H.A.; Magee, J.L.; Hamm, R.N.; Chaterjee, A.; Turner, J.E.; Klots, C.E.; 1985, Calculations of physical and chemical reactions produced in irradiated water containing DNA, *Radiat. Prot. Dosim.*, 13(1–4), 133–136.
2. Charlton, D.E.; 1981, Inner shell ionization produced by the Compton effect, *Rad. Res.*, 88, 420–425.
3. Ribberfords, R. and Carlsson, G.A., 1985, Compton component of the mass-energy absorption coefficient: corrections due to energy broadning of Compton-scattered photons, *Rad. Res.*, 101, 47–59.

4. Hubbel, J.H., Berger, M.J., 1968, Photon attenuation, in *Engineering Compendium on Radiation Shielding*, Vol 1, chap. 4, edited by Jaeger, R.G. et al, Springer-Verlag, Amsterdam.

5. Hubbel, J.H., 1982, Photon mass attenuation and energy-absorption coefficients from 1 keV to 20 MeV, *Int. J. Appl. Radiat. Isot.*, 33, 1269–1290.

6. Storm, E. and Israel, H.I., 1970, *Nucl. Data Tables* A7, 565.

7. Davisson, C.M., 1979, *Alpha-Beta- and Gamma-Ray Spectroscopy*, Vol. 1, cap. 2, Ap. 1, editado por Siegbahn, K., Amsterdam, North-Hollad Publ.

8. Kase, K.R. and Nelson, W.R., 1978, *Concepts of Radiation Dosimetry*, Pergamon Press, New York.

9. Johns, H.E. and Cunningham, J.R., 1983, *The Physics of Radiology*, 4th edition, Charles C Thomas, Springfield, Illinois.

10. ICRU-51, 1993, *Quantities and Units in Radiation Protection Dosimetry*, International Commission on Radiation Units and Measurements, Bethesda, MD.

11. ICRU-60, *Fundamental Quantities and Units for Ionizing Radiation*, International Commission on Radiation Units and Measurements, Bethesda, MD.

12. Papiez, L. and Battista, J.J., 1994, Radiance and particle fluence, *Phys. Med. Biol.*, 39, 1053–1062.

13. Attix, F.H., 1979, The partition of kerma to account for bremsstrahlung, *Health Phys.*, 36, 347–354.

14. Buermann, L., Grosswendt, B., Kramer, H.-M., Selbach, H.-J., Gerlach, M., Hoffmann, M. e Krumrey, M., 2006, Measurement of the x-ray mass energy-absorption coefficient of air using 3 keV to 10 keV synchrotron radiation, *Phys. Med. Biol.*, 51, 5125–5150.

15. Carlsson, G.A., 1985, *Theoretical Basis for Dosimetry, incluído no livro: The Dosimetry of Ionizing Radiation*, Vol. I, editado por: K.R. Kase, B.E. Bjärngard, F.H. Attix, Academic Press.

16. Oliveira, A.D., 2006, Energy and entropy in radiation dosimetry and protection, *Proceedings of full papers, Second European IRPA Congress on Radiation Protection*, Paris, 2006.

17. ICRU-47, 1992, *Measurements of Dose Equivalents from External Photon and Electron Radiations*, International Commission on Radiation Units and Measurements, Washington, DC.

18. ICRP-92, 2003, *Relative Biological Effectiveness (RBE), Quality Factor (Q), and Radiation Weighting Factor (wR)*, Annals of the ICRP, Vol 33(4), Elsevier Science, Oxford, UK.

19. ICRP-103, 2007, The 2007 recommendations of the International Commission on Radiological Protection.

20. Mayneord, W.V. and Clarke, R.H., 1975, Carcinogenesis and radiation risk: a biomathematical reconnaissance, *Br. J. Radiol.*, Suplemento No 12.

21. Shannon, C.E., 1949, *The Mathematical Theory of Communication*, University of Illinois Press Champaign, IL.

22. Oliveira, A.D. and Lima, J.J.P., 2001, The degradation of the energy of primary photons described through the entropy, In *Proceedings of the Monte Carlo 2000 Conference—Advanced Monte Carlo for Radiation Physics, Particle Transport Simulation and Applications*, Lisbon, Portugal. Kling et al., Eds, Springer-Verlag, Berlin, pp. 425–430.

23. Grosswendt, B., 1994, Formation of track entities by photons in water, *Radiat. Prot. Dosim.*, 52(1–4), 237–244.

24. Oliveira, A.D., 2001, The entropy of an x-ray beam, In *Proceedings of the Fifth Regional Congress on Radiation Protection and Safety IRPA*, Pernambuco, Brazil.
25. Chan, H.-P. and Doi, K., 1988, *Monte Carlo Simulation in Diagnostic Radiology, do livro: Monte Carlo Simulation in the Radiological Sciences*, editado por R. L. Morin, CRC Press, FL.
26. Faraggi, M., Gardin, I., Stievenart, J.L., Bok, B.D., Guludec, D., 1998, Comparison of cellular and conventional dosimetry in assessing self-dose and cross-dose delivered to the cell nucleus by electron emmissions of 99mTC, 123I, 111In, 67Ga and 201Tl, *Eur. J. Nucl. Med.*, 25, 205–214.
27. Goddu, S.M., Howell, R.W., Rao, D.V., 1994, Cellular dosimetry: Absorbed fractions for monoenergetic electron and alpha particle sources and S-values for radionuclides uniformly distributed in different cell compartments, *J. Nucl. Med.*, 35, 303–316.
28. Edwards & Lloyd, 1998, Risks from ionising radiation: Deterministic effects, *J. Radiol. Prot.*, 18(3), 175–183.
29. Doll, 1998, Effects of small doses of ionising radiation, *J. Radiol. Prot.*, 18(3), 163–174.
30. Feinendegen et al., 1985, Microdosimetric approach to the analysis of cell responses at low dose and low dose rate, *Radiat. Prot. Dosim.*, 13(1–4), 299–306.
31. Leenhouts and Chadwick, 1985, Radiation energy deposition in water: Calculation of DNA damage and its association with RBE, *Radiat. Prot. Dosim.*, 13(1–4), 267–270.
32. ICRP-60, 1991, Recommendations of the International Commission on Radiological Protection, *Ann. ICRP* 21, (1–3).
33. Pochin, 1980, The need to estimate risks, *Phys. Med. Biol.*, 25(1), 1–12.
34. Lindell, 1996, The risk philosophy of radiation protection, *Radiat. Prot. Dosim.*, 68(3/4), 157–163.
35. Lea, D.E., 1955, *Actions of Radiations on Living Cells*, 2nd Ed, ICRU-51, 1993, Cambridge University Press, Cambridge.
36. Savage, JR. 1993, Update on target theory applied to chromosomal aberrations, *Environ. Mol. Mutagen* 22, 198–207.
37. Nias, A.H.W., 1998, *An Introduction to Radiobiology* (2nd ed.), Wiley, Chichester.
38. Leroy, C. and Rancoita, P.G., 2004, *Principles of Radiation Interaction in Matter and Detection*, World Scientific, Singapore.
39. Alpen, E.L. *Radiation Biophysics* (2nd ed.), Academic Press, San Diego, 1998.
40. Kiefer, J., 1990 *Biological Radiation Effects*, Springer-Verlag, Berlin.
41. Belyakov, O.V., Prise, K.M., Mothersill, C., Folkard, M. and Michael, B.D., 2000, Studies of bystander effects in primary uroepithelial cells using a charged particle microbeam, *Radiat. Res.* 153, 235.
42. Azzam, E.I., de Toledo, S.M., and Little, J.B., 2001, Direct evidence for the participation of gap junction mediated intercellular communication in the transmission of damage signals from -particle irradiated to nonirradiated cells. *Proc. Natl. Acad. Sci. USA* 98, 473–478.
43. Ballarini, F., Biaggi, M., Ottolenghi, A. and Sapora, O., 2002, Cellular communication and bystander effect: A critical review for modelling low-dose radiation action, *Mut. Res.* 501, 1–12.
44. Little, J.B., Azzam, E.I., de Toledo, S.M. and Nagasawa, H., 2005, Characteristics and mechanisms of the bystander response in monolayer cell cultures exposed to very low fluences of alpha particles, *Rad Phys Chem.* 72(2–3), 307–313.

Index